1st International Conference on Agriculture and Rural Development (ICARD 2019)

IOP Conference Series: Earth and Environmental Science
Volume 383

Banten, Indonesia
8 - 9 August 2019

Editor:

Tatang Sutisna

ISBN: 978-1-7138-0251-8
ISSN: 1755-1307

Printed from e-media with permission by:

Curran Associates, Inc.
57 Morehouse Lane
Red Hook, NY 12571

Some format issues inherent in the e-media version may also appear in this print version.

Institute of Physics
Dirac House, Temple Back
Bristol BS1 6BE UK

Phone: 44 1 17 929 7481
Fax: 44 1 17 920 0979

techtracking@iop.org

Additional copies of this publication are available from:

Curran Associates, Inc.
57 Morehouse Lane
Red Hook, NY 12571 USA
Phone: 845-758-0400
Fax: 845-758-2633
Email: curran@proceedings.com
Web: www.proceedings.com

TABLE OF CONTENTS

Preface

All praise is due to Allah, who has bestowed His blessings and allows us for hosting and joining "The 1st International Conference on Agriculture & Rural Development (ICARD. The main objective of this event was to appreciate and share the experience and results of research in agriculture, as well as to improve its understanding and application to develop rural communities.

Agricultural is one of the sectors that significantly influence the community and economic. In the present, the development of agriculture is crucial due to some reasons. First, the potency of agriculture is enormous and having diverse resources. Second, agriculture contribution to national income and exports is quite high. Third, the size of the population who rely on the agricultural sector. Forth, Its role in community food supply very substantial. Fifth, it becomes the basis of economic growth.

Currently, the focus of the agricultural development is to build up a system of food security based on production capability, diversity of food resources, diversity of local institutions and cultures, and to establish globally-oriented agribusiness by improving competitive advantages of local products based on competition and comparative advantage of nature and human resources. Many programs of research and development have been carried out in the field of agriculture in the region that has this potential, particularly in rural areas. The development and economic progress of the community, not only depends on the implementation of a program and budget allocation by the government, but also requires extensive scientific studies and its application in whole sectors, including agriculture.

In Indonesia, research of agriculture is carried out by universities, Indonesian Agency for Agricultural Research and Development (Badan Libang Pertanian), Indonesian Institute of Science (LIPI), National Nuclear Energy Agency of Indonesia (Batan), Agency for the Assessment and Application of Technology (BPPT), and related research institution in other ministries. The results need to be well documented and published to be applied by those who need them. Spreading out the research results are also crucial for further research, knowledge transfer, and reference in compiling scientific papers.

University, as one of the institutions for the implementation of scientific studies, has to carry out its function continuously to research and disseminate various studies in various fields of knowledge. One of them is agriculture in general term, which is very vital sectors. These works are intended to support the

IOP Conf. Series: Earth and Environmental Science **383** (2019) 011001 doi:10.1088/1755-1315/383/1/011001

advancement of science and technology and improve the welfare of society, especially in the countryside.

University of Sultan Ageng Tirtayasa (UNTIRTA), through the faculty of agriculture, has great enthusiasm and attention to strengthen scientific studies and publication of research results. The event of ICARD is a collaboration between the Faculty of Agriculture UNTIRTA and the Faculty of Agriculture, IPB University. One of the outputs this event is scientific publications in Scopus indexed international proceeding. In this respect, we hope this international conference can contribute to the development of the agricultural sector, to support food sovereignty on the people in Indonesia and around the world. Also, it serves as an excellent opportunity to bring together scientists, researchers, and students who are expected to be future researchers. Hopefully, everything we do will benefit all of us.

Thank you very much.

Dean,
Faculty of Agriculture
University of Sultan Ageng Tirtayasa

Prof. Dr. Ir. Nurmayulis, MP

IOP Conf. Series: Earth and Environmental Science **383** (2019) 011002 doi:10.1088/1755-1315/383/1/011002

COMMITTEE
"THE 1st INTERNATIONAL CONFERENCE ON AGRICULTURE AND RURAL DEVELOPMENT (ICARD)"

STEERING COMMITTEE:

Prof. Dr. H. Sholeh Hidayat, M.Pd
Dr. H. Fatah Sulaiman, S.T., M.T
H. Kurnia Nugraha, S.T., M.T
Dr. H. Suherna, S.P., M.Si
Prof. Dr. Ir. Kartina AM, M.P
Prof. Dr. Nurmayulis, Ir., M.P
Dr. Ir. Suwardi, M. Agr

ORGANIZING COMMITTE:

Chairperson : Dr. Susiyanti, S.P, M.P
Secretary : Nur Iman Muztahidin, S.P., M.Sc
Treasurer : Dr. Mustahal, M.Sc
Een, M.Pd

Live Event:
Prof. Dr. Meutia, S.E., M.P

Device and Equipment:
Rifqi Ahmad Riyanto, S.Sc., M.Si

Accomodation:
Abdul Hasyim Sodiq, S.P., M.Si

Proceeding Editorial in Chief:
Tatang Sutisna, S.P., M.Sc

SCIENTIFIC COMMITTEE

Elena Nikitina, Ph.D; Kazan Federal University, Rusia
Christopher D. Elvidge, Ph.D; Earth Observation Group, USA
Assoc. Prof. Daisuke Yasutake, Ph.D; Kyushu University, Japan
Prof. Dr. Hj. Kartina AM., Ir., M.P
Prof. Dr. Nurmayulis, Ir., M.P
Prof. Dr. Meutia, S.E., M.P
Dr. Saed Gorgin, Iran
Dr. Willy Bayuardi Suwarno, S.P., M.Si
Dr. Ir. Suwardi, M.Agr
Assoc. Prof. Nurul Huda, Malaysia
Assoc. Prof. Azura Adam, Malaysia
Dr. Andy Saryoko, S.P., M.P
Dr. Agr.Sc Muharfiza, S.TP., M.Si

Conference Photograph 1st ICARD

IOP Conf. Series: Earth and Environmental Science **383** (2019) 011003 doi:10.1088/1755-1315/383/1/011003

1st International Conference on Agriculture and Rural Development
Serang, 8-9 August 2019

1st International Conference on Agriculture and Rural Development
Serang, 8-9 August 2019

THE 1ST INTERNATIONAL CONFERENCE
AGRICULTURE AND
(ICARD)

IOP Conf. Series: Earth and Environmental Science **383** (2019) 011003 doi:10.1088/1755-1315/383/1/011003

Peer review statement

All papers published in this volume of *IOP Conference Series: Earth and Environmental Science* have been peer reviewed through processes administered by the proceedings Editors. Reviews were conducted by expert referees to the professional and scientific standards expected of a proceedings journal published by IOP Publishing.

IOP Conf. Series: Earth and Environmental Science **383** (2019) 012001 doi:10.1088/1755-1315/383/1/012001

The potency of bio-organic fertilizer containing local microorganism of Cibodas village, Lembang-West Java

A H Sodiq[1*], M R Setiawati[2], D A Santosa[3], D Widayat[2]

[1]Doctoral Program, Faculty of Agriculture, Padjadjaran University, Indonesia
[2]Faculty of Agriculture, Padjadjaran University Jl. Jatinangor Km.12 Sumedang, Indonesia
[3]Department Soil Science and Land Resource, Faculty of Agriculture IPB University Jl. Meranti, IPB Campuss Darmaga Bogor, 16680, Indonesia

*Corresponding author: abdul_hasyimsodiq@yahoo.co.id

Abstract. This experiment was conducted to determine the characteristics of biofertilizers usually called as Local Microorganism (LM) with the raw materials that is easily obtained in the village of Cibodas Lembang. This experiment included ten main raw materials as media for growing LM, including: Bamboo Root, Spring Onion, Broccoli Leaves, Banana Stump, Leaves of Water Hyacinth, Goat Manure, Rabbit Manure, Chicken Manure, Cow Manure and Rabbit Urine. The LM was enriched with material sugar and carbohydrates then incubated for 2 weeks. Parameters observed in the laboratory include IAA hormones, cytokines and gibberellins. The microorganism living in the LM i.e Total microorganism (Total Plate Count Aerob), *Azospirillium, Azotobacter* and *Trichoderma* were also analysed. The results showed that the LM with raw materials from chicken manure showed the highest hormone content in both IAA, Cytokinin (Zeatin), Cytokinin (Kinetin) and Giberelin with the following values: 141.28 ppm, 115.63 ppm, 126.49 ppm, 165.13 ppm, respectively. The same LM showed highest number of microorganisms based on TPC method and *Trichoderma* with a value of 3.75 x 10^8, 6.80 x 10^5 CFU/ml. As for the analysis of the LM goat manure contain the highest number of *Azospirillium* which population of 2.4 x 10^5 CFU/ml and the highest population of *Azotobacter* sp was found in LM of rabbit manure with a value of 6.95 x 10^3 CFU / ml.

Keywords: *Azospirillium, Azotobacter,* biofertilizers, Cytokinins, Gibberellins, IAA Hormones, Local Microorganisms, TPC Aerob, *Trichoderma*

1. Introduction

The changing paradigm of Indonesian society towards understanding healthy living continues to increase. [1] stated that the consumption pattern of balanced nutrition originating from vegetables began to develop since the 1990s. Vegetables are important in fulfilling food needs because they are one of the providers nutrients in the form of fiber, vitamins, proteins and minerals needed by the human body. In order to meet the needs of consumers for vegetables, farmers increase their production and the same time they have to maintain the quality and supply of products. Harvested area of vegetable increased by 3.87%, and production rose to 3.88% in 2016 compared to 2015 [2].

One things that needs to be observed from vegetable crops is the by-product of farmers. Vegetable waste the cultivation of producer in organic waste which can bring problems to the environmental if it is not managed properly. In fact, vegetable waste can be used as an alternative or substitute for inorganic fertilizers. According to [3] the application of biofertilizers to reduce the use of inorganic fertilizers is important to protect the environment from the adverse effects of excessive inorganic fertilizers.

Vegetable waste as well as livestock waste around our environment is a source of raw material for biological fertilizers known as Local Microorganisms (LM). LM is known to contain a number of microbes, macro and micro nutrients, and can produce phytohormones. Microbial communities can play a role in plant growth through several mechanisms, including increasing nutrient absorption into plants and can increase the competitiveness of rooting-transmitted pests and diseases [4].

Media sources for LM can vary according to the availability of raw materials at the farm level. The main components of LM media are sources of carbohydrates from rice washing water, sources of sugar is refined sugar and bacterial sources (fruit & vegetable waste, livestock waste). The diversity of raw materials will have an impact on the variation of microbial and phytohormone contents contained in LM. Therefore research was conducted to determine the characteristics of various kinds of LM with raw materials available in the Cibodas village of Lembang, West Java.

2. Material and Methods

The study was conducted in August to October 2018 in plastic houses (green house) Macakal Farmers Group, Cibodas-Lembang, Bandung. Analysis of functional microorganism living in the LM and analysis of plant growth hormones was carried out at the Environmental Biotechnology Laboratory of Indonesian Center for Biodiversity and Biotechnology (ICBB), Bogor, Indonesia.

In total types of raw materials were used in this experiments, including: bamboo root, broccoli leaves, banana stump, leaves of water hyacinth, goat manure, rabbit manure, chicken manure, cow manure, rabbit urine. The main raw material is then added to the media contained sucrose (3 kg of sugar diluted in 3 liters of water) and source of carbohydrates.

LM is made from of 2 kg wet weight of raw material mashed by chopping up to a maximum size of 10 mm then mixed with 600 ml of sugar solution then soaked in 4 liters from rice washing water in a 10 liter volume plastic container. The container is covered with paper and then incubated for 2 weeks. The pH and temperature are measured every day. After that, LM is filtered and then put in a plastic bottle. Furthermore, to find out the biological and phytohormone properties of LM made by quantification, a series of analyzes are carried out as presented in Tables 1.

Tabel 1. LM Quality Test Parameters

Parameters	Method	Unit
TPC Aerob	ICBB/MU/12.001.4	CFU/ml
Azotobacter	ICBB/MU/12.001.9	CFU/ml
Azospirillium	ICBB/MU/12.001.10	CFU/ml
Trichoderma	ICBB/MU/12.001.6	CFU/ml
Auxin		
IAA	HPLC	ppm
Sitokinin		
Zeatin	HPLC	ppm
Kinetin	HPLC	ppm
Gibberelin	HPLC	ppm

3. Results and Discussion

3.1. Physical Properties LM

The observations pH of LM during the study can be seen in Figure 1. The pH results show a range of values from 3.5 to 5.9. Most of LM showed has a low pH. This is similar to the study of [5], where the pH of liquid organic fertilizer in the range of 3.68 - 4.65. Tthe daily observation of pH was also carried out but no significant change in pH was observed.

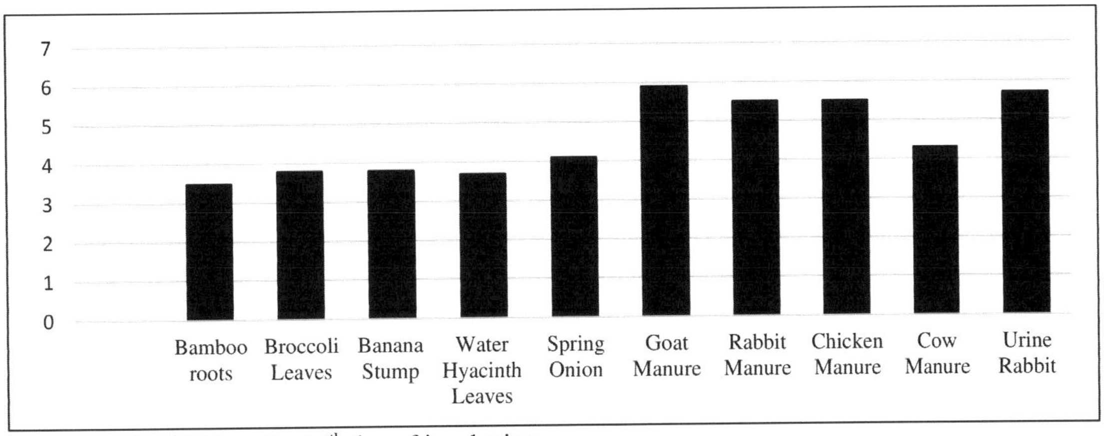

Figure 1. pH of LM on the 14th day of incubation.

Based on Figure 1, the raw material of LM from bamboo roots showed the lowest pH value is 3.5 and the raw material for goat manure LM showed the highest pH value is 5.9. The lowest pH value produced by bamboo root raw material is due to the fact that the material from the plant contains large amounts of organic acid. Microbial activity and organic material derived from the rest of the plant is a cause of acidity in LM. [5] is stated that microbial activity in LM made from plant residues will increase H_2CO_3 levels in LM solution resulting in an increase in H^+ concentration.

The observation of pH LM showed that during of incubation the pH of LM were all in acidic conditions. In LM contains both microbes and various nutrients useful for plants growth. According to [6] organic matter is the main source of macro nutrients such as N, P and S and other essential micro nutrients which are very important for plant growth.

The temperature of LM for 14 days of observation fluctuated (Figure 2). In general, the temperature of several LM increased slightly at the 8th observation, but no difference compared with the temperature of the initial observation. The temperature conditions of the LM are seen to continue to decline until the 14th day, within a value range of 25-27°C.

The temperature range indicated by various LM during the observation is fairly normal, where there is no increase or decrease in extreme temperatures. The temperature of LM is related with microbial activity. The microbial decomposed raw materials and released heat. [7] is stated that when raw material for LM has decomposed completely, microorganisms enter the death phase and the LM temperature will decrease.

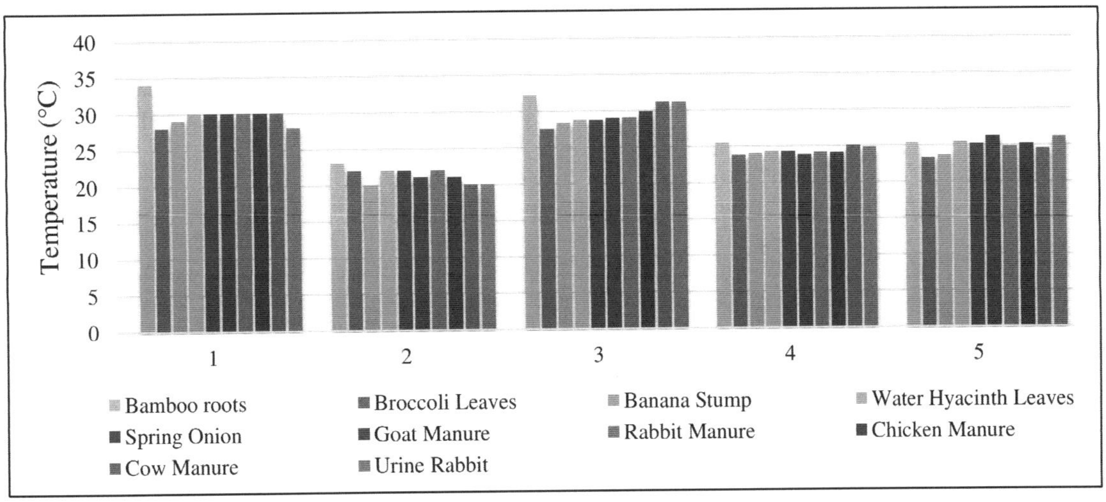

Figure 2. Temperature of LM on 1, 4, 8, 12 and 14 days of incubation

3.2. The Populations of Miccroorganism

Table 2 shows the best results on TPC Aerob analysis is LM with the main raw material of chicken manure with a value of 3.75 x 10^8 CFU/ml, while as for *Trichoderma* the best results are indicated by the main raw material of chicken manure with a value of 6.80 x 10^5 CFU/ml. For the *Azotobacter* population the best results were shown by LM with the main raw material of rabbit manure with a value of 6.95 x 10^3 CFU/ml. Furthermore, in the analysis of *Azospirillium* the best results were shown by the main raw material of goat manure which is 2.4 x 10^5 CFU/ml.

Table 2. Observation results total populations of Mikrob Aerob (TPC Aerob), *Azotobacter, Azospirillium, Trichoderma*

Raw Materials LM	TPC Aerob CFU/ml	*Azotobacter* CFU/ml	*Azospirillium* CFU/ml	*Trichoderma* CFU/ml
Bamboo Roots	1.20 x 10^5	1.3 x 10^3	< 1.8	1.00 x 10^2
Broccoli Leaves	5.90 x 10^6	3.7 x 10^3	2 x 10^3	< 1 x 10^2
Banana Stump	1.15 x 10^5	< 1 x 10^2	780	< 1 x 10^2
Water Leaves	1.14 x 10^5	1.25 x 10^3	400	< 1 x 10^2
Spring Onions	1.04 x 10^5	3.15 x 10^3	< 1.8	< 1 x 10^2
Goat Manure	1.09 x 10^7	4.05 x 10^3	2.4 x 10^5	< 1 x 10^2
Rabbit Manure	6.85 x 10^6	6.95 x 10^3	4.7 x 10^3	< 1 x 10^2
Chicken Manure	3.75 x 10^8	< 1 x 10^2	< 1.8	6.80 x 10^5
Cow Manure	2.48 x 10^5	1.50 x 10^2	< 1.8	3.50 x 10^2
Rabbit Manure	1.34 x 10^5	< 1 x 10^2	< 1.8	< 1 x 10^2

The observation of the microbial population above shows that LM that has been made contains a variety of functional microbes. According to [8], the diversity of microbial populations indicates that many microbes play a role in the process of decomposition of LM organic matter. *Azotobacter* and *Azospirillum* in LM have a very important role in providing nitrogen for plant growth. According to [9], *Azotobacter* and *Azospirillum* are N_2 fixing bacteria that live freely in the soil and also produce plant growth-promoting substances such as giberalin, cytokinin and indolacetic acid.

Population of *Trichoderma* sp. in LM made from raw chicken manure was the highest when it compared to other ingredients. *Trichoderma* sp. known to have antagonistic ability to several pathogenic fungi in vitro. The results of [10], showed that the inoculation of *Trichoderma* sp. able to inhibit the growth of *C. capsici, Fusarium sp*., and *S. rolfsii* fungi in a row of 68.2%, 53.9%, and 35.5%. Therefore LM made from raw chicken manure has the potential as a biofungicide to control the population of plant pathogenic fungi.

3.3. Phytohormone Observation

LM also contains several phytohormones which are useful in plant growth and productivity. [11] states that fitohormones can regulate plant growth. The phytohormon help the plant can also to resist drought condition and heavy metal contamination. In addition, they can protect plant againt high salinity stresses, and help plants grow. Bacteria that are capable of producing gibberellins, namely: *Azotobacter, Azospirillium, Pseudomonas, Acetobacter, Burkholderia* and *Bacillus* [12].

3.3.1. Auxin (IAA)

The function of this auxin hormone is to help in the process of accelerating growth, both root growth and stem growth, accelerate germination, help in the process of cell division, accelerate fruit ripening, reduce the number of seeds in the fruit. This auxin hormone works synergistically with cytokinin and gibberelin hormone.

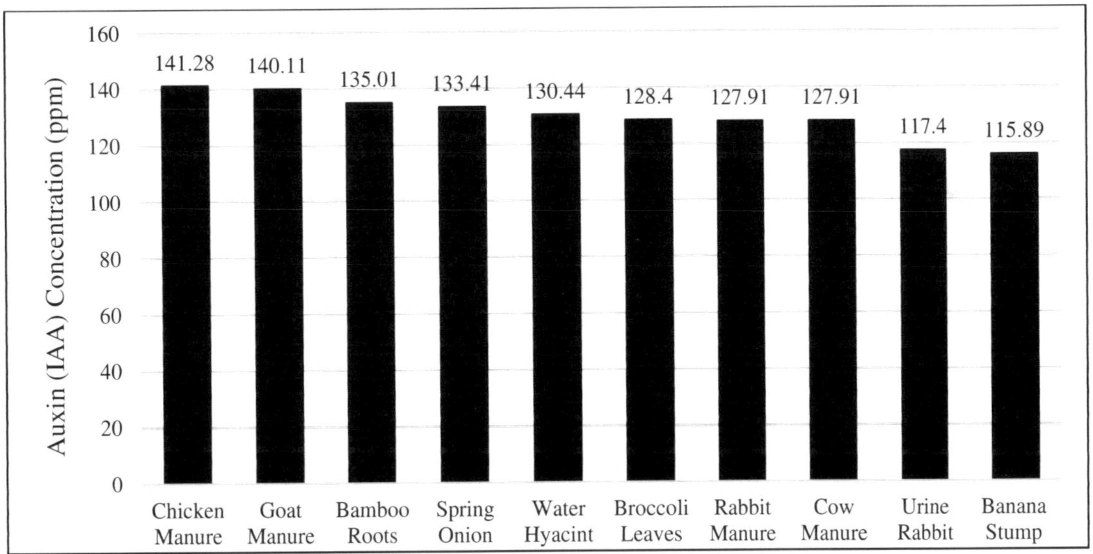

Figure 3. Auxin (IAA) concentration of several sources of LM on 14th days incubations

In Figure 3. The highest concentration of auxin hormone (IAA) is 141.28 ppm resulted from LM chicken manure, which subsequently contributed the second and third best is 140.11 ppm and 135.01 ppm was showed by LM goat manure and bamboo roots. The two best values are produced by raw materials from animal residues. This shows the raw materials from animal organic waste to produce LM with a high composition of the hormone auxin (IAA).

3.3.2. Cytokinins (Zeatin dan Kinetin)

Cytokinins are a group of plant hormones and growth regulators that encourage cell division (cytokinesis) in meristematic tissues. In addition to its main role as a regulator of cell growth and differentiation, cytokinins also influence shoot dominance, edge bud growth, and aging (senescence) leaves. There are two types of cytokines: the type of adenine and the type of phenylurea. Adenine type is represented by kinetin, zeatin, and BA. Phenylurea types, for example, are diphenylurea and tidiazuron (TDZ) which are not formed by plants. Cytokines can work locally or remotely. Usually cytokinins are transported through wooden vessels. In carrying out its physiological functions, cytokines often work in conjunction with auxin.

Based on the analysis of the cytokinin (Zeatin and Kinetin) hormone content the best results was showed by LM chicken manure is 115.63 ppm and 126.49 ppm. These results indicate that chicken manure is still consistently showing the best results in phytohormone analysis. Whereas the second best results are was showed by LM goat manure is 105.23 ppm and 124.36 ppm. The third highest content of Zeatin is found in LM leaves of water hyacinth and kinetin in LM with raw materials from cow manure.

IOP Conf. Series: Earth and Environmental Science **383** (2019) 012001 doi:10.1088/1755-1315/383/1/012001

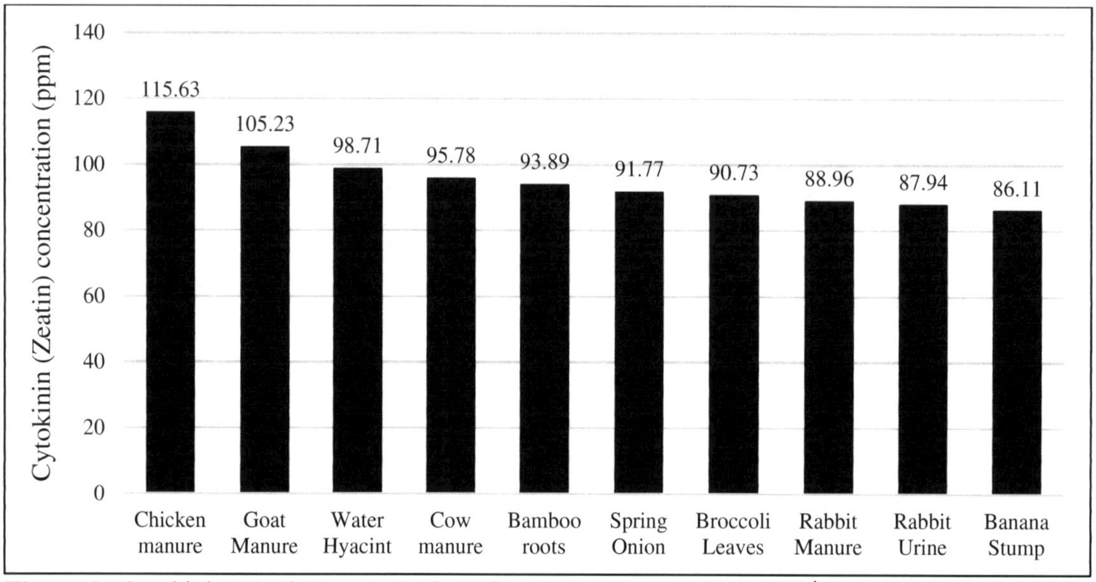

Figure 4. Cytokinin (Zeatin) concentration of several sources of LM on 14th days incubations

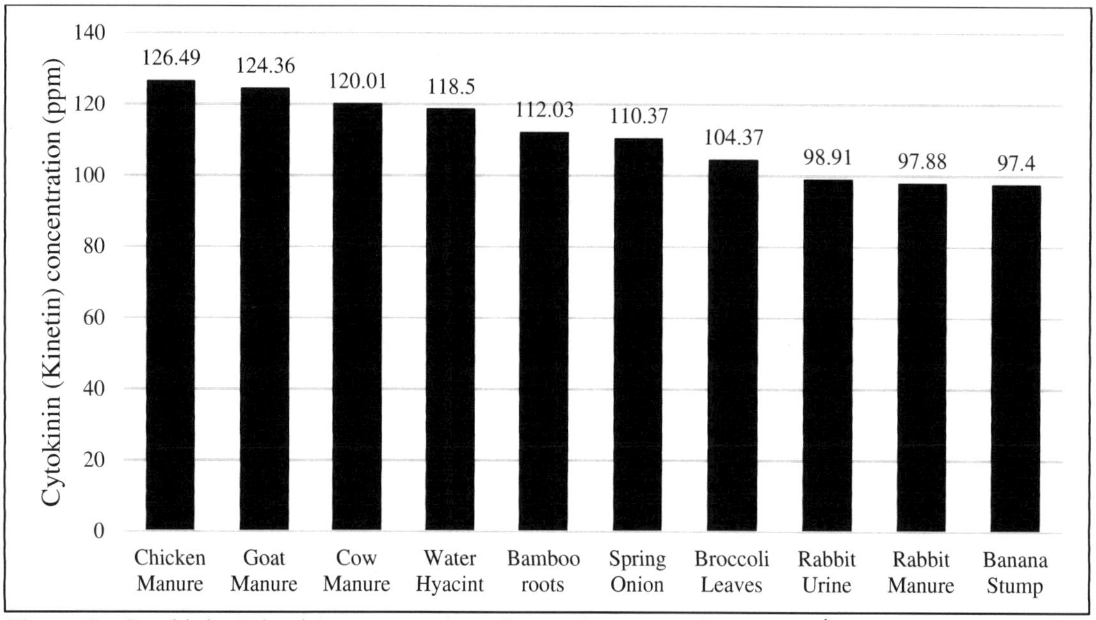

Figure 5. Cytokinin (Kinetin) concentration of several sources of LM on 14th days incubations

3.3.3. Gibberellin

Gibberellin or gibberellic acid are all members of the plant hormone group that have similar functions. GA is needed for most plant life and has the function of regulating germination, stem lengthening, flowering triggering, anther development, seed development and pericarp growth. In addition, this phytohormone also plays a role in responses to stimuli through physiological regulation associated with its biosynthetic mechanism. Gibberellins in plants can be found in two main phases, active gibberellins (bioactive GA) and inactive gibberellins. Biologically active gibberellins (bioactive GA) control various aspects of plant growth and development, including seed germination, extension rods, leaf expansion, and flowers and seed development. Exogenous application of gibberellins can increase the growth of chili plants [13].

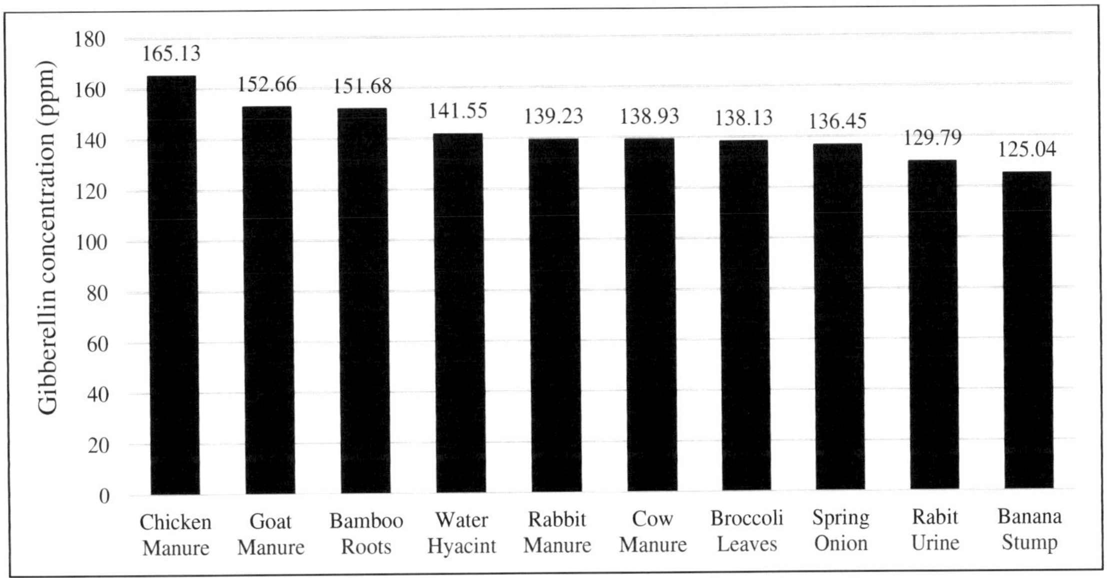

Figure 6. Gibberelin concentration of several sources of LM on 14th days incubations

Based on the results of the analysis of the hormone Gibberellin the best results were again shown by LM with raw material of chicken manure which has value of 165.13 ppm (Figure 6). That can be concluded that the produced with LM raw materials from chicken manure produced highest phytohormone compare with other raw materials.

4. Conclusion

Based on the results showed that the raw materials from chicken manure showed the highest hormone values both on IAA, Cytokinin (Zeatin), Cytokinin (Kinetin) and Gibberellin hormones respectively with the following values of: 141.28 ppm, 115.63 ppm, 126.49 ppm and 165.13 ppm. The LM from chicken manure showed also the highest TPC Aerob and *Trichoderma* values compared to other LM with a value of 3.75 x 10^8 CFU/ml, 6.80 x 10^5 CFU/ml. While the highest population of *Azospirillium* is indicated by the raw material from goat manure, which is 2.4 x 10^5 CFU/ml. Whereas *Azotobacter* sp. the highest population was found in LM with rabbit manure with a value of 6.95 x 10^3 CFU/ml.

5. References

[1] Aswatini, M. Noveria, Fitranita. 2008. Consumption of vegetables and fruit in the community in the context of fulfilling balanced nutrition. J. Indonesian Independence. 3(2): 97-119.

[2] [Kementan RI] 2017. Agricultural Statistics 2017. Jakarta (ID): Agricultural Data and Information Center. Ministry of Agriculture of the Republic of Indonesia. 362 hal.

[3] El-Habbasha, S.F., M.S. Abd El Salam, M.O. Kabesh. 2007. Response of two sesame varieties (*Sesamum indicum* L) to partial replacement of chemical fertilizers. *JABS*. 3(6): 563-571.

[4] Smith, S.E., D.J. Read. 1997. *Mychorizal Symbiosis*. London (UK): Academic Pr.

[5] Handayani, S.H., A. Yunus, A. Susilowati. 2015. Test the quality of liquid organic fertilizer from various kinds of local microorganisms (MOL). *El-Vivo*. 3(1): 54-60.

[6] Smith, J.L., R.I. Rapendick, D.F. Bezdicek, J.M. Lynch. 1993. Soil organic matter dynamics and crop residue management [editorial]. *Soil Microbial Ecology*. Marcel Dekker Inc (US). p: 65-94

[7] Juanda, Irfan, Nurdiana. 2011. Effect and duration of fermentation on the quality of moles (local microorganisms). J. Floratek. 6(1): 140-143.

[8] Batara, L.N, I. Anas, D.A. Santosa, Y. Lestari. 2016. Application of local microorganisms (MOL) enriched with microbes is useful in the cultivation of rice systems of rice intensification (SRI) organic. J. Land and Climate.. 40(1): 71-78.

[9] Hindersah, R., T. Simarmata. 2004. Azotobacter Rhizobacteria Potential in improving soil health. J Nature Indones. 5(2): 127-133.

[10] Alfizar, Marlina, F. Susanti. 2013. The antagonistic abilities of Trichoderma sp. against several types of in vitro pathogenic fungi. *J. Floratek*. 8(1): 45-51.

[11] Widyati, E. 2016. The role of phytohormone in plant growth and its implications for forest management. *Galam*. 2(1): 11-22.

[12] Mansour FA, Aldesuquy HS, Hamedo HA, 2004. Studies on plant growth regulators and enzymes production by some bacteria. Qatar Univ Sci J. 14(2):281-288.

[13] Joo GJ, Gunes A, Turan M, Gulluce M. 2005. Gibberellins producing rhizobacteria increase endogeneous gebberellins content and promote growth of red peppers. J Miicrobiol. 43: 510-515.

The Utilization of Vegetable Waste Silage as Feed Ingredient in Diets for Tilapia *Oreochromis niloticus*

A N Putra[12*], Jaenudin[1], F Sofia[1], Mustahal[1], M B Syamsunarno[1], D Hermawan[1], M Herjayanto[1]

[1]Department of Fisheries, Faculty of Agriculture, University of Sultan Ageng Tirtayasa
Jalan Raya Jakarta KM 04 Pakupatan, Serang, Indonesia 42124.
[2]Indonesia-Center of Excellence for Food Security

*Corresponding author: putra.achmadnp@untirta.ac.id

Abstract. The vegetable waste silage is potential as a feed ingredient for tilapia culture. Silage is a product of the processing of materials by the addition of acid or by anaerobic fermentation. This study aimed to evaluate the effect of vegetable waste silages on the growth of tilapia (*Oreochromis niloticus*). Formic acid is used in this study as organic acid for the preservative. Four diets concentration of formic acid in making silage (0%, 3%, 4%, and 5%) was formulated containing 27% crude protein. The result showed that total digestibility, protein digestibility, and fat digestibility were significantly higher (P<0.05) in 4% formic acid and 5% formic acid than Control. The value of growth was significantly highest (P<0.05) in 5% formic acid (2.96±0.08 %/day), followed by 4% formic acid (2,66±0,14 %/day), 3% formic acid (2,56±0,05 %/day) and control (2,54±0,21 %/day). The utilization of 5% formic acid in making vegetable waste silage showed the best result on growth and feed efficiency than the other treatments.

Keywords: formic acid, silage, tilapia.

1. Introduction

Tilapia (*Oreochromis niloticus*) is a type of fish that has high economic value as a consumption fish and is an important commodity in the freshwater fish business in the world [1]. On the other hand, the intensive tilapia culture faces several problems, including the relatively expensive price of artificial feed that is not followed by the selling price of the product. This certainly has an impact on the income of fish farmers because fish farmers have to incur other production costs such as seeds and medicines. This problem also has an impact on the decline in tilapia production in Indonesia. The annual report of the Directorate of Production, the ministry of marine and fisheries in 2016 states that national tilapia fish production is 1.02 million tons and this number is below the production target targeted by the ministry in 2016, which is 1.65 million tons. The high price of feed is due to high prices of feed raw materials such as fish meal and soybeans meal. Therefore, it's needed local alternative raw materials in the tilapia feed.

Vegetable waste is an alternative material used as raw material for fish feed [2, 3], because it has a high protein content of 15-24% [4]. In addition, the availability of vegetable waste is abundant, inexpensive and does not compete with human needs [5]. Utilization of vegetable waste as feed raw material can also reduce environmental pollution because vegetable waste is waste that pollutes the environment [6]. However, the use of vegetables as feed raw materials is limited by the high durability of vegetable waste and crude fiber. Vegetable waste is easy to rot and voluminous (large) with a crude fiber content of 20.76-29.18%. Therefore, it is necessary to increase shelf life and decrease the value of crude fiber of vegetable waste.

Silage is a technology that is expected to increase shelf life and reduce crude fiber from vegetables. According to [7], the mechanism of action of silage is that the addition of acid will cause the pH to decrease (<4.2), thereby inhibiting the growth of spoilage bacteria and only lactic acid bacteria in the silage material. The acid commonly used in making silage is formic acid [8, 9, 10, 11]. However, there is no research related to the use of silage derived from vegetable waste as feed raw material in tilapia. Therefore, research is needed related to the use of silage from vegetable ingredients as raw material for tilapia fish feed. This study aimed to evaluate the effect of vegetable waste silages on the growth of tilapia.

2. Material and Methods

2.1 Silage preparation

Vegetable waste consisting of cabbage waste, water spinach, spinach, mustard is cut into small pieces to facilitate the fermentation process. Furthermore, vegetable waste is weighed and added formic acid according to treatment. The dosage of formic acid used consisted of 0%, 3%, 4% and 5% (w / w). The mixture of vegetable waste and formic acid is then put into a bucket and sealed tightly to create anaerobic conditions. The fermentation process lasts for 10 days, pH silage is measured every day to ensure the fermentation process occurs in vegetable waste. The silage is then dried in the sun to dry and ground into silage meal.

2.2 Preparation of experimental diets

The feed used is commercial feed with a protein content of 32%. Feed was pollinated into the meal and then chromium oxide (Cr_2O_3) of 0.5% as an indicator of digestibility and 3% tapioca as a feed binder were added to the meal before the pelleting process. Feed formulation refers to [12], i.e. feed formulation consisting of 70% commercial feed and 30% test material, while for control (formic acid 0%) using a vegetable waste meal. The feed is made in the form of a wet pellet using 2 mm size feed molding machine. The formulations of feed are presented in Table 1.

2.3 Experimental design

Tilapia used were male monosex tilapia with an average weight of 10 g with a density of 15 fish/aquarium with a recirculation system. Tilapia is obtained from Ary Farm, Serang, Banten Province, Indonesia. This study is conducted at the Laboratory of Aquaculture, Department of Fisheries, University of Sultan Ageng Tirtayasa, Indonesia. fish were reared in 12 aquariums with a size of 50 x 40 x 30 cm, for 40 days. After the acclimation period is completed (7 days), fish fasted for 24 hours with the aim of eliminating the remaining food in the body. Feeding is done three times a day with satiation. Feces collection was conducted in the day-5 after stocking by siphoning out the faecal material from the bottom of the aquarium. The faecal material was dried at 60°C for 24 h.

2.4 Growth parameters

At the end experiment, the growth parameters of the tilapia were analysed for the specific growth rate (SGR), feed efficiency (FE), survival rate (SR) were calculated using equations described by [13]. Nutrient digestibility (ND) including protein digestibility, fat digestibility, and energy digestibility, and total digestibility (TD) were calculated according to [14].

$$\text{SGR} (\%) = 100 \times (\ln W2 - \ln W1) / \text{time (days)} \quad (1)$$

$$\text{SR} (\%) = \text{NT/N0} \times 100 \quad (2)$$

$$\text{FE} (\%) = 100 \times [\text{wet weight gain (g) / dry feed intake (g)}] \quad (3)$$

Where, W1 and W2 indicate the initial and final weight (g), respectively. NT is the number of fish at the end (individuals), N0 is the number at the beginning (individuals).

$$ND = 100 - [1-a/a' x b'/b] \quad (4)$$

$$TD = 100 - [1-a/a'] \quad (5)$$

Where a is Cr_2O_3 percentage in the feed, a' is Cr_2O_3 percentage in the faeces, b is a nutrient percentage in the feed, b' is the nutrient percentage in the faeces.

2.5 Chemical Analysis

The proximate composition of experimental diets and faecal samples was analysed according to the standard method of [16]. The proximate composition including crude protein, crude lipid, moisture, nitrogen-free extract, and the chromium content of diets and faecal.

2.6 Statistical analyses

Statistical analyses of growth parameters were carried out using the Statistical Package for the Social Sciences (SPSS) program for Windows (v. 16.0). Where significant differences were found, the means within each treatment and among treatments were compared using Duncan of multiple comparisons with a 95% significant level.

Table 1. Formulation of the diet in this study

Ingredients*	Treatments			
	0% formic acid	3% formic acid	4% formic acid	5% formic acid
Commercial feed (%)	67.55	67.55	67.55	67.55
Vegetable waste (A)	28.95	-	-	-
Silage meal (B)	-	28.95	-	-
Silage meal (C)	-	-	28.95	-
Silage meal (D)	-	-	-	28.95
Tapioca	3	3	3	3
Cr_2O_3	0.5	0.5	0.5	0.5
Total	100	100	100	100
Proximate analyses (% dry matter)				
Crude protein (%)	27.91	27.97	27.00	27.80
Crude lipid (%)	4.20	4.26	4.22	4.24
Fiber (%)	7.93	6.15	6.14	6.11
NFE (%)**	39.38	39.40	43.31	41.00
Energy (Kcal/kg feed)***	230.155	230.901	236.957	234.144

Note: * A = commercial feed + vegetable waste, B = commercial feed + vegetable waste silage (3% formic acid), C= commercial feed + vegetable waste silage (4% formic acid), D = commercial feed + vegetable waste silage (5% formic acid).

** Nitrogen free extracts

*** DE: Digestible Energy = carbohydrate: 2.5 kcal DE; protein: 3.5 kcal DE, fat: 8.1 kcal DE [13]

3. Results and Discussions

3.1 Growth parameters of tilapia

Formic acid is routinely used in the manufacture of silage and formic acid salts have been shown to increase disease resistance in tilapia [9]. Application of formic acid in the manufacture of silage has been observed in fingerlings of common carp *Cyprinus carpio* [8, 21], Mozambique Tilapia (*Oreochromis mosaambicus*) [9], discard fish (*Equulites klunzingeri* and *Carassius gibelio*) [10], abalone [11]. In the present study, 5% formic acid treatment can promote best growth compared to control for all parameters. Growth parameters of tilapia fed experimental diet is shown in Table 2.

Table 2. Effect of dietary vegetable waste silage as feed ingredients on growth parameters of tilapia*.

Parameters**	Treatments (%)			
	0% formic acid	3% formic acid	4% formic acid	5% formic acid
FI (%)	337.5±9.04	337±7.22	335±7.07	324±10.32
TD (%)	44.30±4.12^{a}	49.98±4.16^{ab}	55.61±5.92^{b}	55.66±4.15^{b}
PD (%)	78.56±2.63^{a}	81.78±2.30^{ab}	82.13±3.58^{ab}	84.59±1.26^{b}
FD (%)	74.35±1.60^{a}	80.38±1.54^{ab}	81.53±4.52^{b}	84.24±2.80^{b}
ED (%)	59.09±3.04^{a}	65.42±3.01^{ab}	69.22±4.08^{b}	69.48±4.06^{b}
SGR (% day^{-1})	2.54±0.21^{a}	2.56±0.05^{a}	2.66±0.14^{ab}	2.96±0.08^{c}
FE (%)	47.04±2.53^{a}	47.78±3.62^{a}	45.80±6.46^{a}	57.93±2.67^{b}
SR (%)	92.18±3.12	89.06±5.98	87.50±8.83	82.81±7.86

Note: * The value in the same row followed by similar superscript are not significantly different (p>0.05).
** (FW), feed intake (FI), total digestibility (TD), protein digestibility (PD), fat digestibility (FD), energy digestibility (ED), specific growth rate (SGR), feed efficiency (FE), survival rate (SR) of tilapia.

No significant difference (P<0.05) was obtained in feed intake. The value of feed intake in 0% formic acid treatment of 337.5±9.04%, 3% formic acid treatment of 337±7.22%, 4% formic acid of 335±7.07%, and 5% formic acid of 324±10.32%. No significant difference (P<0.05) was obtained in feed intake. This is indicating that the utilization of vegetable waste silage did not affect to feed palatability. Palatability is the level of preference for fish to consume the feed given at a certain time period [17]. The similar result has been reported by [18] in the feed based the other vegetable protein. He noted that the utilization of *Leucaena leucocephala* leaf meal did not affect to the appetite of rohu, *Labeo rohita* (Hamilton). Digestibility describes the fraction of the nutrient or energy in the ingested feedstuff that is not excreted in the feces [19]. Total digestibility is the number of nutrients in the feed that can be digested by fish [13]. The value of total digestibility was significantly higher (P<0.05) in 5% formic acid treatment (55.66±4.15%) and 4% formic acid treatment (55.61±5.92%) compared to control (44.30±4.12%), but there was no significant (P>0.05) difference between 3% formic acid treatment (49.98±4.16%), and 4% formic acid treatment and 5% formic acid treatment. Protein is most feedstuffs that have been properly processed are highly digestible of fish [13]. Protein digestibility showed higher (P<0.05) in 5% formic acid (84.59±1.26%) than control (78.56±2.63%). There are no significant (P>0.05) difference between 4% formic acid treatment, 3% formic acid treatment (82.13±3.58%, 81.78±2.30%, respectively) than control. In the present study, the protein digestibility value (78-84%) of tilapia were lower than the previous study by [20] in tilapia (89-93%).

Fat when administered either alone or in a mixed diet routinely gives digestibility values of 85% to 95% for fish [13]. Fat digestibility showed higher (P<0.05) in 5% formic acid treatment and 4% formic acid treatment (84.24±2.80%, 81.53±4.52%, respectively) than control (74.35±1.60%) and there no difference (P>0.05), between control and 3% formic acid treatment (80.38±1.54%). The similar result was obtained in energy digestibility, the result showed that energy digestibility was significant higher (P<0.05) in 5% formic acid treatment (69.48±4.06%) and 4% formic acid treatment (69.22±4.08%) compared to control (59.09±3.04%) and no significant difference (p>0.05) between control and 3% formic acid treatment (59.09±3.04%). The high energy digestibility value in 4-5% formic acid treatment might be related to the high value of protein digestibility and fat digestibility in this treatment. In this study, nutrient digestibility was significantly higher observed in the treatment of 5% formic acid compared to other treatments. This shows that the dose of formic acid 5% is effective in breaking down the nutrients contained in vegetable waste so that the digestibility of nutrients in the treatment of 5% formic acid is better than other treatments.

The value of specific growth rate was significantly highest (P<0.05) in 5% formic acid (2.96±0.08 %/day), followed by 4% formic acid (2,66±0,14 %/day), 3% formic acid (2,56±0,05 %/day) and control (2,54±0,21 %/day). Feed efficiency in this study showed highest (P<0.05) in 5% formic acid treatment (57.93±2.67%) compared the other treatments, followed by 3% formic acid treatment, control, and 4% formic acid treatment (47.78±3.62%, 47.04±2.53%, 45.80±6.46, respectively). In the present study, the high growth and feed efficiency may be due to the higher nutrient digestibility value in 5% formic acid treatment than the other treatments. The similar result has been reported by [21] in major carp, he found that a higher growth response on major carp feed a 26.4 to 53.1 % silage-based diet than fish fed a fishmeal-based diet. The same results were founded for salmon. [22] noted that salmon fed a 25% silage-based diet grew faster than fish fed a commercial dry pellet. The survival rate of feeding trial in control, 3% formic acid treatment, 4% formic acid treatment, and 5% formic acid treatment (92.18±3.12%, 89.06±5.98, 87.50±8.83, 82.81±7.86, respectively). The result is indicating that the addition of formic acid in making vegetable waste silage as feed did not affect to fish health of tilapia. Similarly, [9] has reported that the utilization of formic acid in making rainbow trout silage not effect on survival rate and immune stimulant of tilapia.

4. Conclusions

Vegetable waste silage can be used as a feed ingredient of tilapia feed. Total digestibility, protein digestibility, and fat digestibility were significantly higher ($P<0.05$) in 4% formic acid and 5% formic acid than Control. The value of growth was significantly highest ($P<0.05$) in 5% formic acid (2.96±0.08 %/day), followed by 4% formic acid (2,66±0,14 %/day), 3% formic acid (2,56±0,05 %/day) and control (2,54±0,21 %/day). The utilization of 5% formic acid in making vegetable waste silage showed the best result on growth and feed efficiency than the other treatments.

References

[1] Food and Agriculture Organization. 2018 *The State of World Fisheries and Aquaculture: Opportunities and Challenges* Roma FAO p 223.

[2] Hertrampf JW, Piedad-Pascual F, 2000 *Handbook on Ingredients for Aquaculture feeds* Los Angles Kluwer Academic Publishers p 624

[3] Lee C-S, Lim C, Gatlin III DM, Webster CD 2015 *Dietray Nutrients, Additives, and Fish Health.* Canada Wiley Blackwell p 355.

[4] Abun, Rusmana D, Saefulhadjar D 2007 The effects of mechanical processing of waste vegetables on digestibility at super native chicken JJ-101 *Journal of Animal Science Padjadjaran University* **7** 81-86.

[5] Superianto S, Harahap AE, Ali A 2018 Nutrition value of cabbage vegetable waste silage with rice bran addition and different duration of fermentation *Indonesia Animal Science Journal* **13** (2) 172-181.

[6] Muktiani A, Achmadi J, Tampoebolon BIM, Setyorini R 2013 The use of vegetable waste silage supplemented with mineral and alginate as feeding for sheep *Jurnal Ilmu dan Teknologi Peternakan* **2** (3) 144-151.

[7] Toppe J, Olsen RL, Penarubia OR, James DG 2018 *Production and Utilization of Fish Silage: a manual on how to turn fish waste into profit and a valuable feed ingredient or fertilizer*. Rome: Food and Agriculture Organization of the United Nations p 16

[8] Ramasubburayan R, Iyapparaj P, Subhashini KJ, Chandran MN, Palavesam A, Immanuel G 2013 Characterization and Nutritional quality of Formic Acid Silage Developed from marine Fishery Waste and Their Potentia Utilization as Feed Stuff for Common Carp *Cyprinus carpio* Fingerlings *Turkish Journal of Fisheries and Aquatic Sciences* **13** 281-289

[9] Goosen NJ, Wet LFD, Gorgens JF 2016 Rainbow Trout Silage as Immune Stimultant and Feed Ingredient in Diets for Mozambique Tilapia (*Oreochromis mosaambicus*) *Aquaculture Research* **47** 329-340

[10] Ozyurt G, Boga M, Ucar Y, Boga EK, Polat A 2017 Chemical, Biactive Properties and In Vitro Digestibility of Spray-Dried Fish Silages: Comparison of Two Discard Fish (*Equulites klunzingeri* and *Carassius gibelio*) Silages *Aquaculture Nutritions* **24** (3) 998-1005

[11] Goosen NJ, Wet LFD, Gorgens JF 2018 Effects of Formic Acid in Abalone Diets that Contain Ingredients Derived from Fish Processing by-product *Aquaculture International* **26** (3) 857-868.

[12] Takeuchi 1988 *Laboratory Work-Chemical Evalualuation of Dietary Nutriens. In Watanabe (Ed) Fish Nutrition and Mariculture* Kanagawa International Fisheries Training. Japan Japan International Coopration Agency (JICA) p 179–233

IOP Conf. Series: Earth and Environmental Science **383** (2019) 012002 doi:10.1088/1755-1315/383/1/012002

[13] National Research Council 1993 Nutrient Requirement of Fish National Academic Press. Washington D.C p 273

[14] Huisman EA 1987 Principles of Fish Production. Department of Fish Culture and Fisheries, Wageningen Agriculture University, Wageningan Netherland p 170

[15] Takeuchi 1988 Labrotary Work-Chemical Evaluation of Dietary Nutriens P.179-233, *In* Watanabe (Ed) Fish Nutrition and Mariculture. Kanagawa International Fisheries Training. Japan International Cooperation Agency (JICA) Japan

[16] Association of Official Agriculture Chemists 1995 Official Methods of Analysis 16th ed Volume II Maryland

[17] Tantiakitti C 2014 Review article: Feed palatability and the alternative potein sources in shrimp feed *Songklanakarin Journal of Science and Technology* **36** (1) 51-55

[18] Bairagi A, Ghosh KS, Sen SK, Ray AK 2004 Evaluation of the nutritive value of *Leucaena leucocephala* leaf meal, inoculated with fish intestinal bacteria *Bacillus subtilis* and *Bacillus circulans* in formulated diets for rohu, *Labeo rohita* (Hamilton) fingerlings *Aquaculture Research* **35** 436-446.

[19] National Research Council. 2011 *Nutrient requirements of fish and shrimp* In The National Academies Press, Washington DC p 376

[20] Borgesi R, Portz L, Oetterer M, Cyrino JEP 2008 Apparent digestibility coefficient of protein and amino acid, biological and enzymatic silage for nile tilapia *Aquaculture Nutrition* **14** 242-248

[21] Ali MZ, Gheyasuddin S, Zaher M, Hossain MA, Islam MN 1994 Evaluation of fish silage prepared from under-utilised marine fishes as protein sources in the diet of major carp (*Cirrhinus mrigala*) *Journal of Aquaculture in the Tropics* **9** 247-254.

[22] Crampton VBN, Watret R 1982 Moist feeds for salmon Fish Farmer **5** 11.

Promising upland rice lines as new shading tolerant varieties with good grain quality

A P Lestari*, Yullianida, R Hermanasari, A Hairmansis

Indonesian Center for Rice Research.Jl Raya 12 Sukamandi Subang West Java Indonesia 41256

*Corresponding author: ap_lestari@yahoo.com

Abstract. The utilization of coconut plantation land to increase rice planting area is a potential effort to maintain food sustainability in Indonesia. The development of shading tolerant rice varieties has been undertaken through plant breeding programs. Preliminary yield trial of upland rice genotypes under coconut plantation with 50% light intensity have been carried out in the village of Cilincing, Tegalbulued District, Sukabumi in the WS 2017. The genotype consists of 36 rice lines 4 check varieties, i.e Inpago 8, Limboto, Situpatenggang, and Jatiluhur. The result showed that the height of the lines ranged from 70.8 cm - 111.1 cm and the number of productive tillers were between 3 - 11.9 tillers. The line B15341-1B-TB-29 showed the earliest flowering time of 83 days after sowing (DAS). The panicle length of rice genotypes ranged from 29.4 cm - 21.8 cm and the average weight of 1000 grains of the line was 25.7 g. The ranges of filled grain per panicle of the lines were 47-157 grains. B14908C-MR-1-25-1-3 had the highest amount of the number of filled grain per panicle. Amylose levels of those genotypes were in the range of 17-22% and the rice texture was sticky to medium. The shape and size of the rice were long to medium. There were six promising rice, namely B14981B-TGB-4, B15344B-TB-30, B15175C-TGB-39, B15209B-MR-17-1, B14908C-MR-1-25-1-3, and B14192E-MR-33 that have the potential to be released as new shade-tolerant varieties.

1. Introduction

The agricultural land conversion into industrial facilities and settlements continues to increase which significantly reduces the agricultural land availability and threatens food supply [6]. It is necessary to extend the rice planting area by using marginal land as dry land in the low or highland. The government's effort to fulfill the food needs utilize suboptimal land maximally because the majority of land in Indonesia was suboptimal such as dry land 148 million Ha (76.5%) from total land in Indonesia [4][9]. The dry land which was suitable for agricultural cultivation was about 76.22 million Ha (52%) [1].

The development of upland rice is directed to the plantation area where upland rice is planted among annual crops such as coconut, oil palm, teak, and rubber. Plantation areas in Indonesia have enormous potential for the extensification of rice production [2]. The area of land that has the potential to overlap upland rice on the land is estimated to reach 2 million ha per year [11]. In such conditions, rice plants often experience a lack of light so production is low. To overcome the stress, rice varieties that are tolerant to shade are required [5].

This research aimed to evaluate the appearance and yield of advanced generation strains of upland rice tolerant to shade and have good quality rice. Lines with high yields and good quality rice have the potential to be released as new shade-tolerant superior varieties.

2. Material and method

2.1. Time and place
The experiment was conducted in Sukabumi, West Java in MT1 2018. The experiment was carried out under coconut plantation with 50% sunlight intensity.

2.2. Materials
A total of 30 lines and check varieties i.e Inpago 8, Limboto, Situpatenggang, and Jatiluhur were used as the experimental material.

2.3. Experiment design
Design of this experiment was randomize complete block design with four replication. The plot size was 2 m x 5 m, spacing was 30 cm x 15 cm. Planting was done with the number of seeds 2-3 grains per hole. Fertilization used were 300 kg/ha NPK + 100 kg / ha Urea. Pests and diseases were controlled optimally.

2.4. Observation
The variables observed were plant height, number of productive tillers, age of flowering plants (50%), age can be harvested (80% yellowing rice), amount of filled grain and empty per panicle, 1000 grain weight (14% moisture content), yield (14% moisture content) and grain quality.

2.5. Data analysis
Data were analyzed using analysis of variance. The difference in average between lines will be tested with the Least Significant Diffrerence (LSD) at the real difference level of 5%.

3. Result

3.1. Field performance
The performance and recapitulation of agronomic components of plant height, number of productive tillers, flowering age, harvest age, and yield of tested lines were shown in Figure 1 and Table 1. The height of the lines ranged from 70.8 cm - 111.1 cm. According to [7] the shade level have significant effect on the plant height and leaves number. Plant height of the check varieties were decline when compared to the characters in their description in [10]. The number of productive tillers were between 3 - 11.9 tillers. Jatiluhur was a check with the highest tillers 6.7 than another check. The earliest flowering age was B15341-1B-TB-29 with an age of 83 DAS. Inpago 8 was the earliest check variety with 84 DAS and the deepest was Limboto with 91.3 DAS. The average flowering age of the lines was between 83 - 99 DAS.

Figure 1. Yield experiment of upland rice lines at Tegalbuleud, Sukabumi WS 2017

The lengths of panicles of the lines were in the range of 29.4 cm - 21.8 cm where the lines with the longest panicle were B15053F-PWR-8 (Table 2). There were 7 lines with not significantly different panicle length from the best check varieties Limboto (26.7 cm), namely B15231-MR-10-1, B15344B-TB-30, B14958-MR-11-25- 4-1, B14908C-MR-1-25-1-3, B14192E-MR-33, and B11908F-TB-1-1.

Table 1. Agronomic characters of upland rice lines, Tegalbuleud, Sukabumi, WS 2017

No	Line	Plant height	No. of productive tiller	Day to flowering	Day to harvest
1	B14168E-MR-2	73.9	4.7	96.3	123
2	B11908F-TB-1-1	77.2	3.0	99.0	126
3	B12825E-TB-2-4	90.7	4.0	96.3	118
4	B11908D-MR-2-2-4	71.9	4.2	91.3	118
5	B11423G-MR-1-7	90.6	4.7	96.0	123
6	B14129E-MR-5	89.6	3.8	98.0	125
7	B14129E-MR-13	103.4	4.4	88.3	115
8	B15053F-PWR-8	70.8	3.4	97.3	124
9	B11957-RS*-2-3-2-18-2-SI-2-MR-2-PN-3-3	81.7	3.8	99.0	126
10	B15175C-TGB-13	82.8	4.6	96.0	123
11	B15175C-TGB-16	79.4	5.0	94.7	122
12	B15175C-TGB-23	75.3	6.3	86.7	115
13	B14957-MR-2-3-2	84.0	5.4	86.3	113
14	B14958-MR-11-25-4-1	95.7	8.8	85.7	113
15	B15114C-TB-22	110.8	7.4	84.3	111
16	B15341-1B-TB-2	95.8	8.4	87.0	114
17	B15341-1B-TB-29	100.4	8.3	83.0	110
18	B14981B-TGB-4	111.1	11.9	84.3	111
19	B15344B-TB-30	108.2	8.0	87.0	114
20	B15344B-TB-34	101.0	6.2	91.7	118
21	B14168E-MR-10-1	85.3	7.0	93.0	120
22	B15175C-TGB-39	85.1	7.1	84.3	111
23	JIRCAS-24	83.2	7.2	90.7	118
24	IR71525-19-1-1	101.6	6.1	90.7	119
25	B15209B-MR-1-3	89.4	6.6	85.7	114
26	B15209B-MR-17-1	90.4	5.2	87.3	114
27	B15408B-MR-4	99.3	4.7	89.7	117
28	B15231-MR-10-1	110.3	5.2	95.3	122
29	B15053F-PWR-2	95.2	4.4	97.3	124
30	B15209B-MR-12-1	91.8	6.0	90.0	117
31	B14908C-MR-1-25-1-3	95.8	8.1	88.3	118
32	B13643G-TB-26	97.2	7.9	91.3	118
33	B14191E-MR-3-61	100.7	6.3	88.7	118
34	B14192E-MR-2	99.3	4.9	88.3	116
35	B14192E-MR-33	103.8	5.1	88.7	116
36	B10580E-KN-28-1-1	86.0	4.4	91.7	120
37	Inpago 8	88.9	4.8	84.0	111
38	Limboto	97.0	5.6	91.3	118
39	Situ Patenggang	104.1	5.0	89.7	117
40	Jatiluhur	85.8	6.7	90.0	117

No	Line	Plant height	No. of productive tiller	Day to flowering	Day to harvest
	Grand mean	92.1	5.9	90.6	118
	Mean of lines	91.9	5.9	90.8	118
	Mean of checks	93.9	5.5	88.8	116
	LSD (5%)	20.9	3.6	8.6	7.9

The average weight of 1000 grains of the lines was 25.7 g, where this value was higher than the average of check varieties with a weight of 24.6 g. The line with the highest weight of 1000 grains was B15175C-TGB-39, where there were also 6 lines with a weight of 1000 grains which were not significantly different from, namely no. B15175C-TGB-23, IR71525-19-1-1, B11908F-TB- 1-1, B15175C-TGB-16, B15175C-TGB-13, and B14192E-MR-2. Limboto variety was the highest 1000 grains compared to other check varieties (26.6 g).

Table 2. Yield and yield components of upland rice lines, Tegalbuleud, Sukabumi, WS 2017

No	Line	Panicle lenght (cm)	No. of filled grain	Weight of 1000 grains(g)	yield (t/ha)
1	B14168E-MR-2	24.5	96	23.4	1.85
2	B11908F-TB-1-1	26.4	99	29.7	1.50
3	B12825E-TB-2-4	25.8	94	26.6	1.34
4	B11908D-MR-2-2-4	21.9	47	26.4	0.77
5	B11423G-MR-1-7	23.4	59	27.1	1.31
6	B14129E-MR-5	25.7	102	25.7	1.77
7	B14129E-MR-13	24.8	88	25.8	1.57
8	B15053F-PWR-8	29.4	114	20.9	1.77
9	B11957-RS*-2-3-2-18-2-SI-2-MR-2-PN-3-3	24.2	84	24.0	1.16
10	B15175C-TGB-13	22.1	53	29.0	0.85
11	B15175C-TGB-16	22.3	56	29.3	1.32
12	B15175C-TGB-23	22.9	65	30.7	1.36
13	B14957-MR-2-3-2	24.6	106	24.3	0.91
14	B14958-MR-11-25-4-1	28.3	120	17.5	1.11
15	B15114C-TB-22	25.3	86	23.7	1.29
16	B15341-1B-TB-2	23.6	76	26.3	1.24
17	B15341-1B-TB-29	25.5	91	26.8	1.70
18	B14981B-TGB-4	26.1	119	27.5	2.08
19	B15344B-TB-30	28.5	144	25.1	2.08
20	B15344B-TB-34	25.1	105	24.6	1.65
21	B14168E-MR-10-1	22.1	87	22.6	1.09
22	B15175C-TGB-39	21.8	57	31.2	1.66
23	JIRCAS-24	23.5	92	24.5	1.72
24	IR71525-19-1-1	24.6	85	30.5	1.26
25	B15209B-MR-1-3	25.6	100	26.8	1.57
26	B15209B-MR-17-1	24.8	95	25.8	2.13
27	B15408B-MR-4	24.2	88	27.2	1.38

No	Line	Panicle lenght (cm)	No. of filled grain	Weight of 1000 grains(g)	yield (t/ha)
28	B15231-MR-10-1	29.2	140	25.1	2.02
29	B15053F-PWR-2	26.1	106	19.9	1.77
30	B15209B-MR-12-1	24.8	68	26.3	1.67
31	B14908C-MR-1-25-1-3	27.6	157	20.1	2.44
32	B13643G-TB-26	23.1	75	24.5	1.37
33	B14191E-MR-3-61	22.9	103	24.5	1.47
34	B14192E-MR-2	22.2	73	28.9	1.40
35	B14192E-MR-33	26.8	125	26.9	2.27
36	B10580E-KN-28-1-1	23.3	100	24.9	1.66
37	Inpago 8	25.2	115	23.1	2.55
38	Limboto	26.7	113	26.6	1.75
39	Situ Patenggang	25.7	116	25.6	1.49
40	Jatiluhur	22.1	107	22.9	1.81
	Grand Mean	24.8	95	25.6	1.58
	Mean of lines	24.8	93	25.7	1.54
	Mean of checks	24.9	113	24.6	1.90
	LSD 5%	3.15	49.1	3.64	1.30

The range of grain content per panicle of the lines was between 47-157 grains where B14908C-MR-1-25-1-3 had the highest amount of grain. This value was higher than the best check varieties Situpatenggang with 116 grain contents.

The highest yield were Inpago 8 varieties with 2.55 t/ha and there were 8 lines with results not significantly diferent with Inpago 8 namely B14168E-MR-2, B15231-MR-10-1, B14981B-TGB-4, B15344B-TB-30, B15209B -MR-17-1, B14192E-MR-33 and B14908C-MR-1-25-1-3. However the selection of strains would be promising lines based not only on results but also by other characters that support high yield.

3.2. Correlation

Correlation character values with rice yields were used as an indicator for selecting improved rice varieties. Values of simple correlation showed that panicle lenght and number of grains per panicle were positively and significantly correlated with grain yield (Table 3). [12] founded that number of tillers per plant, days to maturity, and plant height had significant and positive correlation with grain yield. The selection on the basis of these characters can be helpful for improvement in rice grain yield and quality.

Table 3. Correlation among characters of upland rice lines

	PH	NPT	DF	DH	PL	NFG	WG	Y
PH	1.00000	0.08947	-0.22831	-0.23	0.22768	0.23657	-0.03	0.13601
		0.3332	0.0125	0.0133	0.0128	0.0096	0.7118	0.1402
NPT	0.08947	1.00000	-0.35608	-0.35	-0.01660	0.03061	-0.06	0.09895
	0.3332		<.0001	<.0001	0.8572	0.7400	0.5411	0.2823

	PH	NPT	DF	DH	PL	NFG	WG	Y
DF	-0.22831	-0.35608	1.00000	0.95	0.07141	-0.07013	-0.06	-0.11596
	0.0125	<.0001		<.0001	0.4383	0.4466	0.5213	0.2072
DH	-0.226	-0.353	0.954	1.00	0.06650	-0.06631	-0.06	-0.11628
	0.013	<0.001	<0.001		0.4705	0.4718	0.5137	0.2060
PL	0.228	-0.017	0.071	0.07	1.00000	0.80167	-0.10	0.39089
	0.013	0.857	0.438	0.470		<.0001	0.2877	<.0001
NFG	0.237	0.031	-0.070	-0.07	0.80167	1.00000	-0.13	0.55401
	0.010	0.740	0.447	0.472	<.0001		0.1486	<.0001
WG	-0.034	-0.056	-0.059	-0.06	-0.09785	-0.13267	1.00	0.14125
	0.712	0.541	0.521	0.514	0.2877	0.1486		0.1239
Y	0.136	0.099	-0.116	-0.12	0.39089	0.55401	0.14	1.00000
	0.140	0.282	0.207	0.206	<.0001	<.0001	0.1239	

*PH=plant height, NPT=number of productive tiller, DF=daay to flowering, DH=day to harvest, PL=panicle length, NFG=number of filled grain, WG=weight of 1000 grains, Y=yield

3.3. Grain quality

The percentage of head rice averages above 46% with a range between 35% - 59% (Table 4). Head rice, which is healthy rice grains and defects that have a size greater than or equal to 75% of part of whole rice grains. The highest head rice of check variety was Limboto (56%), followed by Situpatenggang (55%). Some lines with head rice equivalent to check varieties were B15344B-TB-30, B12825E-TB-2-4, B15344B-TB-34, B15175C-TGB-39 and B15053F-PWR-8.

Some lines have milled rice percentage of more than 60%. The highest yield was 67% in lines no. 24 and 31 while other lines with yields above 60% were number B11908F-TB-1-1, B12825E-TB-2-4, B14129E-MR-5, B14129E-MR-13, B14957-MR-2-3-2, B15114C-TB-22, B15341-1B-TB-2, B15231-MR-10-1, B14191E-MR-3-61, B14192E-MR-2 and B14192E-MR-33.

The rice samples used in this experiment were comes from drying treatment. According to [8] the treatment can affect the quality of rice. The use of a drying machine can avoid contamination of sand or other impurities. In addition, the yield loss is less than 1%, lower than the drying method of 1.5-2.0%.

The rice texture test showed that there were 16 medium-textured lines, 15 tender, and 5 lines had hard texture. The hard-texture check was Jatiluhur and medium check varieties were Inpago 8, Limboto, Situpatenggang. Amylose levels of strains ranged from 16.6 (no. 32) to 26.56 (no. 3). The lower amylose content, the texture of the rice is getting fluffier and conversely the higher the amylose content of the rice texture will be more hard.

Medium texture check varieties had amylose content between 21% - 24%. Judging from the size of the rice, the average size of the rice strains tested was long (L), medium (M), medium (M) which meant long, while medium liming or limestone still appeared in the middle of the rice belly. Check varieties were mostly LMM except Inpago 8 with LMS shape and size. There were 6 lines, namely B11908D-MR-2-2-4, B15175C-TGB-13, B15175C-TGB-16, B15344B-TB-34, B15175C-TGB-39 and B14192E-MR-2, were detected as red rice. Line 15 as aromatic. [3] stated that low light during grain filling stage affected physical appearance of rice grains by reducing milled rice rate and increasing the chalkiness rate. In addition, cooking qualities also decreased in response to low light.

Table 4. Grain quality of upland rice lines, Tegalbuleud, Sukabumi WS 2017

No	Galur	Milled rice	Amylose (%)	Rice texture		L	S	C	Description
1	B14168E-MR-2	53	23.84	3.2	medium	L	M	M	
2	B11908F-TB-1-1	62	26.44	4.0	hard	L	M	M	
3	B12825E-TB-2-4	62	26.56	3.8	hard	L	M	M	
4	B11908D-MR-2-2-4	50	19.44	2.0	tender	L	M	M	Red rice
5	B11423G-MR-1-7	54	19.68	2.2	tender	L	M	M	
6	B14129E-MR-5	60	19.52	2.0	tender	L	M	M	
7	B14129E-MR-13	64	21.68	2.8	medium	L	M	M	
8	B15053F-PWR-8	60	23.76	3.0	medium	M	M	M	
9	B11957-RS*-2-3-2-18-2-SI-2-MR-2-PN-3-3	52	24.88	3.8	hard	L	M	M	
10	B15175C-TGB-13	35	22.72	2.8	medium	L	M	M	Red rice
11	B15175C-TGB-16	49	17.52	1.8	tender	L	M	M	Red rice
12	B15175C-TGB-23	42	19.36	2.0	tender	L	M	M	
13	B14957-MR-2-3-2	62	24.00	3.0	medium	L	M	M	
14	B14958-MR-11-25-4-1	58	24.48	3.7	hard	L	M	M	
15	B15114C-TB-22	60	23.04	3.1	medium	L	M	M	Aromatic
16	B15341-1B-TB-2	63	21.52	2.8	medium	L	M	M	
17	B15341-1B-TB-29	49	19.36	2.1	tender	L	M	M	
18	B14981B-TGB-4	58	18.96	2.0	tender	L	M	M	
19	B15344B-TB-30	57	20.72	2.5	medium	L	M	M	
20	B15344B-TB-34	57	22.08	2.8	medium	L	M	M	Red rice
21	B14168E-MR-10-1	45	25.60	3.8	hard	L	M	M	
22	B15175C-TGB-39	56	21.12	2.6	medium	L	M	M	Red rice
23	JIRCAS-24	48	25.52	3.8	medium	L	S	S	
24	IR71525-19-1-1	67	18.64	1.8	tender	L	M	S	
25	B15209B-MR-1-3	37	20.96	2.5	medium	L	M	M	
26	B15209B-MR-17-1	38	17.28	2.1	tender	L	S	S	
27	B15408B-MR-4	55	19.36	2.2	tender	L	S	S	
28	B15231-MR-10-1	66	23.92	3.2	medium	L	M	M	
29	B15053F-PWR-2	59	22.16	2.8	medium	L	M	S	
30	B15209B-MR-12-1	47	18.40	2.0	tender	L	S	S	
31	B14908C-MR-1-25-1-3	67	20.00	2.2	tender	M	M	S	
32	B13643G-TB-26	59	16.64	2.0	tender	L	M	M	
33	B14191E-MR-3-61	66	20.72	2.6	medium	L	M	M	
34	B14192E-MR-2	62	16.96	1.8	tender	L	M	M	Red rice
35	B14192E-MR-33	65	21.52	2.8	medium	L	S	S	
36	B10580E-KN-28-1-1	54	18.64	2.0	tender	L	M	S	
37	Inpago 8	56	21.37	3.0	medium	L	M	S	

No	Galur	Milled rice	Amylose (%)	Rice texture		L	S	C	Description
38	Limboto	65	23.16	3.3	medium	L	M	M	
39	Situ Patenggang	69	24.01	3.2	medium	L	M	M	aromatic
40	Jatiluhur	65	26.59	3.8	hard	L	M	M	

*L=long, M=medium, S=slender/small

Based on the yield components and the grain quality there were 6 lines may be filed as a material for multi-location test to obtain information about the potential and stability of the yield from the various locations, i.e. B14981B-TGB-4, B15344B-TB-30, B15175C-TGB-39, B15209B-MR-17-1, B14908C-MR-1-25-1-3 and B14192E-MR-33. Those lines are expected to become new shade tolerant varieties.

4. References

[1] Abdurachman A, Dariah A and Mulyani A 2008 *Strategi dan teknologi pengolahan lahan kering mendukung pengadaan pangan nasional Jurnal Litbang Pertanian.* Bogor. **27** 43-49

[2] Alridiwirsah Harahap EM Akoeb EN and Hanum H 2018 Growth and production of new superior rice varieties in the shade intensity. *IOP Conf. Series: Earth and Environmental Science* 122 Doi: 10.1088/1755-1315/122/1/012024

[3] Cheng-gang L Jia L Yan W Dan X Chun-bang D Tian L 2015 Low light during grain filling stage deteriorates rice cooking quality, but not nutritional value *Rice Science* **22** (4): 197-206 https://doi.org/10.1016/j.rsci.2015.04.003

[4] Haryono 2012 *Lahan rawa, lumbung pangan masa depan Indonesia.* IAARD Press. Badan Penelitian dan Pengembangan Pertanian.142pp.

[5] Muhidin Jusoff K Elkawakib S Yunus M Kaimuddin Meisanti Gusti Ray K and Baka La Rianda 2013 The Development of Upland Red Rice under Shade Trees *World Applied Sciences Journal* **24**(1): 23-30 doi: 10.5829/idosi.wasj.2013.24.01.13179

[6] Rondhi M Pratiwi PA Handini VT Sunartomo AF and Budiman SA 2018 Agricultural Land Conversion, Land Economic Value, and Sustainable Agriculture: A Case Study in East Java, Indonesia *Land* **7**(4):1-19 https://doi.org/10.3390/land7040148

[7] G R Sadimantara GR Alawyah T Suliartini NWS Febrianti E and Muhidin 2019 Growth performance of two superior line of local upland rice (*Oryza sativa* L.) from SE Sulawesi on the low light intensity *IOP Conf. Series: Earth and Environmental Science* **260** doi:10.1088/1755-1315/260/1/012145

[8] Sasmita P 2007 *Aplikasi tenik kultur antera pada pemuliaan tanaman padi. Prosiding: Apresiasi Hasil Penelitian Padi.* Balai Besar Penelitian Tanaman Padi. Hal:595-609.

[9] Suhaeti RN and Wahyuni S 2018 Farmer's sovereignty in dry land and swamps areas *J. Phys*: Conf. Ser.953 012019

[10] Suprihatno B Daradjat AA Satoto Baehaki SE Widiarta IN Setyono A Indrasari SD Lesmana OS Sembiring H 2009 *Deskripsi Varietas Padi* Subang:Balai BesarPenelitian Tanaman Padi, 105 Hal.105

[11] Toha H M Pirngadi K Permadi K dan A M Fagi 2009 *Meningkatkan dan memantapkan produktivitas dan produksi padi gogo*. Dalam A.A. Daradjat, A. Setyono, A.K. Makarim, dan A. Hasanudin (Eds). Padi Inovasi Teknologi Produksi Buku 2. LIPI Press, Jakarta Hal.167-200.

[12] Wattoo JI Khan AS Ali Z Babar M Naeem M Amanullah M Hussain N 2010 Study of correlation among yield related traits and path coefficient analysis in rice (*Oryza sativa* L.) *African Journal of Biotechnology* **9**(46): 7853-7856

5. Acknowledgments

Authors wishing to acknowledge IAARD Ministry of Agriculture of Indonesia in year 2018 for the financial support and our technical staff for data collecting.

The present status of fixed lift net fishing in Banten Bay Indonesia

A Susanto*,1,2 and H S Nurdin1,2

1Department of Fisheries, Faculty of Agriculture, University of Sultan Ageng Tirtayasa, Jl. Raya Jakarta Km. 4 Panancangan, Cipocok Jaya Kota Serang, Banten 42122 Indonesia
2Indonesia-Center of Excellence for Food Security, University of Sultan Ageng Tirtayasa, Jalan Raya Jakarta Km. 4, Panancangan, Cipocok Jaya, Kota Serang, Banten 42124, Indonesia

*Corresponding author: adisusanto@untirta.ac.id

Abstract. Banten Bay is the main fishing ground for small scale fishing activity in Banten Province. The fishers use several fishing gears, such as gillnet, trap, hook and line, danish seine, and lift net. However, a lift net fishery has a significant contribution to fish production in Archipelagic Fishing Port of Karangantu, Serang City. The purpose of this study is to analyze the status of fixed lift net fishing in Banten Bay. The research was conducted on Mei to July 2019 through survey and field observation. There are more than 150 units of fixed lift net operated in Banten Bay, that have landing place in Karangantu and Panjang Island. The rectangle lift nets (10Wx10L m^2) are fixed at six to ten-meter depth. The fishers used four to six pressure kerosene lamps as a light source before 2010. However, since 2010 the fishers changed the lamp to the electrical light source due to high price, and limited availability of kerosene. In peak season, the lift nets have high productivity (500-1000 kg/trip). The productivity was decreased in the low season (< 100 kg/trip). The main target species in fixed lift net fishing are anchovies and squids with a potential margin more than IDR 60 million per year.

1. Introduction

The fisherman used various fishing gears to utilize the fish resources in Banten Bay. Fishing units in this area dominated by small scale fisheries using < 30 GT fishing vessel. Lift net fisheries have a main contribution to the fish production in the fishing port of Karangantu, as primary landing place in Banten Bay. However, a fixed lift net is the prevailing fishing gears with the main target are anchovy, squid, scad, and mackerel. The fixed lift net (bagan tancap) is the stationary lift net that using an artificial light source to attract and concentrate fish in the catchable area. The fixed lift net fishing in Banten Bay is still developed due to the low investment cost and high productivity.

The fishermen in Banten Bay used pressurized kerosene lanterns as a light source until a mid-year of 2011. The development of the electric light was caused by the transformation of light source application in lift net, from kerosene to the electric lamp, especially a fluorescent lamp [1-2]. The fishers prefer the fluorescent lamp because it is easy to obtain, low cost [1-2] and considered bright enough to attract schooling fish [3]. However, the application of this lamp with various number and wattages for a light source influence the increase of the operational cost. The purpose of this research is to analyze the current status of fixed lift net fishing including a number of gears, and fishing productivity related to the substitution of a light source in Banten Bay.

2. Methods

This research used a mixed-method, i.e. study literature and field survey. The study literature was conducted through review the result of earlier research related to the fixed lift net fisheries in Banten Bay. The survey and field observation was accompanied on Mei to July 2019 in Banten Bay and Archipelagic Fishing Port of Karangantu. The descriptive analysis used to define the present status of a fixed lift net. Moreover, a cost-benefit analysis used to evaluate the economic prospect of the lift net fisheries in Banten Bay.

3. Results and Discussions

The fixed lift net in Banten Bay was made from bamboo in rectangle construction (10Lx10W m^2) as presented in figure 1. The fishers selected a fishing ground based on experience knowledge especially a pattern of sea current, wave or wind energy, and productivity at the earlier fishing season. Usually, the lift net is located at water depth less than 10 m and around a small island in Banten Bay. The lift nets are operated by one fisherman at a dark moon period. There are approximately 20 days of fishing trip every month.

Figure 1. The fixed lift net in Banten Bay

All fishers of fixed lift net in Banten By used the kerosene lamp until at the end of 2011. In each lift net, they usually used four to six kerosene lamps to attract the fish as showed in figure 2 [5]. The high cost of kerosene and limited spare part caused these lamp to become expensive. However, they changed the lamp into the fluorescent lamp as a light source in various number and power. They applied generator as an electric source with four to nine lamps with wattages approximately 45-90 W as presented in figure 3.

Figure 2. Fixed lift net fishing used the kerosene lamp in 2009

Figure 3. Fixed lift net fishing used the fluorescent lamp in 2014

Figure 4 showed the location of the fixed lift net in a different year. There were 101 units in 2011 with a dominant fishing location around the small island and coastal area. However, the number of lift net was increased by approximately 50 per cent in 2018. The fishers move their lift net into deeper location due to of fishing season and monsoon exchange. Generally, the number of fixed lift net in Banten Bay has fluctuation, related to the fishing productivity and profit in recent year. The fisher will be built a new lift net if there is high productivity during operation. The application of fluorescent lamp generates a high proportion of anchovy as the main target catch in lift net. It was related to the light spectrum of a fluorescent lamp which has suitable enough for an adaptation stage of anchovy [6]. As a result, the high productivity of anchovy increases the economic income, thus induce the increase of fixed lift net in Banten Bay.

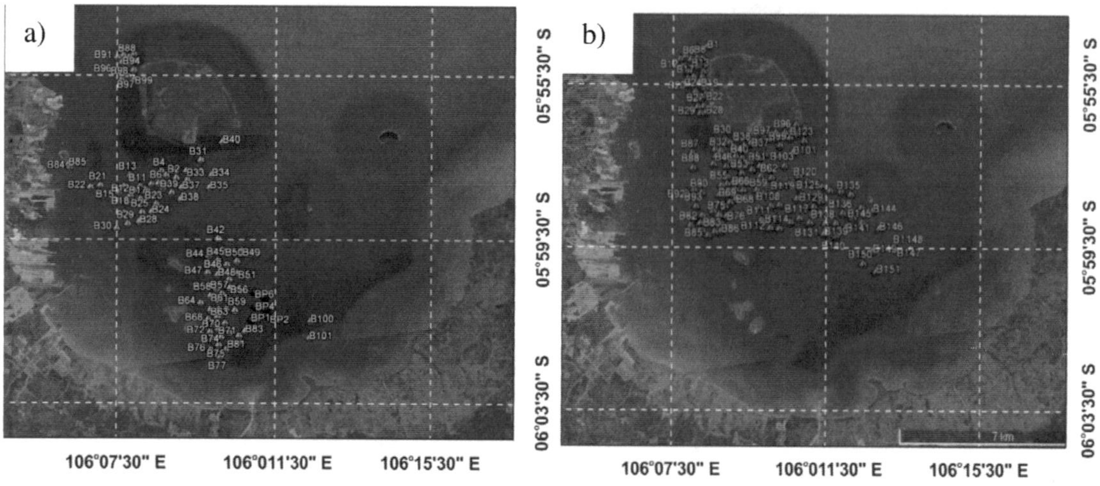

Figure 4. Fishing ground location of fixed lift nets in the different year a) 2011; b) 2018

Fixed lift net fishing in Banten Bay generates high economic income for fishermen. It caused the number of the unit to always fluctuation every year. However, there were increased in the investment and total cost of lift net fishing between 2009 to 2018. It was related to the price of bamboo, fuel cost, and purchase a generator as an energy source for a fishing lamp. Table 1 presented the cost-benefit analysis of fixed lift net fishing in 2009-2018.

Table 1. Cost-benefit analysis of fixed lift net in Banten Bay

Item	Value (USD)	Value (USD)	Value (USD)
Year	2009	2014	2018
Light source	Kerosene lamp	Fluorescent lamp	Fluorescent lamp
Fishing units investment	321.43	642.86	958.57
Total cost	2,140.86	3,178.57	4,707.21
Revenue	3,523.57	5,614.29	9,192.86
Benefit	1,382.71	2,435.71	4,485.64
R/C	1.65	1.77	1.95

The benefit from lift net fishing significantly is influenced by fishing season. There is three seasons in Banten Bay, i.e peak season, middle season and low season. The fishing productivity in peak season reaches 500-1.000 kg per trip. Meanwhile, in the low season, its productivity dropped only 20-50 kg per trip. In conclusion, fixed lift net fishing in Banten Bay is important enough to develop to become profitable small scale fisheries activity. Introduction of energy-efficient fishing light, i.e light-emitting diode (LED) in fixed lift net is a sufficient way to increase efficiency and effectiveness of the fishing operation.

4. References

[1] Hua L T, Xing J 2013 Research on LED fishing light *Res J Appl Sci, Eng and Technol.* **5** (16) 4138-4141

[2] Susanto A, Irnawati R, Mustahal, Syabana M A 2017 Fishing efficiency of LED lamps for fixed lift net fisheries in Banten Bay Indonesia *Turkish Journal of Fisheries and Aquatic Sciences.* **17** 283-291

[3] Khairul W, Mawardi, Riyanto M 2017 Use of blue lights emitting diode (LED) lamp to the catch of floating lift net in Aceh Jaya Regency *Albacore.* **1** (2) 235-243

[4] Hamidi, Baskoro M S, Riyanto M 2017 The use of underwater Light Emitting Diode (LED) with different color: its effect to the catch of boat lift net *Albacore.* **1** (3) 285-296

[5] Lee J W 2010 The influence of moon period to catch composition and fishermen income of fixed lift net in Serang Regency [Thesis]. Bogor Graduate School of Bogor Agricultural University. Bogor [ID] 117 pp

[6] Kondrashev S L, Gnyubkina V P, Zueva L V 2012 Structure and spectral sensitivity of photoreceptors of two anchovy species: *Engraulis japonicus* and *Engraulis encrasicolus Vision Research.* **68** 19-27

Promoting sustainable agricultural management through spatio temporal optimization of food crop land based on pedo-agroclimate at Kalalasi region, Southeast Sulawesi, Indonesia

A M Kandari[1]*, S Baja[2], A Ala[3], Kaimuddin[4], S Kasim[5], Y Taufik[6]

[1]Specifications Agro climatology, Department of Environmental Science, Faculty of Forestry and Environmental Science, Halu Oleo University, Southeast Sulawesi, Indonesia.
[2]Spesifications soil science, Department of Agro technology, Faculty of Agriculture, Hasanuddin University, Makassar, South Sulawesi, Indonesia
[3]Spesifications Agronomy, Department of Agro technology, Faculty of Agriculture, Hasanuddin University, Makassar, South Sulawesi, Indonesia.
[4]Spesifications Agro climatology, Department of Agro technology, Faculty of Agriculture, Hasanuddin University, Makassar, South Sulawesi, Indonesia
[5]Spesification Agroforestry, Department of Environmental Science, Faculty of Forestry and Environmental Science, Halu Oleo University, Southeast Sulawesi, Indonesia.
[6]Spesification Agribusiness, Department of Social Economy, Faculty of Agriculture, Halu Oleo University, Kendari, Southeast Sulawesi, Indonesia.

*Corresponding author: manekandaria@yahoo.com

Abstract. The purposes of this research were to: (1) analyze the suboptimal land availability and its suitability for food crops; (2) to identify the major and prime food crops; (3) analyze the socio-cultural and economic characteristics of farmers in order to develop local based climate change adaptation strategy and to prioritize major food crops using multi-criteria approach ; (4) to promote sustainable crop and land management, through the local specific technologies dealing with the constraints of the Kalalasi Region. As regional based study, this research used overlaid maps of thematic lands, undertaken on suboptimal-land area with dry-climate at Kalalasi Region, Southeast Sulawesi. This study employed a spatial analysis method using GIS. Data collection included data on land biophysics, climate, and farming-based socio-culture and economic, using biophysical and social economic survey method. Land evaluation was carried out using FAO method. The major food crops were identified and analyzed using LQ method. Then, Multi-Criteria Decision Method (MCDM) through an application of AHP was used to promote sustainable crop and land management at the Kalalasi Region. The results of this research showed that: (1) Kalalasi Region had relatively varied characteristics of land biophysic, climate, socio-culture and economic, (2) Potential land availability for the development of food crops was ha, where the land suitability level of S2 (moderately suitable) of ha, S3 (marginally suitable) of ha, and the rest was N1 and N2, (3) there were eight major food crops in the research area, namely: maize, cassava, upland rice, soy bean, sweet potato, ground nut, green bean, and paddy rice, (4) prime or excellent food crops in research area were varied among regions (sub-districts), however, in general, it was concluded that there were four prime food crops, namely: cassava, maize, upland rice, and paddy rice, (5) productive and sustainable crop pattern and land management can be achieved through the application of multiple cropping system, planting diversification, and appropriate planting date for the period of December, January, February, and March, using drought resistant cultivar, and several other efforts for adaptation and conservation, such as mulching and cover crops, to deal with climate change. These were conducted to maintain the land and crop productivity, and the sustainable food availability.

Keywords: Promoting, sustainable agriculture, pedo-agroclimate, suboptimal land

1. Introduction

Land is the basic capital and the main determining factor in the agricultural production system so it needs to be maintained. [1] elucidated that ineffective land use is closely related to climate change, but it can also happen otherwise that incorrect land use maycause climate change.Therefore, sensibility is needed in planning land use so that land optimization is truly effective and sustainable. [2] asserts that the goal of land use planning is to obtain the best use of land, through the achievement of three things, namely (a) efficiency, (b) equality and acceptance, (c) sustainability.

Land use optimization in terms of accurate and effective can be done, by means of land use planning considering many factors, especially land biophysics, applying remote sensing technology and GIS or geographic information system [3]. Furthermore, [4] emphasized that GIS applications have proven to be quite effective in determining crop land suitability. Land use planning by considering many factors or multi criteria results is better, usingnamely an analysis called AHP (Analytic Hierarchy Process) as one method of MCDM (Multiple Criteria Decision Making) [5].

Based on these various explanations, this study aimed to promote Spatio-temporal optimization of food land based on Pedo-agroclimate on suboptimal land with employing a multi-criteria approach, especially in Kalalasi Region District which dominated by marginal land.

2. Methods

This research was spatial based on a regional scale, applying quantitative and qualitative approach integrally (mixed approach) in two stages. Some of the important data used are the result of an overlay between thematic and informational map of previous researches [6].

The LQ value is calculated by using the following equation :

$$\boldsymbol{LQij} = \frac{Xij/Xi.}{X.j/x..} \quad \text{............ [7]}$$

LQij : quotient index sub district location i for food crops j
Xij : average size of food crop harvest j in sub district i
Xi. : average total size of food crops in sub district i
X.j : average total size of food crops j in the Kalalasi Region
X.. : average overall size of food crops in the Kalalasi Region

Notes:

a) LQ > 1; mainstay/basis, haves comparative advantages, the results in addition to fulfilling the needs of said region, it can also be export outside the region
b) LQ = 1; non basis/not mainstay, does not have comparative advantage, only sufficient to fulfill the needs of its own region
c) LQ < 1; non basis/not mainstay, cannot fulfill the needs of its own region therefore needs supplies or imports from outside the region

The R/c rasio is calculated by using the following equation

$$R/C\ rasio = \frac{TR}{TC}\ atau\ \frac{TR}{TN + NTT}$$

R/C rasio : index of farming suitability
TR : Total income; TC : Total cost (Rp)
NT : fixed cost; NTT : inconsistent cost

Notes:
R/C rasio > 1, farming effort yields profit (suitable)
R/C rasio = 1, farming effort break even, input=output
R/C rasio < 1, farming effort incur losses (not suitable)

The workflow scheme of the research can be seen in Figure 1.

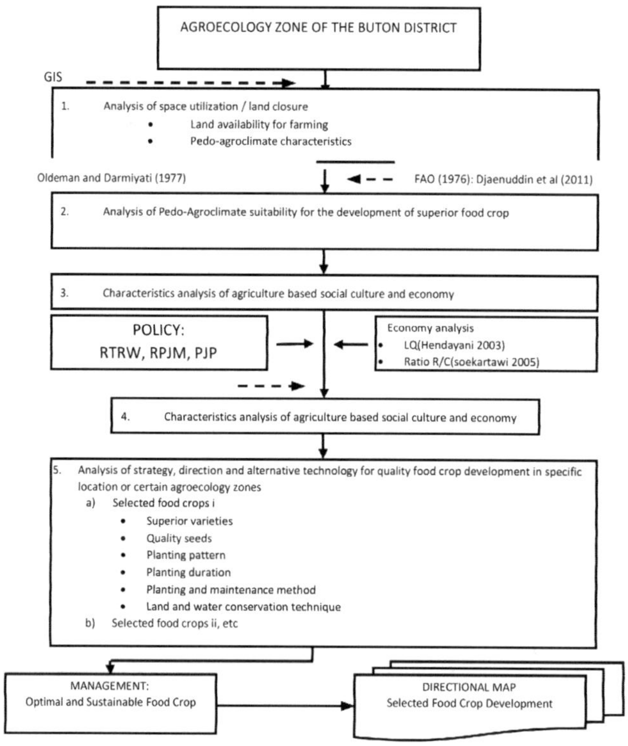

Figure 1. Analysis stage and main analysis parameter

The application of AHP in this research was intended to determine priority of selected food crops, based on factors that consists of 3 criteria: (1) social culture potential, which comprise of three sub criteria, which are policy (KJ), infrastructure (SP), farmer's interest (AP); (2) pedo-agroclimate potential covers two sub criteria which are weighted land suitability (KL) and location quotient (LQ), and (3) social economy potential comprises of two sub cultures which are market opportunity/marketing(PP) and farmer net income or R/C ratio (NI). Whereas the alternative consists of

10 species of food crops which are paddy rice (PS), field rice (PL), corn (JG), cassava (UK), peanuts (KT), green beans (KH), taro (TL) and sorghum (SG). The AHP hierarchy is presented on Figure 2.

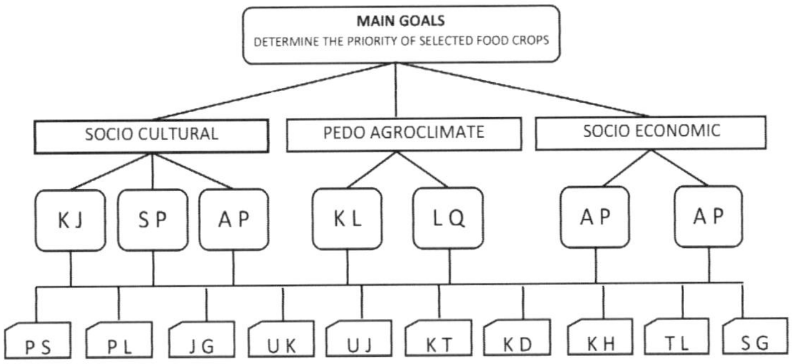

Figure 2. AHP structure determining priority of select food crops in research areas

Note:
- Factors : (1) Socio Cultural, (2) Pedo-Agroclimate, (3) Socio Economic
- Criteria : KJ = Policy; SP= Infrastructure; AP= farmers' interest
  KL = Land suitability; LQ = Location Quotient
  PP = Market opportunity; NI = Net Income (R/C ratio)
- Alternative: PS = Wet rice; PL= Rice field; JG = Corn
  UK = Cassava; UJ = Sweet Potato; KT = Peanuts
  KD = Soybean; KH = Green bean; TL = Taro; SG= Sorghum

3. Results and discussion
3.1. Results
3.1.1. Biophysical characteristic of Pedo-Agroclimate land
Result biophysical characteristic description of the land in the research area is presented in table 1 and can be visually seen on Figure 2.

Table 1 shows the condition of 10 research areas with different sizes of potential land with a total of 74 664.64 ha (35.55%) from the total size of the Kallasi Region (210 030.00 ha). The research areas possessed diversity of land biophysics which consisted of physio-topography and land (Figure 3), along with the climate especially the rainfall (Figure 4), all these factors affected the species and class of food crop suitability, and the coverage of its development.

Table 1. Land Biophysical Characteristics based on Pedo-Agroclimat

No	Satuan lahan dan Karakteristiknya	Satuan lahan dan Karakteristik lahan yang dominan di Wilayah Penelitian									
		Batauga	Sampolawa	Lapandewa	Wabula	Pasarwajo	Wolowa	Siotapina	Lasalimu Sel	Lasalimu	Kapontori
1	Satuan lahan :										
	a. Jumlah satuan lahan (329 unit)	28	20	12	10	31	14	24	52	65	73
	b. Luas lahan potensial (74,664.64 ha)	7.120,29	3.473,56	8.472,45	4.872,97	12.615,67	3.081,27	4.422,33	8.709,94	9.436,40	12.458,86
2	Zona agroekologi	Iax2, IIax2	Iax2 , IVax1	Iax2	Iax2	Iax2	Iax2, Iay2	Iay2, Iax2	IIay2, Iay2	Iay2, IIay2, IVay2	Iax2, IVax1, IVax2
3	Komponen Fisio-topografis :										
	a. Bentuk lahan (landform)	Sis perbukitan	Sis perbukitan	Terumbu karang	Sis perbukitan	Sis perbukitan	Sis perbukitan	Dat piedmont	Sis perbukitan	Sis perbukitan	Sis perbukitan
	b. Geologi tanah	Wapulaka	Sampolakosa	Wapulaka	Wapulaka	Wapulaka	Tondo	Sampolakosa	Wapulaka	Tondo	Wapulaka
	c. Kelerengan (%)	26-40 (Berlereng)	41-60 (agak curam)	41-60 (agak curam)	41-60 (agak curam)	41-60 (agak curam)	41-60 (agak curam)	16-25 (agak berlreng)	9-15 (landai)	41-60 (berlereng)	41-60 (berlereng)
4	Komponen Tanah :										
	a. Jenis tanah	Rendolls	Rendolls	Rendolls	Rendolls	Rendolls	Rendolls	Tropudalsfs	Tropudalfs	Dystropepts	Rendolls
	b. Tekstur tanah	Lempung berpasir	Liat berdebu	Lem liat berpsir	Lem liat berdebu	Lem Liat Berpsir	Lempung berpasir	Lempung	Lempung	Lempung berpasir	Lempung
	c. Struktur tanah	Gumpal membulat	Gumpal bersudut	Gumpal bersudut	Gumpal bersudut	Gumpal bersudut	Remah	Gumpal bersudut	Gumpal bersudut	Gumpal bersudut	Gumpal bersudut
	d. Kedalaman efektif (cm)	20-50 (dangkal)	>50-75 (agak dalam)	20-50 (dangkal)	20-50 (dangkal)	20-50 (dangkal)	20-50 (dangkal)	>75-100 (dalam)	20-50 (dangkal)	20-50 (dangkal)	20-50 (dangkal)
	e. Batuan permukaan (%)	>50-90 (AS berbatu)	>3-15 (C. berbatu)	>15-50 (S. berbatu)	>3-15 (C. berbatu)	>50-90 (AS.berbatu)	>15-50 (S. berbatu)	>3-15 (C. berbatu)	>0.01-0.1 (sedkit berbatu)	>3-15 (C. berbatu)	>15-50 (S. berbatu)
	f. Singkapan batuan permukaan (%)	>50-90 (AS berbatu)	>2-25 (C berbatu)	>25-50 (S. berbatu)	>0.1-2 (A berbaatu)	>2-25 (C berbatu)	>2-25 (C berbatu)	>0.1-2 (A berbatu)	>0.01-0.1 (sedikit berbatu)	>2-25 (C berbatu)	>2-25 (C berbatu)
	g. Tingkat bahaya erosi	sedang	berat	sedang	sedang	sedang	sedang	sangat ringan	sedang	sedang	sedang
	h. Konsistensi tanah	LektBasahKering	LektBasahKering	LektBasahKering	LektBasahKering	LektBasahKering	BashJenuh Kering	LektBasahKering	LektBasahKering	LektBasahKering	LektBasahKering
	i. Kapasitas tukar kation (cmol(+)kg-1	25-40 (tinggi)	17-24 (sedang)	25-40 (tinggi)	25-40 (tinggi)	25-40 (tinggi)	25-40 (tinggi)	17-24 (sedang)	25-40 (tinggi)	25-40 (tinggi)	25-40 (tinggi)
	j. pH	6.6-7.5 (netral)	6.6-7.5 (netral)	7.6-8.5 (agak alkalis)	7.6-8.5 (agak alkalis)	7.6-8.5 (agak alkalis)	7.6-8.5 (agak alkalis)	7.6-8.5 (agak alkalis)	6.6-7.5 (netral)	6.6-7.5 (netral)	6.6-7.5 (netral)
	k. Kejenuhan basa (%)	>70 (sangat tinggi)	51-70 (tinggi)	>70 (sangat tinggi)	51-70 (tinggi)	51-70 (tinggi)	51-70 (tinggi)	51-70 (tinggi)	51-70 (tinggi)	51-70 (tinggi)	51-70 (tinggi)
	l. Kandungan C-organik (%)	< 1 (sangat rendah)	1-2 (rendah)	> 2-3 (sedang)	> 2-3 (sedang)	> 2-3 (sedang)	>3-5 (tinggi)	>3-5 (tinggi)	>3-5 (tinggi)	1-2 (rendah)	>3-5 (tinggi)
	m. Kandungan N-total (%)	0.51-0.75 (tinggi)	0.21-0.50 (sedang)	0.51-0.75 (tinggi)	0.51-0.75 (tinggi)	0.51-0.75 (tinggi)	0.51-0.75 (tinggi)	0.21-0.50 (sedang)	0.21-0.50 (sedang)	0.51-0.75 (tinggi)	0.51-0.75 (tinggi)
	n. Kandungan Phosfor (ppm)	<10 (sangat rendah)	<10 (sangat rendah)	<10 (sangat rendah)	<10 (sangat rendah)	<10 (sangat rendah)	<10 (sangat rendah)	<10 (sangat rendah)	<10 (sangat rendah)	<10 (sangat rendah)	<10 (sangat rendah)
	o. Kandungan Kalium (K-dd, ppm)	41-60 (tinggi)	41-60 (tinggi)	21-40 (sedang)	41-60 (tinggi)	21-40 (sedang)	21-40 (sedang)	41-60 (tinggi)	41-60 (tinggi)	41-60 (tinggi)	41-60 (tinggi)
	p. C/N ratio (%)	<5 (sangat rendah)	<5 (sangat rendah)	<5 (sangat rendah)	<5 (sangat rendah)	<5 (sangat rendah)	5-10 (rendah)	5-10 (rendah)	5-10 (rendah)	5-10 (rendah)	5-10 (rendah)
	q. Kejenuhan Al (%)	<10 (sangat rendah)	<10 (sangat rendah)	<10 (sangat rendah)	<10 (sangat rendah)	<10 (sangat rendah)	<10 (sangat rendah)	<10 (sangat rendah)	<10 (sangat rendah)	<10 (sangat rendah)	<10 (sangat rendah)
	r. Status kesuburan	sedang	sedang	sedang	sedang	sedang	sedang	sedang	sedang	sedang	sedang
5	Komponen Iklim :										
	a. Suhu udara (oC)	26-28	26-28	26-28	26-28	26-28	28.1-30	28.1-30	28.1-30	28.1-30	28.1-30
	b. Kelembaban udara (%)	82	94	94	94	94	87	87	87	87	83
	c. Curah hujan tahunan (mm th-1)	1.989,70	1.747,20	1.747,20	1.747,20	1.646,50	1.464,50	1.464,50	1.464,50	1.464,50	1.552,40
	d. Tipe iklim Oldeman	C3	C3	C3	C3	C3. E3	C3, E3	D3	D3	D3, E3	C3
6	Neraca Air Lahan										
	a. Periode deficit	Juli-Nov	Juli-Okt	Juli-Okt	Juli-Okt	Juli-Nov	Juli-Okt, Nov	Juli-Jan	Juli-Nov	Juli-Nov	Juli-Nov
	b. Periode surlus	Des-Apr	Nov-Apr	Nov-Apr	Nov-Apr	Des-Apr	Nov,Des-Apr	Feb-Jun	Des-Apr; Feb-Jun	Des-Apr; Feb-Jun	Des-Apr
7	Komponen hidrologi :										
	a. Kapasitas drainase	Baik	Baik	Baik	Baik	Baik	Baik	Baik	Baik	Baik	Baik
	b. Tingkat bahaya banjir	Tanpa banjir (F0)	Tanpa banjir (F0)	Tanpa banjir (F0)	Tanpa banjir (F0)	Tanpa banjir (F0)	Tanpa banjir (F0)	Tanpa banjir (F0)	Tanpa banjir (F0)	Tanpa banjir (F0)	Tanpa banjir (F0)
8	Komponen Pengg lahan	PLK, SB	SB	SB	HLKS, SB	PLKCS	PLKCS	PLKCS	PLK, PLKCS	PLKCS	HLKS

Figure 3 show the result of analysis land of general water scale in the research area, consists of four categories according to the information of climate factors from the local rain station particularly the potential evapotranspiration in the area of Betoambari, Kaisabu, Kapontori, and Lawele station (Figure 5).

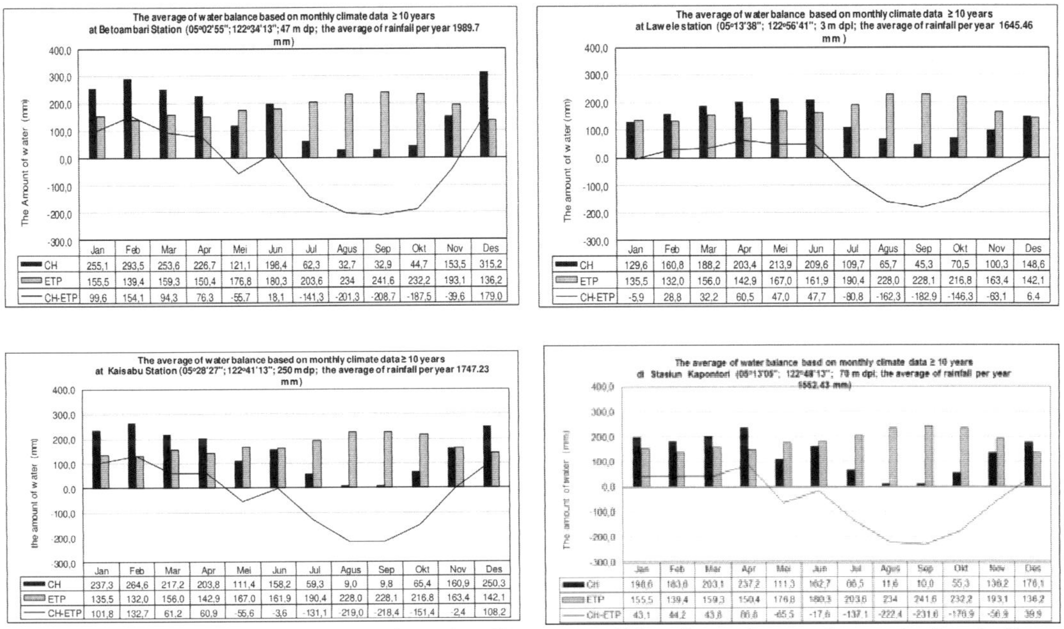

Figure 3. General water scale based on the climate condition from observation at the local climate station

Figure 3 shows that the research area around the Betoambari station namely the Batauga sub district, had experienced humid/wet months for 6 months (Dec, Jan, Feb, Mar, Apr, June), and 4 dry months (July-October). In the area of Kaisabu station, namely Pasarwajo, Wabula, Lapendewa, Sampolawa sub district and part of Wolowa and Kapontori sub district experienced 5 months of wet and dry months which are, Des, Jan, Feb, Mar, and Apr (wet months) and Mei, July-November (dry months), and wet months on June and November. In the area of Kapontori station which covers Kapontori sub district and part of Wolowa, Lasalimu, and Pasarwajo sub district undergo wet and dry months for 5 months each, Des, Jan, Feb, Mar, and Apr (wet months) and May, July-November (dry months), and wet months on June and November. In the coverage area of Lawele station which consisted of Lasalimu, South Lasalimu, Siotapina sub district and part of Wolowa sub district, shows that rainy season and dry season of the Lawele station area.

Figure 4. Some of the examples of the lands biophysical condition in the research area

IOP Conf. Series: Earth and Environmental Science **383** (2019) 012005 doi:10.1088/1755-1315/383/1/012005

Each occured for 5 months which are February-April (rainy season), and July-November (dry season), and humid condition happens for two months December and January. Based on these descriptions, it generally shows that the research area have lower yearly rainfall than its potential evapotranspiration therefore it's classified as a dry climate area, consequently farming activities must prioritize rainfall information usage cautiously for dry land agriculture, mainly to determine planting period, planting pattern, planning of irrigation system and selection of suitable variety.

Drainage capacity of the research area varies and consists of 4 categories: good, medium, slightly poor and poor. Space utilization and domination varies as well

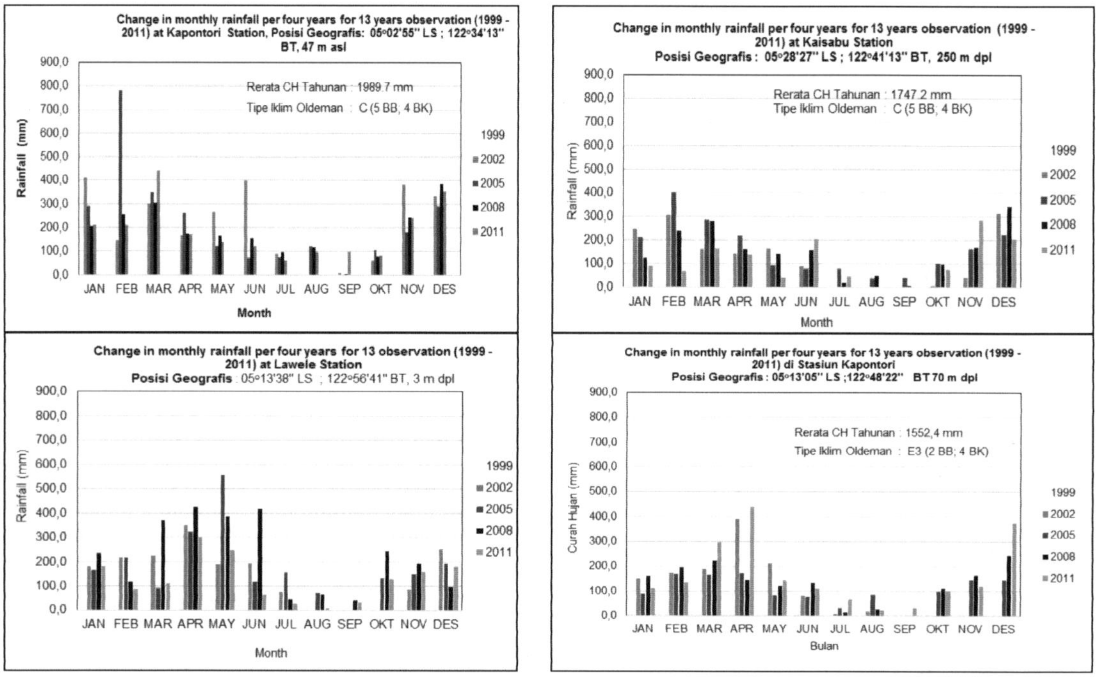

Figure 5. Monthly Rainfall fluctuation (mm) every 4 year based on observation at the Betoambari, Kaisabu, Lawele and Kapontori station in the period of 1999-2011.

Figure 5 shows that monthly rainfall rate every 4 year for the last 13 years in the Buton district have been subject to fluctuation and experiences shift in value every year. These circumstances imply on space utilization, farm planning, arrangement of planting pattern, determining planting period and crop productivity that the farmer developed.

3.1.2. Assessment of food crop soil suitability

Results of the suitability grade of actual land and food crop development potential evaluation in the research area are presented on table 2.

Table 2. Suitability grade of actual land and the potential of 10 species of food crop in the research area

RESULT OF ANALYSIS OF SUITABLE LAND FOR FOOD CROPS

Researh area (sub-district)	Actual Land Suitability for Food Crop Development in the Research Area									
	WET RICE	RICE FIELD	CORN	Cassava	S.Potato	Peanuts	Soybean	G. Beans	Taro	Sorghum
Batauga	N2	N2	N2	S3 N	N2	N2	N2	N2	N2	N2
Sampolawa	S3 N2	S3 N2	N1 N2	S2 S3 N	S3 N1 N2	S3 N1 N2	N1 N2	S3 N1 N2	S3 N1 N2	S3 N1 N2
Lapandewa	S3 N2	S3 N2	N1 N2	S3 N	N1 N2	S3 N2	N2	N1 N2	N1 N2	S3 N2
Wabula	N2	N2	N2	N	N2	N2	N2	N2	N2	N2
Pasarwajo	N2	N2	N2	N	N2	N2	N2	N2	N2	N2
Wolowa	N2	N2	N2	N	N2	N2	N2	N2	N2	N2
Siontapina	S3 N1 N2	N1 N2	N1, N2	S2 S3 N	N1 N2	S3 N1 N2	N1 N2	S3 N1 N2	S3 N1 N2	S3 N1 N2
Lasalimu Selatan	S3 N1 N2	N1 N2	N1 N2	S2 S3 N	N1 N2	S3 N1 N2	N1 N2	S3 N1 N2	S3 N1 N2	S3 N1 N2
Lasalimu	S3 N1 N2	N1 N2	N1 N2	S2 S3 N	N1 N2	S3 N1 N2	N1 N2	S3 N1 N2	S3 N1 N2	S3 N1 N2
Kapontori	S3 N1 N2	N1 N2	S3 N1 N2	S2 S3 N	N1 N2	S3 N1 N2	N1 N2	S2 S3 N1 N2	S3 N1 N2	S3 N1 N2

No.	Research areas (Sub-districts)	Potential Land suitability for food crops in the research areas									
		WET RICE	RICE FIELD	CORN	Cassava	S. Potato	Peanuts	Soybean	G. Beans	Taro	Sorghum
1	Batauga	N2	N2	N2	S3 N	N2	N2	N2	N2	N2	N2
2	Sampolawa	S2 N2	S3 N2	S3 N1 N2	S1 S2 N	S3 N1 N2	S2 S3 N2	N1 N2	S2 N1 N2	S3 N2	S3 N1 N2
3	Lapandewa	S2 N2	S3 N2	N1 N2	S2 N	N1 N2	S3 N2	N1 N2	N1 N2	S3 N2	S3 N2
4	Wabula	N2	N2	N2	N	N2	N2	N2	N2	N2	N2
5	Pasarwajo	N2	N2	N2	N	N2	N2	N2	N2	N2	N2
6	Wolowa	N2	N2	N2	N	N2	N2	N2	N2	N2	N2
7	Siontapina	S2 S3 N1 N2	N1 N2	N1, N2	S2 S3 N	N1 N2	S3 N1 N2	N1 N2	S2 S3 N1 N2	S3 N1 N2	S3 N1 N2
8	Lasalimu Selatan	S2 S3 N1 N2	N1 N2	N1 N2	S1 S2 S3	N1 N2	S2 S3 N1 N2	S3 N1 N2	S2 S3 N1 N2	S3 N1 N2	S3 N1 N2
9	Lasalimu	S2 S3 N1 N2	N1 N2	N1 N2	S2 S3 N	N1 N2	S3 N1 N2	S3 N1 N2	S2 S3 N1 N2	S3 N1 N2	S3 N1 N2
10	Kapontori	S2 S3 N1 N2	N1 N2	S3 N1 N2	S2 S3 N	N1 N2	S3 N2	N1 N2	S2 S3 N1 N2	S3 N1 N2	S3 N2

Table 2 shows that from the 10 food crop species that were evaluated whether it has long been planned or just been planned, the highest suitability grade is S2 (suitable enough) next is S3 grade (marginal fit), and the rest are N1 grade (temporarily unsuitable) and N2 (permanently unsuitable). Suitability grade varies relatively varies, whether it's between plants in the same region or between the same species of plants on different regions. This distinctness indicates that growth and productivity potential of plants in each region will be diverse

3.1.3. Determining mainstay/basis food crop

Results of the LQ value calculation from 10 different species of food crops in the research area based on the size of the harvest in the last six years (2006-2011) are presented on Table 3.

Table 3. Average harvest size and Location Quotient (LQ) value result of 10 species of food crop on the research area

No.	Reserach areas (sub-districts)	Average harvest area (ha) of food crops (ha) in the study area										Total
		Wet Rice	Rice fields	Corn	Cassava	S. Potato	Peanuts	Soybean	G.Beans	Taro	Sorghum	
1	Batauga	-	113,83	270,21	455,17	32,61	2,50	-	-	-	-	874,31
2	Sampolawa	-	147,65	412,20	377,77	57,22	272,98	-	4,26	-	-	1.272,07
3	Lapandewa	-	-	518,96	142,03	1,00	8,17	1,67	-	-	-	671,83
4	Wabula	-	-	93,16	63,83	0,17	12,33	-	-	-	-	169,50
5	Pasarwajo	-	275,94	628,03	167,87	53,75	36,58	2,28	13,26	-	-	1.177,68
6	Wolowa	-	109,60	130,98	84,30	6,32	15,00	1,17	6,83	-	-	354,18
7	Siontapina	-	449,78	122,36	71,07	26,28	17,54	156,90	8,55	-	-	852,46
8	Lasalimu Selatan	297,00	747,41	371,59	65,43	59,51	28,47	314,10	29,58	-	-	1.913,09
9	Lasalimu	458,83	214,70	252,81	50,97	28,24	15,11	9,67	15,73	-	-	1.046,05
10	Kapontori	657,17	257,96	234,79	167,37	94,04	5,61	20,23	3,80	-	-	1.440,96
Total of Harvest areas		1.413,00	2.316,86	3.035,07	1.645,79	359,13	414,27	506,00	82,00	-	-	9.772,12
Average of harvest areas		141,30	231,69	303,51	164,58	35,91	41,43	50,60	8,20	-	-	977,21
Deviation standard of harvest areas		242,89	226,00	177,45	140,83	30,26	82,00	104,45	9,30	-	-	514,54

No.	Type of food crops	The average grade of Food crops Location Quotient (LQ) in research areas (10 sub-districts)									
		Batauga	Sampolawa	Lapandewa	Wabula	Pasarwajo	Wolowa	Siontapina	Lasalimu S	Lasalimu	Kapontori
1	Wet rice	-	-	-	-	-	-	-	1,70	4,71	4,87
2	Rice field	1,51	0,76	0,00	0,00	1,69	1,98	3,44	2,37	1,09	1,15
3	Corn	0,77	0,94	2,10	1,75	1,44	1,13	0,41	0,54	0,63	0,47
4	Cassava	1,64	1,02	0,87	1,23	0,54	0,77	0,30	0,12	0,19	0,39
5	Sweet Potato	1,70	1,52	0,06	0,04	1,39	0,64	0,90	1,02	0,94	1,93
6	Peanuts	0,05	6,41	0,41	2,88	1,02	1,36	0,57	0,48	0,49	0,15
7	Soybean	-	0,00	0,23	-	0,04	0,13	5,56	4,36	0,48	0,65
8	Green beans	-	0,49	-	-	1,79	1,59	1,45	2,61	2,26	0,42
9	Taro	-	-	-	-	-	-	-	-	-	-
10	Sorghum	-	-	-	-	-	-	-	-	-	-

Note:
LQ : index whether or not it's a mainstay/basis food crop in a region
LQ > 1; mainstay/basis, haves komparatif advantages, the results in addition to fulfilling the needs of said region, it can also be export outside the region
LQ = 1; non basis/not mainstay, does not have komparatif advantage, only sufficient to fulfill the needs of its own region
LQ < 1; non basis/not mainstay, cannot fulfill the needs of its own region therefore needs supplies or imports from outside the region

Table 3 points out that harvest size of each plant varies and not every plant based on its harvest size can become a mainstay plant in a region, being classified as a mainstay plant in a region does not necessarily make it a mainstay plant in a different region. These cases can happen due to LQ value closely relates to the plants ability to grow and reproduce, related to the degree of land suitability for plant development and harvest size, the larger the size of suitable land the production amount also increased, in consequence the LQ value increase or larger than 1.

3.1.4. Social culture and economy characteristic identification

Results of Social culture and economy characteristic identification in the research area are presented on Table 4.

Tabel 4. The result of recapitulation of socio, cultural and economic aspects in the research areas

No	Karakteristik Sosial Budaya	Rekap Hasil Deskripsi Karakteristik Sosial Budaya yang Dominan di Wilayah Penelitian									
		Batauga	Sampolaw a	Lapandew a	Wabula	Pasarw ajo	Wolow a	Siotapina	Lasalimu Sel	Lasalimu	Kapontori
1	Identitas petani										
	a. Umur (tahun)	30-72 (53.7)	34-72 (49.5)	35-63 (52.3)	37-69 (58.2)	36-56 (43.5)	40-62 (50.5)	39-69 (54.3)	45-60 (54.3)	25-61(43.2)	34-66 (49.4)
	b. Lama bertani (tahun)	3-36 (20.3)	12-40 (23.1)	11-42 (26.4)	15-39 (29.1)	10-31 (19.8)	8-25 (15.6)	10-40 (23.9)	9-32 (19.6)	10-30 (17)	4-40 (18.9)
	c. Tingkat pendidikan	SD	SD	SD	SD	SD	SD	TS/TT SD	SD	SD/SLTP	SLTA
2	Kepemilikan lahan :										
	a. Luas Kepemilikan (ha)	> 0.5-1	> 0.5-1	> 0.5-1	> 0.5 - 1	> 0.5 - 1	≤ 0.5	≤ 0.5	≤ 0.5	> 0.5 - 1	≤ 0.5
	b. Status kepemilikan (ha)	Hak milik	Hak milik	Hak milik	Hak milik	Hak milik	Hak milik/pinjam	Hak milik	Hak milik	Hak milik/pinjam	Hak milik
3	Luas dan pola tanam exixting :										
	a. Lua tanam (ha)	> 0.5-1	> 0.5-1	> 0.5-1	> 0.5-1	> 0.5-1	≤ 0.5	≤ 0.5	≤ 0.5	≤ 0.5 / >0.5-1	≤ 0.5
	b. Pola tanam	Monokultur	Monokultur	Monokultur	Monokultur	Monokultur	Campuran	Campuran	Campuran	Campuran	Monokultur
4	Animo/preferensi petani	Jagung/U kayu	Jagung/U kayu	Jagung/U kayu	Jagung/U kayu	Jagung/U kayu	Jagung/U kayu	Jagung U kayu	P ladang/Jagung	P ladang/Jagung	P sawah/Jagung
5	Kebiasaan/tradisi petani :										
	a. Nama acara ritual	Haroa	Haroa	Haroa	Pe Bante	Pe Gandaa	Kacingkaha	Tamburie	Pesta panen	Tidak aktif	Tidak aktif
	b. Nama Lembaga ritual	Parabela/Parica	Parabela/Parica	Parabela/Parica	Parabela/Parica	Parabela/Parica	Parabela	Parabela	Parabela/Kel Kerja	Tidak ada	Kel Kerja/Kel Tani
	c. Waktu tanam	DES/JAN	DES/JAN	DES/JAN	MEI	DES/JAN	DES/JAN	DES/JAN	JAN	JAN	JAN/FEB
	d. Upaya adaptasi	Rubah W.tanam	Rubah P tanam	Rubah P tanam	Rubah P tanam	Rubah P tanam	Tidak ada	Tidak ada	Tdk ada/R W tanam	Tidak ada	Rubah P tanam
	e. Petunjuk perubahan w.tanam	Pergeseran MH	Pergeseran MH	Pergeseran MH	Pergeseran MH	Pergeseran MH	Pergeseran MH	Tidak ada	Air Irigasi	Hak milik/pinjam	Air Irigasi
	f. Petunjuk perubahan p.tanam	Pergeseran MH	Tidak aktif	Pergeseran MH	Pergeseran MH	Pergeseran MH	Pergeseran MH	Tidak ada	Air Irigasi	Harga	Tidak aktif
	g. Petunjuk w.panen	Warna buah	Warna buah	Warna buah	Warna buah	Warna buah	Umur tanaman	Umur panen	Umur panen	Umur panen	Warna buah

Notes :
R/C ratio : index of farming suitability
R/C ratio > 1 farming yields profit
R/C ratio = 1 farming breaks even
R/C ratio < 1 farming incur loss

3.1.5. Determining priority of select food crop

Results of AHP sub criteria weighting based on pair comparison between sub criteria in its relation with determining select food crop priority in the research area are presented in Table 5 and its spread in Figure 6.

Table 5. Recapitulation Results of AHP sub criteria weighting based on pair comparison between sub criteria in its relation with determining select food crop priority in the research area are presented in and its spread (Picture.6) in the research area.

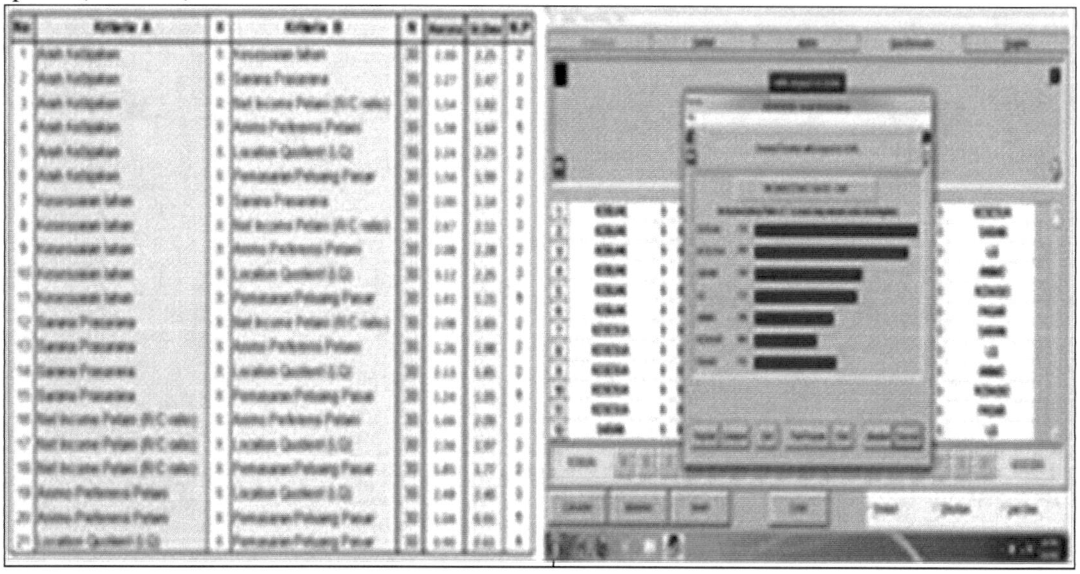

AHP results in determining select food crop priority on each region by involving by applying ample criteria are presented on Figure 7.

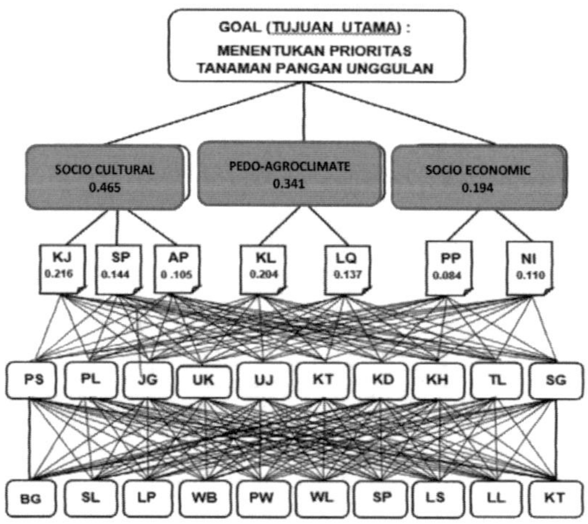

Figure 6. Hierarchy of determining select food crop priority

Figure 6 shows that social culture reached first place with a value of 0.465, next in second place is the pedo-agroclimate criteria with a score of 0.341, and the third or last place is social economy criteria with a score of (0.194). In sub criteria level, policy direction sits in first place (0.216), then in sequence is weighted land suitability in second place (0.204), infrastructure in third place (0.144), LQ fourth place (0.137), farmer net income (R/C ratio) fifth place (0.110), farmer interest/preference sixth place (0.105), and market opportunity/marketing seventh place (0.084)

AHP result according to Figure 7, we identified the spread of value of select food crop priority ranking in each research region (sub district) as presented on table 6, and diagrammatically on Figure 8.

Table 6. Recapitulation spread of weighted select food crop priority in research areas and results of priority ranking on each research area based on the whole AHP sub criteria that are taken into consideration

No.	Jenis Tanaman Pangan	Batauga	Sampolawa	Lapandewa	Wabula	Pasarwajo	Wolowa	Siotapina	Lasalimu S	Lasalimu	Kapontori	Nilai Statistik		
												Max	Min	Stdev.
1	Padi Sawah	0.060	0.063	0.069	0.075	0.071	0.065	0.083	0.122	0.141	0.156	0.156	0.060	0.035
2	Padi Ladang	0.128	0.121	0.076	0.082	0.139	0.150	0.137	0.127	0.118	0.119	0.150	0.076	0.024
3	Jagung	0.155	0.144	0.189	0.171	0.144	0.135	0.128	0.116	0.112	0.113	0.189	0.112	0.026
4	Ubi Kayu	0.199	0.170	0.200	0.196	0.143	0.158	0.141	0.121	0.138	0.143	0.200	0.121	0.029
5	Ubi Jalar	0.138	0.127	0.108	0.100	0.130	0.121	0.106	0.108	0.112	0.107	0.138	0.100	0.012
6	Kacang Tanah	0.094	0.133	0.116	0.131	0.116	0.113	0.099	0.100	0.101	0.102	0.133	0.094	0.014
7	Kedelai	0.074	0.077	0.084	0.083	0.076	0.082	0.129	0.124	0.091	0.094	0.129	0.074	0.020
8	Kacang Hijau	0.077	0.084	0.085	0.083	0.110	0.103	0.101	0.105	0.109	0.087	0.110	0.077	0.012
9	Talas	0.075	0.081	0.073	0.079	0.071	0.073	0.076	0.077	0.078	0.079	0.081	0.071	0.003
10	Sorghum	.	.	.	.	.	.	.	.	.	.	.	.	.
	Maximun	0.199	0.170	0.200	0.196	0.144	0.158	0.141	0.003	0.141	0.156	0.200	0.121	0.035
	Minimum	0.060	0.063	0.069	0.075	0.071	0.065	0.076	0.077	0.078	0.079	0.079	0.060	0.003
	In Consintensis Index	0.05	0.05	0.05	0.05	0.05	0.05	0.05	0.05	0.05	0.05	0.050	0.050	

Urutan Prioritas Unggulan	Hasil perengkingan prioritas tanaman pangan unggulan berdasarkan seluruh sub kriteria AHP yang di pertimbangkan di masing-masing wilayah penelitian									
	Batauga	Sampolawa	Lapandewa	Wabula	Pasarwajo	Wolowa	Siotapina	Lasalimu Sel	Lasalimu	Kapontori
Prioritas 1	Ubi Kayu	Ubi Kayu	Ubi Kayu	Ubi Kayu	Jagung	Ubi Kayu	Ubi Kayu	Padi Ladang	Padi Sawah	Padi Sawah
Prioritas 2	Jagung	Jagung	Jagung	Jagung	Ubi Kayu	Padi Ladang	Padi Ladang	Kedelai	Ubi Kayu	Ubi Kayu
Prioritas 3	Ubi Jalar	Kacang Tanah	Kacang Tanah	Kacang Tanah	Padi Ladang	Jagung	Kedelai	Padi Sawah	Padi Ladang	Padi Ladang
Prioritas 4	Padi Ladang	Ubi Jalar	Ubi Jalar	Ubi Jalar	Ubi Jalar	Ubi jalar	Jagung	Ubi Kayu	Jagung	Jagung
Prioritas 5	Kacang Tanah	Padi Ladang	Kacang Hijau	Kedelai	Kacang Tanah	Kacang Tanah	Ubi Jalar	Jagung	Ubi Jalar	Ubi Jalar
Prioritas 6	Kacang Hijau	Kacang Hijau	Kedelai	Kacang Hijau	Kacang Hijau	Kacang Hijau	Kacang Hijau	Ubi Jalar	Kacang Hijau	Kacang Tanah
Prioritas 7	Talas	Talas	Padi Ladang	Padi Ladang	Kedelai	Kedelai	Kacang Tanah	Kacang Hijau	Kacang Tanah	Kedelai
Prioritas 8	Kedelai	Kedelai	Talas	Talas	Padi Sawah	Talas	Padi Sawah	Kacang Tanah	Kedelai	Kacang Hijau
Prioritas 9	Padi Sawah	Padi Sawah	Padi Sawah	Padi Sawah	Talas	Padi Sawah	Talas	Talas	Talas	Talas
Prioritas 10	.	.	.	.	.	.	.	.	.	.
Inconsistency	0.05	0.05	0.05	0.05	0.05	0.05	0.05	0.05	0.05	0.05

Table 6 and the diagram on figure 7 indicates that the distribution of select food crop priority value in the research area are relatively homogeneous with a very low deviation standard (0.003 ≤ St.D ≤ 0.035), both between regions and between food crop species. This fact is relevant with the index inconsistency value which is quite small (0.05) from priority value of superior food crop on the research area.

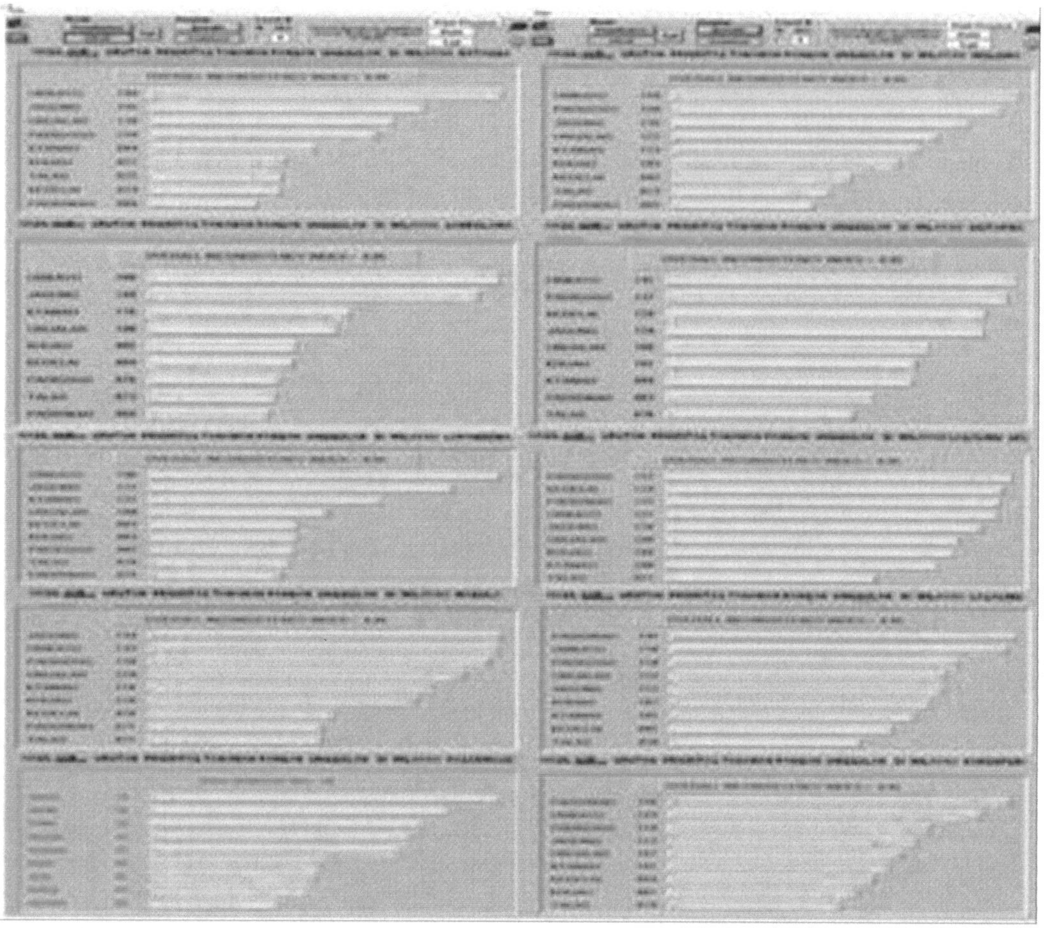

Figure 7. Weight diagram of select food crop priority based on all considered sub criteria. Source: Kandari, 2014.

Table 6 shows that the some ranking of select food crop priority (1-9) in each research region (sub district) are the same and some of them differ. For instance on priority 1, mainstay food crop in Batauga region is cassava same with Sampolawa, Lapendewa, Pasarwajo, Wolowa, dan Siotapina, while in Kapontori and Lasalimu the mainstay food crop is paddy rice, in South Lasalimu field rice and on Wabula is corn. These facts indicates the existence of varying roles of each criteria that are classified by its sub criteria in the process of determining select food crop priority

3.1.6. Land resources management in an effort to guarantee sustainable food crop availability

Land resources management on each research regions are based upon the types of limiting factors found. Result of the research shows there are many combined limiting factors on each research area, which are generally divided into 7 factors: (1) condition of the slope/terrain (s), (2) level of erosion hazard (e), (3) water availability (w), (4) surface rock (p), (5) rooting media and condition (r), (6) nutrient retention (f), (7) nutrient availability (n).

Based upon these limiting factors, land resources management activities needed to be done are generically summarized in these efforts [12]:

a) Management of land based on soil and water conservation principle
b) Utilization of water and climate resources optimally
c) Management of forest and food vegetation

d) Coaching of human resources wisely
e) Choosing commodity according to agro ecology

4. Conclusions

Based on result and discussion, such points can be concluded:

1. The Kalalasi Region has land biophysical diversity, which consists of physio-topography, soil and climate that affects the types and class of food crop land suitability, and the coverage of its development
2. From the extent of the Kalalasi Region mainland 210,030.00 ha, there is only 74,664.64 ha (35.55%) suboptimal land that has the potential of food crop cultivation with highest suitability grade of S2 (suitable enough) specifically 5,096.52 ha, S3 grade (according to margin) 44,521.38 ha, and the rest are N1 grade (temporarily unsuitable) and N2 (permanently unsuitable)
3. Select food crops that are economically worth to cultivate in the Kalalasi Region, comprises of cassava (66,205.33 ha), soy (14,423.34 ha), field rice (14,296.50), sweet potato (13,758.67 ha), peanut (13,478.90 ha), corn (13,428.27 ha), green bean (12,339.57 ha), and paddy rice (11,278.87 ha)
4. From the social and cultural aspect farmer in the research area as a community are classified as subsistent farmer, and have strong cultural asset in plant cultivation system
5. Need to follow up through field experiments for the advancement of science and technology, especially the development of superior food crops in each region by making conservation efforts and site-specific technology input.

5. References

[1] Rounselvell M.D.A., and D.S. Reay. 2009. *Land Use and Climate Change in UK.* Land Use Policy 26S, S160-S169. Elsevier.
[2] Food and Agriculture Organization. 1976. *A Framework for Land Evaluation.* FAO Soils Bulletin No.32. Food and Agriculture Organisation of the United Nations, Rome. Rome.
[3] Baja, S. 2012a. *Perencanaan Tata Guna Lahan dalam Pengembangan Wilayah. Pendekatan Spasial dan Aplikasinya.* Penerbit ANDI, Yogyakarta. 378p.
[4] Wirosoedarmo R., A. T. Sutanhaji, E.Kurniati dan R. Wijayanti. 2011. *Evaluasi Kesesuaian Lahan Untuk Tanaman Jagung Menggunakan Metode Analisis Spasial.* Journal AGRITECH, Vol. 31, No. 1, Februari 2011. Fakultas Teknologi Pertanian, Jurusan Keteknikan Pertanian, Universitas Brawijaya, Malang.
[5] Baja, S. 2012b. *Metode Analitik Evaluasi Sumber Daya Lahan* : Aplikasi GIS, Fuzzy Set, dan MCDM. Penerbit: IDENTITAS. Universitas Hasanuddin, Makassar. 242p.
[6] RePPProT, 1990. *The Land Resources of Indonesian:* a National Over View Regional Physical Planning Programme for Transmigration, Jakarta. Ministry of Transmigration and Land Resources, Departement NRI, UK. OverSears Development Administration.
[7] Hendayana R. 2003. *Aplikasi Metode Location Quetient (LQ) dalam Penentuan Komoditas Unggulan Nasional.* Informatika Pertanian 12 (1) : 658-675.

IOP Conf. Series: Earth and Environmental Science **383** (2019) 012006 doi:10.1088/1755-1315/383/1/012006

Phytochemical analysis of beneng taro (*Xanthosoma undipes* K.Koch) leaves: cultivation as raw material for biopesticides for eco-friendly agriculture

A A Fatmawaty, N Hermita*, D Hastuti, Kartina AM, S Hilal

Department of Agroecotechnology, Faculty of Agriculture, Sultan Ageng Tirtayasa University, Jl. Raya Jakarta Km.4 Pakupatan Serang, Fax. +62254 828529

*Corresponding author: nuniekhermita@untirta.ac.id

Abstract. Beneng taro (*Xanthasoma undipes* K.koch) is a local wild taro which grows spreading the land area of Gunung Karang in Pandeglang District, Banten Province. Recently, the demand of Beneng taro tubers is increasing due to improvements in its utilization as a food crops. Therefore, Beneng taro is now being cultivated widely by the farmers. However, the use of Beneng taro is still focus on its tubers, while its leaves has not been utilized and only become an agricultural waste. The aim of research was to find out the chemical content of cultivated Beneng taro leaves as a raw material for biopesticide. The research was conducted from July to September 2018. The leaves sample of cultivated Beneng taro were taken from farmer's field in Juhut Village, Karangtanjung Sub-district. Phytochemical analysis was accomplished in the laboratory. Further, chemical content in the form of alkaloids, flavonoids, saponins, tannins, and oxalic acid were observed. The research result showed that there were negative content for alkaloids, flavonoids, saponins in Beneng taro leaves. The cultivated Beneng taro leaves positively containing tannins, and oxalic acid as much as 3448.20 and 8361.52 mg/kg, respectively. The chemical content of tannins in Beneng taro leaves indicated the potential use of Beneng taro as a raw material for biopesticide.

Keywords: Beneng taro, biopesticide, cultivation, leaves, phytochemical

1. Introduction

Beneng taro (*Xanthasoma undipes* K.koch) is a local wild taro which grows spreading the land area of Gunung Karang in Pandeglang District, Banten Province. Recently, the demand of Beneng taro tubers is increasing due to improvements in its utilization as a food crops. Therefore, Beneng taro is now being cultivated widely by the farmers surround Gunung Karang. The protein and carbohydrates content found in Beneng taro tubers is relatively high, thus makes this crop to be use as wide variety of food products. On the other hand, the Beneng taro also containing oxalic acid, the chemical that become a problem in food. The cultivated Beneng taro, however, reported has much lower oxalic acid than wild Beneng taro [1].

Recently, the variability of products made from Beneng taro is increasing, such as chips, brownies, beneng noodles, beneng ice cream, and many more [2]. The main plant organ of Beneng taro plants that has been commonly use is tubers. However, the leaves of this plant have not been utilized to date, thus becomes only organic wastes. To optimize the use of Beneng taro plants and to reduce the agricultural wastes, it is necessary to utilize Beneng taro leaves as a raw material for biopesticide. In general, biopesticides is defined as pesticides whose basic ingredients come from plants or plant parts such as roots, leaves, stems, or fruits [3]. Additionally, the characteristics of biopesticide is to kill, to repel, act as anti-feedant, toxicant, and as growth inhibitors of insect.

Indonesia is a country that has the largest biodiversity in the world, second only to Brazil. Plants are storehouse of various chemical compounds that are rich in active ingredients, including secondary metabolic products. These compounds plays an important roles in the process of interacting or competing, including protecting themselves from interference or from competitors. Secondary metabolite products can be used as active ingredients of biopesticides.

Refer from various function of plants that have active ingredients on leaves as biological pesticides, it is important to know the chemical content of Beneng taro leaves and its potential use as raw material for biopesticide. The aim of research was to find out the chemical content of cultivated Beneng taro leaves as a raw material for biopesticide.

2. Material and Methods

The research was conducted from July to September 2018. Data collection was carried out by means of exploratory surveys which included interview and direct sampling from the field. Interviews were aimed at community leaders, users, or those who are familiar with Beneng taro plants. The leaves samples of cultivated Beneng taro were taken from farmer's field in Juhut Village, Karangtanjung Sub-district, Pandeglang Regency, Banten Province. A Phytochemical analysis was accomplished in the Bio-pharmachy Study Center Laboratory, Bogor, West Java. Further, chemical content in the form of alkaloids, flavonoids, saponins, tannins, and oxalic acid were observed. The results obtained were then analysed using quantitative descriptive analysis.

3. Result and Discussion

Based on the results of phytochemical analysis presented in Table 1, it is shown that there are negative results for flavonoids, alkaloids, and saponins for cultivated Beneng taro leaves. These results are different from previous research that show the results of phytochemical tests for Bogor taro containing anti-nutrients, including calcium oxalate, saponins, tannins, and flavonoids. The presence of these compounds shows that taro leaves and stems had the potential to be larvacides [4].

Table 1. Phytochemical content of Cultivated Beneng Taro leaves

Sample	Content	Result
Beneng Taro leaves	Flavonoid	Negative
	Alkaloid	Negative
	Saponin	Negative
	Tannin	3448.20 mg/kg
	Oxalic Acid	8361.83 mg/kg

Beneng taro leaves are positive in containing tannin with a yield of 3448.20 mg/kg. This shows the potential for Beneng taro leaves to be used as an ingredient for biopesticides. It was reported that the ingredients of *nginang* (Indonesian traditional chewing) in the form of betel, tobacco, gambier, and lime betel have the potential to be used as basic ingredients for environmental-friendly insecticides because they contain chemical compounds in the form of tannins of 67.65%, alkaloids of 1.43 %, and essential oil by 0.2%. These compounds can reduce the intensity of plants diseases such as leafhoppers, stink bugs, etc. by inhibiting the growth of adult insects and their larvae.

[2] reported that active substances that have the potential to be used as vegetable insecticides include alkaloids, terpenoids, phenolics, tannins, and other secondary chemicals. These chemical compounds affect the insect in the nervous or muscle systems, hormonal balance, reproduction, anti-feeding behavior, and respiratory system.

This research showed that the content of cultivated Beneng taro leaves have the potential of tannin compounds. The tannin compounds taste bitter which react with proteins, amino acids, and alkaloids which contain many hydroxyl and carboxyl groups to form strong and complex bonds with other proteins and macromolecules, making it very bitter in taste, thus not liked by insects that become pests on plants. The presence of saponin and tannin compounds in maja fruit is one reason why it is highly recommended as one of the ingredients of vegetable pesticides.

In addition to the chemical content of tannin compounds found in cultivated Beneng taro leaves, oxalic acid content is also present in the leaves with a value of 8361.83 mg/kg. Oxalate is one of the plant metabolites that have unique role in plants. In plants, oxalate can be in the form of oxalic acid or in the form of calcium oxalate crystals [5]. Oxalic acid and its water-soluble salts can be dangerous, because these compounds are toxic. High levels of oxalic acid can cause death from digestive symptoms (abdominal cramps and vomiting) which are rapidly followed by circulatory failure and rupture of blood vessels [6].

The distribution of oxalic acid in plant parts is uneven. In general, the leaf part contains more oxalic acid than that in the stalk, whereas in *Poligonaceae*, the oxalic acid content in the petiole is almost double than that in the stalk. In addition, young leaves contain less oxalic acid compared to old leaves. For example, in *Chenopodiaceae* leaves, the content of oxalic acid can be doubled during the aging process [7]. According to [8] calcium oxalate crystals in taro are found in two forms which are druse (round shape) and raphide (like fine needles), which is about 80 to 85 percent of the total calcium oxalate content.

Furthermore, the research on the treatment of taro tuber extract on mice behaviour showed that mice were run after ate the taro tuber. This is, presumably, because taro tubers have an unpleasant odor which can be felt or tasted by mice after eating it. Taro is a weed plant that has a sharp odor, bitter taste, and gives itching sensation. The smell and itching stimulant of taro tubers are difficult to wash off, and to remove them repeated wash are necessary. This odor and itchiness is produced by calcium oxalate crystals and saponins which have a bitter taste that usually mice do not like [9]. In addition, according to [10] there are 2-3% essential oils, 20-60% starch, resin, oxalic acid in ginger rhizome extract (Zingiber officinale). The results of previous research in the use of red ginger rhizome extract with a concentration of 20% were very effective against the mortality of *xylostella plutella* larvae in cabbage plants. Various results of previous research that have been described concerning the content of oxalic acid in plants that provide an effect as a natural pesticide, whose characteristics are including reducing pest attacks, although it does not directly kill the target Plant Pest Organisms.

4. Result and Discussion

The research concluded that the chemical content in cultivated Beneng taro leaves are tannin and oxalic acid compounds. Thus, cultivated Beneng taro leaves is potentially to be use as a raw material for bio pesticide. Further research is needed to examine the Beneng taro leaf extracts and its effectiveness in controlling pest in the field.

5. References

[1] Kartina AM, Ranthy Pancasasti, Wawan Ichwanuddin , dan Nuniek Hermita. 2015. Pengaruh Elevasi Terhadap Kadar Asam Oksalat Talas Beneng (*Xanthosoma undipes* K.Koch) Di Sekitar Kawasan Gunung Karang Provinsi Banten. Laporan Penelitian MP3EI.

[2] Setyowati M, Hanarida I, dan Sutoro. 2007. Karakteristik Umbi Plasma Nutfah Tanaman Talas (*Colocasia esculenta*). Buletin Plasma Nutfah Vol.13 No.2. Balai Besar Penelitian dan Pengembangan Bioteknologi dan Sumberdaya Genetik Pertanian. Bogor.

[3] Kardinan, Agus. 2008. Prospek Tanaman Aromatik Dalam Menanggulangi Permasalahan Nyamuk Dan Lalat. Warta Penelitian dan Pengembangan Tanaman Industri, Volume 14 Nomor 1, Hal. 25-23.

[4] Widhyastini Manik I dan Hutagaol Ricson P. 2013. Pemanfaatan talas Bogor (*Colocasia esculenta* (L) Schoot) Sebagai Larvasida Nyamuk. Jurnal Sains Natural Universitas Nusa Bangsa. Vol. 4, No. 2, Juli 2014, 92 – 97 hal.

[5] Franchesi, V.R., dan Nakata, P.A, 2005. Calcium oxalate in plants: formation and functions. Annual Review of Plant Biology 56:41-71.

[6] Herlena Fitriani, Nurlailah, Dinna Rakhmina, 2016. Kandungan Asam Oksalat Sayur Bayam. Jurusan Analis Kesehatan Poltekkes Kemenkes Banjarmasin. Medical Laboratory Technology Journal 2 (2), 51-55 hal.

[7] Kiantoro, Andi. 2011. Pembuatan asam oksalat dari limbah kulit pisang dengan pengaruh waktu dan konsentrasi asam nitrat. Politeknik negeri sriwijaya. Palembang.

[8] Kurdi W. 2002. Reduksi Kalsium Oksalat pada Talas Bogor (Colocasia esculenta (L.) Schott) sebagai Upaya Meningkatkan Mutu Keripik Talas. [Skripsi]. IPB, Bogor.

[9] Dian Yustisia dan Ismail, 2017. Efek Beberapa Ekstrak Tanaman Terhadap Hama Tikus Sawah (*Rattus argiventer*). Program Studi Agroteknologi, STIP Muhammadiyah Sinjai. Jurnal Agrominansia, 2 (1) Juni. ISSN 2527 - 4538 41.

[10] Mursito B. 2003. Sehat di usia lanjut dengan ramuan tradisional. Penebar Swadaya. Jakarta.

6. Acknowledgement

We thank The Director of LPPM of Sultan Ageng Tirtayasa University for his generous support and experiment facilities. This study was supported by the Faculty Agriculture, Sultan Ageng Tirtayasa University.

Performance of improved upland rice cultivars under farmer's field condition in Banten province, Indonesia

A Saryoko*, S Kusumawati, and A Pohan

Assessment Institute for Agricultural Technology (AIAT) Banten, Indonesian Agency for Agricultural Research and Development (IAARD), Jl. Ciptayasa km.01 Ciruas Serang Banten Indonesia

*Corresponding author: andysaryoko@yahoo.com

Abstract. The use of improved upland rice (*Oryza* spp.) cultivars would be an alternative approach for improving rice production in Indonesia. The objectives of study were to identify cultivars with high yield potential and to determine its related traits. Five improved upland rice cultivars, namely Inpago 8, Inpago 9, Inpago 10, Inpago 11 and Inpago 12, released in 2011 to 2017 were observed under farmer's field condition in Lebak district, Banten Province, Indonesia from September 2018 to January 2019. Grain yield, yield components, and biomass production and partitioning were measured at maturity stage. Grain yield of five cultivars showed higher performance than regional average yield performance. Grain yield performance was closely correlated with panicle number per m^2. The Inpago 9 and Inpago 12 were greater in grain yield due to greater number of panicle as compared to other cultivars. Superior performance of Inpago 9 and Inpago 12 were also due to greater biomass production and better harvest index (HI). On the other side, smaller value of panicle number per m^2 and HI of Inpago 11 resulted in lower performance in grain yield. In conclusion, panicle number, biomass production, and partitioning were some good traits in selecting adaptive cultivars. In addition, cultivars Inpago 9 and Inpago 12 are considerable to be chosen by the farmers in Lebak district, Banten Province to improve upland rice production.

1. Introduction

Rice is the most important food crop after wheat and maize. Rice production system can be classified into two types, lowland and upland rice system [1][2]. In lowland rice, field is usually flooded for whole or part of growing season [1]. In upland rice system, rice is grown in rainfed fields that prepared and seeded under dry condition [3], and in some cases supplemented with irrigation water [1].

Rice is number one of staple food crops, being eaten by almost population in Indonesia. Rice is mostly produced from lowland rice system, contributed to more than 95% of total rice production in Indonesia. Nowadays, rice production is being faced by the reduction of lowland area due to land-use change from agricultural use to non-agricultural use. On the other side, dryland area is highly potential to be explored in terms of rice production improvement [4].

Banten Province is one of the rice producers in Java Island, Indonesia, produced for more than 2 billion rice per year, contribute for more than 1.6 million tons of rice in 2018, contributed to near 3 % of national rice production [5]. Upland rice production in Banten Province is relatively low, ranged from 2 to 3 tons per hectare and vary between area [6]. The dryland area in Banten, besides, has a large potential to be used for improving upland rice production, combining with high-yielding upland rice cultivars and improvement in agricultural practices.

In the last decade, Indonesian Rice Research Centre – IAARD, has been released new high-yielding upland rice cultivars that can produce more than seven tons rice per ha with the average of four to five tons per ha [7]. The use of high-yielding upland rice cultivars is expected to improve rice production in upland rice system and enhance its contribution to total rice production. The aim of study was to identify cultivars with high yield potential and to determine its related traits

2. Material and Methods

2.1. Cultivars and agronomical practices

The field experiment was conducted from September 2018 to January 2019 under farmer's field condition in Lebak district, Banten Province, Indonesia (Lat. 6.8°S, Long. 106.2°E). On 20 September 2018 seeds of five improved upland rice cultivars, namely Inpago 8, Inpago 9, Inpago 10, Inpago 11 and Inpago 12, released in 2011 to 2017, were directly seeded. Three to five seed per hole were arranged with a double row with a spacing of 25 cm between rows, 12.5 cm between plant within rows, and 50 between two rows. The cultivars were arranged into a plot with 4 m wide and 13 long for each cultivar, resulting in 6 double rows for each. The soil was fertilized with 11.5 g m^{-2}, 5.4 g m^{-2} and 4.5 g m^{-2} of N, P_2O_5 and K_2O, respectively. The N fertilizer in the form of urea was applied three times at 10, 25 and 35 days after sowing (DAS) as much as 50, 25 and 25 % respectively. The P_2O_5 and K_2O in the form of SP36 and KCl, respectively, were applied at 10 DAS. Regional recommended management programs were employed for irrigation, weed and pest control to optimize growth condition.

2.2. Agronomical traits, plant growth, and yield components

Plant height was measured weekly from 4 weeks after planting (WAP) to 11 WAP. Two plants per plot were collected at 6, 8 and 11 WAP and oven-dried at 80° C for 48 hours to measure the above-ground biomass (TDW). Total leaf area was measured at 6, 8 and 11 WAP using the similar plant samples. Fresh leaf sample was taken and specific leaf area was measured using digital image technique. The leaf further oven-dried at 80 °C for 48 hours and leaf dry weight was recorded. Total leaf area was calculated from specific leaf weight and total leaf dry weight.

Canopy size and development were measured using a digital imaging technique [8] [9][10][11] using ImageJ (National Institute Health, USA) with some modifications, and expressed as a percentage of soil coverage by the rice canopy. For this purpose, plots were maintained as weed-free as possible during the growth stage.

At maturity stage, four plants per cultivar were harvested, and yield and yield components were observed. Panicle number per plant was counted, and panicle length and seed number per panicle were observed from 5 panicles per plant. Seed number was counted and weighted, and 1000 seed weight (1000SW) and seed moisture content was measured. Grain yield per plant was calculated as a multiplication of grain number, 1000SW, panicle number per plant at 14% moisture content. The remaining plant organs (leaves and straws) were dried at 80° C for 48 h. Harvest index (HI) was calculated as a ratio of dry grain yield (Y) and total biomass at maturity (TDW), as follows:

$$HI = Y / TDW \quad (1)$$

2.3. Cultivars preference

Cultivar preference in terms of plant performance and grain quality was evaluated by collecting data using structured questioner. In total 47 respondents, consisted of farmers, rice consumers, rice traders were involved. For the plant performance, respondents were provided a 110 days old standing crop of each cultivar, and asked to give a score of preference for each trait i.e. plant structure, plant height, panicle number, panicle length, days to maturity, grain number per panicle, and resistance to pest and disease. For grain quality, respondents were provided grain rice, husked-grain rice and steamed rice to be evaluated in terms of grain shape, size and color, and of steam rice eating quality (aroma, color, and stickiness).

2.4. Data Analysis

Variability among cultivars on plant performances, growth and yield components were analyzed using an analysis of mean. As for cultivars preference data, descriptive statistic was applied.

3. Results and Discussion

3.1. Plant growth and development

The growth in plant height is shown in figure 1. The plant height for all cultivars at 4 WAP was relatively similar. The variation in plant height occurred starting from 7 WAP and later. Cultivar Inpago 11 consistently showed taller in plant height as compared to other cultivars. The characteristics of most upland rice cultivars in Indonesia is indicated with height of plant (> 120 cm), especially when it compared to lowland cultivars (around 100 cm) [7]. The cultivar Inpago 8 was consistently the shortest as compared to other cultivars. The TDW of five tested cultivars were increased significantly from 6 WAP to 8WAP (figure 2). The TDW was also significantly increased from 8 WAP to 11 WAP, but not for Inpago 9. The characteristics of Inpago 9 was earlier in maturity (109 days) especially as compared to Inpago 10 (115 days) [7]. The cultivar Inpago 11 was the highest in TDW, and TDW was correlated with plant height (R^2= 0.56).

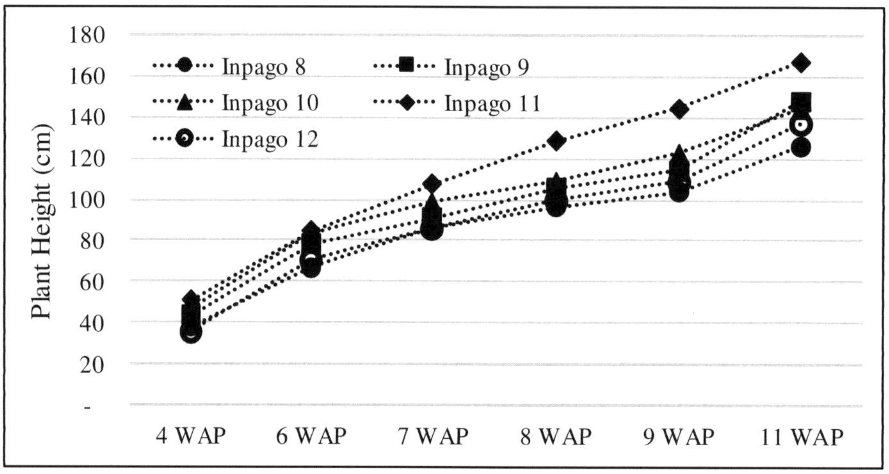

Figure 1. Growth of plant height of five upland rice cultivars from 4 to 11 week after planting.

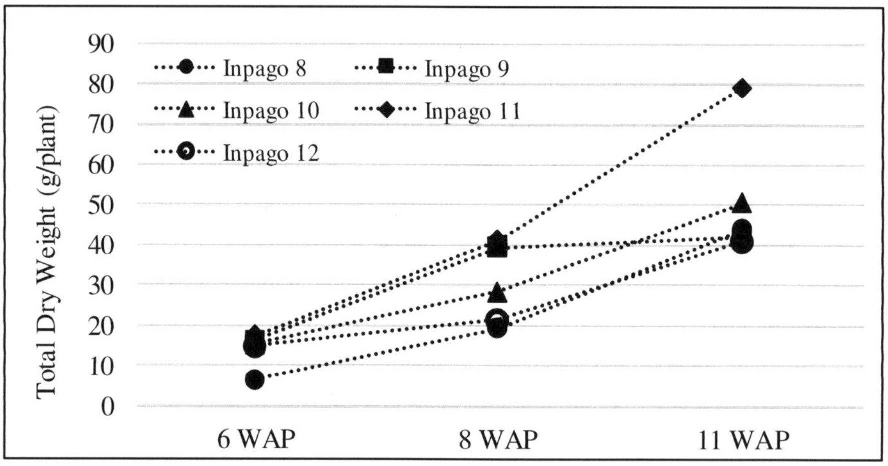

Figure 2. Growth of total dry weight (TDW) of five upland rice cultivars from 6 to 11 week after planting.

The variation in growth of total leaf area from 6 WAP to 11 WAP was various among cultivars (figure 3). At the 6 and 8 WAP, cultivar Inpago 8 was the smallest in total leaf area, but tend to be moderate at 11 WAP. The growth of total leaf area of Inpago 11 from 8 WAP to 11 WAP was the highest resulted in greater total leaf area as compared to other cultivars.

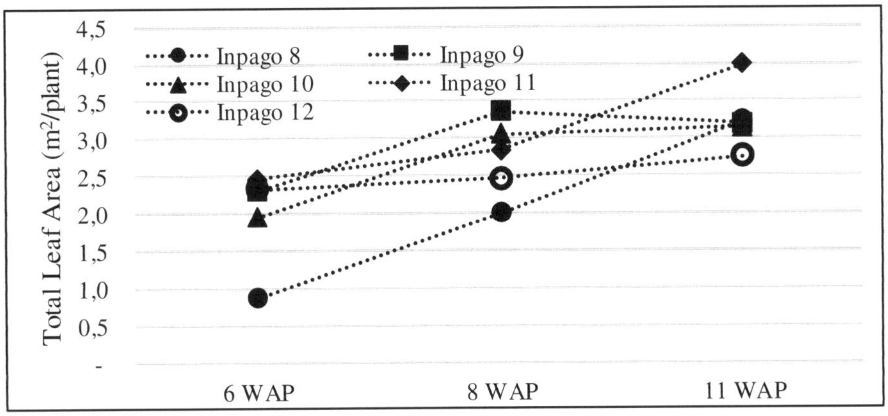

Figure 3. Growth of total leaf area of five upland rice cultivars from 6 to 11 week after planting.

For the canopy growth, in terms of canopy coverage, were measured from 3 WAP to 7 WAP (figure 4). Canopy coverage expressed the ability of plant to intercept light for photosynthetic activities [12][8]. Larger variation in canopy coverage occurred at 4 WAP. Canopy coverage of tested cultivars reached a fully covered at 7 WAP.

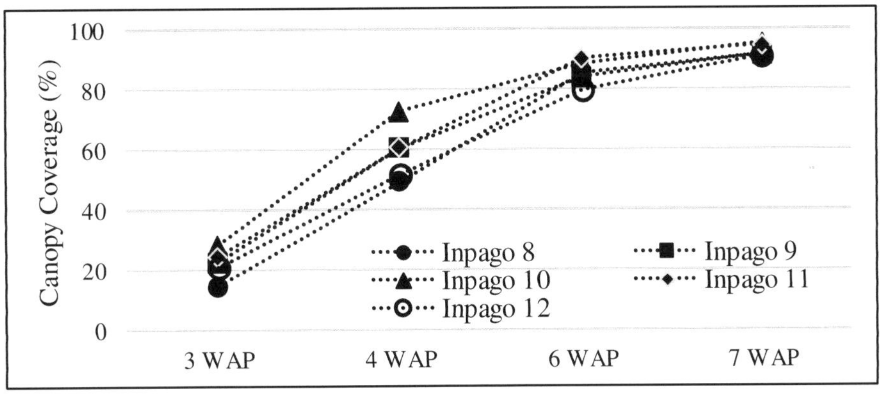

Figure 4. Growth of canopy coverage of five upland rice cultivars from 3 to 7 week after planting.

3.2. Yield and yield components

Grain yield and biomass accumulation performance of five tested cultivars were varied (table 1). Grain yield was ranged from 510 – 821 g m^{-2}, and the TDW ranged from 1,362 – 1,894 g m^{-2}. Inpago 9 and Inpago 12 were the greatest in grain yield with the yield of 821 g m^{-2} and 771 g m^{-2}, respectively. The biomass accumulation in terms of TDW of Inpago 8 was the highest followed by Inpago 9 with the TDW of 1,894 and 1,874 g m^{-2}, respectively. In addition, the HI, ratio dried grain yield and TDW, of Inpago 9 was the highest (0.517) followed by Inpago 12 (0.514).

The yield components that commonly used to explain yield performance were also varied. Panicle number was ranged from 8.0 – 16.6 per plant. The Inpago 12 were the greatest in panicle number, followed Inpago 9 with the panicle number around 16 per plant. The greater number of filled seed number was found in Inpago 11 and Inpago 12 with the filled seed number per panicle of 129.6 and 123.8, respectively. Bigger seed size was identified in Inpago 8 (31.0 g 1000seed^{-1}) followed by Inpago

9 (30.2 g 1000seed^{-1}). Total filled seed number per panicle length of Inpago 12 was the highest with the seed number per panicle length of 5.5 seed cm^{-1}.

Table 1. Yield and yield components of five upland rice cultivars

Cultivar	Yield			Yield Components				
	Yield (g/m^2)	TDW[a] (g/m^2)	HI[b]	Panicle Number (per plant)	Panicle Length (cm)	Filled Seed Number (per panicle)	Sterile Seed Number (per panicle)	1000SW[c] (g 1000seed^{-1})
Inpago 8	693	1,894	0.512	12.5	26.8	94.2	68.1	31.0
Inpago 9	821	1,874	0.517	16.0	23.0	103.7	29.6	30.2
Inpago 10	623	1,796	0.511	13.3	26.5	103.9	44.3	25.4
Inpago 11	510	1,362	0.509	8.0	27.4	129.6	23.6	27.0
Inpago 12	771	1,711	0.514	16.5	22.4	123.8	35.0	25.9

[a] TDW = total dry weight
[b] HI = harvest index
[c] 1000SW = 1000 seed weight

A positive correlation was found between grain yield and panicle number per plant (figure 5) with R^2 0.88. There are three groups of cultivar based on yield and panicle number performance. The first group consist of Inpago 9 and Inpago 12 with greater yield and panicle number. The second group consist of Inpago 8 and Inpago 10 with moderate yield and panicle number. The last group is Inpago 11 with inferior yield and panicle number.

Grain yield of five cultivars (ranged from 510 – 821 g m^{-2}) showed higher performance than regional average yield performance (2 to 3 tons per ha ≈ 200 to 300 g m^{-2}). Grain yield performance was closely correlated with panicle number per m^2. The Inpago 9 and Inpago 12 were greater in grain yield due to greater number of panicle as compared to other cultivars. Superior performance of Inpago 9 and Inpago 12 were also due to greater biomass production and better harvest index (HI). On the other side, smaller value of panicle number per m^2 and HI of Inpago 11 resulted in lower performance in grain yield.

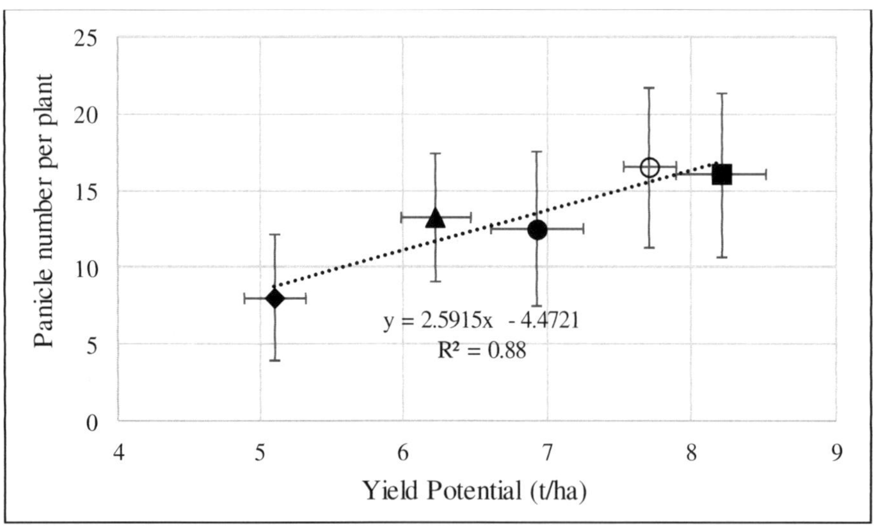

Figure 5. Correlation between panicle number per plant with yield potential of 5 upland rice cultivars.

3.3. Cultivar preference

Cultivar preference was evaluated from two aspects, plant performance in the field, and grain quality. The main purpose of cultivars preference evaluation is to know the acceptability of cultivars by the users (rice growers, rice consumers, and rice traders) in accompanied by yield performance. Figure 6 and 7, respectively, show the cultivar performance in the aspect of plant performance in the field and grain quality for respective 5 upland rice cultivars.

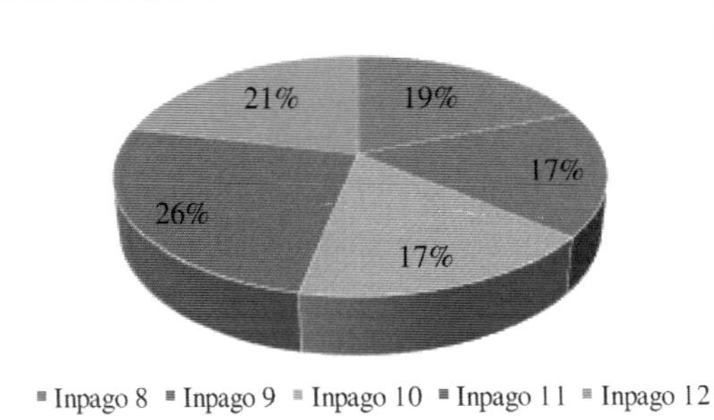

Figure 6. Cultivars preference in plant performance in the field of five upland rice cultivars

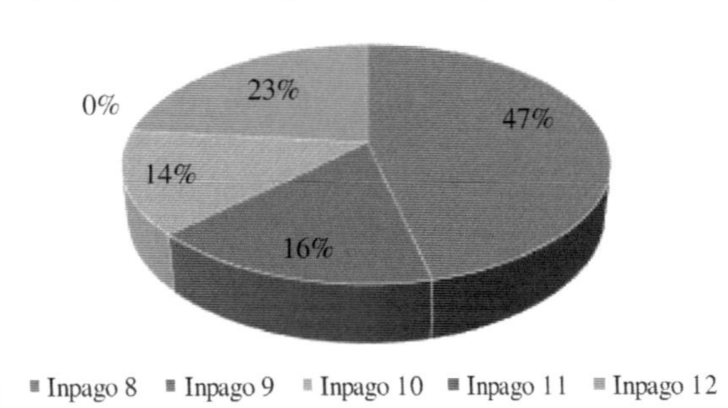

Figure 7. Cultivars preference in grain quality of five upland rice cultivars

For the field plant performance, cultivars preference is moderately distributed for each cultivars, ranged from 17 % to 26 %. The Inpago 11 was the most preferable cultivars due to plant vigor (plant height), panicle length and seed number that superior as compared to other cultivars (Table 1). For the other aspect, however, Inpago 8 was the most preferable cultivars (figure 7) in terms of grain quality. The Inpago 8 was preferable due to its good eating quality (stickiness), grain shape and colour and seed size. The rice consumers tend to choose cultivars with large size with long grain shape.

4. Conclusion

Grain yield of upland rice cultivars, Inpago 8, Inpago 9, Inpago 10, Inpago 11 and Inpago 12 showed higher performance than regional average yield performance. Grain yield performance was closely correlated with panicle number per m^2. The Inpago 9 and Inpago 12 were superior in grain yield due to greater biomass production and better harvest index (HI). Panicle number, biomass production and partitioning are some good traits for select adaptive cultivars. The Inpago 9 and Inpago 12 are considerably to be chosen by the farmers in Lebak district, Banten Province to improve upland rice production.

5. References

[1] Saito K, Asai H, Zhao D, Laborte A G and Grenier C 2018 Progress in varietal improvement for increasing upland rice productivity in the tropics *Plant Production Science* **1008** 1–14

[2] Saito K 2016 Plant Characteristics of High-Yielding **286** 276–86

[3] Arraudeau M A 1995 Upland Rice : Challenges and Opportunities in a Less Favorable Ecosystem *GeoJournal* **35** 325–6

[4] Abdurachman, A., Dariah, A., Mulyani A 2008 Strategi dan teknologi pengelolaan lahan kering mendukung pengadaan pangan nasional *Jurnal Litbang Pertanian* **27** 43–9

[5] BPS 2019 Luas Panen, Produksi, dan Produktivitas Padi Menurut Provinsi, 2018

[6] BPS 2015 Provinsi Banten dalam Angka Tahun 2015 632

[7] Wahab, M. I., Satoto, Suprihanto, Guswara, A. S 2018 *Deskripsi varietas unggul baru padi* (Badan Penelitian dan Pengembangan Pertanian)

[8] Purcell L C, Ball R a, Reaper J D and Vories E D 2002 Radiation use efficiency and biomass production in soybean at different plant population densities *Crop Science* **42** 172–7

[9] Bajgain R, Kawasaki Y, Akamatsu Y, Tanaka Y, Kawamura H, Katsura K and Shiraiwa T 2015 Biomass production and yield of soybean grown under converted paddy fields with excess water during the early growth stage *Field Crops Research* **180** 221–7

[10] Kawasaki Y, Tanaka Y, Katsura K, Purcell L C and Shiraiwa T 2016 Yield and dry matter productivity of Japanese and US soybean cultivars *Plant Production Science* **19** 257–66

[11] Saryoko A, Fukuda Y, Lubis I, Homma K and Shiraiwa T 2018 Field Crops Research Physiological activity and biomass production in crop canopy under a tropical environment in soybean cultivars with temperate and tropical origins *Field Crops Research* **216** 209–16

[12] Van Roekel R J, Purcell L C and Salmerón M 2015 Physiological and management factors contributing to soybean potential yield *Field Crops Research* **182** 86–97

6. Acknowledgement

We would like to thank Tana Rukmana of AIAT Banten, and Usep for his generous support with conducting the experiments.
This study was supported by the IAARD, Ministry Agriculture, Indonesia [grant number 560.4/HM.230/H.1/07/2018.K].

Morphology changes of rice root nucleus under iron stress

A Polosoro[1]*, W Enggarini[1], T Hadiarto[1], and N Ohmido[2]

[1]Indonesian Center for Agricultural Biotechnology and Genetic Resource Research and Development (ICABIOGRD), Jl. Tentara Pelajar No. 3A, Cimanggu, Bogor, Indonesia.
[2]Graduate School of Human Development and Environment, Kobe University, 3-11 Tsurukabuto, Nada-ku, Kobe 657-8501 Japan.

*Corresponding author: aqwin@pertanian.go.id, +62-818-0890-9220

Abstract, Understanding the response of the Root Apical Meristem (RAM) region under environmental stress is a long-standing issue in plant biology. Our previous study successfully gave new evidence in the epigenetic level that there is a heterochromatinization, especially in the proximal meristem region in rice under environmental stresses (high iron and salinity). These changes are suspected to be responsible for the reduction of the growth rate of RAM. On this study, we used sensitive (Ciherang), moderate (Nipponbare) and tolerant (Mentong and Siam Arjuna) rice varieties to see the meristematic cell response under high iron content environment. For the methodology, we use the nucleus size as the indicator for cell phase. Normal meristematic cells are dominant in the interphase stage (spread to G0, S, and G1 phase). However, our observation and quantitative analysis showed that iron stress reduced the S, G0, and cell cycle phase and pushed the cell to form G0-like phase. There is no significant difference between tolerant and sensitive rice varieties response, this proves that this response is a general mechanism in abiotic stresses.

Keyword: rice, iron stress, interphase, RAM

1. Introduction

Rice (Oryza sativa) is an essential crop for most Asian countries, majority inbreeding species that successfully supply as primary nutrition over half of the world's population [1]. An understanding of the physiological, developmental, and morphological mechanisms in cultivated rice is crucial for developing new varieties, especially in the aspects of quality, reliability, and sustainability as the world's food supply. The rapid growth of the human population, remarkably in developing countries push the researchers to accelerate the breading process to get more option in stress-tolerant, resource-use efficient, and highly productive varieties.

One of the main problems in rice cultivation is heavy metal-contained soil by natural or artificial contamination. In high concentration, heavy metal can decrease rice production significantly. Most iron (Fe)-polluted land is found in developing countries including Malaysia, India, Myanmar, Thailand,

Laos, Cambodia, Vietnam, Bangladesh, the Philippines, Sri Lanka and Madagascar, which are rice-producing countries [2,3]. For example, approximately 60% of the paddy fields in Central Africa and West Africa suffer from Fe poisoning. However, Indonesia suffers worse, the large scale of iron toxicity spread in 1 million ha of intensified tidal and peatland areas in the provinces of Southern Kalimantan and Southern Sumatra [4].

Iron is a crucial precursor in several biochemical and cellular mechanisms. This metal also has the primary function as electron acceptors and donors, cofactors in the enzymatic process and necessary in biological pathways, such as photosynthesis in chlorophyll, DNA synthesis, nitrogen metabolism and hormone synthesis [5]. Typically, the soil contains two types of iron oxidation forms; the divalent (Fe^{2+}) and trivalent (Fe^{3+}) which is related to several environmental conditions such as pH, redox potential (Eh) and water status. Mostly iron will be stable in high oxidation stage in the aerobic environment, while the divalent state is stable in anaerobic condition.

The plant growth regulator abscisic acid (ABA) has a crucial role in plant growth, development, biotic, and abiotic adaptation. Its function not only promotes seed differentiation and dormancy but also contributes to organizing the immune system under abiotic stresses [6]. ABA signaling pathways have been elucidated rapidly in much detail just only over a couple of years late. A significant example has been shown in progress at the discovery of ABA signaling complex in Arabidopsis, which regulates gene expression [7–9]. Our previous study also successfully gave new evidence in the global epigenetic level that there is a heterochromatinization, especially in the proximal meristem region in rice under environmental stresses (high iron and salinity) and also by exogenous ABA addition [10]. These changes are suspected to be responsible for the reduction of the growth rate of RAM and also cell transformation into a transient phase [11].

In the present study, the meristematic cell response was analyzed under high iron content environment. High iron content environment was set for treating germinated rice roots. Several rice cultivars, including sensitive, moderate, and tolerant varieties. The nucleus size population were counted based on size range and plotted on the curve for evaluating the meristematic cell response,. The result provides the population changes of nucleus specimens in the proximal meristem region.

2. Materials and methods

2.1. Plant materials

The experiment used several rice varieties, which was characterized by high iron condition as sensitive Ciherang (ICABIOGRAAD-Indonesia), moderate Nipponbare (Kobe University-Japan), and tolerant Siam Arjuna and Mentong (ICABIOGRAAD-Indonesia).

2.2. Treatments

To treated plant samples, the seeds were germinated in the growth chamber (NK System Biotron, Japan) at 25–30 °C under a 12-h photoperiod. 3-5-day-old seedlings were treated by dipping in 150 ppm (mg/L) iron sulfate ($FeSO_4$) for 6 h.

2.3. Slide preparation

RAM tissues were cut and fixed for 10 min under vacuum condition in 4% PFA buffer (paraformaldehyde in MTSB buffer), followed by 20 min in room temperature. To remove the fixative agent, the roots were washed twice in sterile distilled water for 10 min. To remove the cell wall, the samples were treated by an enzyme mixture of 1%Cellulose Onozuka RS [Yakult Co., Japan] and 2% Pectolyase Y-23 [Kikkoman, Japan] for 1 h and washed twice for 10 min each with 1x PBS buffer. The proximal meristem tissues were separated from the elongation zone, squashed on the glass slide, and covered by a cover glass. The cover glass was detached after freezing in liquid nitrogen and immediately transferred to cold 1x PBS buffer for 10 min. For the nucleus imaging, the sample stained by 0.5 ng/μL

DAPI solution for 10 min in the darkroom and washed by distilled water for couple time. A cover glass was placed on the samples after the addition of a drop of Vectashiled. The specimens were observed by a fluorescence microscope (BX60, Olympus-Japan), and the images were taken by a cooled CCD camera (SPOT RT3, Spot Imaging Solitons).

2.4. Data analysis

The nucleus size was converted to the actual wide by a pixel-size table and clustered by each 2 μm^2 range for counting the population. The histogram was projected to the line curve based on each population percentage and nucleus size.

3. Results

3.1. Nucleus size changes

Significant changes in the nucleus population of the four rice cultivars were observed during 24 h of treatment with 150 ppm iron ($FeSO_4$). Visual observation of rice root proximal meristem nucleus (fig. 1) shows the changes of global nucleus dimension in all rice varieties under iron treatment. In general, nucleus specimens were reduced in size after the treatment, and the nucleus population became more homogenous in the smaller size. Moreover, telophase specimens became challenging to find in the treatment sample. However, this result should be validated based on several statistical techniques to make more objective conclusions.

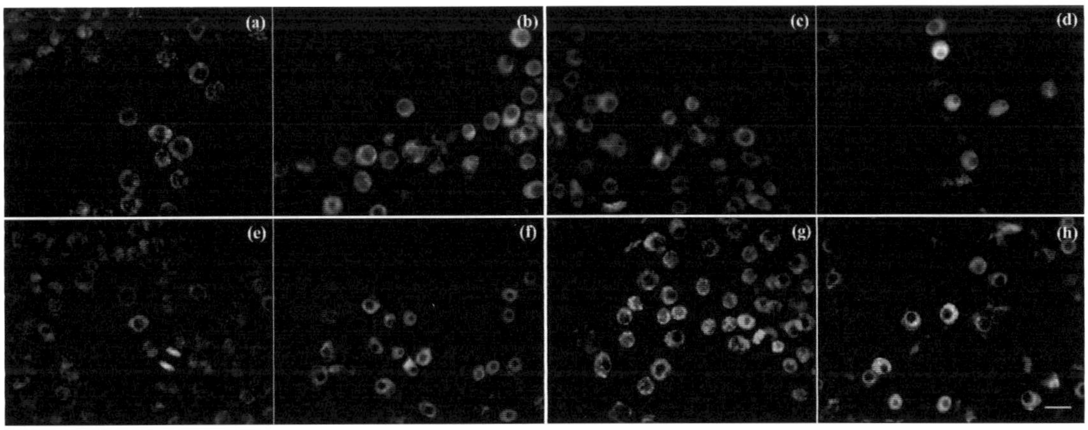

Figure 1. Visual observation of root proximal meristem nucleus of (a) Ciherang control, (b) Ciherang treatment, (c) Nipponbare control, (d) Nipponbare treatment, (e) Mentong control, (f) Mentong treatment, (g) Siam Arjuna control, and (h) Siam Arjuna treatment under high iron content medium. bar = 10 um

We used box and whisker plot and counting the mean to investigate the nucleus size data distribution and more objective observation,. Nucleus size data were taken from the nuclei of 225 Ciherang, 174 Nipponbare, 221 Mentong and 160 Siam Arjuna in the proximal meristem region. All controls have nucleus size of approximately 9 μm^2, and they were reduced to 7 μm^2 after 6 hours of iron treatment. Moreover, the data distribution also dramatically changed after the treatment. All 4th quartile of the control nucleus specimens has a wide distribution, from 11 μm^2 to 18 μm^2, gone after iron exposure. The nucleus size distribution also spread to a narrower dimension within range 3-12 μm^2 from 3-18 μm^2 before the treatment. However, there are no significant changes in the minimal size nucleus, which is still distributed around 3-4 μm^2. We also found 50% of nucleus size population (in quartile 2nd and 3rd) spread on 5-8 μm^2, which was far smaller compared to the control which was about 7-11μm^2. This showed that the nucleus transformed into one size specimen.

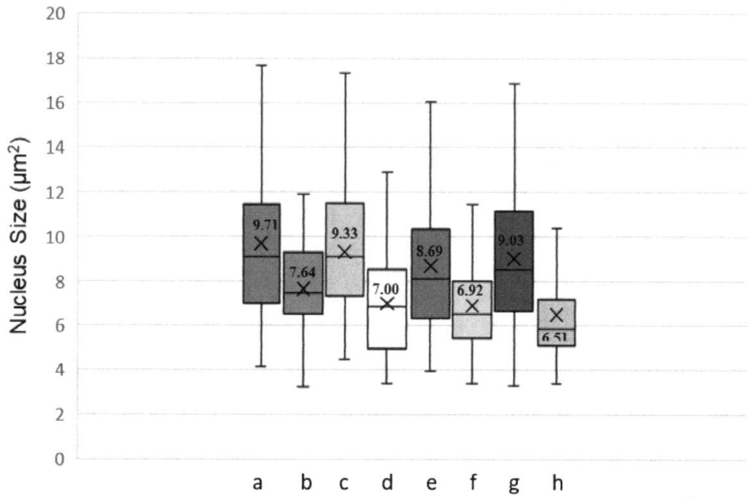

Figure 2. Nucleus size data distribution compared between varieties and treatment. The lower bar indicates the extreme lower quartile; the upper bar, the extreme upper quartile; the lower line of the box, the lower quartile; the upper line of the box, the upper quartile; and the cross shows the median. (a) Ciherang control, (b) Ciherang treatment, (c) Nipponbare control, (d) Nipponbare treatment, (e) Mentong control, (f) Mentong treatment, (g) Siam Arjuna control, and (h) Siam Arjuna treatment

3.2. Nucleus phase arrest

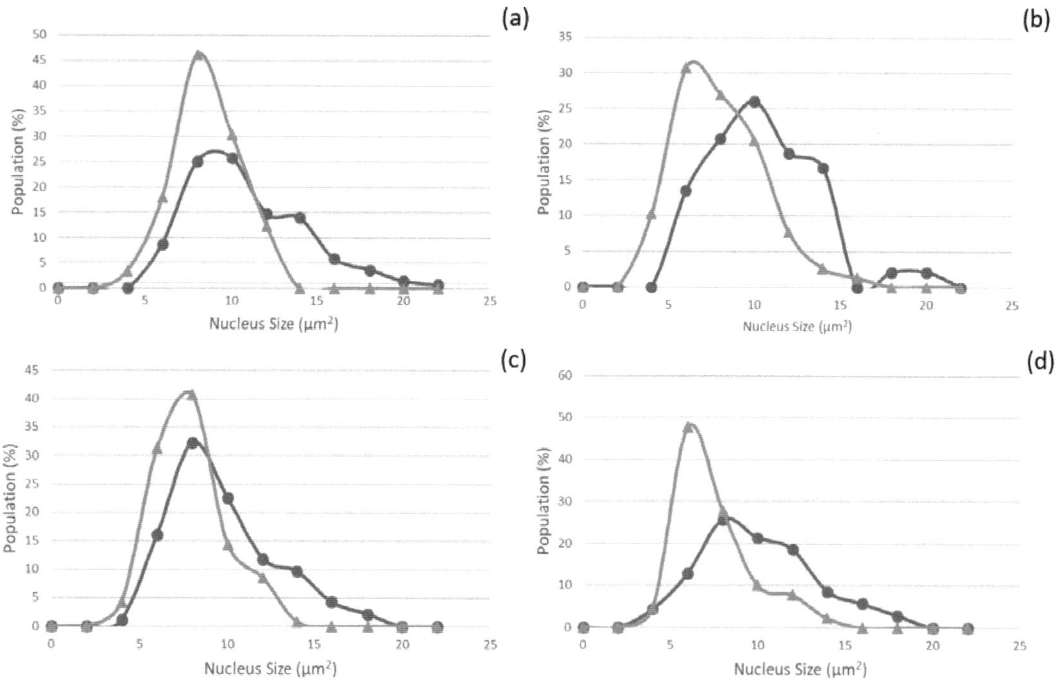

Figure 3. The proximal nucleus population based on the nucleus size on the rage 2 µm² in a). Ciherang, b). Nipponbare, c). Mentong and d). Siam Arjuna. circle = control and triangle = iron treatment.

We utilized a histogram plot from all nucleus size population data from the range of every 2 µm² size difference to analyze the population of nucleus specimens. The results (figure 3) successfully imitated flow cytometry histogram results in the meristematic nucleus. Entirely untreated variety showed two peaks at 7-10 μm^2 and 13-15 μm^2 that were estimated as G1 and G2 checkpoints. However, 6 hours of iron exposure successfully erased the second peak and increased the population of G1-like checkpoint in all cultivars. The G1 checkpoint climbed from just only 25% to 35-45 % proximal nucleus population. The single peak curve demonstrated that the nucleus on the proximal region arrested on the G1 checkpoint.

4. Discussion

4.1. Effect of iron stress on rice meristematic activity

In this study, the high iron content environment not only successfully changes the average size of the rice nuclei but also changes the nucleus composition from heterogenous size to smaller homogenous size nuclei. Several studies have also been performed to investigate the effects of environmental stresses on the RAM activities, the stress can initiate early maturation of RAM [12], reduction of cell division [13–15], reduction of nucleus size [13,15], nucleus heterochromatinization [10], and transformation of nuclei into a transient state [11].

Using the box and whisker plot diagram, we can identify extinction of 25% of nucleus specimens (11 μm^2 to 18 μm^2 wide) during 6-hour treatment. The histogram analysis also detects the erasure of G2 phase and increment of the G1 population, but there are no significant differences between varieties. The G1 phase arrest dominates in the canonical cell cycle in several biological processes such as meristematic quiescence, dormancy, and terminal differentiation. Meristematic quiescence is in substance-related to non-cell-autonomous regulation of meristem cell identity, and especially by the effect of ubiquitin-dependent proteolysis, which also associated with reactive oxygen species (ROS), abscisic acid (ABA), and auxin [16]. All varieties also show domination of G1-like phase under iron stress, and it is also supported by our previous research that demonstrated the link between heterochromatinization of proximal meristem with an increment of ABA, ROS, and also IAA negative effect [10].

4.2. G1 arrest and nucleus heterochromatinization

The association between G1 arrest and heterochromatinization might be apparent since global heterochromatinization in the meristematic nucleus during stress condition will decrease. Histone acetylation is needed for the specification and activation of replication origins during the transition at the G1 and in S phase. In G2 phase, a deposition of CENH3 was discovered in the centromeric region, but the mechanism still unknown [17]. During mitosis, euchromatinization occurs globally, which is induced by histone modification, mainly phosphorylation. Finally, at the end of the cell cycle, heterochromatinization spread to avoid re-replication or endoreduplication (E) and re-entry into S phase [18]. Heterochromatinization will close DNA accessibility and affects gene transcription and DNA replication. As a result, protein synthesis and DNA replication will be interrupted, and it will reduce the S and G2 nucleus population and raises of G1 nucleus specimens.

4.3. G1 arrest as a transient phase (Meristematic quiescence)

In the case of the cell culture, growing the plant cell in the salt medium will give stress condition that leads to cellular reprogramming and development of a transient, dedifferentiated, stem cell-like state. This transition is related to chromatin decondensation or heterochromatinization that usually accompanied by DNA transposition and/or recombination [11]. This transient phase is categorized as meristematic quiescence, a transient state of the suppressed cell cycle in undifferentiated cells. It is also classified as embryonic quiescence, including organizing centre (OC) and the quiescent centre (QC) in the same state [16].

ROS and redox signaling pathways are proved as an essential player to maintain the balance of quiescence and proliferation in the RAM and SAM [19]. In the case of G1 arrest in QC cells, antioxidants ascorbate and glutathione have an important role which is dependent on the polar auxin maximum in the QC. Therefore, the ascorbate and glutathione are both highly oxidized in the QC, but on the contrary in the proximal meristem [20]. Our previous experiment and this paper also proved the robust evidence about the connection between G1 arrest and ROS increment in proximal meristem tissues.

5. Conclusion

In conclusion, our experiment demonstrates the response of rice root nucleus to high iron exposure by suppressing the growth rates and arresting the nucleus in the G1 phase. The mitotic activity reduction probably is a result of global heterochromatinization and also a mechanism for defending meristematic cell under environmental stress. However, there is no significant difference between tolerant and sensitive rice varieties response, which indicates that this response is a general mechanism in abiotic stresses. Future research should evaluate the G1 arrest state actual status by using more integrated parameters such as cell physiology and epigenetics for discovering the real mechanism of G1 arrest under environmental stresses.

6. References

[1] Zhao K, Tung C, Eizenga G C, Wright M H, Ali M L, Price A H, Norton G J, Islam M R, Reynolds A, Mezey J, Mcclung A M, Bustamante C D and Mccouch S R 2011 a rich genetic architecture of complex traits in Oryza sativa *Nat. Commun.* **2** 467
[2] Asch F, Becker M and Kpongor D S 2005 A quick and efficient screen for resistance to iron toxicity in lowland rice *J. Plant Nutr. Soil Sci.* **168** 764–73
[3] Tam Aung 2006 *Physiological mechanisms of iron toxicity tolerance in lowland rice Master of Science Tam Aung* (Universität zu Bonn)
[4] Ismunadji 1990 Alleviating iron toxicity in lowland rice *Indon. Agric. Res. Dev. J.* **12** 67-72.
[5] Jeong J and Connolly E L 2009 Iron uptake mechanisms in plants: Functions of the FRO family of ferric reductases *Plant Sci.* **176** 709–14
[6] Cutler S R, Rodriguez P L, Finkelstein R R and Abrams S R 2010 *Abscisic Acid: Emergence of a Core Signaling Network* vol 61
[7] Lee S C and Luan S 2012 ABA signal transduction at the crossroad of biotic and abiotic stress responses *Plant, Cell Environ.* **35** 53–60
[8] Raghavendra A S, Gonugunta V K, Christmann A and Grill E 2010 ABA perception and signalling *Trends Plant Sci.* **15** 395–401
[9] Yoshida T, Mogami J and Yamaguchi-Shinozaki K 2014 ABA-dependent and ABA-independent signaling in response to osmotic stress in plants. *Curr. Opin. Plant Biol.* **21** 133–9
[10] Polosoro A, Enggarini W and Ohmido N 2019 Global epigenetic changes of histone modification under environmental stresses in rice root *Chromosom. Res.*
[11] Grafi G, Florentin A, Ransbotyn V and Morgenstern Y 2011 The Stem Cell State in Plant Development and in Response to Stress *Front. Plant Sci.* **2** 1–10
[12] Ji H, Liu L, Li K, Xie Q, Wang Z, Zhao X and Li X 2014 PEG-mediated osmotic stress induces premature differentiation of the root apical meristem and outgrowth of lateral roots in wheat *J. Exp. Bot.* **65** 4863–72
[13] Fernandez-Marcos M, Sanz L, Lewis D R, Muday G K and Lorenzo O 2011 Nitric oxide causes root apical meristem defects and growth inhibition while reducing PIN-FORMED 1 (PIN1)-dependent acropetal auxin transport *Proc. Natl. Acad. Sci.* **108** 18506–11
[14] Sacks M M, Silk W K and Burman P 1997 Effect of Water Stress on Cortical Cell Division Rates within the Apical Meristem of Primary Roots of Maize. *Plant Physiol.* **114** 519–27
[15] West G, Inzé D and Beemster G T S 2004 Cell Cycle Modulation in the Response of the Primary Root of Arabidopsis to Salt Stress *Plant Physiol.* **135** 1050 LP – 1058
[16] Velappan Y, Signorelli S and Considine M J 2017 Cell cycle arrest in plants : what distinguishes quiescence , dormancy and differentiated G1 ? 495–509
[17] Lermontova I, Rutten T and Schubert I 2011 Deposition, turnover, and release of CENH3 at Arabidopsis centromeres *Chromosoma* **120** 633–40
[18] Raynaud C, Mallory A C, Latrasse D, Jégu T, Bruggeman Q, Delarue M, Bergounioux C and Benhamed M 2014 Chromatin meets the cell cycle *J. Exp. Bot.* **65** 2677–89

[19] Considine M J and Foyer C H 2013 Redox Regulation of Plant Development *Antioxid. Redox Signal.* **21** 1305–26
[20] Jiang K, Meng Y L and Feldman L J 2003 Quiescent center formation in maize roots is associated with an auxin-regulated oxidizing environment *Development* **130** 1429 LP – 1438

Characterization flour of two seaweeds (*Gracilaria spp.* and *Kappaphycus alvarezii*) for reducing consumption of wheat flour in Indonesia

A Munandar[1]*, D Surilayani[1], S Haryati[1], M H Sumantri[2], R P Aditia[1], and G Pratama[1]

[1] Fisheries Department, Faculty of Agriculture, Sultan Ageng Tirtayasa University, Jl. Raya Jakarta km. 04 Pakupatan, Serang City, Banten Province, Indonesia.
[2] Graduate of Fisheries Department, Faculty of Agriculture, Sultan Ageng Tirtayasa University, Jl. Raya Jakarta km. 04 Pakupatan, Serang City, Banten Province, Indonesia.

*Corresponding author: aris.munandar@untirta.ac.id

Abstract. Dependence on wheat flour in Indonesia can be reduced by looking for substitute ingredients which had high nutritional value, one of which was seaweeds. The objective of this research was to determine the characteristics of flour of two seaweeds (*Gracilaria* spp. and *Kappaphycus alvarezii*). Seaweed flours have produced using a hammer mill. The comparative of characteristics on two seaweed flours with proximate composition, crude fiber, and L* parameter. The proximate content of seaweed flours from *Gracilaria* spp. and *K. alvarezii* were moisture 11,83% and 17,56%, ash 15,08% and 16,90%, fat 0,32% and 0,13%, and protein 8,77% and 4,07%, respectively. The value of crude fiber and L* parameter in seaweed flour *K. alvarezii* (5,43% and 80,50%) were higher than *Gracilaria* spp. (4,77% and 49,90%). The best of seaweed flour was *Gracilaria* spp. flour, because of the content almost similar to wheat flour.

Keywords: Seaweed flour, *Gracilaria* spp., *Kappaphycus alvarezzii*

1. Introduction

In Indonesia, consumption of wheat flour is increasing every year. In 2013, its needs reached 5,351 million or increased by around 4,06% compared to 2012 [1]. One of the causes of the increasing need for wheat flour is the increasing number of food products that use wheat flour as the main basic ingredient. Indonesia is the second-largest wheat importing country in the world after Egypt with an average import volume of over 5 million tons per year. In 2013/2014, the volume of Indonesia's wheat imports reached 7,392 million and increased in 2014/2015 by 7,7 million. The increase in imports was due to the lack of domestic production [2].

Dependence on wheat flour in Indonesia must be reduced by finding substitution materials that have high nutritional value. One of the substitution materials that can be used mainly for making meal was seaweed [3]. Seaweed is an important commodity in Indonesia so it can be optimized into seaweed flour. Seaweed contains carbohydrates (sugar or vegetable gum), protein, less fat, and ash which are mostly sodium and potassium salt compounds [4]. Seaweed that can be used were *Gracilaria* spp. and *K. alvarezii* [5].

K. alvarezii was included in seaweed which has commercial value and export commodity. This type of seaweed is one of carragenophytes which is carrageenan-producing seaweed. The processed products from K. alvarezii are widely used as emulsifiers, gel formers, stabilizers, and thickeners [6]. *Gracilaria* spp. produce agar which can form a gel so that it can function as a gelling agent and also as a thickener and stabilizer [7]. Therefore, we desire to attempt the characteristics of flour of two seaweeds (*Gracilaria* spp. and *K. alvarezii*).

2. Material & Methods

2.1 Preparation of materials

The seaweed of *Gracilaria* spp. and *K. alvarezii* were collected from Lontar Beach in July 2016. The areas were in Serang, Banten, Indonesia. In general, the process of making flour includes cleaning and washing, soaking, reducing the size, drying, grinding and sifting. Seaweed flours have produced using a hammer mill [4].

2.2 Analysis of sample

The comparative of characteristics on two seaweed flours with proximate composition, crude fiber, and L* parameter. Proximate and crude fiber measurement used the concept of AOAC [8]. Measurement of the L* parameter using a whiteness meter. The sample was put into the tool, at the place provided for the sample [4].

3. Result & Discussion

3.1 L parameter*

The color produced in seaweed flour was influenced by the quality of raw materials. *Gracilaria* spp. was blackish green so that the color of seaweed flour was less than the maximum. Because there are still impurities in the raw material and contain color pigments. It was different from *K. alvarezii* which has a white color so that the value was higher when compared to *Gracilaria* spp. Red seaweed has a phycobiliprotein pigment consisting of red phycoerythrin and blue phycocyanin [9]. In this condition, this seaweed can make pigment adjustments to the lighting quality so that it can cause various colors on the thallus. The colors formed include dark red, pink, blond, brown, yellow and green [7].

3.2 Crude fiber

The crude fiber content of Gracilaria spp. and K. alvarezii were 4,77% and 5,43%, respectively. It was closely related to the content of polysaccharides. Polysaccharides are composed of hydrocolloids making up cell walls and intercellular space-filling materials including carrageenan and agar [3]. The content of carrageenan and agar in seaweed plays a role as does gluten. The function of gluten as a form of skeletal structure in food dough [4]. Carrageenan and agar can also function to form gels in food [10]. Therefore, seaweed flour can be used as an alternative to wheat flour substitution.

3.3 Proximate composition

The proximate content of seaweed flours from *Gracilaria* spp. and *Kappaphycus alvarezii* were moisture 11,83% and 17,56%, ash 15,08% and 16,90%, fat 0,32% and 0,13%, and protein 8,77% and 4,07%, respectively. High mineral salts in seaweed raw materials will make water levels high. The surface of the seaweed has white salt crystals. The presence of hygroscopic salt crystals can result in increased water content of seaweed during storage. Carrageenan is hydrophilic so it was easy to absorb and dissolve in water [9]. Carrageenan was stored the longer the more water vapor is absorbed. The content of water content in seaweed flour affects shelf life. The higher the water content of seaweed flour, the more susceptible to microbes during storage [4].

Seaweed protein levels are different from one area to another because it was influenced by the habitat and type of seaweed itself. *Gracilaria* spp has a higher protein content when compared with *K. alvarezii* in several studies. Because the content of phycobiliprotein in *Gracilaria* spp was higher than *K. alvarezii*

[7]. fat and ash in both seaweed there was no significant difference. The proximate content has shown in Table 1.

Table 1. Seaweed flour parameter

Parameter	Seaweed flour	
	Gracilaria spp.	*K. alvarezii*
L* parameter (%)	49,90	80,50
Crude fiber (%)	4,77 ± 0,54	5,43 ± 0,47
Moisture (%)	11,83 ± 0,18	17,56 ± 0,15
Ash (%)	15,08 ± 0,48	16,90 ± 0,11
Fat (%)	0,32 ± 0,25	0,13 ± 0,01
Protein (%)	8,77 ± 0,20	4,07 ± 0,12

4. Conclusion

The best of seaweed flour was *Gracilaria* spp. flour, because of the content almost similar to wheat flour, but the color was not good.

5. References

[1] [APTINDO] 2014 *Industri Tepung Terigu Nasional Indonesia*. Jakarta: APTINDO

[2] [USDA] United State Department of Agriculture 2015 *Indonesia Grain and Feed Annual Report 2015*. USA: USDA

[3] Alvarenga N B, Lidon F C, Belga E, Patricia M, Guerreiro S, Carvalho J M, Canada J 2011 Characterization of gluten-free bread prepared from maize , rice and tapioca flours using the hydrocolloid seaweed agar-agar *Recent Res. Sci. Technol.* **3** 8 64–68

[4] Mamat H, Matanjun P, Ibrahim S, Amin S F M, Hamid M A, Rameli A S 2014 The effect of seaweed composite flour on the textural properties of dough and bread *J Appl. Phycol.* **26** 2 1057–1062

[5] Kumoro A C, Johnny D, Alfilovita D 2016 Incorporation of microalgae and seaweed in instant fried wheat noodles manufacturing: Nutrition and culinary properties study Incorporation of microalgae and seaweed in instant fried wheat noodles manufacturing: nutrition and culinary properties study *Int. Food Res. J.* **23** 2 715–722

[6] Larasanti F, Pratama R I, Haetami K, Rostini I 2019 The addition of seaweed flour (Kappaphycus alvarezii) and carrageenan on the preference level of freshwater pomfret meatballs *Sci. News Pacific Reg.* **3** 30–40

[7] Buschmann A H, Hernández-González M C, Varela D 2008 Seaweed future cultivation in Chile : perspectives and challenges *Int. J. Environ. Pollut.* **33** 4 432–456

[8] [AOAC] 2005 *Official Methods of Analysis*, 18th ed. Mayland: AOAC

[9] Karthikeyan K, Shanmugam M 2016 Grain yield and functional properties of red gram seaweed extract powder maufactured from *Kappaphycus alvarezii Int. J. Recent Adv. Multidiscip. Res.* **3** 3 1353–1359

[10] da Silva R L, Barbosa J M 2009 Seaweed meal as a protein source for the white shrimp Litopenaeus vannamei *J Appl. Phycol.* **21** 2 193–19

6. Acknowledgement

We are grateful to University of Sultan Ageng Tirtayasa for shared the grant research with scheme “Penelitian Madya”.

IOP Conf. Series: Earth and Environmental Science **383** (2019) 012010 doi:10.1088/1755-1315/383/1/012010

Empowerment of farmers in increasing production of environmentally friendly soybean plants: case study in Pandeglang regency Indonesia

A Mulyaningsih*, A T Sumantri

[1]Department of Agribisnis, Faculty of Agriculture, Universitas Sultan Ageng Tirtayasa, Jl. Raya Jakarta Km 4 Pakupatan Serang Banten Indonesia

*Corresponding author: asihmulya@ymail.com

Abstract. Soybean is one of the food commodities needed as a supplement to food nutrition because it is rich in vegetable protein, carbohydrates and fats. This study aims to analyze the empowerment of soybean farmers in Pandeglang Regency. The population in this study were soybean farmers in two sub-districts in Pandeglang Regency, Banten Province, namely Carita and Panimbang Subdistricts, which are centers of soybean plants. The number of samples is 150 farmers, soybean, two sub-districts in Pandeglang Regency, Banten Province. The research activities were carried out for three months (April to June 2019). Data is collected and processed using descriptive statistics and inferencing statistics. Data analysis with Different Mann Withney Test. The results showed that there were significant differences in the empowerment of farmers in the two sub-districts in Pandeglang District, where the empowerment of soybean farmers in Carita District was lower than that of Soybean Farmers in Panimbang District. For this reason, it is necessary to increase the empowerment of soybean farmers in Carita Subdistrict by increasing technical capabilities in managing soybean farming, the ability to overcome problems in soybean farming, to improve soybean farming planning capabilities. In addition, it is also necessary to increase the perception of soybean farmers that soybean commodities are potential commodities to be developed.

Keywords: empowerment, soybean.

1. Introduction

Soybean is an important food commodity in Indonesia because it is widely used as a source of vegetable protein for public consumption and the food industry in Indonesia. Soybean seeds contain many nutrients such as protein 30-50%, fat 15-25% and several important nutrients, such as vitamins (phytic acid) and lecithin [7]. The multipurpose nature of soybean causes high demand for soybeans in the country. In addition, the benefits of soybeans as a cheap source of protein make soybeans increasingly in demand. As the population increases, the demand for soybeans in the country has the potential to increase every year. Estimated domestic demand for soybeans per year is 2.4 million - 2.6 million tons, of which 70% - 80% these needs are met from imported soybeans. On the other hand, the ability to produce soybean domestically there is a tendency to decline from year to year so that the level of dependence on soybean imports is also high. Therefore efforts must be made to increase domestic soybean production.

Based on research carried out by several parties, various technologies have been found to increase function soybean production in Indonesia, but only a small portion is applied by farmers. This is because farmers consider the application of technology unprofitable because the price of soybeans in the market is low due to the presence of cheap imported soybeans. Therefore farmers only use makeshift technology so that production is not optimal and causes the productivity of national soybeans to be relatively low.

Based on the problems faced by farmers, it is necessary to find an alternative how so that the low input technology applied by farmers does not reduce soybean production, such as the use of drought resistant cultivars, little fertilization and weeds that are resistant to weeds. Finding cultivars that are resistant to weeds is an important thing because weeds can increase production costs and can reduce soybean yield. This is considered important because weeds are one of the contributors to the decline in soybean yields.

The ability of farmers to implement environmentally friendly innovations in addition to being seen from the technical and environmental aspects, can also be seen from the ability of farmers to plan and evaluate their farming in order to obtain business efficiency. The ability to adapt in the form of fighting power and adaptive power is generally still low. This can be seen from submission to unfavorable conditions, such as carrying out cultivation activities through hereditary methods, surrender to the conditions of infertile land, limited availability of water, pest attacks, low production and other problems encountered. Adaptation to various types of changes that occur, including market appetite for products that are safe for chemical residues, is still not anticipated by farmers in conducting their farming. Support for increasing farmer empowerment can be seen from the synergistic partnership capability in the form of building communication, interacting and collaborating as an important process in exchanging experiences, ideas, changing knowledge and views. The ability of farmers to cooperate with other groups and organizations is expected to overcome problems of limited capital and technology, improvement of product quality and marketing problems.

Based on the results of previous studies emphasizing the aspects of cultivation techniques, overcoming the negative impacts of conventional agriculture and evaluating the level of application of technology in programs to increase soybean production [15], [4], [16], [11], [13], [8], [6], [10]. The results of research related to farmer empowerment focus more on aspects of developing farmers' capabilities through institutions or farmer groups [1], [3], [19], [14], [18], not yet seeing the learning process of farmers as subjects that have the potential to be improved its ability through the power it has. The purpose of this study was to analyze the empowerment of soybean farmers in Pandeglang Regency.

2. Materials and Methods

This study used a survey method by distributing questionnaires. Qualitative analysis is done by in-depth interviews to explore or explore (exploratory) and understand the experiences of informants in the concept of empowerment. The format of questions is arranged in an ordinal scale with response categories arranged in a matrix form and consists of four answer choices, namely: always, often, rarely, never [12]. This study uses a probability sample design, meaning sampling is based on the idea that the entire population unit has the same opportunity to be sampled [2], [9]. The study was conducted for three months from April to June 2019 at the soybean production center in two sub-districts in Banten Province, namely Carita and Panimbang Districts. The sample in this study were 150 farmers from two sub-districts where each sub-district was 75 soybean farmers.

This study uses a probability sample design, meaning sampling is based on the idea that the entire population unit has the same opportunity to be used as a method [2], [9]. To see the difference in the level of empowerment of farmers, the data were analyzed using a Mann Withney different test conducted using SPSS 21 program assistance (Statistical Product and Service Solution 21).

Mann Withney

$$U_1 = n_1 n_2 + \frac{n_1(n_1+1)}{2} - R_1 \tag{1}$$

$$U_2 = n_1 n_2 + \frac{n_1(n_1+1)}{2} - R_2 \quad (2)$$

Information:
n_1 = number of samples 1
n_2 = number of samples 2
R_1 = number of steps in sample 1
R_2 = number of levels in sample 2

3. Results and Discussion

Community empowerment are elements that enable people to survive and in a dynamic sense develop themselves and achieve progress. Empowering the community is an effort to increase the dignity of the people who are not able to escape from the trap of poverty and underdevelopment.

Table 1. Average distribution and results of different tests of soybean farmers in Carita and Panimbang Districts based on indicators of farmer empowerment

Empowerment of farmers	Category	Percentage (%)			U-Test
		Carita	Panimbang	Total	
Technical ability	Less fortunate	7,40	33,80	20,60	0,000**
	Quite capable	23,10	34,30	28,70	
	Able	49,50	28,20	38,90	
	Very capable	19,90	3,70	11,80	
Average		2,82	2,02	2,42	
Ability to overcome problems	Less fortunate	18,50	21,30	19,90	0,254
	Quite capable	48,10	49,50	48,80	
	Able	26,40	25,00	25,70	
	Very capable	6,90	4,20	5,60	
Average		2,22	2,12	2,17	
Planning ability	Less fortunate	4,20	17,60	10,90	0,000**
	Quite capable	48,60	47,20	47,90	
	Able	45,80	33,80	39,80	
	Very capable	1,40	1,40	1,40	
Average		2,44	2,19	2,32	
Adaptability	Less fortunate	6,00	46,80	26,40	0,000**
	Quite capable	39,80	36,60	38,20	
	Able	50,90	15,30	33,10	
	Very capable	3,20	1,40	2,30	
Average		2,51	1,710	2,11	

Empowering is enabling and empowering the community [5]. The empowerment of farmers in this study is measured by the ability of technical abilities, ability to overcome problems, planning ability, adaptability. The empowerment of soybean farmers is used as information on soybean farmers in Carita and Panimbang Districts. Based on the results of the Mann Whitney difference test there are differences in the empowerment of soybean farmers in Carita District with soybean farmers in Panimbang District.

3.1. *Technical ability*

Technical capability is the capability of farmers related to the technical management of soybean farming which includes: the ability to rotate varieties, the ability to maintain plants in the form of balanced fertilization and integrated pest control, to the ability to manage irrigation. In addition, the technical ability to apply the principles of environmentally friendly farming is also an important part.

Understanding of balanced fertilization, use of chemical pesticides, intermittent irrigation, utilization of straw as organic fertilizer as well as the replacement of soybean seeds can be understood by farmers, namely the level of error in answering low statements. This is because farmers have always listened to information on integrated crop management technology and in fact most have been applied in farming activities.

Based on Table 1, there are significant differences between the technical capabilities possessed by farmers in Carita and Panimbang Districts. It appears that the percentage of respondents who have technical capabilities in Carita District is better than Kedele farmers in Panimbang District. This is indicated by the presentation value of the technical capabilities of farmers in Carita District in the category of being able to manage soybean farming technically with a value of 49.5 percent while the tennis ability of soybean cultivation in Panimbang District is quite capable with a value of 34.3 percent. The average technical capability of soybean farmers in Carita District is higher than that of farmers in Panimbang District.

The complaints of farmers in the application of environmentally friendly innovations are in the use of organic fertilizers and vegetable pesticides. This innovation requires an outpouring of labor and time to make it because there is not much available compared to the use of chemical fertilizers and pesticides. The management of soybean farming in both sub-districts in Pandeglang Regency has been using conventional technology, such as planting soybean seeds at close distances, flooded irrigation, and using poor quality seeds. This was done based on the assumption that the technology had been carried out for generations and the results could be enough to support their lives so far..

3.2. Ability to overcome problems

The ability to overcome problems is an individual's ability which is also very important for farmers. The many problems in farming that are faced are often a burden and can interfere with farming activities. This individual ability, is the ability possessed by each individual, including farmers. The ability to overcome farming problems includes: (1) the ability to overcome scarcity of production facilities; (2) the ability to make organic fertilizers with available ingredients; (3) the ability to jointly control in one stretch; and (4) the ability to cooperate in cleaning irrigation channels independently. It can be seen in Table 1 that the level of ability to overcome problems in soybean farming in the total Carita Subdistrict reaches 2.22 percent, or more when compared to the number of respondents in Panimbang sub-district of 2.12 percent.

The results of the real difference test showed that there were significant differences between the ability to overcome the problems that were owned by farmers in Carita District and farmers in Panimbang District in the aspect of soybean farming. Common problems in aspects of soybean cultivation are more in handling pests and diseases, water management and balanced fertilization. Water problems are caused more by the availability of water and the frequent improvement of irrigation so that water closures occur. Some of these problems occur in both sub-districts. This certainly interferes with farming activities, so that effective and efficient use of water through intermittent irrigation systems can help farmers.

It can be seen in Table 1 that soybean farmers in both sub-districts were quite able to overcome the problem. However, there are significant differences between soybean farmers in Carita District and soybean farmers in Panimbang Subdistrict, where farmers in Panimbang sub-district are better able to overcome problems than farmers in Carita District. The farmers' understanding of the concepts of soybean management, such as: the function of providing organic material (manure or compost) to the land, the benefits of conducting varieties rotation and the use of superior seeds, the benefits of intermittent irrigation, the use of decomposers and vegetable pesticides and fertilization balanced and specific locations classified as soybean farmers in Carita District are classified as capable of managing soybean farming technically. The existence of a good understanding of these concepts, it turns out, has not been fully implemented in managing the farm. The management of farming by most soybean farmers is still done conventionally and relies on habits originating from their parents.

3.3. Planning ability

Planning capability is a personal ability related to regular and planned farming efforts, ranging from land management, breeding, planting, maintenance, harvesting to post-harvest handling. All of them are intended to be well planned, so that they can obtain optimal results. Planning capabilities in this study include: (1) planning of farmland area; (2) financing planning; (3) farming records; and (4) taking into account the risks faced.

Table 1 shows that the average planning ability score is quite good in both Carita District and Panimbang District, but there are differences where farmers in Carita District are more capable than farmers in Panimbang District. Farming planning capabilities are influenced by individuals or personal, so generally very minimal is done in farming activities. Knowledge of planning generally can only be obtained from non-formal education activities such as: training, mentoring or other activities. Farmers generally do not do farming planning or farming records. His farming activities are generally based on habits. This shows that the level of ability of farmers, especially related to farming planning capabilities is still very low. These conditions can be obstacles in an effort to increase production.

3.4. Adaptability

Adaptation ability is the ability of each individual farmer to adapt to the environment. The adaptation ability is very important in farming activities, considering that agricultural activities are very dependent on natural conditions, such as climate, season and other natural conditions. The average score of adaptability of farmers in Carita Subdistrict is higher than that of petabi in Panimbang District (Table 1). Farmers in Carita sub-district are able to adapt to the environment while farmers in Panimbang sub-district are less able to adapt to the environment. Environmental adaptability becomes capital in farming activities, so that production results can be optimized. This is in line with [17] that the ability of farmers to know, master, understand and implement the technical principles of farming management will increase the independence and success of farmers.

Adaptation ability of farmers in Carita Subdistrict is the ability to overcome the attack of plant pests. The reason for farmers who are difficult to adapt is that they are accustomed to cultivating soybean plants conventionally and feel that they will find it difficult to try ways that are environmentally friendly, with perceptions of being busy, complicated and having no time.

Adaptation actions taken by farmers if there is a crop failure, ideally utilizing savings from the proceeds in the following planting seasons. But most farmers do not have savings to anticipate such an emergency. Most then rely on loans from other parties as a source of farming funding and the cost of daily living. This action certainly has further consequences, namely the emergence of debt, especially if the loan is flowering. This will be a new burden if farmers cannot pay it off. Adaptation actions carried out by respondent farmers still require costs. The condition of lack of farming capital that occurs in respondent farmers, especially if they experience crop failure, makes adaptation actions not optimal. One of them is the scale of farming that is managed to be minimized from the scale of previous farming. This implies the risk of production that is not optimal in the following season. The conventional adaptation actions taken by these respondents have also not been able to guarantee the return of lost business capital and the accumulation of income. In the long term, of course this can have an impact on the socio-economic conditions of farmers and their families.

Adaptation to the use of superior seeds was addressed by most farmers in both districts with captive breeding of local farmers and exchanging seeds among fellow farmers. Farmers replant their seeds 2-3 times, then replace or make a rotation of varieties to break the cycle of pest disease. For farmers who do not get superior seeds directly from Balai Benih exchange seeds so that they can plant superior seeds in the next planting season.

3.5. Level of Empowerment of Soybean Farmers

The level of empowerment is the ability possessed by farmers in the form of a connection of individual personal abilities in the form of perceptions of technical abilities in soybean farming, problem solving abilities, planning ability, and adaptability. Helpless farmers are farmers who have the ability to get

IOP Conf. Series: Earth and Environmental Science **383** (2019) 012010 doi:10.1088/1755-1315/383/1/012010

technical management in kelede farming, are able to overcome problems faced in farming, are able to plan farming management, and are able to adapt to soybean farming.

Community empowerment are elements that enable people to survive and in a dynamic sense develop themselves and achieve progress. Empowering the community is an effort to increase the dignity of the people who are not able to escape from the trap of poverty and underdevelopment. In other words, empowering is enabling and empowering the community [5].

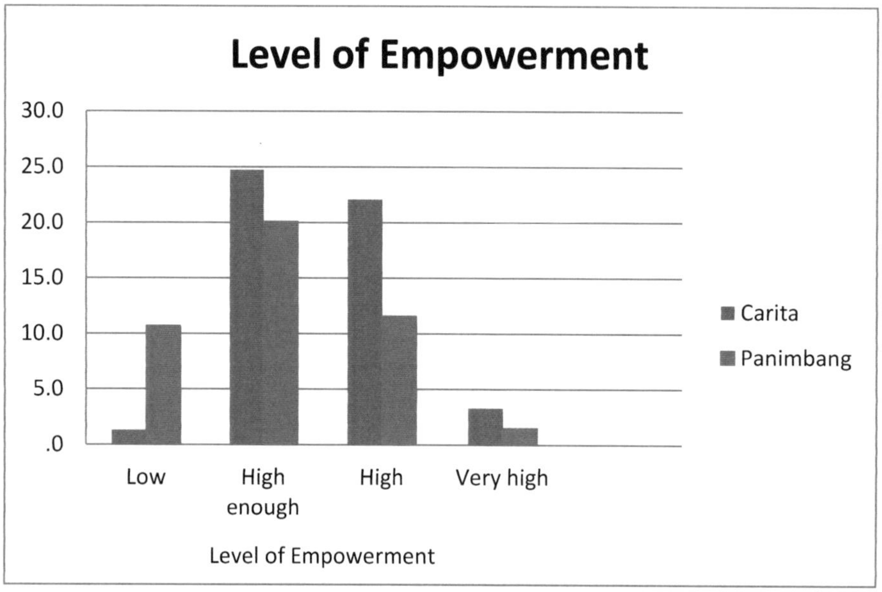

Figure 1. The level of empowerment of soybean farmers

Based on Figure 1 it can be seen that the level of empowerment of soybean farmers in Carita and Panimbang Districts is quite high. However, the empowerment of farmers in Carita District is higher than the empowerment of soybean farmers in Panimbang District. This is because soybean plants are plants that are generally still relatively new applied in Pandeglang Regency. Where Pandeglang Regency is generally the main crop is rice. Marketing soybean plants still have difficulties compared to rice plants. However, soybean plants have good prospects because generally the people of Banten Province favor processed products from soybeans, namely tofu and tempeh. For this reason, soybean farmers need to increase their empowerment with extension activities.

Conclusion

The level of empowerment of eco-friendly soybean farmers is quite high both farmers in Carita District and farmers in Panimbang District. However, there are differences where soybean farmers in Carita Subdistrict are from the aspect of technical ability, problem solving ability, planning ability and adaptability in managing soybean farming.

Recommendation

It is necessary to increase the empowerment of soybean farmers that are environmentally friendly by providing counseling to soybean farmers so that farmers can apply the cultivation of soybean crops that are environmentally friendly so that soybean production increases while maintaining natural sustainability.

References

[1] Anantanyu S. 2012. Model partisipasi petani lahan kering dalam konservasi lahan. Jurnal

Ekonomi Pembangunan 13(2): 218-234. Surakarta.

[2] Bungin B. 2006. *Metodologi Penelitian Kuantitatif – Komunikasi, Ekonomi, dan Kebijakan Publik serta Ilmu-Ilmu Sosial lainnya*. Jakarta: Prenada Media Grup.

[3] Fatchiya A. 2010. Tingkat kapasitas pembudi daya ikan dalam mengelola usaha aquakultur secara berkelanjutan. *Jurnal Penyuluhan*. 6(1): 11 – 18.

[4] Hesti L, Hamid A, Suyatno A. 2014. Evaluasi penerapan teknologi pada program Sekolah Lapangan Pengelolaan Tanaman Terpadu (SLPTT) Padi di Kecamatan Benua Kayong Kabupaten Ketapang. Jurnal Litbang Pertanian. 20(3): 101-112.

[5] Hendratmoko C, Marsudi, Hidup. 2010, Analisis tingkat keberdayaan sosial ekonomi nelayan tangkap di Kabupaten Cilacap. Jurnal Dinamika Sosial Ekonomi 6 (1): 1-17.

[6] Hermanto. 2009. Reorientasi kebijakan pertanian dalam perspektif pembangunan berwawasan lingkungan. Jurnal Analisis Kebijakan Pertanian. 7(12): 369 - 383.

[7] Irawan B. 2006. Fenomena anomali iklim El nino dan La nina: Kecenderungan jangka panjang dan pengaruhnya terhadap produksi pangan. Bogor: Pusat Analisis Sosial Ekonomi dan Kebijakan Pertanian. *Forum Penelitian Agro Ekonomi* (FAE). 24 (1): 28-45.

[8] Istiantoro, Bambang AN, Soeprobowowati TR. 2013. Analisis faktor-faktor sosial ekonomi yang mempengaruhi pengendalian hama dan penyakit padi sawah ditinjau dari sistem pertanian berkelanjutan. *Jurnal Ekosains*. 5(2): 16-20.

[9] Kerlinger FN, Lee HB. 2000. *Foundation of Behavioral Research*. Second edition. London: Holt, Rinehart and Winston. Inc.

[10] Las IK, Subagyono, Setiyanto AP. 2006. Isu dan pengelolaan lingkungan dalam revitalisasi pertanian. *Jurnal Litbang Pertanian*. 25 (3): 106-114

[11] Muhidin, Leomo S. 2008. Sistem produksi padi ramah lingkungan. Jurnal WartaWiptek. 16 (1): 14-21.

[12] Neuman LW. 2006. *Social Reserch Methods Qualitative and Quantitative Approachs*. Boston: Pearson

[13] Pirngadi K. 2009. Peran Bahan Organik dalam Peningkatan Produksi Padi Berkelanjutan Mendukung Ketahanan Pangan Nasional. Jurnal Pengembangan Inovasi Pertanian. 2(1): 48 – 64.

[14] Ruhimat IS. 2014. Model peningkatan kapasitas petani dalam pengelolaan hutan rakyat: Studi di Desa Rangrang, Kalimantan Selatan. *Jurnal Penelitian Kehutanan Wallacea* 4(1): 11 – 21.

[15] Saptana, Ashari. 2007. Pembangunan pertanian berkelanjutan melalui kemitraan usaha. Jurnal Litbang Pertanian. 26(4): 123 – 130.

[16] Setiawan I. 2012. Dinamika Pemberdayaan Petani: Sebuah Refleksi dan Generalisasi di Jawa Barat. Bandung (ID): Widya Padjajaran.

[17] Suprayitno AR, Sumardjo, Gani DS, Sugihen BG. 2012. Motivasi dan partisipasi dalam pengelolaan hutan kemiri di Kabupaten Maros Provinsi Sulawesi Selatan. Jurnal Penyuluhan. 8(2): 184-199.

[18] Yumi, Sumardjo, Darwis SG, Sugihen BG. 2012. Model pengembangan pembelajaran petani dalam pengelolaan hutan rakyat lestari (Kasus di Kabupaten Gunung Kidul, Provinsi Daerah Istimewa Yogyakarta dan Wonogiri, Provinsi Jawa Tengah). *Jurnal Penyuluhan*. 8(2): 17 – 27.

[19] Yunita, Sugihen BG, Asngari PS, Susanto D, Amanah S. 2012. Strategi peningkatan kapasitas rumah tangga petani padi sawah lebak menuju ketahanan pangan rumah tangga (Kasus di Kabupaten Ogan Ilir dan Ogan Komering Ilir, Sumatera Selatan). *Jurnal Penyuluhan*. 8(2): 10-16.

Foraging behavior of stingless bee (*Geniotrigona thoracica* and *Heterotrigona itama*) on star fruit trees (*Averrhoa carambola* L.)

S S Asma, N A Adam*, M Y Siti Najwa , N Asilah , T S Syukri, and W N Asiah

Department of Plant Protection, Faculty of Agriculture, Universiti Putra Malaysia, 43400 UPM Serdang Selangor Darul Ehsan, Malaysia

*Corresponding author: nur_azura@upm.edu.my

Abstract. Star fruit produce flowers and fruits nearly the whole year round and it require cross-pollination for good fruit set and yields. An inadequate pollination rate is one of the factor that contributes to low yield. One of the effective way of increasing the pollination rate is by having insect pollinators such stingless bees in the star fruit farm. Malaysia is rich with native pollinator such as stingless bee, but there still lack of study related on their foraging behaviour. Therefore, the objectives were to study the foraging behaviour of *Geniotrigona thoracica* and *Heterotrigona itama* on star fruit trees. The colonies of *G. thoracica* and *H.itama* were selected for observation of activity foraging of stingless bee. Foraging bees was enumerated by using a manual tally counter to determine the number of individuals that flew out and into their hive within a 10 minute period each hour for each colonies. Then 10 of foragers' *G. thoracica* and *H.itama* were selected randomly and marked using insect colouring on their thorax. The period for each flower that visited by *G. thoracica* and *H.itama* was recorded by using stopwatch. The number of flower visited by *G. thoracica* and H.itama also recorded in 3 minutes. Result showed the peak time foraging activity of both stingless bees were in between 8.00am until 10.00am. The foraging activity for each hours showed there was not significant differences, $p > 0.05$. Then the highest number flower visited by *G.thoracica* in 3 minute is 31 flowers while the lowest number flower visited in 3 minute is 7 flowers. However, for *H.itama* the highest number visited flowers are 27 and lowest number visited flowers are 8, for the longest average visited period for each flower is 14.76 seconds and the shortest period is 3.57 seconds. Meanwhile for *G.thoracica* longest average visited period for each flower is 12.88 seconds and the shortest average visited period for each flower is 2.72 seconds. In conclusion, this experiment showed that stingless bees have the characteristic for effective pollinator and have the potential to be used as a pollinator for star fruit trees. However, temperature, humidity and weather gave big influence on the foraging activity of stingless bees on star fruit plant.

Keywords: stingless bee, *Geniotrigona thoracica*, *Heterotrigona itama*, pollination, pollinators

1. Introduction

Star fruit (Averrhoa carambola L.) belongs to the family Oxalidaceae. Star fruit's flower is a complete flower but heterodistylous. It is difficult to self- pollinate and needs pollinators for development of fruits. One of the effective ways is by introducing the stingless bee as pollinator. Stingless bee is a good candidate as a pollinator that is a highly diverse and abundant group of eusocial bees, originated from tropical and subtropical region [1]. Pollinators are known to provide symbiosis process with the ecosystem function. This study provided the guideline for grower to manage the population of pollinators in proper ways.

2. Materials and Method

2.1. Sampling site

The site of this experiment was in Integrated farm, Faculty of Agriculture, University Putra Malaysia with the latitude 2.9917°N; longitude 101.7163° E.

2.2. Sampling method

2.2.1. Preparation materials

The star fruit tree was pruning 1 month before experiment started. It was to encouraged growth of buds flower. Each tree of star fruit was fertilized with NPK Blue in once 2 weeks.

2.2.2. Observation active time of foraging

The 3 colonies of *G. thoracica* and *H.itama* were selected for observation of activity foraging of stingless bee. Foraging bees was enumerated by using a manual tally counter to determine the number of individuals the fly out and in their hive within 10 minutes period each hour.

2.2.3. Observation pollination service

10 of forager's *G. thoracica* and *H.itama* were selected randomly and marked using insect colouring on their thorax. The period for each flower that visited by stingless bees were recorded by using stopwatch. The number of flower visited by stingless bees also recorded in 3 minutes.

2.3. Data Analysis

For data analysis of number *G.thoracica* and *H.itama,* stingless bee flew out and into hive by using One way analysis of variance were used with $P < 0.05$ to determine the difference between number stingless bee flew out and into hive and time.

3. Results and Discussion

The study on the number of stingless bee (*G.thoracica* and *H.itama*) that flew out from the hive and the number of stingless bee that flew into the hive related to the time of day is presented in Figure 1, Table 1.1 and 1.2.

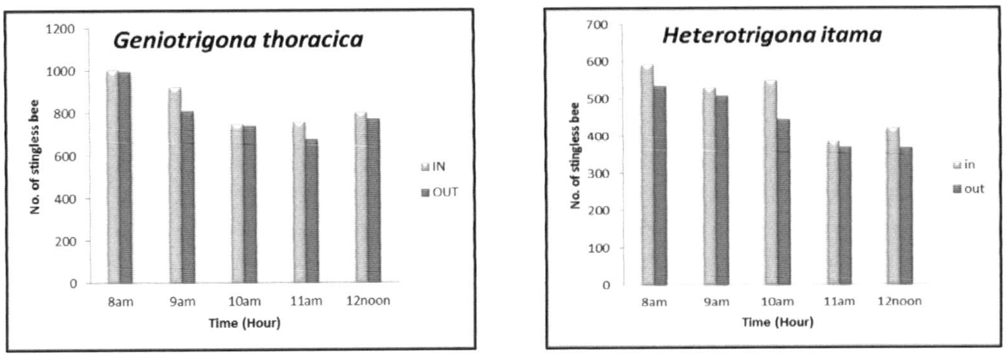

Figure 1. Number of stingless bee (*G.thoracica* and *H.itama*) that flew out and into hive

The highest and lowest number for *G.thoracica* flew out from hive were 995 individuals (9.00 am) and 675 individuals (11.00 am), respectively. Furthermore, the highest and lowest number individual *G.thoracica* flew into hive were 1005 individuals (8.00 am) and 749 individuals (10.00 am), respectively. On the other hand, respectively, for *H.itama* the highest and lowest number individual flew out from hive were 535 individuals (8.00 am) and 368 individuals (12.00 noon). Also, the highest and lowest number *H.itama* flew into hive were 593 individuals (8.00 am) and 386 individuals (12.00 noon), respectively.

Table 1.1: Number of stingless bee (*G.thoracica*) that flew out and into hive

Time (Hours)	No. of stingless bee that flew out from hive	No. of stingless bee that flew into hive
0800 – 0900	331.67 ± 39.77ᵃ	335.00 ± 40.10ᵃ
0900 – 1000	270.33 ± 74.37ᵃ	308.33 ± 44.99ᵃ
1000 - 1100	247.33 ± 58.52ᵃ	249.67 ± 49.33ᵃ
1100 – 1200	225.00 ± 26.08ᵃ	253.67 ± 53.49ᵃ
1200 – 1300	257.33 ± 20.99ᵃ	268.33 ± 48.33ᵃ

Note: Value with the same letter were not significantly different t P = 0.05

Table 1.2: Number of stingless bee (*H.itama*) that flew out and into hive

Time (hours)	No. of stingless bee that flew out from hive	No. of stingless bee that flew into hive
0800 – 0900	178.33 ± 39.99ᵃ	197.67 ± 58.17ᵃ
0900 – 1000	169.33 ± 36.38ᵃ	177.33 ± 34.46ᵃ
1000 – 1100	148.00 ± 40.95ᵃ	183.33 ± 30.12ᵃ
1100 – 1200	123.33 ± 36.52ᵃ	128.67 ± 29.24ᵃ
1200 – 1300	122.67 ± 22.05ᵃ	140.67 ± 28.85ᵃ

Note : Value with same letter were not significantly different at P = 0.05

Based on the tables shown, there were no significant differences ($p > 0.05$) for number of both stingless bees that flew out from hive and the number of both stingless bees that flew into the hive between times. Thus, the range of time from 8.00 am to 1.00 pm did not influence the activity of stingless bee. The number of both stingless bees that flew into the hive and flew out from the hives is relatively decrease by time for foraging activity. However, *G.thoracica* has slight increase at 12.00noon for individuals that flew into and flew out the hive. [6] stated that some foraging patterns do exist with a pollen-collecting peak in the early morning and a nectar-collecting peak in the late morning.

The total flowers that had been visited by *G.thoracica* were 583 flowers. Then the highest number flower visited in 3 minutes by *G.thoracica* was 31 flowers while the lowest number flower visited in 3 minute was 7 flowers. Total numbers of flowers visited by *H.itama* were 463 flowers where the lowest visited flowers were 8, while the highest numbers of flowers visited were 27. Respectively, the longest and shortest average visited period for each flower for *G.thoracica* were 12.88 and 2.72 seconds. Meanwhile, for *H.itama* the longest and shortest average visited period for each flower were 14.76 and 3.57 seconds, respectively. This mean that *H.itama* spent more time in one flower than *G.thoracica*, this might be because of the size proboscis of *H.itama* shorter than *G.thoracica*.

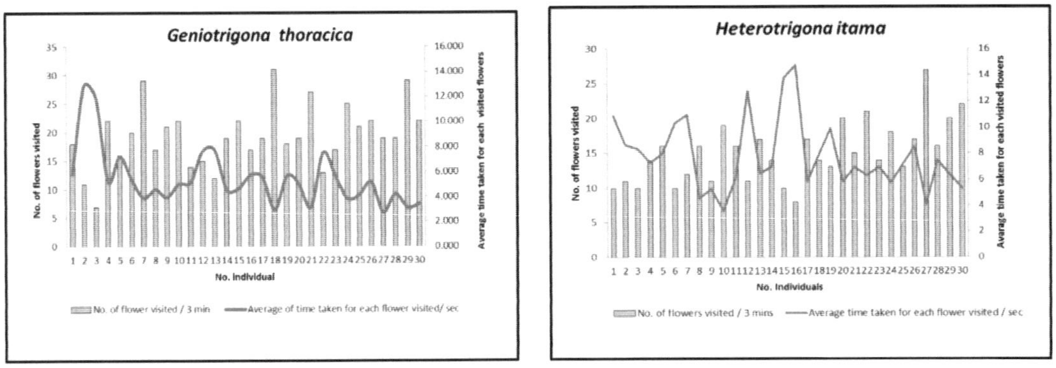

Figure 2. The number of flower that visited by stingless bees *G.thoracica* and *H.itama* in 3 minute and average time taken visited for each flower

The peak hour of stingless bees for foraging activities was around 8.00am to 10.00am. Increasing individuals flew in and out was because of the routine activity housekeeping bees such as brings out the trash or brings out the dead individuals besides foraging activities. Usually stingless bee starts their foraging activity at early morning because of favorable temperature. Temperature, light intensity, and relative humidity were the factors that most influenced the foraging activities [3].

Energy preservation and achieved nutrient are the factor that can influence the time foraging of stingless bee. Early morning is the time of highest foraging activity because of stingless bee need to fulfil their energy requirement while at mid-day temperature is increasing. The optimal level of relative humidity and temperature within between 70% to 90% and approximately 20ºC is the favorable condition for stingless bee starts their foraging activities [2].

Stingless bees remained active in the morning to collect pollen and nectar. Since starfruit flowers were almost closed between 1400 hours to 1800 hours [7]. Stingless bees tend to show flower constancy, repeated visit to the same flower. This is one of the characteristic for effective pollinator [5]. Frequency and duration of visit by pollinators to nectariferous depend on nectar production level [4, 5].

4. Conclusion

Foraging activity of this bee species highly overlapped with the opening time of the flowers. Stingless bees tend to showed flower constancy and a single flower-visit was sufficient for efficiently transferring pollen to the stigma, thus promoting pollination. Stingless bees start their foraging activity at early morning because that time has favorable temperature and prefers foraging in the hot sunny day than the rainy season due to the small size of body beside rain can be disturbing their flight.

5. References

[1] Heard, T. A. (1999). The role of stingless bees in crop pollination. Annual Review of Entomology, 44(1), 183-206.

[2] Hilário, S.D., Imperatriz-Fonseca, V.L. and Kleinert, A. M. P. (2013). Flight Activity And Colony Strength In The Stingless Bee Melipona Bicolor (Apidae, Meliponinae). Revista Brasileira de Biologia. 60(2): 299–306.

[3] Iwama, S. A., (1977), Influência de fatores climáticos na atividade externa de Tetragonisca angustula (Apidae, Meliponinae). Bol. Zool., 2:189-201.

[4] Mohd Norowi, H., Sajap, A. S., Rosliza, J., Mohd Fahimie, J., & Suri, R. (2008). Conservation and sustainable utilization of stingless bees for pollination services in agricultural ecosystems in Malaysia. Department of Agriculture, Malaysia.

[5] Putra, D. P., Dahelmi, S. S. M., & Swasti, E. (2016). Pollination in chili pepper (Capsicum annuum L.) by Trigona laeviceps and T. minangkabau (Hymenoptera, Meliponini).

[6] Roubik, D. W., (1989), Ecology and natural history of tropical bees. Cambridge Tropical Biology Series, 514p.

[7] Saúco, V. G., & Tindall, H. D. (1993). Carambola cultivation (No. 108). Food & Agriculture Org.

6. Acknowledgements

This research was funded by Fundamental Research Grant Scheme (FRGS) 07-01-16-1794FR and Trans-disciplinary Research Grant Scheme (TRGS) TRGS/1/2016/UPM/01/5/2 from Ministry of Education, Malaysia.

Selection of potential genotypes and traits evaluation of honeydew (*Cucumis melo* L.)

B A Iskandar[1]*, W B Suwarno[1,2], E Gunawan[2], and S K Saptomo[3]

[1] Department of Agronomy and Horticulture, Faculty of Agriculture, IPB University, Jl. Meranti Kampus IPB Dramaga, Bogor 16680, Indonesia
[2] Center for Tropical Horticulture Studies, Institute of Research and Community Development (LPPM), IPB University, Jl. Raya Pajajaran Kampus IPB Baranangsiang, Bogor 16141, Indonesia
[3] Department of Civil and Environmental Engineering, Faculty of Agricultural Technology, IPB University, Gedung Fateta Kampus IPB Darmaga, Bogor 16680, Indonesia.

*Corresponding author: bunga.aprini@gmail.com

Abstract. Quality improvements of honeydew (*Cucumis melo* L.) can be carried out by developing new superior hybrid varieties. This study aimed to identify potential genotypes having good fruit quality for breeding improved hybrid honeydew varieties. The experiment was conducted from March to May 2019 at the Tajur II experimental station of PKHT, IPB University, Bogor. The experiment was arranged in a single-factor randomized complete block design with three replicates. There were six test genotypes from S2 generation, namely IPB HD1, IPB HD2, IPB HD3, IPB HD4, IPB HD5, IPB HD6; three inbred lines, namely SMM, SMF, and OMM; and also one check variety (Alisha F1). The results indicated that genotypes had significant effects on all observed quantitative characters, namely leaf width, leaf length, stem diameter, internode length, days to harvest, fruit length, fruit weight, and total soluble solids. The SMM genotype had the highest average fruit weight as well as the other characters, i.e. leaf width, leaf length, stem diameter, and fruit length. There were also IPB HD1 and HD4 having the highest average of flesh thickness and sugar content, respectively. These three genotypes are predicted to have good potential for developing improved hybrid honeydew varieties.

Keywords: genetic diversity, fruit quality, melon breeding

1. Introduction

Honeydew (*Cucumis melo* L.) is a nutritious fruit and popular in many countries, including Indonesia. Honeydew contains 92% water; 1.5% protein; 0.3% fat; 6.2% carbohydrate; 0.5% fiber; and 357 IU of Vitamin A [1]. Honeydew is a healthful food choice as it is rich in ascorbic acid, β-carotene, folic acid, potassium, as well as another bioactive compound [2].

Consumer preferences for a good honeydew determined by the sugar content, taste or flavor, fruit texture, and nutrition content [3]. This indicates that fruit quality is important. The quality of fruit products can be divided into two, external quality which is fruit appearance, and internal quality which is taste and nutrition [4]. Both are important factors affecting the price and customers' decision about buying products.

One important problem is the availability of domestic superior varieties. This limits the variability of honeydew fruit types in the market, as well as encourages seed imports. Imported honeydew varieties may not suitably adapt with the environmental conditions at the production centers in Indonesia. Additionally, the continuous supply of imported seed cannot be guaranteed, and the price of imported seeds is unfortunately expensive [5].

Quality improvements can be carried out by developing improved hybrid varieties. Hybrid varieties have both quality and uniformity traits which benefit the farmers and consumers. At the same time, the seed industries also get a benefit as the farmers would require to buy hybrid seeds every planting time. If the seed is replanted, the uniformity of the fruits, as well as the plants' resistance to pest and diseases, will be altered because of genetic segregation. Breeding improved honeydew varieties is potential because the genetic diversity of melon is large [6]. Developed varieties are expected to reduce imported seed and increase the variability of superior varieties in the market. The challenge is producing high quality of domestic honeydew varieties that can compete with imported qualities. With that, this study aimed to identify potential genotypes having good fruit quality for breeding improved hybrid honeydew varieties.

2. Materials and Methods

The experiment was conducted from March to May 2019 at the Tajur II experimental station of PKHT, IPB University, Bogor. There were six test genotypes from S2 generation, namely IPB HD1, IPB HD2, IPB HD3, IPB HD4, IPB HD5, and IPB HD6; three inbred lines, namely SMM, SMF, and OMM; and one check variety (Alisha F1). The experiment was arranged in a single-factor randomized complete block design with three replicates.

The experiment began with land clearing. Soil tillage was conducted, and manure and chemical fertilizers were applied. Fertilizer doses were 1 kg of manure, 25 g Urea, 30 g KCl, and 20 g TSP per plant. Additionally, 250 g of $CaCO_3$ per square meter was applied. The beds were made of 50 cm high and covered with plastic mulch. The seedlings were planted 15 days after nursery. Plant spacing was 60 cm x 60 cm.

Plants were maintained includes watering, binding of plants and fruit, lateral pruning except for the 8th to 12th branches, also pests and disease control if needed. Supplementary nutrients of NPK 27:7:7 fertilizers were applied in a solution (5 g l^{-1}) at 7, 14, 21 days after planting, also NPK 16:16:16 (10 g l^{-1}) at 28, and 35 days after planting. KNO_3 fertilizer was applied in a solution (1 g l^{-1}) when the fruits were formed.

Morphological traits were observed based on *Descriptor for Melon* (*Cucumis melo* L.) [7] with necessary modifications. Quantitative traits observed were: leaf width, leaf length, stem diameter, internode length, days to harvest, fruit length, fruit weight, fruit diameter, flesh thickness, skin thickness, and total soluble solids. Qualitative traits observed including stem color, leaf color, leaf shapes, leaf lobes, leaf glossiness, flower color, fruit shape, unripe and ripe fruit skin color, net presence, groove presence, flesh color, flesh flavor, internal aroma, external aroma, and after taste. Leaf observation was carried out on the leaves in the 8th branch.

3. Result and Discussion

3.1. General Condition

Generally, the condition of the crop was good from the beginning until harvesting time. However, there were several symptoms of diseases detected, including bacterial wilt and *Cucumber Mozaic Virus*. Additionally, abnormal plants were also found. Some pests were also identified, i.e. grasshopper, red bugs, and *Aulocophora sp* in low intensity intrusion. *Aulocophora sp.* attacks every stage of cucurbits; it fed under cotyledon by biting holes [8]. Fruit flies (*Bactrocera dorsalis*) identified with medium to high intensity intrusion at fruit development stage. Fruit flies attack was considerably high due to the presence of other vegetable plants such as tomatoes and chili around the experiment area. Pest and disease incursions were controlled by pesticides applications.

3.2. Qualitative Traits

3.2.1. Stem, Leaf, and Flowers Traits

Qualitative traits observed at vegetative stage were stem color, leaf color, leaf shape, leaf lobes, leaf glossiness, and flower color. All observed genotypes possessed the same traits, i.e. green stem color, dark green leaf color, and bright yellow flower color (Figure 1). This shows that these traits cannot be used as markers for differentiating the genotypes. The dark green leaf color indicated that all genotypes grew on an optimum environment with sufficient N nutrient.

Figure 1. (a) Green stem color, (b) Dark green leaf color, and (c) Bright yellow flower color.

The observation results have shown that all genotipe had *entire* leaf shape, but had diversity on leaf lobes (Figure 2). Several genotypes had *shallow* leaf lobes, namely IPB HD1, HD2, HD4, HD5, HD6, and SMM. While IPB HD3, OMM, SMF, and Alisha F1 had *intermediate* leaf lobes.

Figure 2. The diversity of leaf shapes and leaf lobes on observed genotypes

3.2.2. Fruit Traits

The fruits were developed from manual self- or cross-pollination. The variability of fruit traits from each genotype can be seen in Figure 5. These pictures represent the dominant type of fruits, as there were also segregants within each genotype. The dominant fruit types were determined to be superior in appearance. Mostly inodorus melons have greenish yellow skin, with white or light yellow flesh [9].

IOP Conf. Series: Earth and Environmental Science **383** (2019) 012012 doi:10.1088/1755-1315/383/1/012012

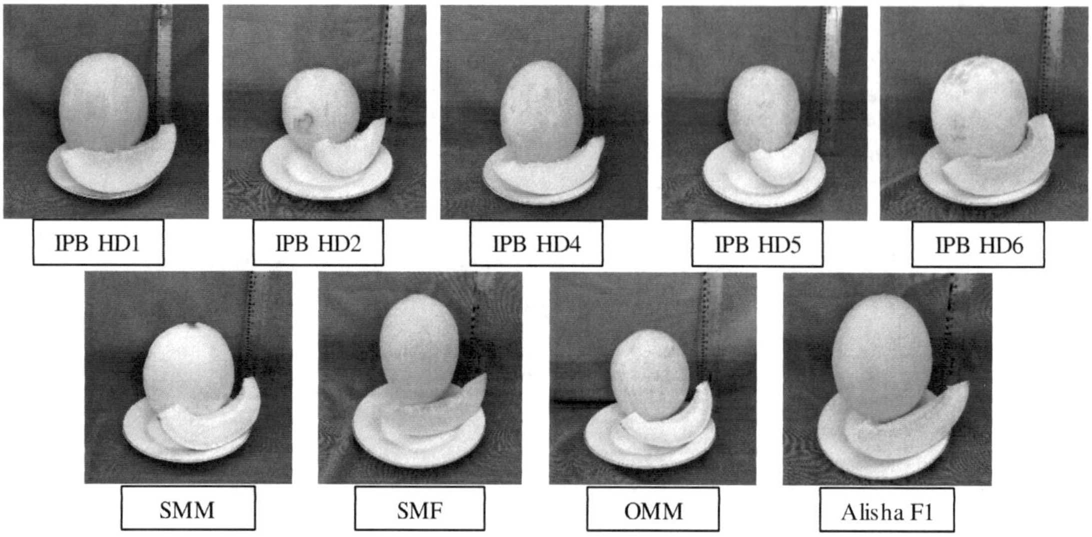

Figure 3. The fruit diversity of all observed genotypes

Observed fruit qualitative traits were fruit shapes, unripe and ripe fruit color, flesh color, and flesh texture (Table 1). There was no self-pollinated fruit sample for IPB HD3 genotypes because of the pollination and fruit development failures. The fruit shapes of IPB HD2 and IPB HD6 were *round*; IPB HD1, HD5, SMM, OMM, and Alisha F1 were *flattened*; and IPB HD4 and SMF were *elliptical.*

Mostly, skin color of unripe fruit is greenish and will turn to yellow when the fruit is ripe. However, ripe fruits of IPB HD6 and SMM had a white skin color. Moreover, the flesh color and flesh texture were different among the genotypes. IPB HD1, HD6, and SMF had orange flesh, while IPB HD1, HD4, SMM, and OMM, had white flesh. Meanwhile, IPB HD5 had different flesh color, which is pale green. Nowadays crunchy fruit texture on demand, and we found that five genotypes have crunchy fruit texture, namely IPB HD1, HD2, HD5, OMM, and Alisha F1.

Table 1. Fruit shape, unripe and ripe fruit color, flesh color, and flesh texture in several honeydew genotypes

Genotype	Fruit shapes	Fruit color		Flesh color	Flesh texture
		Unripe	Ripe		
IPB HD1	Flattened	Green	Yellow	Orange	Crunchy
IPB HD2	Round	Green	Yellow	White	Crunchy
IPB HD3[a]	-	-	-	-	-
IPB HD4	Elliptical	Green	Yellow	White	Soft
IPB HD5	Flattened	Green	Yellow	Pale Green	Crunchy
IPB HD6	Round	White	Light Yellow Green	Orange	Spongy
SMM	Flattened	White	Light Yellow	White	Spongy
SMF	Elliptical	Green	Yellow	Orange	Soft
OMM	Flattened	Green	Yellow	White	Crunchy
Alisha F1	Flattened	Green	Yellow	White-orange	Crunchy

[a]There was no fruit from self pollination that can be observed.

The other qualitative fruit traits that observed were external aroma, internal aroma, groove presence, net presence, flesh flavor, and after taste. Honeydew categorized in non-climacteric fruits has smooth skin surface, also lacking of musky odor and flavor [10]. This statement is in agreement with this experiment, which all of the genotypes likely have absent external aroma. Groove and net appeared with low intensity on several genotypes, shown in Table 2. Internal aroma appeared in all observed genotypes except IPB HD1 and IPB HD5. Honeydew are common for having sweet, delicately flavored, juicy flesh [9]. Our experiment indicate that all observed genotypes had slightly sweet to sweet flesh with no musky after taste. This indicates that all genotypes have good internal qualities.

Table 2. External aroma, internal aroma, groove presence, and net presence, flesh flavor, and after taste in several honeydew genotypes

Genotype	External aroma	Groove	Net	Internal aroma	Flesh flavor	After taste
IPB HD1	Absent	Absent	Absent	Absent	Intermediate	Absent
IPB HD2	Absent	Absent	Present	Present	Sweet	Absent
IPB HD3	-	-	-	-	-	-
IPB HD4	Present	Present	Present	Present	Sweet	Absent
IPB HD5	Absent	Present	Present	Absent	Intermediate	Absent
IPB HD6	Absent	Absent	Present	Present	Sweet	Absent
SMM	Present	Absent	Absent	Present	Intermediate	Absent
SMF	Present	Present	Present	Present	Sweet	Absent
OMM	Absent	Absent	Present	Present	Sweet	Absent
Alisha F1	Absent	Absent	Present	Present	Sweet	Absent

From the observation results of qualitative fruit traits (Table 1 and Table 2), we found that IPB HD1 and IPB HD4 have similarity on several fruit traits with Alisha F1. Furthermore, the fruits from inbred lines, namely SMM, SMF, and OMM, had the same traits as the previous experiment [11]. A previous experiment showed that SMM had white fruit flesh and skin surface, SMF had *elliptical* fruit shapes and orange colored flesh, while OMM had yellow skin and white colored flesh [11]. These indicate that the fruits were true to type.

3.3 Quantitative Traits

Genotypic effects were significant for all observed traits except for fruit diameter, flesh thickness, and skin thickness (Table 3). Meanwhile, the replication did not have significant effects on all observed traits except leaf width traits and stem diameter. Coefficient of variation (CV) of all traits were less than 20%, indicating that the experiment was reliable. The trait with the smallest CV was the day to harvest (2.02 %), and the one with the highest CV was fruit weight (19.49 %).

Table 3. Variance analysis recapitulation of quantitative traits in several honeydew genotypes

Traits	Mean square of replication	Mean square of genotypes	Coefficient of variation (%)
Leaf width (cm)	3.67^{*}	7.35^{**}	4.79
Leaf length (cm)	1.10^{ns}	4.50^{**}	4.97
Stem diameter (mm)	0.60^{**}	0.60^{**}	3.03
Internode length (cm)	0.69^{ns}	3.62^{**}	12.24
Days to harvest (dap)	0.03^{ns}	6.30^{**}	2.02
Fruit length (cm)	2.07^{ns}	5.52^{**}	7.93
Fruit weight (g)	43254.49^{ns}	100198.60^{*}	19.49
Fruit diameter (cm)	0.27^{ns}	1.73^{ns}	6.92
Flesh thickness (cm)	0.00^{ns}	0.15^{ns}	9.28
Skin thickness (cm)	0.00^{ns}	0.00^{ns}	17.67
Total soluble solids (°brix)	1.90^{ns}	5.04^{**}	9.04

* Significant at $P < 0.05$
** Significant at $P < 0.01$
^{ns} Not significant

3.3.1. Stem, Leaf, and Days to Harvest Traits

Genotype means for leaf width, leaf length, stem diameter, internode length, and days to harvest are shown at Table 4. IPB HD1 had the highest average of leaf width (20.6 cm), but not significantly different from IPB HD4 (19.3 cm) and SMM (19.6 cm). The IPB HD4 and SMM had the highest average of leaf length. The highest average on stem diameter was shown by SMM, indicating that SMM had better plant performance than the other genotypes. IPB HD1 and IPB HD2 had significantly longer internode than the others. Genotypes having the shortest average days to harvest were SMM and Alisha F1 (62 days after planting, DAP). This result is different from the last experiment, where SMM had a longer day to harvest (76 DAP) [11], which can be due to the difference in environmental conditions.

Table 4. Mean of quantitative traits: leaf width, leaf length, stem diameter, internode length, and days to harvest

Genotype	Leaf width (cm)	Leaf length (cm)	Stem diameter (mm)	Internode length (cm)	Days to harvest (DAP)
IPB HD1	20.6a	14.0b	8.0d	9.4a	65abc
IPB HD2	17.4b	12.9cb	8.0d	7.0b	64bc
IPB HD3	16.9bc	13.8b	8.9ab	6.4b	66a
IPB HD4	19.3a	15.7a	8.4cd	9.1a	66ab
IPB HD5	17.4b	12.3cd	8.0d	6.6b	65ab
IPB HD6	15.5c	11.7d	8.7bc	6.4b	66ab
SMM	19.6a	15.2a	9.3a	7.5b	62c
SMF	16.9cb	13.3cb	8.5bcd	6.7b	64bc
OMM	16.9cb	13.1cb	8.8abc	6.8b	64abc
Alisha F1	17.1cb	13.2cb	8.0d	6.6b	62c

Notes: Means followed by the same letters in the same column are not significantly different based on DMRT at $\alpha = 0.05$

3.3.2. Fruit Traits

Every plant breeding products are expected to comply with the consumers' demand. Fruit quality is important to be considered when choosing a potential genotype. Many consumers prefer sweet taste, while the farmers also look for a large fruit size. There were no significant difference among genotypes on fruit diameter, flesh thickness, and skin thickness in this experiment. The IPB HD5 had the smallest average of fruit length and weight, 11.8 cm, and 664.3 g respectively (Table 5). The fruits of SMM and SMF were among the longest. The results also showed that SMM did not have a significant difference in fruit weight with SMF, IPB HD1, IPB HD4, and Alisha F1. Fruit sizes and quality are determined by cell enlargement, where cell enlargement is strongly altered by water and nutrients.

The SMM genotype appeared to have similar weight and total soluble solids with the check variety, Alisha F1 (Table 5). The observed genotypes have a range of TSS from 9.3 to 13.7°Brix. This result was considerably good because one suggested that the TSS for commercial honeydew should be equal or higher than 8-10°Brix [12]. IPB HD4 had the highest average TSS, but not significantly different from that of IPB HD2, HD3, HD5, HD5, SMF, and OMM. Sugar content, indicated by total soluble solids, is related to days to harvest [13]. Fruits that are harvested longer tend to have a sweeter taste. The longer the harvest days, the sweeter the fruit gets until reach the limit, which is leading to alcoholic fermentation and then rot.

Table 5. Mean of quantitative traits: fruit length, fruit weight, fruit diameter, flesh thickness, skin thickness, and total soluble solids

Genotype	Fruit length (cm)	Fruit weight (g)	Fruit diameter (cm)	Flesh thickness (cm)	Skin thickness (cm)	Total soluble solids (°Brix)
IPB HD1	14.2ab	1176.1ab	12.8	3.2	0.1	9.3d
IPB HD2	13.2bc	887.0bc	12.4	3.1	0.1	13.0abc
IPB HD3	14.1ab	903.9bc	11.9	2.9	0.1	12.3abc
IPB HD4	14.9ab	1000.7abc	12.3	3.0	0.1	13.7a
IPB HD5	11.8c	664.3c	10.9	2.7	0.1	12.0abc
IPB HD6	13.0bc	897.8bc	12.4	2.8	0.1	12.0abc
SMM	16.2a	1287.3a	13.7	3.2	0.1	11.0cd
SMF	16.0a	933.2abc	11.7	2.7	0.1	13.0abc
OMM	14.7ab	884.0bc	12.0	2.6	0.1	13.3ab
Alisha F1	14.9ab	1167.1ab	12.9	3.2	0.1	11.3bc

Notes: Means followed by the same letters in the same column are not significantly different based on DMRT at $\alpha = 0.05$

3.3 Correlation among Quantitative Traits

Table 6 depicts the correlation coefficients among the observed traits. There were positive correlations between several fruit traits, i.e fruit length, fruit weight, fruit diameter, and flesh thickness. Fruit weight and flesh thickness were positively related to leaf width, leaf length, and internode length. Total soluble solid had negative, small significant correlation with leaf width, while had no significant correlation with internode length, fruit weight, fruit diameter, and flesh thickness. Indirect selection can be performed on other traits having a significant association with the trait of interest.

Table 6. Correlation coefficient among traits

	LW	LL	SD	IL	DTH	FL	FW	FD	FT	ST
LL	0.787**									
SD	-0.086	0.189								
IL	0.837**	0.657**	-0.155							
DTH	-0.116	-0.077	-0.095	0.069						
FL	0.302	0.561**	0.316	0.227	-0.301					
FW	0.433*	0.460*	0.230	0.402*	-0.307	0.744**				
FD	0.361	0.400*	0.318	0.342	-0.353	0.593**	0.927**			
FT	0.457*	0.369*	-0.059	0.409*	-0.149	0.376*	0.791**	0.812**		
ST	-0.104	-0.020	0.268	-0.086	0.182	-0.096	0.065	0.100	0.105	
TSS	-0.387*	0.023	0.001	-0.253	0.241	0.220	-0.122	-0.050	-0.090	0.013

Notes : LW=Leaf width; LL=Leaf length; SD=Stem diameter; IL=Internode length; DTH=Days to harvest; FL= Fruit length; FW=Fruit weight; FD=Fruit diameter; FT=Flesh thickness; ST=Skin thickness; TSS=Total soluble solids (°brix)

* Significant at $P < 0.05$

** Significant at $P < 0.01$

4. Conclusion

The genotype had significant effects on all observed quantitative characters, namely leaf width, leaf length, stem diameter, internode length, days to harvest, fruit length, fruit weight, and total soluble solids. All observed genotypes have similar vegetative traits, specifically green stem color, dark green leaf color, and bright yellow flower color. The SMM, IPB HD1, and IPB HD4 have good fruit traits as compared with Alisha F1, and therefore are considered potential for developing improved hybrid honeydew varieties.

5. References

[1] Daryono B S, Maryanto S D, Nissa S and Aristya G R 2016 *Biogenesis* **4** 1-9
[2] Lester G E and Hodges D M 2008 *Postharvest Biology and Technology* **48** 347-54
[3] Lester G E 2008 *Journal of Agricultural and Food Chemistry* **56** 3694-98
[4] Hariyadi P 2009 *Food Review Indonesia* **4** 16-9
[5] Sobir and Siregar F D 2014 *Berkebun Melon Unggul* (Jakarta: Penebar Swadaya)
[6] Robinson R W, Decker-Walters D S 1997 *Cucurbits* (Wallingford, Oxon, U.K. ; New York, N.Y: CAB International)
[7] International Plant Genetic Resources Institute 2003 *Descriptors for Melon* (*Cucumis melo* L.) (Rome: International Plant Genetic Resources Institute)
[8] Rashid M A, Khan M A, Arif M J and Javed N 2014 *Journal of Entomology* **7** 38-54
[9] Ajuru M G, Okoli B E 2013 *International Journal of Modern Botany* **3** 15-9
[10] Nunez-Palenius H G, Gomez-Lim M, Ochoa-Alejo N, Grumet R, Lester G and Cantliffe D J 2008 *Biotechnology* **28** 13-55
[11] Huda A M 2015 Thesis: Karakteristik morfologi beberapa genotipe melon (*Cucumis melo* L.) grup cantalupensis dan inodorus (Bogor: Institut Pertanian Bogor) p 12-5
[12] Supapvanich S, Boon-Lha and Mhernmee N 2011 *Kasetsart J. (Nat. Sci.)* **45** 874-82
[13] Kusumaliski N 2015 Thesis: Pengembangan metode deteksi kematangan melon (*Cucumis melo* L.) dengan respon impuls akustik (Bogor: Institut Pertanian Bogor)

Analysis of the socio-economic impact of renewable energy hybrid electricity utilization for rural community development (case study: Pantai Anyar, Yogyakarta special region, Indonesia)

D Rachmawatie[1]*, E Rustiadi[2], Aad Fauzi [3], Bambang Juanda[4]

[1)] Universitas Muhammadiyah Yogyakarta
[2)] Department of Regal Development, Faculty of Agriculture, Bogor Agricultural University
[3)] Department of Resource Economics and Environment, Faculty of Economy and Management, Bogor Agricultural University
[4)] Department of Economic Science, Faculty of Economy and Management, Bogor Agricultutal University

*Corresponding author: dessyvalue@gmail.com

Abstract. The aim of this research were to: (1) study on socio-economics from Renewable Energy Hybrid Power Plant (PLTH) electricity development for rural communities, (2) find out information related to average households beneficiaries of electricity (pre and post the utilization of electrical energy hybrid electricity at Baru Beach. The research was conducted at Baru Beach, Yogyakarta Special Region on November – December, 2017. The data used were primary and secondary data. The research was conducted by surveying 68 respondents who are direct beneficiaries of hybrid electric energy utilization activities in Baru Beach. The analytical method used in this part of the study is the visualization analysis of the Baru Beach map area for 2007-2017, analysis of household income, and *Paired Sample T-test*. The results showed that: (1) based on changes in the map of Pantai Baru area in 2007-2017, changes in road changes, the access and connectivity of this region to other regions have become much better. In the analysis of income obtained, post PLTH was developed at Baru Beach. The income analysis was found that, post PLTH was developed in Baru Beach, there was an increase in the average income in the non-agricultural sector and household income for PLTH electricity beneficiaries, but there was a decrease in the average income value in the agricultural sector, (2) based on *Paired Sample T-Test* analysis show there is a significant difference the average household income post PLTH developed at Baru Beach with a significance level of 0.041. Thus the results of this research conclude that there is any a socio-economic impact from the utilization of renewable electricity not only for the beneficiaries of electricity in rural communities, but also for the development of local area.

Keywords: electricity beneficiaries, renewable energy hybrid power plant, rural community development.

IOP Conf. Series: Earth and Environmental Science **383** (2019) 012013 doi:10.1088/1755-1315/383/1/012013

1. Introduction

Energy access is a prerequisite for the realization of a country's socio-economic development. The availability of energy is enough to open up opportunities for people to carry out productive activities and work more effectively, so they can contribute to economic and social progress. The economic and social progress that occurs will drive demand for increasingly quality energy. The availability of electricity is very important, because it can trigger the emergence of productive economic activities. Enough lighting at night provides opportunities for children to learn, read books and do school work. People can access information via television or access the internet. Besides that, there is also access to electricity, besides being able to support activities in the primary sector (agriculture, fisheries and livestock) that already exist in the village, but also the presence of electricity can trigger the emergence of secondary sectors, such as: trade and tourism in the village. So that access to electricity is not just to provide access to electricity to the village community, but also in the broader sense that the emergence of electricity in rural areas is important because it can encourage existing economic growth and move potential sectors to be developed in the village.

Although most literature agrees that electricity access is necessary for poverty reduction, the causal chains suggested for this dependence are various. This is to be expected given that researchers must deal with multi-factorial systems, with energy only one of a linked set of variables that determine poverty outcomes. The most frequently reported causal chains linking electricity access to poverty impacts are: (1) the improved productivity or output of enterprises using electricity for their activities [12,16,15,18,19,20,21,22,37], (2) the freeing up of time for paid work, which would increase labour supply [5,8,12,16,18,28,31,32,34].

Besides that, the availability of access to electricity in remote areas that is empirically empirical can also have significant social and economic impacts. For example: in supporting human life [23,35], improving educational facilities [29], increasing water supply [9], enabling gender and youth empowerment [33,25], and enables the creation of productive activities in the village [26,4,7]. This statement is confirmed by the results of the research by [3,6,17,36] which explains that, access to electricity is usually always in tandem with sustainable activities that allow contributions to improving education, entrepreneurship, gender equality, and increasing income at the local level, as well as social and economic benefits for development in rural [27]. The results of research conducted by [13] also state that, in an electrification program if supported by other possible activities in the countryside, in the rural context it can provide positive feedback in the future. Another economic benefit of electrification in the countryside is the virtuouse circle of development driven by investment activities in other infrastructure services in the countryside, such as: water supply, street lighting, health services, education which in turn will have an impact on increasing electricity affordability and increasing investment others in the countryside [17].

The development of renewable hybrid electric energy held in Ngentak Hamlet was carried out in 2008 located in Pantai Baru, Poncosari Village, Srandakan District, Bantul Regency, Special Region of Yogyakarta. Before the existence of Hybrid Electric Energy Development activities at the New Coast At that time, Ngentak Hamlet was one of the hamlets in Batul Regency which was categorized as a lagging village, this area was originally a closed area from outside communities, this was due to the inaccessible network in this region, in addition to the limited road infrastructure to the region, this region is increasingly difficult to reach by the local community and surrounding communities.

Lack of access to electricity is seen as a major constraint to economic growth and increased welfare in developing countries. However, having good access to electricity requires more than the existence of physical connection or a source of electrical generation. Numerous attributes (such as duration, affordability, safety and convenience, among others), drive the extent to which people and enterprises can make use of electricity and affect their ability to improve lives and livelihoods. Geographically Pantai Baru in Ngentak Hamlet is directly facing the Indian Ocean, this causes Pantai Baru to have abundant wind energy and solar energy potential. The development of a Hybrid Power Plant (PLTH) is a realization of a series of Regional Innovation Systems (SIDa) in Bantul Regency, initiated by the Ministry of Research and Technology, National Aeronautics and Space Agency

(LAPAN), Agency for the Assessment and Application of Technology (BPPT) and the Ministry of Maritime Affairs and Fishery.

The productive use of electricity offers the potential for sustained poverty reduction (Omri 2014) as well as inequality in accessing electricity, especially for rural communities living in remote areas. The three main pillars of the development of renewable electricity in Pantai Baru are through: technology, community empowerment, and institutions. In this study, researchers focused more on aspects of community empowerment in an effort to develop renewable hybrid electric energy through business activities developed at Pantai Baru. Thus, some of the research questions to be analyzed in this study are: (1) how the impact of changes in the development of renewable hybrid electricity (PLTH) at Pantai Baru on the socio-economic conditions of PLTH electricity beneficiaries and local village communities (2) whether there are significant changes to the income of household beneficiaries by comparing conditions before and after the existence of PLTH.

Based on the research questions that have been formulated, the objectives of this study are as follows:

1) Identified the impact of changes in the development of renewable hybrid electricity (PLTH) on the socio-economic conditions of PLTH electricity beneficiaries and local village communities;
2) Analyze the income of household PLTH electricity beneficiaries before and after PLTH was developed at Pantai Baru.

2. Methodology

The research were designed as survey research, with 68 respondents who are PLTH electricity beneficiaries. The data used are primary and secondary data. Primary data is obtained from the results of direct observations in the field, and structured and in-depth interviews through questionnaires of research questionnaires aimed at: PLTH electricity beneficiaries, tourist visitors, and local community leaders.

The analytical method used in this study are: (1) descriptive analysis and analysis of household income. Descriptive analysis is used as supporting information from the quantitative analysis used. Analysis of household income obtained from income obtained from the agricultural and non-agricultural sectors is reduced by total costs, and (2) Paired Sample T-Test to find out whether there are significant differences between household income of PLTH electricity beneficiaries at conditions before and after PLTH were developed at Pantai Baru.

2.1. Household Income Analysis on Agriculture Sector

Analysis of household income obtained from the agricultural sector is formulated as follows:

$$TR_t = Pt\ x\ Qt \tag{1}$$

Where,

TR_t = Total farm income (Rupiah/month)
Pt = Selling price per unit of output (Rupiah)
Qt = Total output (tons)

In farming, it is also necessary to calculate the amount of total production costs needed to produce output in one production period. Mathematically written as follows:

$$TC_t = TFC_t + TVC_t \tag{2}$$

Where,

TC_t = Total productsion costs originating from the agriculture sector (Rupiahs/month)
TFC_t = Total fixed costs originating from agriculture sector (Rupiahs/month)
TVC_t = Total variabel costs originating from agriculture sector (Rupiahs/month).

Income originating from agriculture sector are obtained from from the difference in total revenues originating from sales on farm with the total cost on farm production. Mathematically, it is written as follows:

$$I_t = TR_t - TC_t \quad (3)$$

Where,
I_t = Income originating from the agriculture sector (Rupiahs/month)
TR_t = Total revenue from farming (Rupiahs/month)
TC_t = Total costs incurred from farming (Rupiah)

2.2. Businesses Analysis Non-Agriculture Sector

Business analysis in the non-agriculture sector is is calculated based on wages received by PLTH electricity beneficiaries from non-farming activities (I_n), namely wages received outside of tourism and farming businesses, such as: private employment / PNS, PLTH honorary staff, teachers, and tourism businesses that are calculated based on the wages received by respondents per month. Therefore, the revenue and total costs of non-farming are calculated by the formula:

$$TR_n = Pn\ x\ Qn \quad (4)$$

Where,
TR_n = Total revenue originating from non-agriculture sector (Rupiahs/month)
TC_n = Total costs originating from non-agriculture sector (Rupiahs/month)

Revenues earned from non-agriculture sector are calculated using the formula:

$$I_n = TR_n - TC_n \quad (5)$$

Where,
I_t = Income originating from non-agriculture sector (Rupiahs/month)
TR_t = Total revenue originating from non-agriculture sector (Rupiahs/month)
TC_t = Total costs originating from agriculture sector (Rupiahs/month)

Household income obtained by the beneficiaries of electricity is calculated by the formula:

$$I_H = \sum_{i=1}^{n} I_t + \sum_{k=1}^{n} I_n \quad (6)$$

Where,
I_{RT} = Household income (Rupiahs/month)
I_t = Income originating from agriculture sector (Rupiahs/month)
I_n = Income originating from non-agriculture sector (Rupiahs/month).

In this analysis taking into account the inflation value for the years 2007 - 2017. The compound factor method is carried out from the value of the household income. Then the value was analyzed using by *Paired Sample T-Test* to compare the value of household income for the year. Calculated by the formula:

$$FV = PV\ (1 + i)^t \quad (7)$$

Where,
FV = F*uture value* of 2017 household electricity beneficiaries (Rupiahs/month)
PV = P*resent value* of 2007 household electricity beneficiaries (Rupiahs/month)
i = Average inflation rate for DIY Province in 2007 - 2017 (percent) = 5.44% (BPS Province of DIY. 2017).
t = Year to be corrected (year of 2007 until 2017 = 10 years)

3. Results and Discussion

3.1. The Socio-Economic Impact of Electric Energy Development for Rural Community Development
Socio-economic impacts are impacts or benefits arising from social, economic and infrastructure aspects for the development of PLTH electricity in Baru Beach. The aspects that will be discussed in this section are: the condition of road infrastructure along the Pantai Baru area, the household income of PLTH electricity beneficiaries, the number of visitors to the Pantai Baru tourism, and the Regional Original Revenue (PAD) of Bantul Regency originating from Pantai Baru.

3.2. Road Infrastructure Throughout Baru Beach Before and After PLTH Electric Energy Development

Year of 2007 Year of 2017

Picture 1. Visualizing the Condition of Road Infrastructure along Baru Beach in 2007 and 2017

Based on a map visualization analysis at the same location in Pantai Baru, it is not only access roads along the Pantai Baru area that are much better, but also the width of the road along the area is now much wider. In addition, the improvement of road infrastructure is important, because this region connects the southern part of Bantul Regency with Kulon Progo Regency, so that improved access to the road facilitates interconnection between regions.

3.3. Household Income from PLTH Electricity Beneficiaries Before and After PLTH Developed at Baru Beach
Based on the results of the analysis show that, before PLTH was developed at Pantai Baru (in 2007), the average income in the agricultural sector was Rp 2,262,367 per month, the average income in the non-agricultural sector was Rp 630,119 per month, and the average household income of the beneficiaries before PLTH was developed at Pantai Baru in the amount of Rp 4,038,102 per month. After PLTH was developed in Pantai Baru, the livelihoods of the local people not only relied on the agricultural and non-agricultural sectors, but also the tourism sector. household income before PLTH was developed at Baru Beach, with the average income obtained by PLTH electricity beneficiaries from the agricultural sector amounting to Rp 1,388 897, the average income in the non-agricultural sector was Rp 3,600,074 / month, and the average total household income after PLTH was developed on Pantai Baru that is equal to Rp 4,988,981 / month. Thus, after PLTH was developed at this location, there was an increase in the average income in the non-agricultural sector and household income from PLTH electricity beneficiaries, but there was a decrease in the average value of income in the agricultural sector.

Thus, it can be concluded that household income before PLTH was developed (in 2007) after PLTH was developed (in 2018) after it was corrected with the average inflation rate for 2007 - 2018 indicating that the household income of PLTH electricity beneficiaries increased with the average household income in 2007 amounting to Rp 4,038,102 / month, while the average household income in 2017 is Rp.4,988,981 / month. This means that in 2017 there was an increase in household income of PLTH electricity beneficiaries around 24.54% or in other words that there was an increase in household income compared to household income in 2007.

3.4. Different Household Income Test Results Before and After PLTH Developed at Pantai Baru

Based on the results of different pairing tests also showed that, the significance value (2-tailed) of the income before and after the existence of PLTH was 0.041 (<0.05), it can be concluded that there was a significant difference between the average household income before and after the PLTH. Thus, the conditions before and after the existence of PLTH at Pantai Baru are not the same as Zero or in other words that, the hypothesis is accepted. Add standard deviation explanation.

Table 1. Different Paired Test Results Average Income of Electric Beneficiary Households Before and After the Availability of PLTH

Household Income	Mean	N	Std. Deviation	Std. Error Mean
Before PLTH existed	4.038.101	68	3.315.972,2	402.120,69
After PLTH existed	4.988.970	68	1.799.727,3	218.248,99

Source: Result of analysis. 2019.

In otherwise, based on the results of observations in the field it was found that, since the development of PLTH at Pantai Baru proved to be able to provide alternative business income for hamlet women to help the family economy. This can be seen from 79.41% of business managers on the beach are women. The results of the analysis also show that the value of the standard deviation of household income from the beneficiaries of electricity after the existence of PLTH developed in Pantai Baru has decreased from Rp. 3,315,972.2 to Rp. 1,799,727.3. This shows that after the existence of PLTH in Pantai Baru the income gap of beneficiary households has decreased, meaning that the income of household beneficiaries of PLTH electricity is more evenly distributed (Table 1).

3.5. Number of Tourist Visitors and New Coast Regional Income

Based on data from the [2], it shows that, from 2012 to 2018, the number of tourist visitors who visit New Beach in general continues to increase, with average tourist visitors in 2012 - 2018 total 284,579 visitors / year, both tourist visitors originating from Yogyakarta and from outside Yogyakarta. This has an impact on the increase in the value of regional original income obtained from the regional levies obtained from Pantai Baru in 2018 obtained at Rp. 5,655,731,250/year.

Table 2. Number of Tourist Visitors and Regional Retribution from the Sector of Tourism Baru Beach in 2012 - 2018

Year	Number of Visitors (Person / year)	Regional Retribution (Rupiahs/year)
2012	128.648	259.000.000
2013	129.055	319.682.250
2014	141.573	400.798.750
2015	163.109	496.234.750
2016	140.050	499.227.500
2017	558.740	2.536 .114.750
2018	730.875	5.655.731.250
Rerata	284.579	1.863.398.464

Source: Local Regional Income of Bantul Regency Tourism Sector 2012 – 2018. Bantul Regency Tourism Office. 2018.

4. Conclusion

Based on the results of the analysis of this study shows that the presence of PLTH has had a positive impact both from many aspect: there are the access of infrastructure and better electricity. In addition, the presence of PLTH in Pantai Baru has contributed to the Original Regional Income (PAD) of Bantul Regency. Based on the analysis of household income, it shows that the average income of household beneficiaries of electricity when compared to the conditions before and after the PLTH in Pantai Baru shows an increase. Likewise, based on the results of the paired sample t-test, there is a significant difference in the income of PLTH electricity beneficiary households (before and after) PLTH in Pantai Baru. The analysis also shows that the income earned by PLTH electricity beneficiaries has a decreased standard deviation, meaning that the household income of PLTH electricity beneficiaries is more evenly distributed (well distributed).

5. Reference

[1] Adams S, Edem K, Alfred A. 2018. *Renewable Energy and Non Renewable Energy, Regime Type and Economic Growth*. Journal Renewable Energy. 125 (2018) 755 – 767.

[2] Bantul Regency Tourism Office. 2018. *Local Regional Income of Bantul Regency 2012 – 2018*. Bantul Regency Tourism Office. Province of Yogyakarta Special Region.

[3] Bastakoti, BP. 2003. *Rural Electrification and Efforts to Create Enterprises for The Effective Use of Power*. Applied Energy 76(1-3); 145-155.

[4] Bazilian M., Nakhooda S., Van der Graaf T. 2014. *Energy Governance and Poverty*. Journal of Energy Resource Social Science. 1(2014) 217-225.

[5] Chowdhury, S. K. (2010). Impact of infrastructures on paid work opportunities and unpaid work burdens on rural women in Bangladesh. Journal of International Development 22(7): 997-1017.

[6] Colombo E, Bologna S, Masera D. 2013. *Renewable Energy for Unleashing Sustainable Development*. Switzerland (SW): Springer International Publishing.

[7] Day R, Walker G, Simcock N. 2016. *Conceptualising Energy Use and Energy Poverty Using a Capabilities Framework*. Energy Policy. 93: 255-264.

[8] Dinkelman, T. (2008). "The Effects of Rural Electrification on Employment: New Evidence from South Africa."

[9] Emec S, Bilge P, Seliger G. 2015. *Design Production Systems With Hybrid Energy and Water Generation of Sustainable Value Creation*. Clean Technology Environment Policy. 17(7): 1807-1829.

[10] Halicioglu F. 2009. *An Economic Study of CO_2 Emissions, Energy Consumption, Income and Foreign Trade in Turkey*. Energy Policy. 37(3): 1156 – 1164.

[11] [IEA] International Energy Agency. 2015. *World Energy Outlook*. [internet]. [diacu 2015 Desember 2]. Tersedia dari: htpp: //www.worldenergyoutlook.org/wen.2015/.asp

[12] Grimm, M., Hartwig, R. and Lay, J. (2013). Electricity access and the performance of micro and small enterprises: evidence from West Africa. The European journal of development research The European journal of development research Vol:25 Iss:5 (2013) Pg:815-829

[13] Kanagawa M, Nakata T. 2008. *Assesment of Access to Electricity and The Socio Economic Impact in Rural Areas of Developing Countries*. Energy Policy. 36(6): 72-82.

[14] Khandker, S. R., Barnes, D. F. et al. (2009a). Welfare impact of rural electrification: a case study from Bangladesh. Policy Research Working Papers, World Bank.

[15] Khandker, S. R., Barnes, D. F. et al. (2009b). Welfare impact of rural electrification: Evidence from Vietnam. Policy Research Working Paper 5057. World Bank

[16] Khandker, S. R., Samad, H. A. et al. (2012). Who benefits most from rural electrification?: evidence in India. Washington, The World Bank. 6095.

[17] Khandker SR, Barnes DF, Samad HA. 2013. *Welfare Impact of Rural Electrification: a Panel Data analysis From Vietnam*. Economic Development and Cultural Change. 61(3): 659-692. https://www.journals.uchicago.edu/doi/IQ/1086/669262.

[18] Kooijman-van Dijk, A. L. and Clancy, J. 2010. Impacts of Electricity Access to Rural Enterprises in Bolivia, Tanzania and Vietnam. Energy for Sustainable Development 14(1): 14-21.
[19] Kooijman-van Dijk, A. L. 2012. *The role of Energy in Creating Opportunities for Income Generation in The Indian Himalayas*. Energy Policy 41(0): 529-536.
[20] Maleko, G. C. (2005). *Impact of Electricity Services on Microenterprise in Rural Areas in Tanzania.*
[21] Meier, P., Tuntivate, V. et al. 2010. *Peru: National Survey of Rural Household Energy Use.* ESMAP Special Report 007/10. August 2010, World Bank. Washington DC.
[22] Mulder, P. and Tembe, J. 2008. *Rural Electrification in an Imperfect World: A Case Study From Mozambique*. Energy Policy 36(8): 2785-2794.
[23] Nielsen BF, Santos ALR. 2013. *Designing for Multiple Stakeholder Interest Within the Humanitarian Market: The Case Off-Grid Energy Devices*. International Journal Learn Change. 7: 49-67.
[24] Omri A. 2014. *An International Literature Survey on Energy Economic Growth Nexus: Evidence from Country Specific Studies*. Renewable Sustainable Energy Rev. 38: 951 – 9.
[25] Permana AS, Aziz NA, Siong HC. 2015. *Is Energy Efficient? A Study of Gender Household Energy Consumption and Family Decision Making in Indonesia.* Energy Resource Social Science. 6:78-86.
[26] Ramani KV, Heijndermans E. 2003. *Energy, Poverty, and Gender*. Washington DC (US): The World Bank.
[27] Riva F, Helen A, Elias H, Sonali P. 2018. *Energy for Sustainable Development. Electricity Access and Rural Development: Review of Complex Socio Economic Dynamics and Causal Diagrams for More Appropriate.* Journal Energy for Sustainable Development. 43: 203 – 223.
[28] Sadanand, A. (2013). Rural electrification and employment in poor countries. World development, 43, 252-265.
[29] Sovacool BK, Ryan SE. 2016. *The Geography of Energy : Leaders, Lag – Gards, and Lessons for Achieving Primary and Secondary School Electrification*. Renewable Sustainable Energy Review. 58: 107-123.
[30] Stern DJ, CJ Cleveland. 2004. *Energy and Economic Growth*. Encyclopedia Energy. 2: 35 – 51.
[31] UNDP and ESMAP (2004). *The Impact of Energy on Women's Lives in Rural India.* UNDP and World Bank Energy Sector Management Assistance Program (ESMAP).
[32] UNDP and World Bank (2002). *Rural Electrification and Development in the Philippines: Measuring the Social and Economic Benefits*. UNDP and World Bank Energy Sector Management Assistance Program (ESMAP).
[33] [UNIDO] United Nation of Industrial Development Organization. 2015. *Renewable Energy Unit What We Do.Vienna* [id]: Energy and Climate Change Branch.
[34] Van de Walle, D., Ravallion, M., Mendiratta, V. and Koolwal G. 2013. *Long-term Impacts of Household Electrification in Rural India.*
[35] Vargaz RM., Bruce A, Watt M., Mogollon L.G, Alvarez H.R. 2017. *Community in Panama, a Sustainable Assessment of The Boca de Lura PV-Wind Hybrid Power System.* World Renewable Energy Congress XVI (5-9 Feb). Western Australian (ID): Murdoch University
[36] Winther T, Matinga MN, Ulsrud K, Standal K. 2017. *Women's Empowerment Through Electricity Access: Scoping Study and Proposal for A Framework of Analysis*. Journal of Development Effectiveness. 9(3):389-417.
[37] World Bank (2008). *Project Assessment Report Lao People's Democratic Republic Southern Provinces Rural Electrification Project*. Credit 3047 - Sector Thematic and Global Evaluation Division, Independent Evaluation Group (World Bank).

The effectiveness of acetic acid solution agains brown planthopper pest (*Nilaparvata lugens* Stal.) of ciherang rice (*oryza sativa* L).

D Hastuti*, M F Islamy , Rusmana

Department of Agroecotechnology Department, Agriculture Faculty, University of Sultan Ageng Tirtayasa, Jl. Raya Jakarta km 4, Kampus Pakupatan Untirta Serang, Banten, Indonesia

*Corresponding author: dewihastuti@untirta.ac.id

Abstract. One of the main pests that attack rice crops, is brown planthopper (*Nilaparvata lugens* STAL.). This study aimed to determine the potential of acetic acid solution as an insecticide to control brown planthopper and their effects on rice plant Ciherang variety. The research was conducted in the plastic house at Beberan village, Ciruas, Serang District and the Biotechnology Laboratory of Agriculture Faculty; University of Sultan Ageng Tirtayasa and Greenhouse of Indonesian Rice Research Institut Subang; West Java in April to June 2016. The research arranged out experimentally and used a completely randomized design (CRD), which consisted of 7 treatments and 4 replications. The data obtained were statistically analyzed by analysis of variance and Duncan's New Multiple Range Test at 5% level. The results showed that acetic acid can control the brown plant hopper, the highest mortality namely at contretation acetic acid solution 30%. Rice plants showed symptoms of poisoning in acetic acid concentration of 25% and 30% and Acetic Acid no effect on high rendering rice crop.

Keywords: *Acetic Acid, Concentration, brown plant hopper, and Ciherang Rice variety.*

1. Introduction

According to Banten Central Statistics Agency data [3], rice production is predicted to increase 6.21%. The area of rice harvest in 2015 rose 2.21% to 369,631 ha with productivity rising 5.35%

to 57.20 quintals/ha. But the area of paddy fields fell by 28.64% to 17.67 ha as well as productivity decreased by 0.67% to 37.06 quintals/ha.

Efforts to increase rice production are faced with obstacles and problems, including pest attacks. Losses caused by pests account for 13% of total production. One of them is the brown plant hopper pest which in the mid-1970s had caused a national disaster in rice cultivation [1]. In 1960-1970 there was an explosion of brown plant hopper pests in West Java and Central Java, the area of rice plantations attacked by brown plant hopper reached 1.5 million ha with yield losses of more than 2.3 million tons. In the period of 2000-2005, the area of rice crop damaged by this pest reached an average of 20,000 ha / year [5].

According to [4], the brown plant hopper pest is not a familiar name for farmers. Pests that have the Latin name Nilavarpata lugens are the most difficult to overcome among other pests. Besides its very small size, this type of pest is also very fast population, even a female plant hopper when laying eggs can produce 100-500 eggs and also at a temperature of 250C. Wereng is also able to live up to the age of 30 days.

Wereng has a type of mouth piercing tool sucking plant fluids. The part that pierces into plant tissue is a pair of mandible stylets and one pair of maxillary stylets. Brown planthopper attacks in the field fluctuate, ranging from mild to reaching the peak of its development during an explosion that causes puso / burning (hopperburn). Stem worms directly attack rice plants by sucking plant cell fluids on the phloem tissue so that the plants become dry. While the non-direct attack ie planthopper can transfer three viruses that are harmful to rice plants, namely the vacuum dwarf virus, type 1 dwarf virus, and type 2 grass dwarf virus [2].

Facing these obstacles, most Indonesian farmers use chemical insecticides. Unwise use of chemical insecticides has a negative impact on the environment and humans. In Asia in recent years, *N. Lugens* has developed and become resistant to chemical insecticides [10]. This technology is quite popular because its effects can be seen in a relatively short time after application and chemical insecticides are easy to obtain when needed. However, this technology is relatively expensive for farmers, and dangerous for humans, animals, non-targeted species, and the environment if the application is not in accordance with procedures. Other technologies that can be used as components of pest control are alternative insecticides, one of which is vinegar [7].

Based on empirical data, the results of tests conducted by Karawang farmers of the Subur Farmers Group (2015), the use of acetic acid can eradicate the attack of planthopper pests on rice plants. Then in a preliminary test where a solution of acetic acid in a volume of 150ml / L can eradicate planthopper pests 75%. From the test of 25 rice plants infused with 100 planthopper, acetic acid can eradicate 78 nymphs of planthopper (LD50).

In fact, the use of chemical insecticides has not been able to overcome the problem of brown planthopper pests themselves. Farmers need to find alternative insecticides that can increase the mortality of planthopper pests. Based on this, research is needed on the effectiveness of the use of Acetic Acid solutions on planthopper pests (*Nilaparvata lugens* stal) in rice plants (Oryza sativa L) Ciherang variety.

2. Materials and methods

This research was carried out from April to June 2016 at the Beberan Village Plastic House, Serang Regency, Banten Province, also at Biotechnology Laboratory, Faculty of Agriculture, University of Sultan Ageng Tirtayasa, and Greenhouse Sukamandi Rice Research Center.

In this study, the tools used were buckets, aspirators, labels, mylar plastic cages with a diameter of 15 cm, height 50 cm with gauze covers, scissors, hoes, bamboo, plastic, 50% paranet, test tubes, nameplate, mini sprayer, tools write and documentation (camera).

The materials used are Ciherang Varieties rice seeds, soil, vinegar which contain acetic acid at a concentration of 25%, 2-3 instar brown plant hopper which is propagated in the greenhouse of Sukamandi Subang Rice Research Center. The experimental design used was a Completely Randomized Design (CRD) applied to a homogeneous environment, the same plant species and the same plant varieties. The treatment consisted of one factor with 7 levels each repeated 4 times so that there were 28 experimental units. Each experimental unit consisted of 5 individual rice plants so that there were 149 total number of individual rice plants, the need for leafhoppers for insecticide testing was 20 individuals (IRRI, 2002). So that the number of planthopper needed is 560 2-3 planthopper nymphs.

This study used acetic acid solution as an alternative insecticide to the mortality of brown plant hopper pests in rice plants. Wereng is propagated in the Subang Paddy Plant Center. While the rice varieties used were Ciherang varieties.

The treatment of acetic acid in this study consisted of 7 factors, namely:

α 1: 0% (0 ml Acetic Acid / L)
α 2: 5% (50 ml Acetic Acid / L)
α 3: 10% (100 ml Acetic Acid / L)
α 4: 15% (150 ml Acetic Acid / L)
α 5: 20% (200 ml Acetic Acid / L)
α 6: 25% (250 ml Acetic Acid / L)
α 7: 30% (300 ml Acetic Acid / L)

To find out the difference between treatments, α 5% variance (F Test) was carried out and if there were no differences α further 5% DMRT test was performed.

3. Results and discussion

3.1. Percentage of Slope Mortality (N.lugens) (%)

Treatment of acetic acid (A) with various concentrations there was a significant difference in the percentage of mortality of *N. Lugens* to the control treatment. This can be seen in observations 2 hours to 10 hours after application. (Table 1).

Table 1. Percentage of *N. Lugens* mortality at various concentrations of acetate acid (%)

Concentrations of Acetate Acid %	**Average mortality of *N. Lugens* after time of application [hour (h)]**				
	2 h	**4 h**	**6 h**	**8 h**	**10 h**
0	1.75 **e**	1.10 **c**	1.25 **d**	1.31 **a**	1.31 **a**
5	6.17 **c**	1.35 **b**	1.10 **d**	1.10 **a**	1.10 **a**
10	6.83 **c**	1.39 **b**	1.25 **d**	1.10 **a**	1.10 **a**
15	7.44 **b**	1.45 **b**	1.45 **c**	1.25 **a**	1.25 **a**
20	8.32 **a**	1.10 **c**	1.50 **bc**	1.10 **a**	1.10 **a**
25	8.75 **a**	1.10 **c**	2.02 **a**	1.10 **a**	1.10 **a**
30	8.30 **a**	2.05 **a**	1.66 **b**	1.45 **a**	1.45 **a**

Average	**6.79**	**1.36**	**1.34**	**1.20**	**1.20**

Note: This was the result of transformation (SQRT + 0.05).

The results of variance in Table 1 show that acetic acid has an effect on the mortality of brown plant hopper pests at 2 hours, 4 hours and 6 hours after application. Based on the 5% DMRT test results at observation 2 hours after application showed that 25% concentration produced the highest percentage of mortality (8.75%), not significantly different from the concentration of 30% (8.30%) and concentration of 20% (8.32%). Concentration of 20% (8.32%) was significantly different from a concentration of 15% (7.44%), a concentration of 10% (6.83%), a concentration of 5% (6.17%) and a concentration of 0% (1.75). Concentration of 15% (7.44%) was significantly different from the concentration of 10% (6.83%), concentration of 5% (6.17%), concentration of 0% (1.75%). The 10% concentration was not significantly different from the concentration of 5% (6.17%), and it was significantly different from the concentration of 0% (1.75%). While the concentration of 0% (1.75%) which shows the lowest percentage of mortality.

Based on the 5% DMRT test results observed 4 hours after application showed that the concentration of 30% produced the highest percentage of mortality (2.05%), significantly different from the concentration of 25% (1.10%), concentration of 20% (1.10%), concentration of 15% (1.45%), a concentration of 10% (1.39%), a concentration of 5% (1.35%) and a concentration of 0% (1.10%). Concentration of 15% (1.45%) was not significantly different from the concentration of 10% (1.39%) and concentration of 5% (1.35%). While the lowest percentage of mortality at a concentration of 0% (1.10%) was not significantly different from the concentration of 25% (1.10%) and a concentration of 20% (1.10%).

Based on the results of the 5% DMRT test on observations 6 hours after application showed that the concentration of 25% produced the highest percentage of mortality (2.02%). Significantly different from the concentration of 30% (1.66%), concentration of 20% (1.50%), concentration of 15% (1.45%), concentration of 10% (1.25%), concentration of 5% (1.10%) and concentration of 0% (1.25%) . Concentration of 30% (1.66%) was significantly different from concentration of 20% (1.50%), concentration of 15% (1.45%), concentration of 10% (1.25%), concentration of 5% (1.10%) and concentration of 0% (1.25%). Concentration of 20% (1.50%) was not significantly different between concentration of 30% (1.66%) and concentration of 15% (1.45%), significantly different from concentration of 10% (1.25%), concentration of 5% (1.10%) and concentration of 0% (1.25%). While the lowest percentage of mortality at a concentration of 5% (1.10%) was not significantly different from a concentration of 10% (1.25%) and a concentration of 0% (1.25%).

So it was concluded that the higher the concentration of acetic acid, the higher the percentage of mortality. This can be understood because pure or concentrated solution of acetic acid is very corrosive and can cause painful burns [9], so the nature of this acetic acid solution can damage skin tissue resulting in mortality of brown plant hopper.

Based on the results of observations for 2 hours, 4 hours, 6 hours, 8 hours, and 10 hours the average percentage of mortality in Table 1 shows that the most effective time for the mortality of brown plant hopper pests is that of observation 2 hours after application. (presented in Figure 3) and tends to decrease in the observations 6 hours, 8 hours, and 10 hours after application, this decrease is caused by the number of mortality of brown plant hopper which has decreased in the previous hours.

Based on observations (Table 2.) acetic acid gives a different effect on the phytotoxicity of rice plants. Phytotoxicity score 0 indicates no damage and rice plants showed normal growth during vegetative periods, while score 2 showed that rice plants were poisoned with an abnormal shape or leaf color.

Table 2. Effects of acetic acid application on scoring of phytotoxicity of rice plants

Concentrations of Acetate Acid %	Number of leaf	Damaged leaf	Phytotoxicity score
0	20	0	0
5	22	0	0
10	21	0	0
15	22	1	0
20	23	1	0
25	22	3	2
30	21	3	2

Based on Lasmini and Wahid in [8] At the concentration of acetic acid 0%, 5%, 10%, 15% and 20% the phytotoxicity score of rice plants is 0, with a toxicity level of 0% (at a concentration of 0%, 5% , 10%), 4.55% (at a concentration of 15%), 4.35% (at a concentration of 20%). Whereas at the concentration of acetic acid 25% and 30% the phytotoxicity score of rice plants was 2 with 13.64% and 14.29% poisoning levels. It is caused at a concentration of 0% to 20% acetic acid solution is not too concentrated so it does not affect plant phytotoxicity rice, but at concentrations of 25% to 30% concentrated acetic acid solution that affects phytotoxicity in rice plants.

3.2. Rice plant height

One parameter observed is plant height which is the most easily observed growth indicator when given treatment. Stem height is a plant size that is often observed both as an indicator of growth and as a parameter used to measure environmental influences or the treatments applied [6].

Table 3. Average Growth in Rice Plant Height (cm) for 4 Observations

Concentrations of Acetate Acid %	Observation [Week After Planting (WAP)]			
	1	**2**	**3**	**4**
0	27.9	31.0	34.9	39.3
5	28.0	32.4	36.1	41.0
10	25.5	30.5	33.8	38.9
15	28.0	31.6	36.4	39.9
20	29.9	31.1	34.1	38.8
25	25.4	31.4	34.1	37.6
30	26.3	34.6	37.3	41.5
KK (%)	**10.30**	**6.29**	**7.80**	**8.01**

Note: All numbers are no different

Based on variance shows that acetic acid has no significant effect on the growth of rice plant height. Descriptive observation results at 1 week after application of normal rice plant height at a concentration of 0% but at a concentration of 25% rice plants experienced a low plant height compared with other concentrations. At 2 WAP plant height was equally good at both 0% and other concentrations as well as 3 weeks after application and 4 weeks after application. This is due to the influence of the open environment so that acetic acid compounds can be mixed with the outside air so that the growth of rice plants that applied acetic acid has no effect on plant height even though plant height is different from control plant height where the application is not given acetic acid solution. This can be understood because according to Banteng (2010) in [6], that when acetic acid is applied into the soil, acetic acid will evaporate into the air and naturally damaged in the atmosphere by sunlight, so it does not affect the soil nutrients needed for rice growth.

4. Conclusions

In the present study it can be concluded as follows:

1. Acetic acid influences the control of brown plant hopper (*Nilaparvata lugens* STAL).
2. The most effective concentration of acetic acid solution to the mortality of brown plant hopper is at a concentration of 30% and acetic acid has no effect on phytotoxicity in rice plants to a concentration of 20%. At a concentration of 25% to 30% can affect the phytotoxicity of rice plants but does not affect the height growth of rice plants.

5. References

[1] Amirhusin, B. 2004. Perakitan Tanaman Transgenik Tahan Hama. Jurnal Litbang Pertanian, 23(1), 2004.
[2] Baehaki, SE. 2011. Strategi Fundamental Pengendalian Hama Wereng Batang Coklat Dalam Pengamatan Produksi Padi Nasional. Balai Besar Penelitian Tanaman Padi. Subang. Pengembangan Inovasi Pertanian, 2011 : 63 75.
[3] BPS Banten.Produksi Tanaman Pangan Di Provinsi Banten (kuintal/ha) Diakses Melalui http://banten.bps.go.id/linktable dinamis/view/id/20. Diakses Senin 19 januari 2019 jam 13:00 WIB.
[4] Hadi. 2009. Biologi Insekta Entomologi. Graha Ilmu.
[5] Kartohardjono, A. 2011. Penggunaan Musuh alami Sebagai Komponen. Pengendalian Hama Padi Berbasia Ekologi. Balai Besar Penelitian Tanaman Padi. Subang.
[6] Pujisiswanto, H., P. Yudono, E. Sulistyaningsih, dan B. H. Sunarminto. 2014. Pengaruh Asam Asetat Sebagai Herbisida Pratumbuh Terhadap Perkecambahan Jagung. Jurnal Penelitian Pertanian Terapan Vol. 15 (1): 61-67
[7] Puslitbang Hutan Bogor No KT.9/VIII/P3KKPHH-6/2011. Cuka Dapat di Manfaatkan Sebagai Biopestisida.
[8] Riskitavani, D. V., dan Kristanti Indah P., 2013. Studi Potensi Bioherbisida Ekstrak Daun Ketapang (*Terminalia catappa*) terhadap Gulma Rumput Teki (*Cyperus rotundus*). FMIPA Institut Teknologi Sepuluh Nopember, Surabaya. *Jurnal Sains dan Semi Pomits, Vol. 2 No. 2.* Hal 59-63.
[9] Shakhashiri. 2008. *Acetic Acid dan Acetic Anhydride. General Chemistry.* www.Scifun.Org
[10] Wang YH, CF Gao, YC Zhu, J Chen, WH Li, YL Zhuang, DJ Dai, W J Zhou, CY Ma, dan JL Shen, 2008. *Imidacloprid susceptibility survey and selectionrisk assessment in Þeld populations of Nilaparvata lugens(Homoptera: Delphacidae). J Econ Entomol 101: 515-5p22.*

Fractionation distribution of metal Al, Fe, Mn, and microbial population in acidic soil horizon

D Firnia[1*], S Anwar[2], D Andreas[2], B Nugroho[2], D P Tedjo[2]

[1]Doctoral Program, Department of Soil Science and Land Resources, Faculty of Agriculture, IPB University, IPB Darmaga Campus, Bogor 16680
[2]Department of Soil Science and Land Resources, Faculty of Agriculture, IPB University, IPB Darmaga Campus, Bogor 16680

*Corresponding author: dewi.firnia@untirta.ac.id

Abstract. Acidic mineral soil generally has mild acidic to high acidic reaction or pH ≤ 5.5, with very low exchangeable bases that can exchanged and rich in oxide as well as Al and Fe hydroxides. Acidized soil can lead to increased toxicity from aluminum, iron, and manganese element to the plants and deficiency of phosphorous, alkaline cation (calcium, potassium, and magnesium), and molybdenum element as well as soil microbial population. This study aimed to observe the fractionation distribution of metal aluminum (Al), iron (Fe), and manganese (Mn) and soil microbial population in acidic soil horizon. Al fraction parameters included exchangeable Al (Al-exc), non-crystalline Al (Al-o), inorganic non-crystalline Al (Al-po), Organically Bound Al (Al-p), Low- and Medium- Stability Complexes with Organic Matter (Al-Cu) and High-Stability Complexes with Organic Matter (Al-pCu), exchangeable fractionation of Fe and Mn (Fe/Mn-exc), Fe and Mn from Bound to Carbonates/Acid-extractable, Bound to Fe-Mn Oxides/Reducible, Bound to Organic Matter Fe Mn/Oxidizible, Fe- Mn Residual and soil microbial population (total microbes and *Acidithiobacillus* bacteria). The results, distribution of Al fraction on the upper horizon of Jasinga and Lebak were dominated by stabilized organic amorphous Al fraction with fluctuative decrease of organic matter (Al-pCu), while the lower horizon tended to be inorganic amorphous Al (Al-op). The fractionation distribution of Fe fractionation on Jasinga and Lebak was dominated by Oxidizible Fe fraction. Mn fraction distribution on Jasinga and Lebak were dominated by Mn-residual fraction in each acidic soil horizon. The total population of Microbes and *Acidithiobacillus* .sp in each soil horizon decreased along with the alteration of soil chemical properties in each soil depth horizon.

Keywords: Acidic soil, Fraction of Al, Fe, and Mn, Soil microbes

1. Introduction

Tropical soil is usually acidic and has a high concentration of Al [22]. This acidic land has 67% area of the total land in Indonesia with the order *Entisols, Inceptisols, Ultisols, Oxisols, and Spodosols*. Considering the area and distribution, acidic mineral soil has great potential for agricultural development. However, acidic mineral soil nowadays has not been utilized. Acidic mineral soil generally has acidic to highly acidic pH, namely ≤ 5.5, followed with poor exchangeable base content, besides rich in oxide and hydroxide of Al and Fe [15]. Acidic soil can lead to increased toxicity of Al, Fe, and Mn to plants, phosphorous deficiency, reduced alkaline cation (calcium, potassium, and magnesium) and molybdenum [17]. Acidic soil affects soil chemical and biological characteristics, including the availability of nutrients and metal toxicity (which affects the microbial community in many aspects [18].Beneficial microbes and crops prefer a nearly neutral pH with the range of 6 to 7, thus resulting increased soil benefits accompanied by microbe types and activity shifts in the soil.High level of Al on acidic soil as in *Oxisol, Ultisol and Inceptisol* soil causes low growth of plant roots. The study conducted showed that 90% of acidic soil source was hydrolyzed Al. The solubility of Al in soil solution depends on the pH solution and the existence of components that can release Al. Therefore, Al stability in the soil solid phase is one of the key factors to avoid this element dissolvement. [8] reported that Al solubility of 9.25 mmol/L or more could lead to the mortality of cotton plant roots. [28] stated that 1 ppm Al level in soil solution would result decreased soybean production.

Excess concentration of Fe and Mn in the soil will poison the plant, whereas low concentration will suffer the plant from the element deficiency content. Fe element plays a role in the enzyme system in plant metabolism, while Mn serves as an activator of various enzymes, i.e. phosphate transfer and Krebs cycle enzyme. In addition, Mn is an important part of the chloroplasts and participates in oxygen-generating reactions [14]. [11] stated that the deficiency of Fe element will cause chlorotic in the younger parts of the plant, while older tissues remain green. This symptom first arises as a chlorosis among leaf bones, which critically disperses to the entire leaf followed with white coloration of the leaf.

The fractionation of Al, Fe, and Mn aims to observe specifically the reactive mineral fraction affecting the availability of nutrients in the soil. According to [1], the availability of Al in soil can be discovered by extraction method, namely: (1) ammonium chloride (NH_4Cl) extraction, which aims to estimate Al-dd [13], (2) oxalic acid extraction, which aims to estimate amorphous Al (Al--O) [3] , (3) sodium pyrophosphate (Al-P) extraction, which aims to estimate organic amorphous Al [2, (4) 0.3M $CuCl_2$ extraction, which aims to estimate moderate to medium bond of organic Amorphous Al (Al-Cu) [7], (5) strong bond of with organic Al (Al-pCu), the difference of Al-p and Al-Cu [21, and (6) Al-0p fraction, which is the difference of Al-o and Al-p fraction to suspect inorganic amorphous Al in the soil [7]. [19] reported that extraction techniques were gradually conducted to separate various forms of Fe, Mn, Cd, Co, Cu, Ni, Pb, and Zn, which were distinguished in five fractions, i.e. exchangeable, bonded to carbonates (acidic extract), bonded to Fe/Mn Oxide (reduced), organic bound fraction (oxidized), and residual fraction (residual). The form of Fe and Mn in the soil based on [19] and [24] are differentiated based on the availability and mobility in the soil, namely: exchangeable fraction (dd), acetic acid extracted (asm) as a less/low available form, oxidized fraction (oks), reduced fraction (red) as unavailable form, and residual fraction (res) as unavailable form for plants. This study aimed to observe the distribution of Al, Fe, and Mn fractionation as well as microbial population in acidic soil horizon.

2. Methods

Soil samples were taken from two locations, namely Neglasari Village, Jasinga, Bogor and Maja Lebak, Banten. Soil samples were taken on each horizon in the land profile. This study was a descriptive study by conducting a sample analysis of the soil without giving any treatments, then presenting and elaborating the result in the form of data and writing. Al fraction parameters included the chlorium oxalate acid extracting material (1 g of soil sample shaken for 4 hours with 100 mL 0.2 M ammonium oxalate acidized into pH 3 with oxalate acid), sodium pyrophosphate (1 g of soil sample shaken for 16 hours with 100 mL 0.1 M $Na_4P_2O_7$), copper chloride (5 g of soil samples shaken for 5 minutes with 100 mL 0.5 M $CuCl_2$, stopped for 12 hours, and shaken again for 30 minutes), and ammonium chloride (5 g of soil samples shaken for 5 minutes with 100 mL 1 M KCl, then filtered after 12 hours at extracted Al measured with *Atomic Absorption Spectrophotometry* (AAS) Extracted Al by oxalate ammonium acid (Al-o) gives non crystalline-Al content [3]; Extraction with sodium pyrophosphate [2] shows organic Al bound (Al-p); Extraction with copper chloride [5]; [7]; [9] provides stabilized Al formed as low or medium complex with organic matter (Al-Cu); Extracted Al by KCl is considered as exchangeable Al [13], the value of Al-Cu from Al-p (Al-pCu) provides highly stabilized Al complex with organic matter [21], Al-o from Al-p (Al-op) provides estimate of the total Al formed as non-crystalline inorganic Al.

Sequential extraction of soil testing stages for the fractionation of Fe and Mn was based on [19]; [24] to estimate exchangeable Fe and Mn fraction (10 mL 1M CH_3COONH_4 pH 7.0 shaken for 1 hour), Fe and Mn fraction of extracted acids (20 mL 1M CH_3COONa pH 5.0 shaken for 5 hours), reduced Fe and Mn fraction (20 mL 0.04 M NH2OH · HCl 25% CH_3COO H 96^{0}C shaken for 5 hours), oxidized Fe and Mn fraction (5 mL 0.02 M HNO_3 + 5 ml 30% H_2O_2, 85^{0}C pH 2.0 shaken for 2 hours + 5 ml 30% H_2O_2 85^{0}C pH 2.0 shaken for 3 hours + 10 mL 3.2 M CH_3COONH_4 in 20% HNO_3 0.5 hours) and residual Fe and Mn fraction (2 mL $HClO_4$ + 10 mL HF, then evaporated to almost dry dissolved in 6 M HCl). All final solutions of the extract are further analyzed using Atomic Absorption Spectrophotometry (AAS). Total soil microbial was calculated using dish agar method. Pouring method was done by pouring 20 ml media over 1 ml inoculum that has been inserted into a sterile Petri dish. Furthermore, Petri dish was rotated by hand over the table surface, then cooled down a few minutes to make the solid media hardened. The media used was nutrients for soil microbial analysis (24 g/L of media). Dilution 10^3, 10^4, 10^5, and 10^6 were used for the total number of microbes analysis. *Acidithiobacillus ferrooxidant* isolation was performed by inserting 10 g of soil samples into 90 ml of sterilized physiological solution (0.85% NaCl), then shaken for 15 minutes using a shaker. Solution was taken 1 ml and inoculated Leathen liquid media. Incubation was done for 14 days or until the media color turned into rusty yellow at room temperature by shaken using 250 rpm shaker. Color turned on the media was observed every day. Media color turned to yellow or rusty yellow indicated the media contained *Acidithiobacillus* sp. Data results obtained were analyzed in a descriptive method.

3. Result and Discussion

3.1 Aluminum Fractionation in Acidic Soil

The analysis result of concentration fractionation and fraction percentage of Al in each acidic soil horizon is presented in Table 1 and Figure 1.

Based on Table 1 and Figure 1, the fraction concentration of Al-dd in Jasinga and Lebak had fluctuative increase on soil horizon depth. Concentration of Al-exc in Jasinga was 8.26 – 12.64%, while in Lebak was 8.00 – 11.49%. Al-exc distribution of Jasinga soil accumulated at a soil depth of 37-70 cm while Lebak soil Al-exc distribution accumulated at a soil depth of 0-15 cm.

Table 1. Fraction concentration of Al in each acidic soil horizon

Depth (cm)	Fraksi											
	Al-exc		Al-o		Al-o				Al-p			
					Al-p		Al-po		Al-Cu		Al-pCu	
	ppm	%	ppm	%	ppm	%	ppm	%	ppm	%	ppm	%
Jasinga												
0-35	1,800	12.24	12,900	87.76	8,100	55.10	4,800	32.65	1,062	7.22	7,038	47.88
35- 70	2,200	12.64	15,200	87.36	7,600	43.68	7,600	43.68	933	5.36	6,667	38.32
70 - 100	2,000	11.56	15,300	88.44	6,100	35.26	9,200	53.18	867	5.01	5,233	30.25
100 -120	1,900	8.26	21,100	91.74	8,600	37.39	12,500	54.35	849	3.69	7,751	33.70
Lebak												
0-15	1,000	11.49	7,700	88.51	4,000	45.98	3,700	42.53	566	6.51	3,434	39.47
15-40	1,400	9.93	12,700	90.07	4,800	34.04	7,900	56.03	668	4.74	4,132	29.30
40-70	1,200	11.32	9,400	88.68	4,400	41.51	5,000	47.17	592	5.58	3,808	35.92
70-120	1,100	8.09	12,500	91.91	4,400	32.35	8,100	59.56	510	3.75	3,890	28.60

Al-exc= exchangeable; Al-o= Amorphous Al; Al-p= Organic amorphous Al; Al-po= Inorganic amorphous Al; Al-Cu = Organic amorphous Al with low to moderate bound; Al-pCu= Organic amorphous Al with strong bound

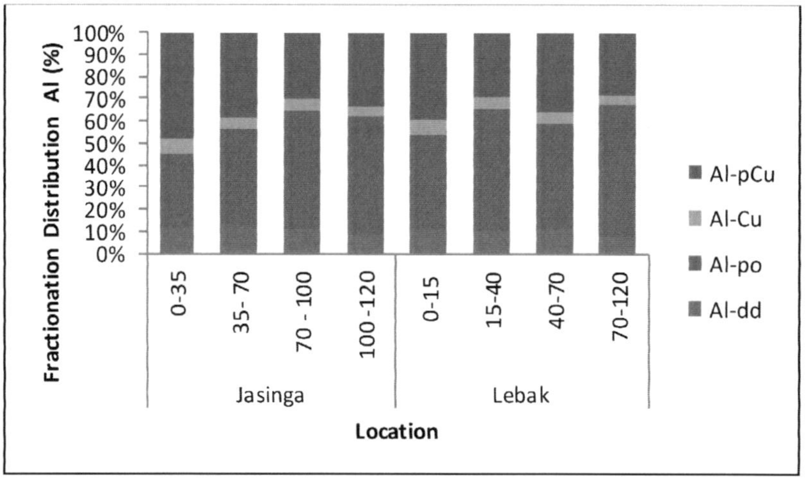

Figure 1. Fraction distribution percentage of Al in each acidic soil horizon, (Al-dd= Al-exc= exchangeable; Al-o= Amorphous Al; Al-p= Organic amorphous Al; Al-po= Inorganic amorphous Al; Al-Cu = Organic amorphous Al with low to moderate bound; Al-pCu= Organic amorphous Al with strong bound.

Al-o fraction is the total Aluminium that binds to organic amorphous and inorganic compounds in the soil. Al-o concentration in Jasinga and Lebak was 87.36-91.74% and 88.51-91.91%, respectively, with fluctuative increase distribution on soil horizon depth. The distribution of the Al-o fraction of Jasinga soil accumulated in the soil depth of 100-10 cm and for lebak soil at a depth of 70-120 cm. High concentration of Al-o reflected high content of amorphous material. According to [23], Al-o fraction was found in amorphous and humous Al. Concentration of Al-p in Jasinga was 35.26 – 55.10% with decreased distribution on soil horizon depth, while in Lebak was 32.35 – 45.98% with declined fluctuative distribution. Distribution of the Al-p fraction of Jasinga soil accumulated at a soil depth of 0-35 cm and Lebak soil accumulated at a soil depth of 0-15 cm. The accumulation of Al-p fraction on the horizon over the soil is related to the content of organic matter where the Al-p fraction is related to the content of Al which is organic.

Al-po fraction concentration in Jasinga was 32.65 – 54.35% with fluctuative increase on soil horizon depth, while in Lebak was 42.53 – 59.56% also with fluctuative increase distribution. The distribution of the Al-po fraction of Jasinga soil accumulated at a depth of 100-120 cm and Lebak soil at a depth of 70-120 cm. This was influenced by the amount of organic matter and texture in the soil sample. Al-Cu fraction is weak and moderately bound Aluminium formed as organic compounds. Al-Cu concentration in Jasinga was 3.69 – 7.22% with reduced distribution on soil horizon depth, while in Lebak was 3.75 – 6.51% with fluctuative decrease on soil horizon depth. The distribution of Al-Cu fraction of Jasinga soil accumulates in the soil depth of 0-35 cm and Lebak soil at a soil depth of 0-15 cm.. Al-Cu fraction is closely related to the organic matter in the soil, the more horizon discovered, the lower organic soil observed, resulting lower concentration of Al-Cu. According to [5], the high content of organic carbon would increase Al-P and Al-Cu fraction in the soil. Al-pCu fraction is aluminum that strongly binds with the organic compounds in the soil [1]. The concentration of Al-pCu fraction in Jasinga was 30.25 – 47.88% with fluctuative decrease distribution on soil horizon depth, while in Lebak was 28.60 – 39.47% also with fluctuative decrease distribution. Distribution of Al-pCu fraction of Jasinga soil accumulated in the soil depth of 0-35 cm and Lebak soil at a soil depth of 0-15 cm. This shows that the deeper soil horizon, the lower Al-pCu formed as lower organic matter discovered.

3.2 Iron Fractionation in Acidic Soil

The analysis result of Fe fraction in each acidic soil horizon is presented on Table 2 and Figure 2. The result showed that Fe concentration generally increased on each fractionation stage in acidic land soil.

Based on Table 2 and Figure 2, Fe-exc concentration in the upper horizon of soil was lower than the down horizon with fluctuative increase in the distribution pattern on soil horizon depth. The concentration of Fe-exc fraction concentration in Jasinga was 0.26-1.04%, while in Lebak was 0.09-0.50%. Distribution of Fe-dd fraction on dried soil accumulated at a depth of 37-70 cm and soil swamp at a depth of 40-70 cm. Fe-exc concentration is influenced by soil texture with high clay minerals. [4] reported acidic condition of soil texture dominated by vermiculate clay minerals can be transformed to release Fe ion in the soil.

Fe-chlorolic acid (Fe-asm) concentration was higher than Fe-exc fraction. Fe-asm concentration in Jasinga was 0.51-3.31% with fluctuative increase on soil horizon depth, while in Lebak was 0.23 – 1.59% also with fluctuative increase distribution. The distribution of Fe-asm fraction of Jasinga soil accumulates in the soil depth of 100-120 cm and Lebak soil at a soil depth of 15-40 cm. Fe-asm fraction concentration is related to soil texture with high clay minerals. Fe-reduced fraction (Fe-red) aims to see metal compound bound to organic compounds. Fe-red concentration in Jasinga was 5.69 – 11.85 with fluctuative increase on soil horizon depth, while in Lebak was 3.92 – 7.93% with fluctuative decrease distribution on soil horizon depth. The distribution of Fe-red fraction in Jasinga soil accumulates at a soil depth of 100-120 cm and Lebak soil accumulates at a soil depth of 40-70 cm.

Table 2. Fraction concentration of Fe in each acidic soil horizon

	Fraksi										
	Fe-exc		Fe-asm		Fe-red		Fe-Oks		Fe-res		Total
Depth (cm)	ppm	%	ppm	%	ppm	%	ppm	%	ppm	%	ppm
					Jasinga						
0-35	16.19	0.26	32.32	0.51	359.46	5.69	5251.60	83.12	658.16	10.42	6317.73
35- 70	35.05	1.04	29.78	0.88	335.00	9.94	1723.50	51.13	1247.18	37.00	3370.50
70 - 100	20.16	0.46	33.50	0.77	335.00	7.68	1453.30	33.31	2520.9	57.78	4363.06
100 -120	11.12	0.75	49.22	3.31	176.60	11.85	684.90	46.10	564.3	37.98	1485.65
					Lebak						
0-15	12.97	0.09	32.51	0.23	811.38	5.71	9360.40	65.93	3981.3	28.04	14198.54
15-40	36.88	0.45	129.19	1.59	426.14	5.24	4478.80	55.02	3069.45	37.71	8140.05
40-70	21.63	0.50	68.35	1.59	341.40	7.93	3395.30	78.83	480.53	11.16	4307.21
70-120	15.11	0.20	33.02	0.42	307.62	3.92	3409.90	43.51	4071.3	51.94	7837.84

Fe- exc = exchangeable; asm = acid extracted; red = reduced; oks = oxidized; res = residual

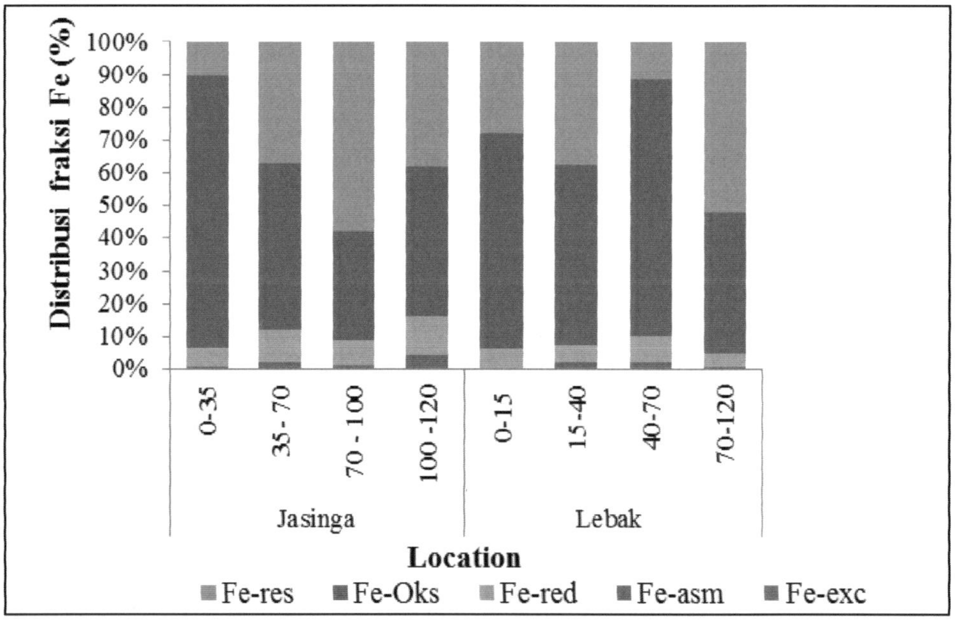

Figure 2. Fraction distribution percentage of Fe in each acidic soil horizon, (exc = exchangeable; asm = acid extracted; red = reduced; oks = oxidized; res = residual)

Fe-oxidized fraction (Fe-oks) in each soil horizon was observed with fluctuative decrease in soil horizon depth. Fe-oks concentration in Jasinga was 33.31 – 83.12% with fluctuative decrease distribution, while in Lebak was 43.51 – 78.83% also with fluctuative decrease distribution. The distribution of the Fe-ox fraction of Jasinga soil accumulates at a depth of 0-35 cm and lebak soil in a soil depth of 40-70 cm. Fe-oks fraction is associated with organic compound level in soil, as decreased Fe-oks concentration happened because of deeper soil horizon resulted in lower organic matter content. Residual fraction is the last stage of metal fractionation. Fe-residual (Fe-res) concentration in Jasinga was 10.42-37.00% with fluctuative decrease distribution on soil horizon depth, while in Lebak was 11.16%-51.94% with fluctuative increase on soil horizon depth. Distribution of Fe-res fraction of Jasinga soil accumulated at a depth of 70-100 cm and Lebak soil at a depth of 70-120 cm. Based on Table 3, Figure 3, and Figure 4, Fe fraction in Jasinga was dominated by oxidized Fe (53.42%), residual Fe (35.80%), reduced Fe (8.79%), acid extracted Fe (1.37%), and interchangeable Fe (0.63%) fraction. Fe fraction in Lebak was dominated by oxidized Fe (60.82%), residual Fe (32.21%), reduced Fe (5.70%), acid extracted Fe (0.96%), and exchangeable (0.31%) fraction. Oxidized Fe fraction indicates the presence of metal that is easily oxidized by peroxide, showing the metal fraction obtained can bind to organic compounds and sulfide [6]. Oxidized Fe fraction is a fraction conditions where Fe are less available, while Fe residual indicates unavailable Fe fraction formed as stabilized crystalline Fe.

3.3 Manganese Fractionation in Acidic Soil

Manganese fraction analysis results in each soil horizon are presented in Table 3 and Figure 3. Mn-exc fraction concentration in Jasinga was 0.05 – 0.26% with fluctuative decrease distribution on soil horizon depth. Mn-exc fraction concentration in Lebak was 0.10 – 0.34% also with fluctuative decrease distribution on soil horizon depth. The distribution of Mn-dd fraction in Jasinga soil accumulated in the soil depth of 0-35 cm and Lebak soil at a soil depth of 0-15 cm. According to [24] Mn concentration is related to the amount of organic matter and type of vegetation on the soil as Mn has weak capability to bind to organic compounds [29].

Table 3. Fraction concentration of Mn in each acidic soil horizon

Depth (cm)	Fraksi Mn										
	Mn-exc		Mn-asm		Mn-red		Mn-Oks		Mn-res		Total
	ppm	%	ppm	%	ppm	%	ppm	%	ppm	%	ppm
Jasinga											
0-35	212.09	0.26	64.83	0.08	88.18	0.11	1016.70	1.26	79539.88	98.29	80921.67
35- 70	59.22	0.08	13.23	0.02	31.02	0.04	80.64	0.11	74343.75	99.75	74527.86
70 - 100	41.72	0.05	9.67	0.01	27.74	0.03	64.08	0.08	79595.63	99.82	79738.83
100 -120	55.26	0.08	22.16	0.03	19.41	0.03	111.02	0.16	70625.63	99.71	70833.47
Lebak											
0-15	267.47	0.34	109.38	0.14	19.55	0.02	4961.20	6.25	74003.38	93.25	79360.97
15-40	179.73	0.22	44.46	0.05	201.76	0.24	2779.80	3.35	79853.13	96.14	83058.87
40-70	144.66	0.18	25.37	0.03	184.46	0.23	2171.90	2.68	78545.63	96.88	81072.02
70-120	103.42	0.10	53.26	0.05	322.86	0.33	5538.60	5.61	92711.25	96.88	98729.39

exc = exchangeable; asm = acid extracted; red = reduced; oks = oxidized; res = residual

Acid extracted Mn (Mn-asm) fraction concentration was lower than Mn-dd fraction. Mn-asm concentration of Jasinga was 0.01%-0.08% with fluctuating decrease distribution on soil horizon depth, while in Lebak was 0.03%-0.14% with fluctuative decrease distribution on soil horizon depth. The distribution of the Mn-asm fraction in Jasinga soil accumulated in the soil depth of 0-35 cm and Lebak soil at a soil depth of 0-15 cm. Fe-asm concentration is related to the amount of organic matter in the soil as lower organic matter discovered due to higher soil layer observed. Reduced Mn fractionation (Mn-Red) aims to observe the metal compound that binds to organic compounds. Mn-red concentration in Jasinga was 0.03%-0.11% with decreased distribution on soil horizon depth, while in Lebak was 0.02%-0.33% with fluctuative increase distribution on soil horizon depth. The distribution of Mn-red fraction in Jasinga soil accumulates in the soil depth of 0-35 cm and Lebak soil at a soil depth of 70-120 cm.

Oxidized Mn (Mn-oks) fraction was observed in each acidic soil horizon with fluctuative decrease distribution. Mn-oks concentration in Jasinga was 0.08%-1.26%, while in Lebak was 2.68%-6.25%. The distribution of Mn-ox fraction in Jasinga soil accumulates in the soil depth of 0-35 cm and Lebak soil at a soil depth of 0-15 cm. Mn-oks is associated with the level of soil organic matter as organic matter concentration decreases due to deeper soil horizon observed.

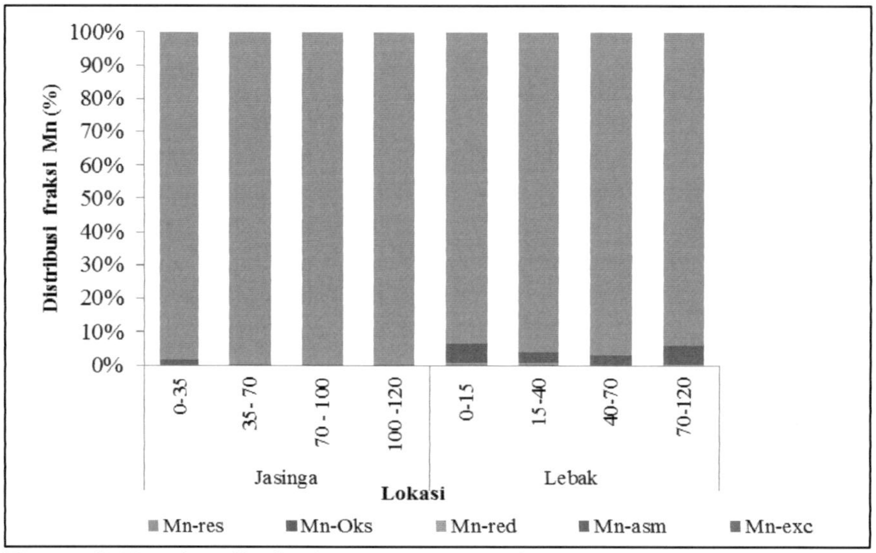

Figure 3. Fraction distribution percentage of Mn in each acidic soil horizon(exc = exchangeable; asm = acid extracted; red = reduced; oks = oxidized; res = residual

Residual fraction is the last stage of metal fractionation. Residual Mn (Mn-res) concentration in Jasinga was 98.29%-99.82% with fluctuative increase on soil horizon depth, while in Lebak was 93.25%-96.88% with increased distribution on soil horizon depth. [24] reported that only about 12-41% of Mn-res fraction in soil that could be extracted. Soil originated from Jasinga and Lebak was dominated by residual Mn fraction. According to [24] Mn concentration is related to the amount of organic matter and type of vegetation on the soil as Mn has weak capability to bind to organic compounds [29]. Residual (res) fraction aims to suspect highly stabilized crystalline mineral that is unavailable to plants, binding to aluminosilicate part as the last stage of metal speciation [26].

3.4 Soil Microbial Population

Table 4 shows that the total population of microbes and *Acidithiobacillus* sp. in each acidic soil horizon decreased as there was an increased value of soil pH and C-organic content. [25] stated that the growth and development of bacteria, one of which is affected by pH. Most bacteria grow in neutral pH range and fewer bacteria can grow at H $\leq$ 3, i.e. Acidophilic bacterial group bacteria, such as *Acidithiobacillus ferrooxidant* and *Acidithiobacillus thiooxidant* that can grow at pH 1.

Table 4. Total microbial and *Acidithiobacillus* in each acidic soil horizon

Depth (cm)	Total microbial		*Acidicthiobacillus*,sp
	(cfu/g)	Log cfu/g)	x 10^2 Apm.g^{-1}
	Jasinga		
0-35	800794.10	5.90	2.5
35 - 70	306240.43	5.49	2.5
70 - 100	515808.09	5.71	0.6
100 -150	408891.53	5.61	0.4
	Lebak		
0-15	189415.04	5.28	0
15-40	32258.06	4.51	2.5
40-70	70652.17	4.85	0
70-120	22160.67	4.35	0

The existence of microbes in soil assists the biochemical process of organic matter as microbes produce enzymes to decompose the organic matter for nutrient source of plants. Microbes also has symbiotic relationship with plant roots that can aid the development and stability of the soil through the exudate production, secondary metabolic, and organic inputs as adhesives between organic and inorganic soil. Increased organic matter can be attributed to the increased microbial biomass and activity, producing extracellular polysaccharides that serve as a good soil sealing agent. *Acidithiobacillus* bacteria could conduct reduction reaction due to increased pH above the required pH, which was 3–4. According to [27] microbial population dynamics are caused by nutrient factors, humidity, soil aeration, temperature, pH, agricultural practices, fertilization, pesticide usage, and the addition of organic materials. [10] studied *Acidithiobacillus ferrooxidant* is an acidophilic microbe that grows in low pH environment.

4. Conclusion

Distribution of Al fraction of upper layer soil horizon taken from Jasinga and Lebak was dominated by highly stabilized amorphous Al fraction complex with fluctuative decrease of organic compound (Al-pCu) and amorphous inorganic Al fraction (Al-op) in the down layer. Furthermore, the distribution of Fe and Mn fractionation in acidic soil sample taken from Jasinga and Lebak sequentially in each soil horizon composed of: Fe-exc < Fe-asm < Fe-Red < Fe-res < Fe-Oks, while Mn fraction concentration sequentially composed of: Mn-exc < Mn-asm < Mn-Red < Mn-oks < Mn-res. In addition, the total population of microbes and *Acidithiobacillus* sp. in each acidic soil horizon were decreased along with changes in soil chemical properties in each acidic soil horizon

5. References

[1] Alvarez E, Fernandez-Sanjurjo M, Nunez A, Seco N, Corti G .2012 Aluminium fractionation and speciation in bulk and rhizosphere of a grass soil amended with mussel shells or lime.*Geoderma* **173**:322–329

[2] Bascomb, C.L., 1968. Distribution of pyrophosphate extractable iron and organic carbon in soils of various groups. Journal of Soil Science 19, 251–268

[3] Blakemore, L.C., 1978. Exchange complex dominated by amorphous material ECDAM). In: Smith, G.D. (Ed.), The Andisol proposal. Soil Bureau, DSIR, New Zealand, pp. 21–22.

[4] Dubikova M, Cambier P, Sucha V, Caplovicova M. 2002. Experimental in soil acidification. *Appl. Geochem*. **17**: 245-257

[5] Drabek, O., Boruvka, L., Mladkova, L., & Kocarek, M. 2003. Possible method of aluminum speciation in forest soils.. *Journal of Inorganic Biochemistry*, **97**, 8–15. doi:10.1016/S0162-0134(03)00259-9.

[6] Gasparatos, D., C. Haidouti, Adrinopoulus and O, Areta, 2005, Chemical Speciation and Bioavailability of Cu, Zn, and Pb in Soil from The National Garden of Athens, Grecee, *Proceedings* : *Internasional Conference on Environmental Science and Technology*, Rhodes Island, Greece, 1-3 September 2005

[7] Garcia-Rodeja E, Novoa JC, Pontevedra X, Martínez-Cortizas A,Buurman P. 2004. Aluminum fractionation of European volcanic soils by selective dissolution techniques. *Catena* **56**:155–183

[8] Hue, G.R.C. and F. Adams. 1986. Effect of organic acids on aluminum toxicity in subsoils. *Soil Sci. Soc. Am. J.* **50**:28-34.

[9] Juo, A.S., Kamprath, E.J., 1979. Copper chloride as an extractant for estimating the potentially reactive aluminum pool in acid soils. *Soil Science Society of America Journal* **43**, 35–38

[10] Kelly DP, Wood AP. 2000. Reclassification of some species of *Thiobacillus* to the newly designated genera *Acidithiobacillus* gen nov., *Halothiobacillus* gen. nov., and *Thermithiobacillus* gen nov. *Int. J. Syst. Evol. Microbiol.* **50**:511-516

[11] Leiwakabessy, F. M,, U. M. Wahjudin, dan Suwarno. 2003. Bahan kuliah Kesuburan Tanah. Jurusan Tanah. Fakultas Pertanian. Institut Pertanian Bogor

[12] Lin Y, Su P. 2010. Behavior of aluminum adsorption in different compost-derived humic acids. *Clean.* 38:916–920. doi:10.1002/clen.201000034

[13] Peech, L., Alexander, L.T., Dean, L.A., 1947. Methods of Analysis for Soil Fertility Investigations. USDA, Circ. No. 757.

[14] Soepardi, G. 1983. Sifat dan Ciri Tanah. Jurusan Tanah. Fakultas Pertanian, Institut Pertanian Bogor. Bogor.

[15] Subagyo H, Suharta N, Siswanto AB. 2000. *Tanah Tanah pertanian di Indonesia. Sumberdaya Lahan Indonesia dan Pengelolaannya*. Pusat Penelitian Tanah dan Agroklimat. Badan Penelitian dan Pengembangan Pertanian. Departemen Pertanian.Hal. 21-65.

[16] Sutherland RA. 2002. Comparison between non-residual Al, Co, Cu, Fe, Mn, Ni, Pb and Zn released by a three-step sequential extraction procedure and a dilute hydrochloric acid leach for soil and road deposited sediment. *J. Applied Geochemistry*. **17**: 353–365

[17] Shazana M, Shamshuddin J, Fauziah C, Syed Omar S. 2013. Alleviating the infertility of an acid sulfate soil by using ground basalt with or without lime and organic fertilizer under submerged conditions. *Land Degrad Dev*. **24**:129–140

[18] Sylvia DM, Fuhrmann JG, Hartel PG, Zuberer DA, eds. 2005. Principles and Applications of Soil Microbiology (No. QR111 S674 2005). Upper Saddle River, NJ, Pearson Prentice Hall

[19] Tessier, A., Campbell, P.G. and Bisson, M.1979. Sequential extraction procedure for the speciation of particulate trace metals. *AnalChem*.**51**:844–851

[20] Tisdale, S.L., and W.L. Nelson, and J.D. Beaton. 1985. Soil Fertility and Fertilizer. 4th ed. The Macmillan Co., New York. 367 p.

[21] Urrutia, M., García-Rodeja, E., Macías, F., 1988. Aplicación de disoluciones no tamponadas para la extracción de Al "activo" ligado a la materia orgánica en suelos de Galicia. Anales de Edafología y Agrobiología 47, 1289–1301 (in Spanish)

[22] Vitorello, V. A., Capaldi, F. R., & Stefanuto, V. A. 2005. Recent advances in aluminum toxicity and resistance in higher plants. Brazilian Journal of Plant *Physiology*, **17**, 129–143. doi:10.1590/S1677-04202005000100011.

[23] Wada,K.1986. Allophane and imogolite. In: Dixon JB, Weed SB (eds) Minerals in soil environments. SSSA, Madison

[24] Walna B, Spychalski W, Ibragimow A. 2010. Fractionation of iron and manganese in the horizons of a nutrient-poor forest soil profile using the sequential extraction method. *J. Polish of Environ.* Vol. **19.5**: 1029-1037.

[25] Waluyo L. 2009. Mikrobiologi Lingkungan. Malang: UMM Pr Wang WX, Yan QL, Fan W, Xu Y. 2002. Bioavailability of sedimentary metals from a contamined bay. *J. Marine Ecology Progress Series.* Vol. **240**: 27-38.

[26] Wang C, Zhao XQ, Aizawa T, Sunairi M, Shen RF. 2013. High Aluminum Tolerance of *Rhodotorula sp.* RS1 is Associated with Thickening of the Cell Wall Rather than Chelation of Aluminum Ions. *Pedosphere.* **23**:29-38

[27] Widyati E. 2013. Rhizosphere microbe community dinamic and it contribution on plants growth. Tekno Hut Tan. **6**: 55-64

[28] Wright, R.J., 1989. Soil aluminum toxicity and plant growth. Commun. *Soil Sci. Plant Anal.* 50-56

[29] Xue N, Seip HM, Guo J, LiAO B, Zeng Q. 2006. Distribution of Al, Fe, Mn pools and their correlation in soils from two acid deposition small catchments in Hunan, China. *J. Chemosphere*. **65**: 2468-2476

IOP Conf. Series: Earth and Environmental Science **383** (2019) 012016 doi:10.1088/1755-1315/383/1/012016

Consumption patterns of processed cold meat after listeriosis *(Listeria Monocytogens)* outbreak: A case study of North-West University (NWU) students, Mmabatho Campus, South Africa

T Dikgole[1], U Luvhengo[1]*, and J N Lekunze[2]

[1] Department of Agricultural Economics and Extension, North-West University, Mmabatho, 2735 South Africa.
[2]Research Manager, Faculty of Economics and management Sciences, North West University Business School, South Africa.

*Corresponding author: luvhengousapfa@gmail.com

Abstract. The aim of this study was to analyse the effect of a foodborne disease outbreak (Listeria Monocytogens) and factors which affect the consumption patterns of processed cold meat products by students at North-West University, Mmabatho Campus, South Africa. A well-structured questionnaire was used as a data collection instrument, 375 registered students who enrolled for postgraduate and undergraduate studies were randomly selected within two clusters. The sample size was calculated from the total population of 12864 students using RASOFT calculator. Data was captured and analysed using Microsoft Excel and Statistical Package for Social Science (SPSS) version 25 computer software. Probit Regression Model was employed to identify factors influencing consumption patterns of students after the Listeria Monocytogens outbreak. Factors such as gender, residence, allowance (Income), meal preference, regulation of meals time, price of alternatives and academic level were found to be statistically significant and they do influence consumption patterns of processed cold meat by both undergraduate and postgraduate students in North-West University Mmabatho Campus. The study concluded that disease outbreaks (Listeria Monocytogens) do have a negative impact on student's consumption patens and switching behaviour. Findings further revealed that disease outbreak led to the majority of students spending more time preparing meals and this may have a negative effect on their academic and/or extramural activities.

Keywords: Processed cold meat, students' consumption patterns, food-borne disease outbreak, factors influencing consumption

1. Introduction

Worldwide, Individuals have different consumption patens which form part of our living and lifestyle in general [1]. Furthermore, disease outbreak is not a new concept especially in agro-food chains whereby the effects can lead to food insecurity or famine [2]. Processed meat products such as ham, bacon, sausages and other deli meats are used in the preparation of the majority of convenient meals because these products have already been prepared for easy and safe consumption and provide a variety of choice to consumers. According to [3], majority of the aforementioned convenient foods are produced through intensive technical processing and there is an increase in demand which is driven by the demand for easy access, affordability, collection, preparation and overall convenience.

Tertiary students are faced with various lifestyle challenges being on campus and away from home. They are faced with a trade-off between academics and a healthy lifestyle due to the "limited" number of hours in a day which often leads to the consumption of takeaways and ready-made meals to try and save time for academic obligations and other activities they might be a part of including church, cultural and sports activities. Processed meat constitutes a great part of these ready-made and take-away meals that students consume, and this is because among other reasons they are relatively cheaper than other meat products. [4] further stated that, as students, it becomes increasingly difficult to keep up with regular meal time, therefore, less time is spent preparing meals than eating them, as a result, convenient meal such as sandwiches, hotdogs, sausages, and bunny chow amongst many others form a perpetual diet of many students. Furthermore, food consumption and preparation by students is influenced by several factors such as food preference, perception of healthy eating, peer influence, lifestyle characteristics, residence type, income/allowance, gender and beliefs among others [5].

The 2017–18 South African listeriosis outbreak in South Africa resulted from contaminated processed meats produced by Enterprise Foods, a subsidiary of Tiger Brands, in Polokwane. As of 12 March 2018, there have been 183 deaths and 973 confirmed infections. It is the world's worst ever listeriosis outbreak [4]. The occurrence and effects of foodborne disease outbreaks on consumers have been studied worldwide. However, fewer studies have outlined the impact of the foodborne disease on students, especially in South African tertiary institutions. Thus the main aim of this study was to analyse the effect of a foodborne disease outbreak (*Listeria Monocytogens*) on the consumption patterns of processed cold meat products by students in North West University, Mmabatho South Africa.

2. Materials and Methods

2.1. *Study Area*

The study was conducted at North-West University Mafikeng Campus, Mmabatho, North-West province of South Africa. The University was formally known as the University of Bophuthatswana prior to democracy and voted by ballot by the people of Mafikeng to have changed to be called the University of North West after the democratic change. In 2004 the merger between Potchefstroom University for Christian Higher Education, University of North West and Sebokeng Campus of Vista took place collaborating and forming the current North-West University with three different campuses. The Mafikeng campus is in Mmabatho Unit 5, it approximately has 12 864 registered students with seven faculties which are the faculty of Natural Science and Agriculture, Faculty of Economic and Management Science, Faculty of Education, Faculty of Humanities, Faculty of Health science, faculty of Theology and the Faculty of Law. The survey and interviews were conducted from the period June-July 2018.

2.2. *Population and sample size determination*

The population of the study refers to the study of a group of people or individuals taken from the general population of approximately 12,864 who share a common characteristic, such as age, sex, health conditions. This group can be identified for different reasons such as to analyse factors affecting consumption patterns of students after disease outbreak [6]. For the purpose of this study, the population included 12,864 students registered at the North-West University, Mafikeng campus

Mmabatho, South Africa. [6] state that sampling involves taking a representative selection of the population and using the data collected as reach information. Therefore, for this research, a sample size of 374 registered students taken from the sample population of 12,864 students using the raosoft calculator software was used.

2.3. Methods of data collection
Primary data was used for this research with a well-structured questionnaire. Students were questioned on different aspects based on the study objectives; student's consumption patterns before listeriosis outbreak, effect of listeriosis on the consumption of processed cold meat by students and the factors which affect student consumption patterns.

2.4. Methods of data analysis
SPSS version 25 computer software was used for analysis and Probit Regression Model for the analysis of factors affecting consumption of processed red meat as the main objective. Frequency counts and percentages were used to summarize the data and socioeconomic characteristics of students. Furthermore, tables and figures were also used to categorise variables which significantly contribute towards change in consumption patens by students.

2.5. Probit model
Factors influencing consumption patterns processed red meat products by students after disease outbreak were estimated using some economic, social and demographic variables of the student's time frame of response to the outbreak were also be analysed. The Probit Model was used due to the fact that the dependent variable (Y_i) is binary with values being yes (1) if consumption was affected and no (0) if consumption of processed cold meat was not affected by the outbreak. The model is stated by equation 1 below as:

$$Y_i = \eta + \beta i \sum_{i=1}^{10} Zi + \varepsilon i \quad (1)$$

Consequently, the Probit regression model can then be predicted as flows equation (2) below:

$$yi^* = \beta_0 + \beta_1 X_1 + \beta_2 X_2 + \dots \beta_2 X_p \quad (2)$$

The independent variables ($Zi's$) measured the following parameters; whether students eat processed meat or not, regulatory of meal time, resident type, income/allowance, meal preference, gender, academic level, extra-curricular activities, age, religion, price of alternative products and the error term. η βi are parameters and prior expectations are explained in table 1 below.

Table 1. Definition of variables

Explanatory variables	Variable name
Gender	1 for male, 0 for female
Religion	1 if religious, 0 if not religious
Age	Number of years
Allowance	In Rands
Residence	(1 for off-campus living, 0 for those living on campus)
ACL	Academic level (1 for undergrad, 0 for postgraduate)
MPL	Meal preference (1 for cooked food, 0 for processed cold meats)
RMT	Regulatory meal time (1 for yes, 0 for no)
ECA	Extra-curricular activities (1 for yes, 0 for no)
PAP	Price of Alternative Products influence on consumption (1 for yes, 0 for no)
ALI	Academic level influence on consumption (1 for yes, 0 for no)
RIOC	Residence influence on consumption (1 for yes, 0 for no)
ECAI	Extra-curricular activities influence of consumption (1 for yes, 0 for no)

3. Results and Discussion

This section discussed the major findings of the study. SPPSS 20 software was used to analyse the collected data. Before the discussion of the findings, the multi-collinearity test was conducted using the Variance Inflation Factor (V.I.F) to ensure that independent variables are not correlated, furthermore, the test for heteroscedasticity was conducted using the Glejser test. Thereafter the significance of socio-demographic characteristics of student's consumption patterns of processed cold meat after the listeriosis outbreak using descriptive statistics and probit model.

3.1. Multi-collinearity test amongst parameters

3.1.1. Variance inflation factor

Variance Inflation Factor (V.I.F) measures the amount of multicollinearity in a set of multiple regression variables. It provides an index that measures how much variance, the square of the estimates standard deviation, of an estimated regression coefficient is increased because of collinearity. According to table 2 below, the mean VIF =1.44, which is > 10 meaning there is no multicollinearity. Therefore it was concluded that there was little correlation which makes the model unbiased.

Table 2. Variance inflation factor

Variable	Variance Inflation Factor	Tolerance
Gender	1.065	0.939
Age	1.882	0.531
Residence	1.325	0.755
Religion	1.037	0.965
ACL	1.682	0.595
Allowance	1.433	0.698
MLP	1.092	0.915
RMT	1.086	0.921
ECA	2.315	0.423
PAP	1.060	0.944
ALI	1.320	0.757
RIOC	1.286	0.778
ECAI	2.232	0.448
Mean VIF = 1.44		

Source: IBM SPSS Statistics Viewer (2018)

3.1.2. Test for heteroscedasticity

Heteroscedasticity is used to examine whether there is a difference in the residual variance of the observation period to another period of observation. This study tested for heteroscedasticity the Glejser test in SPSS and the results are shown in table 3 below. Based on output coefficients the obtained value of Sig >0.05, therefore, it can be concluded that there is no heteroscedasticity problem.

Table 3. Test results output Glejser

Variables	Unbiased coefficients		Standardized coefficients	T
	B	Std Error	Beta	
Gender	-0.019	0.052	-0.020	- 0.370
Age	-0.017	0.013	-0.096	-1.328
Residence	-0.133	0.058	-0.136	-2.281
Religion	0.420	0.245	0.092	1.75
ACL	-0.081	0.081	-0.068	-1.005
Allowance	-3.294E-5	0.000	-0.148	-2.362
MLP	0.108	0.65	0.091	1.657

RMT	0.051	0.079	0.045	0.815
ECA	-0.053	0.188	-0.053	-0.667
PAP	-0.108	0.062	-0.031	-0.574
ALI	0.009	0.058	0.009	-0.149
RIOC	-0.052	0.081	0.033	-0.900
ECAI	0.034	0.053	0.033	0.418

Source: IBM SPSS Statistics Viewer (2018).

3.1.3. Probit model

The Probit model was used to analyse factors that influence student's consumption patterns of processed cold meat. The results from the Probit model in Table 4 below shows that, out of 10 variables that were regressed to check their influence on students' consumption patterns of processed cold meat, their variables were not statistically significant at 5 and 10 percent level of significance, and the variables were: Age, Religion and Extracurricular activity enrolment. However, the coefficients for seven parameters were statistically significant at 5 and 10 percent level of significance as shown in table 3 below.

Table 4. Results of the Probit model Analysis on factors affecting students' consumption of processed cold meat

Parameter	Estimate	Std. Error	T	Significance
Gender	-.492	.349	-1.411	*
Age	.021	.075	.278	NS
Religion	.747	1.173	.637	NS
Residence	-.351	.349	-1.006	*
Allowance	.000	.000	-1.650	*
Meal Preference	.646	.636	1.017	*
RMT	-.238	.419	-.567	**
ECA	.431	.680	.633	NS
PAP	2.245	3.282	-1.183	**
ALI	.518	.438	.231	*
Intercept	-2.959	3.917	-.755	**
	Chi-Square		Sig	
Pearson Goodness of Fit Test	161.193		0.252	

Source: Authors computation (2018)
Significance: * at 10% and ** at 5%

3.2. Gender

Table 5. Frequency of consumption and gender

Gender	Frequency	Consumption after the outbreak	Percentage
Female	175	60	16.9
Male	180	12	3.4
Total	355	72	20.3

Source: Field Survey with questionnaires (2018).

From table 5 above, the parameter for gender is with a negative sign and statistically significant at 10% level. This implies that gender has a significant influence on the consumption pattern of processed cold meats by students in the study are. Table 5 above further revealed that out of the 175-female sampled 60 female student's consumptions of processed cold meat continued but it had decreased while the remaining 115 stopped consuming and out of 180 male students sampled only 12 changed their consumption pattern. After data collection it was discovered that male students consumed processed cold meat more than female students regardless of the news of the outbreak, this indicated that females where more concerned about food safety than males while males are more concerned about hunger satisfaction and convenience of food than safety and literature of studies similar to this one made the same observation. [7] mentioned that females are more interested in food safety and not hunger satisfaction, which seems to be the opposite for males. It was also discovered that females were more concerned than males over food safety risks [8].

3.3. Residence

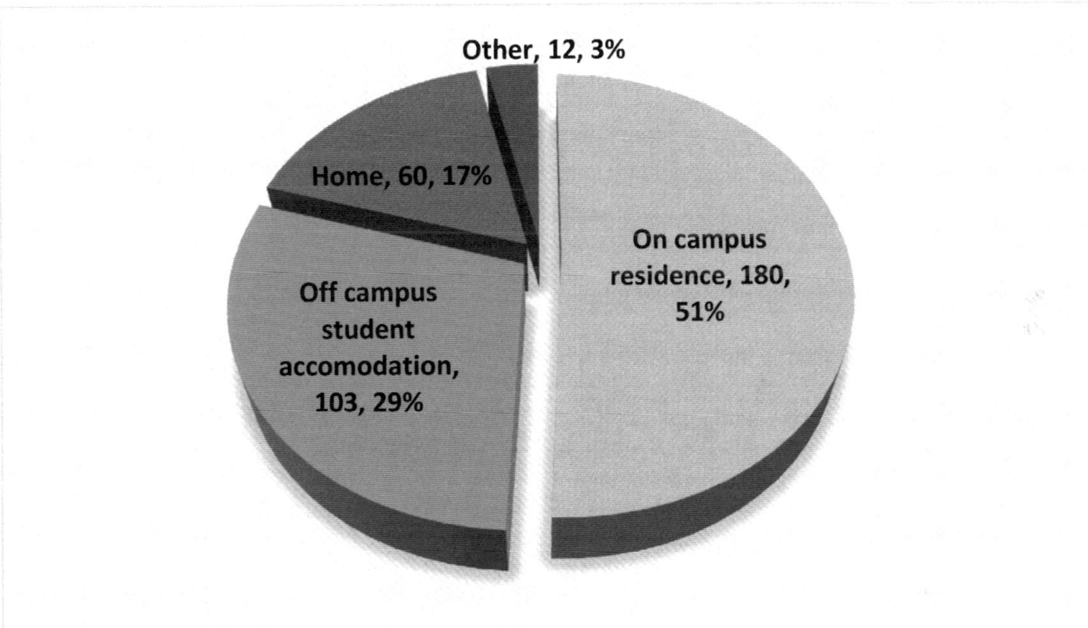

Figure 1. Student residential composition
Source: Authors composition (2018)

Figure 1 above revealed that, the different residences that students lived in and 180 sampled students lived on campus residence and out of this 180, 188 changed their consumption of processed cold meat after the news of the outbreak and used their spare time in between classes to cooking so they avoid buying food from cafeteria and other food stalls where most of these meats were sold and they could do this because they did not have to walk or travel to campus as a result they could go back to their residences after class while waiting for the next class and cook while 29% lived off-campus students' accommodation such as communes, student flats and off-campus student accredited accommodation, 16.9% students lived at home with their families reported that they only consume processed cold meat when they are on campus because at home they find proper cooked nutritious food and 3.4% indicated other forms of residence. This observation shows that students' residence had a significance in the consumption patterns of students. Furthermore, the parameter for the residence was found to be negative and statistically significant at the 10% level (Table 4). Students living off campus choose different food from those students living off-campus. [9] hence the survey carried out discovered that students who lived on campus residence consumed less processed meals after the news of the outbreak. The same cannot be said about students who reside off campus and had to spend most of their days on campus waiting for classes to end before they could go home therefore while waiting they would purchase food at the cafeteria or at the food stalls outside which serve mostly processed cold meat, this observation showed that even after the outbreak because of time and hunger these students would consume what they could quickly eat and proceed with the day's activities. Students revealed that when living in students residences they made time and cooked together with their peers while socializing, which increased chances of eating healthier food as compared to convenient food such as processed cold meats [11].

3.4. Meal Preference

Parameter for meal preference is with a positive sign and statistically significant at 10% level (Table 4). Participants where asked what type of meals they prefer between cooked meals and processed cold meats and those who said they preferred cooked meals said so because of various reasons such as health and taste preferences while those who indicated that they preferred processed cold meat said so because of their convenience and taste. Research has shown that the most important factors that influenced food selection are taste, cost, nutrition, convenience and pleasure amongst others [11]. Preference of meals influenced how much of processed cold meat is consumed by the students, it correlates with the level of nutritional knowledge that students had as some felt strongly about nutrition and thus preferred cook meals over processed cold meat simply because of the level of nutrition that they got from home-cooked meals. It was also recorded in previous studies that people who are actively involved with food preparation such as those who love to cook are usually more concerned about food safety and quality [13]. The consumption of processed cold meats by students was changed after the outbreak and as a result student had to resort to alternative sources of food.

3.5. Students allowance (disposable income)

Table 6. Income/allowance distribution of respondents

Income/Allowance	Frequency of processed cold meat Consumption	Percentage
R0 – R300	45	13.4%
R350 -R600	24	6.8%
R650- R900	33	9.4%
R950 – R1200	29	8.2%
R1250- R1500	152	42.8%
R1550- R1800	20	5.6%
R1850-R2100	16	4.0%
> R2150	26	9.8%

Source: IBM SPSS Statistics viewer (2018).

Parameter for allowance was found to be positive and statistically significant at 10% level (Table 6). This indicates that allowance that students get to influence their consumption of processed cold meat product. Table 4.5.4 showed that 283 of the respondents got < R1550 and made up 80.6% of the sampled respondents, these 283 respondents consumed proceed cold meat 211 times more than those individuals who got > R1550 allowance monthly. This could be because processed cold meats are cheaper as compared to fresh meat and with the limited budget that students live on it becomes more economical to purchase processed cold meat, for example one can consume a 1kg loaf of Palony longer (3-4weeks eating a slice every day) than a 1kg roll of Wors or 1kg beef. This observation also shows that the outbreak did not influence student's consumption of processed cold meat because they did not have enough to change with substitutes, therefore, most of the respondents stated that they only changed brands and not products. [13] observed that it was mostly individuals with high incomes who perceived food safety to be important than individuals from other categories, mostly because high-income earners have the luxury of substituting anything deemed to be hazardous with the next best alternative. The cost of food is among the highest factors influencing the choice of food consumed amongst college students [14].

3.6. Regular Meal Time

Parameter for regulatory meal time is with a negative sign and statistically significant at 5%(See table 4). This indicates that the regulation of meals has an influence on the type of food consumed by the students. Only 23% of the surveyed students had a regular meal time while 76% of the students said they did not have a regular meal time (refer to table 7) because of their academic obligations and schedules which were not constant , this resulted in these students grabbing something to eat quickly while going through their day and most of the foods they had at the disposal while busy were take outs with processed cold meat products such as hot dogs, bunny chow and sandwiches. Factors placing students at risk nutritionally are financial risks, skipping meals, snacking and frequently eating fast foods [15] which put students at a higher risk of being diagnosed with *Listeria Monocytogenes* during the outbreak. Factors such as skipping of meals and not having a regular meal time, snacking in-between classes and frequent consumption of fast food are among factors influencing healthy eating among college students [16].

Table 7. Frequency counts of variables

Variable	Sample number	Frequency	Percentage
Meal preference			
Cooked meals	355	75	21%
Processed cold meat	355	280	79%
Regular mealtime			
Yes	355	83	23%
No	355	272	76%
Academic level influence			
Yes	355	239	33%
No	355	116	67%

Source: IBM SPSS Statistics Viewer (2018).

3.7. Price of Alternatives

Figure 2. Graph of alternative products vs price of procced cold meat
Source: Pick n Pay Supermarket Megacity (May 2018).

Parameter for the price of alternative products is with a positive sign and statistically significant at 5% level (refer to table 4). There is a positive relationship between the consumption of processed cold meat products and price of alternative products , figure 2 shows the relationship between processed cold meat prices and prices of fresh meat , it is evident that the latter is more expensive as compared to the former and the utility thereof also differs , the former provides more utility for long periods (weeks or days) which it can be consumed while fresh meat provides maximum utility for shorter periods in terms of weeks or days it can be consumed. Reported preference for alternative products is determined by food price [14].

3.8. Academic Level Influence (ALI)
Parameter for academic level influence is with a negative sign and statistically significant at 10% level (Table 4). This indicates that the level of the academic year has an influence on the consumption of processed cold meat, only 33% of students surveyed said that their academic level does not influence consumption when asked, this 33 % constituted of students in their first year while 67% said that their academic level influences their consumption. When asked why they said their academic level influenced their consumption most of them indicated that as the level went up the workload also did so there is a time difference between the time they had for the first year and the time which they have in their 2nd or 3rd years of study it was also found out that the faculty which they were enrolled in influenced their consumption in terms of time as some students noted that they had practical's and theoretical work in their courses. It was found that eating and physical activities of University students are presumed to change over their University experience because there are different obligations with the academic levels and these obligations influence students' consumption /eating behaviour [15]. Reasons that influence food choices amongst college students are lifestyle changes such as academic obligations and convince of food due to lifestyle changes where students are academically pressured and eat a lot of convenient meals such as processed cold meats [17].

3.9. Non Significant Parameters
Findings further revealed that parameter for age, religion, extra-curricular activities and their influence of consumption were statistically insignificant and had no relationship with the consumption of processed cold meats. However, the research conducted by [2] revealed that older respondents tend to perceive food safety as an important factor in consumption and purchasing by consumers. From the study by [2], the parameter for age was statistically significant as it was a determinant in the response to food safety by individuals. [4] supported this view by stating that concerns about food risks tend to increase with age which made age a significant variable in studying public perceptions to food risks.

4. Conclusions and Recommendations

The results obtained from this study showed that there are socio-demographic factors which influence the consumption of processed cold meat by students. From the study parameters such as gender, income, residence, regulatory meal time, meal preferences, price of alternatives and academic level have been identified to influence consumption of prospected cold meat before, during and after a disease outbreak while other parameters were redundant and not statistically significant.

The research findings of this study further highlight that, civil society is concerned about food-related diseases and that some individuals still make decisions on food consumption, preparation and purchase which are not deemed healthy and safe. For University students who are considered to be more alert and cautious in this regard the study showed that this is not always the case due to the lifestyles which they lead when away from home and the pressure of having to be independent, they overlook some things.

Department of health and food processing companies should design, formulate and implement special public education programmes with the aim of informing consumers the importance and significance of proper food handlings practices to avoid contamination of food by bacteria and other microbes.

Government and food safety policymakers must try and to at least help consumers understand the information given on food handling practices that must be adopted to ensure on the using simple words or terminologies and help consumers understand at least basic and advanced nutritional aspects of food and thus consider food safety and handling important information to practice as daily routines before consumption of a food product.

Policymakers of food safety should work on and try to eliminate reported difficulties that are encountered by consumers in adapting to changes in consumption and purchase of recalled food during an outbreak like being product /brand specific and clear on what can and cannot be consumed by making sure that information regarding the phenomenon is clear, simple and reaches most consumers so that they do not rely on rumours as a lot of confusion was caused by the recall that students also lost faith in other brands due to the value chain process where contamination occurs.

Phenomenon such as food recalls and disease outbreaks not only affect the producer negatively but also affects the lifestyles of consumers. The costs borne by the consumer is a shift in their overall lifestyle due to time and expenses lost from the outbreaks. Consumption is a primary activity which all consumers engage in and thus need to be taken seriously in terms of information dissemination.

5. References

[1] Nelis, Delphine, Fanny Weytens, and Pauline Dupuis. 2011. Increasing Emotional Competence Improves Psychological and Physical Well-Being, Social Relationships, and Employability. 11(2): 354–66.

[2] Baker, G.A (2003). Food safety and fear: Factors affecting consumer response to food safety risk. International Food and Agribusiness Management Review, Vol 6, Issue 1.

[3] Leroy, F .and Degreef, F. (2015). Convenient meat and meat products. Societal, Technological Issues. Appetite, Vol 94, Pages 40-46.

[4] Meagher, K.D., 2018. Public perceptions of food-related risks: a cross-national investigation of individual and contextual influences. Journal of Risk Research, pp.1-17.

[5] Fitzgerald, A, Henry, C and Kelly, C. (2008). Factors influencing the food consumption ofchildren and adolescents: a qualitative investigation. Proceedings of The Nutrition Society, Vol 67, Issue No OCE7.

[6] Frey, L,. R., Carl, H., B., and Gary, L., K. (2000), Investigating Communication: An introduction to research methods, (2nd ed). Boston: Allyn and Bacon.

[7] Baiardi, D., Puglisi, R. and Scabrosetti, S., 2016. Individual attitudes on food quality and safety: Empirical evidence on EU countries. Food quality and preference, 49, pp.70-74.

[8] Liu, P. and Ma, L., 2016. Food scandals, media exposure, and citizens' safety concerns: A multilevel analysis across Chinese cities. Food Policy, 63, pp.102-111.

[9] Racette, S.B., Deusinger, S.S., Strube, M.J., Highstein, G.R. and Deusinger, R.H., 2005. Weight changes, exercise, and dietary patterns during freshman and sophomore years of college. Journal of American college health, 53(6), pp.245-251.

[10] Deliens, T., Clarys, P., De Bourdeaudhuij, I. and Deforche, B., 2014. Determinants of eating behaviour in university students: a qualitative study using focus group discussions. BMC public health, 14(1), p.53.

[11] Deshpande, S., Basil, M.D. and Basil, D.Z., 2009. Factors influencing healthy eating among college students: An application of the health belief model. Health marketing quarterly, 26(2), pp.145-164.

[12] Gaskell, G., Hohl, K. and Gerber, M.M., 2017. Do closed survey questions overestimate publicperceptions of food risks?. Journal of Risk Research, 20(8), pp.1038-1052.

[13] Dosman, D.M, Adaowics, W.L, and Hrudey, S.E. (2001). Socio-economic determinants ofhealth and food safety-related risk perceptions. Risk Analysis, Vol 21, No 2.

[14] Deric, D.H., Godfrey, E. and Quaicoe, O., 2017. Food Sources and Preferences of College Students. European Scientific Journal, ESJ, 13(10).

[15] Driskell, J.A., Kim, Y.N. and Goebel, K.J., 2005. Few differences found in the typical eatingand physical activity habits of lower-level and upper-level university students. Journal of the American Dietetic Association, 105(5), pp.798-801.

[16] Plotnikoff, R.C., Costigan, S.A., Williams, R.L., Hutchesson, M.J., Kennedy, S.G., Robards, S.L., Allen, J., Collins, C.E., Callister, R. and Germov, J., 2015. Effectiveness of

interventions targeting physical activity, nutrition and healthy weight for university and college students: a systematic review and meta-analysis. International Journal of Behavioral Nutrition and Physical Activity, 12(1), p.45.

[17] Lupi, S., Bagordo, F., Stefani, A., Grassi, T., Piccinni, L., Bergamini, M. and Donno, A.D., 2015. Assessment of lifestyle and eating habits among undergraduate students in northern Italy. Annali dell'Istituto superiore di sanita, 51, pp.154-161.

Determination of potential level of Indonesian rhizomes as an antioxidant based on phenolic compound and antioxidant activity

D A Sandrasari, M Sabariman and I N Azni

Department of Food Technology, Faculty of Food Technology and Health, Sahid University, Indonesia
E-mail dinysandra@gmail.com

Abstract. This study was aimed to determined of the potential level of rhizomes from Indonesia as an antioxidant. The calculated of potential level rhizome based on the total phenolic compound and antioxidant activity.The phenolic compound was determined using the Folin-Ciocalteau method, antioxidant activity was determined by the DPPH (1-diphenyl-2-picrylhydrazyl) method expressed in IC50. Potential level was calculated by comparing the value of total phenolic compound and antioxidant activity. The results showed that red ginger (*Zingiber officinale* var Rubrum) had the highest a phenolic and flavonoid content. Therefore that red ginger, *Zingiber officinale* var Rubrum is the most potential level as an antioxidant followed by *Curcuma longa*, *Curcuma zedoaria*, *Curcuma zanthorrhiza*, and *Curcuma aeruginosa* Roxb.

1. Introduction

Plants such as herbs have long been used in traditional medicine in various cultures throughout the world. Rhizome is one of the traditional medicinal plants that have been used in Indonesia for treating diabetes, high blood pressure, cancer, and many other illnesses. In Indonesia, many of rhizomes are often commonly provide as a medicinal traditional to health-promoting effects to relieve certain illnesses such as nausea, motion sickness, stomach-ache, asthma, diarrhoea, digestive disorder, vomiting, rheumatism, swelling, common cold and cough. Many phytochemicals of these rhizomes have been reported to have interesting biological activities include antifungal, antioxidant, and anti-inflammatory activities.

Free radicals or reactive oxygen species (ROS) during metabolism and other activities beyond the antioxidant capacity of a biological system gives rise to oxidative stress [1]. Antioxidants are substances or compounds that inhibit oxidation. Antioxidants are having scavenging power of free radicals produced inside the human body which are responsible for many of the metabolic disorders such as atherosclerosis, arthritis, cancer, and gastrointestinal disorders. [2]. In oxidation processes happening in biochemical reactions inside the human body, free radicals are produced and enzymes such as superoxidase dismutase, catalases and hydroperoxides act as natural antioxidants against them [3]. When antioxidants are consumed or fortified in food, they further slowdown the oxidation processes by fixing the stray electrons which are responsible for causing oxidative stress leading to cellular damage resulting in aging and various diseases such as cancer, diabetics, rheumatoid arthritis, cardiovascular diseases, chronic inflammation, and stroke [4].

Phenolic compounds from plants belong to a class of bioactive components with antioxidant activities [5-7]. The antioxidant activity of plants has been demonstrated in many recent studies [8,9]. Flavonoids represent one of the most studied classes of phenolic compounds containing carbohydrate units important for their biological activities [10]. Flavonoids exhibit a wide range of biological effects (antibacterial, antiviral, anti-inflammatory and anti-allergic) by reducing lowdensity lipoproteins in plasma, inhibiting platelet aggregation, scavenging free radicals, and preventing cell proliferation [11].

Ginger contains a component of flavonoids which can increase prostaglandin which is a defensive factor of the stomach. In addition, ginger also contains acetone and methanol can protect the stomach by reducing stomach acid and prevent irritation of the gastric mucosa. Curcuma rhizome contains active compounds including terpenoids, alkaloids, flavonoids, phenols and curcuminoids which can function as antioxidants and antimicrobials. Turmeric as an anticholesterol, anti-cancer, antimutagenic, white turmeric functions as an antioxidant, anti-inflammatory, anticancer, antibacterial and immunomodulatory. The objective of this work was to estimate the total phenolic and flavonoid content from five rhizome from Indonesia and to determined the level of potential plant species of rhizomes as an antioxidant agent.

2. Materials and Methods

2.1. Plant Material

Fresh rhizomes of *Zingiber officinale* var Rubrum, *Curcuma longa*, *Curcuma zedoaria*, *Curcuma Curcuma zanthorrhiza,* and *Curcuma aeruginosa* Roxb were purchased from Bogor, Indonesia (table 1). The rhizomes were washed, sliced and dried in a hot air oven at 50°C for 72 h.

Table 1. Botanical name, common name and traditional uses.

Botanical name	Common name	Traditional uses	References
Zingiber officinale	Red ginger	Arthritis, pneumonia, tuberculosis, tootache, cough, bronchitis, diarrhea, headaches	[12]
Curcuma longa	Tumeric	Hematologic, flatulence, infection	[13]
Curcuma zedoaria	White tumeric	Carminative, blood purifier, infection, fever, and heart disease	[14]
Curcuma xanthorrizha	Ginger	Stomach disorders, diarrhea, fever, liver damage	[15]
Curcuma aeruginosa Roxb	Black tumeric	Cough, asthma, stomach disorders, intestinal, antihelmintic	[16]

2.2. Solvent and chemicals

All solvents used were of analytical grade. Folin-Ciocalteu reagent, sodium carbonate, trichloroacetic acid, ethanol (EtOH), ascorbic acid, 1,1-Diphenyl-2-picrylhydrazyl (DPPH), gallic acid, and sodium chloride were purchased from Sigma-aldrich (Germany).

2.3. Extraction and yield of samples

The equal amount of different rhizomes powder (20g) was extracted with EtOH (200mL) for 2 days at room temperature (25°C) with continuous magnetic stirring to prevent oxidation by air and shielding from sunlight. The extraction was carried out so that the components were completely extracted and not oxidized. Then the solutions were filtered through double filter paper (Whatman™ No. 1). Fresh EtOH was added into the used plant material and the process was repeated three times. The filtered solutions were dried on a rotary evaporator under reduced pressure at 40°C. The yield of all extracts was recorded and kept in the refrigerator at 4°C for experimental analyses.

2.4. Total phenolic contents

Total phenolic content was estimated as gallic acid equivalents (GAE) per gram of dried plant extract, according to the Folin-Ciocalteu phenol reagent method [17]. First, a standard curve was generated using gallic acid as a standard. Different concentrations of gallic acid were prepared in 80% methanol, and their absorbance values were measured at 765 nm. For sample measurement, 0.5 mL (1/10 dilution) of Folin-Ciocalteu phenol reagent and 1000 mL of distilled water were added to 100 μL of plant extract. The solutions were mixed and incubated at room temperature for 1 min. After 1 min, samples were combined with 1500 mL of 20% sodium carbonate (Na_2CO_3) solution, mixed, and incubated for an additional 120 minutes. Absorbance at 765 nm was measured. Data presented are average values of four measurements for each sample.

2.5. Total flavonoid contents
Total flavonoid contents (TFC) were estimated according to the method [18]. Quercetin was used to make a calibration curve. 100μL of samples (1000μg/mL) were put in the microplate and 100μL of 2% $AlCl_3$ was added. The reaction was mixed and stand at room temperature for 15min. The absorbance of the mixture was measured at 430 nm by using a Spectrophotometer. The TFC was expressed as quercetin equivalents in mg per g extract. The experiment was carried out in triplicate.

2.6. DPPH radical scavenging assay
The DPPH radical scavenging assay was estimated according to the procedure [19]. 40μL of DPPH and 80μL of sodium acetate buffer (0.1M, pH=5.5) were added into 80μL samples at different concentrations (10, 25 and 50μg/mL). The mixture solution was incubated at room temperature in the dark for 30min. The absorbance was measured at 517nm by using a Spectrophotometer (Winooski, USA). Trolox was used as positive control in this assay. The % DPPH radical scavenging activity was calculated using the formula:

$$\% \text{ scavenging capacity } (A_{control} - A_{sampel})/A_{control} \times 100 \quad (1)$$

where control is the absorbance of control without test sample and sample is the absorbance of the sample. The experiment was carried out in triplicate. The % radical scavenging activity was plotted against the corresponding extract concentration to obtain the IC_{50} value.

2.7. Determination of potential level
Determination of potential level was calculated based on phenolic content and antioxidant activity using the formula :

$$\text{Potential level} = \text{total phenolic content} / \text{antioxidant activity} \quad (2)$$

where antioxidant activity asumsed has a same density of extract. Rhizomes that have the highest potential value are expressed as the most potential plants to overcome digestive disorders.

3. Results and discussion
3.1. Total Phenolic and flavonoid content
Total phenolic and flavonoid of rhizomes extract are shown in Table 2. The amount of total phenol was determined with the Folin-Ciocalteu reagent. Gallic acid was used as a standard compound and the total phenols were expressed mg GAE/g extract. Zingiber officinale had the greatest levels of phenolic content (400.20 mg GAE/g extract) whereas Curcuma aeruginosa had the lowest level of phenolic (51.49 mg GAE/g extract). This value from this research is significantly lower in compared to research of [20] that polyphenolic content in ginger extract nearly 2 times (871 mg GAE/g ginger extract) higher than our sample. Composition and quantity of the phenolic are vary significantly according to different intrinsic and extrinsic factors, including plant genetics and cultivars, soil and growing conditions, maturity state and harvest conditions. Extracted process also has a significant effect on the composition and properties of the final extract.

The amount of flavonoid content was determined with the Quercetin reagent. Quercetin was used as a standard compound and the total flavonoid were expressed as mg QE/g extract. Similar trend with total phenolic that was observed in total flavonoid. The maximum flavonoid content was found in the ethanol extract of *Zingiber officinale* (400.20 mg GAE/g extract) and the minimum also was found of *Curcuma aeruginosa* (33.36 mg QE/g extract).

Table 2. Total phenolic and flavonoid content of rhizomes.

Sample	Total phenolic content (mg GAE/g extract)	Total Flavonoid content (mg QE/g extract)	Antioxidant Activity, IC_{50}
Zingiber officinale	400.20 $\pm$ 0.1	268.20 $\pm$ 0.02	3.96 $\pm$ 0.02
Curcuma longa	285.00 $\pm$ 0.5	79.36 $\pm$ 0.01	24.06 $\pm$ 0.05
Curcuma zedoaria	195.10 $\pm$ 0.3	58.69 $\pm$ 0.03	40.00 $\pm$ 0.03
Curcuma xanthorrizha	139.16 $\pm$ 0.2	101.60 $\pm$ 0.05	167.03 $\pm$ 0.04
Curcuma aeruginosa Roxb	51.49 $\pm$ 0.6	33.36 $\pm$ 0.02	406.52 $\pm$ 0.02

3.2. Antioxidant activity

In this research, the antioxidant activity of ethanol extracts of rhizomes were investigated by using DPPH radical scavenging assay. The DPPH antioxidant assay is based on the ability of DPPH, a stable free radical, to decolourize in the presence of antioxidants. The colour of DPPH radical will turn to yellow from purple when the odd electron of DPPH radical becomes paired with hydrogen from antioxidant compound. DPPH has the advantage of being unaffected by certain side reactions, such as metal ion chelation and enzyme inhibition brought by various additives. A freshly prepared DPPH solution exhibits a deep purple colour generally disappears when antioxidants present in the medium. Thus the antioxidants present in the extract can quench DPPH free radicals by donating hydrogen atom or by electron transfer and convert them to the colourless product (2,2-diphenyl-1-picrylhydrazyl) or a substituted analogous hydrazine, resulting in a decreasing absorbance at the 517 nm. The antioxidants present in the rhizome extract can acts as radical scavengers may protect the cells against various diseases such as cancer, neurodegenerative and digestive disorder.

The result of antioxidant activity from five rhizomes extract are tabulated in Table 2. A low IC_{50} value indicates a strong ability to scavenge free radicals. Z. officinale exhibited the highest antioxidant activity both assays, with IC_{50} values of 3.96 μg/mL (DPPH) and the lowest antioxidant activity was Curcuma aeruginosa Roxb (406.52 μg/mL).

The relationship between antioxidant activity and total phenolic content was found strong correlation. It has been reported that phenolic compounds are responsible for antioxidant activity due to their capacity by donating electrons or a hydrogen atom to free radical. The high value of total phenol in Z. officinale with low antioxidant activity showed that Z. officinale had high antioxidant activity.

3.3. Determination of potential level

Determination of potential level of rhizomes as an antioxidant agent calculated by comparing phenolic content with antioxidant activity. The rhizome that has a highest value showed the most potential rhizome as antioxidant agent. The result of this research presented in table 3.

Table 3. Potential level of rhizomes as an agent antioxidant.

Sample	Total phenolic content (mg GAE/g extract)	Antioxidant Activity, IC_{50}	Potential level
Zingiber officinale	400.20 $\pm$ 0.1	3.96 $\pm$ 0.02	101.06
Curcuma longa	285.00 $\pm$ 0.5	24.06 $\pm$ 0.05	11.85
Curcuma zedoaria	195.10 $\pm$ 0.3	40.00 $\pm$ 0.03	4.87
Curcuma xanthorrizha	139.16 $\pm$ 0.2	167.03 $\pm$ 0.04	0.83
Curcuma aeruginosa Roxb	51.49 $\pm$ 0.6	406.52 $\pm$ 0.02	0.13

Table 3 presented that *Z. officinale* has a highest potential level followed by *C. longa, C. zedoaria, C. xanthorrizha* and the lowest is *C. aeruginosa*. Based on this research known that *Z. officinale* is rhizome that is most potential as an agent of antioxidant. It is known that the antioxidant activity of plant extracts containing polyphenol components is due to their capacity to be donors of hydrogen atoms or electrons and to capture the free radicals. The ginger extract showed antioxidant effect in inhibiting DPPH radical, IC_{50} was 3.96 μg/ml.

4. Conclusion

In the present study showed that red ginger (*Zingiber officinale* var Rubrum) had the highest phenolic and flavonoids content so *Zingiber officinale* var Rubrum is the most potential level as an antioxidant, followed by *Curcuma longa, Curcuma zedoaria, Curcuma zanthorrhiza*, and *Curcuma aeruginosa* Roxb.

5. References

[1] Zheng W, Wang, S Y. 2001. Antioxidant activity and phenolic compounds in selected herbs. *J. Agric. Food Chem.* **49** 5165–5170

[2] Astley S B. 2003. Dietary antioxidants past, present and future. *Trends Food Sci. Technol.* **14**, 93–98

[3] Ighodaro O M and Akinloye O A. 2018. First line defence antioxidants-superoxidase dismutase (SOD, catalase (CAT), and glutathione peroxidase (GPx) : Their fundamendal role in the entire antioxidant defence grid. *Alexandria Journal of Medicine*. **54** 287-293

[4] Shanmugam S, Usha Rani V, Pradeep B V. 2018. Antioxidant activity of rhizome extract of *Coleus forskohlii* Briq. *Asian journal of pharmaceutical and clinical research.* Vol11(11) 275-279

[5] Kirca A and Arslan E. 2008. Antioxidant capacity and total phenolic content of selected plants from Turkey, Int. *J. Food Sci. Technol.* **43** 2038-2046

[6] Ivanova D, Gerova D, Chervenkov T, Yankova T. 2005. Polyphenols and antioxidant capacity of Bulgarian medicinal plants. *J. Ethnopharmacol.* **96** 145-150

[7] Wojdylo A, Oszmianski J, Czemerys R. 2007. Antioxidant activity and phenolic compounds in 32 selected herbs. *Food Chem.* **105** 940-949

[8] Zheng W and Wang SY. 201. Antioxidant activity and phenolic compounds in selected herbs, *J. Agric. Food Chem.* **49**, 5165-5170

[9] Shan B, Cai Y Z, Sun M, Corke H. 2005. Antioxidant capacity of 26 spice extracts and characterization of their phenolic constituents, *J. Agric. Food Chem.* **53** 7749-7759

[10] Yang X, Sun Y, Xu Q, Guo Z. 2006. Synthesis and immunosuppressive activity of L-rhamnopyranosyl flavonoids. *Org. Biomol. Chem.* **4** 2483-2491

[11] Pairul, Susianti, Nasution H.S. 2017. Jahe (Zingiber Officinale) Sebagai Anti Ulserogenik. *Medula*. **7**(5). 42-46

[12] Chen C Y, Cheng K C, Chang A Y, Lin T Y, Hseu Y C, Wang H M. 2012. 10-Shogaol, an Antioxidant from Zingiber officinale for Skin Cell Proliferation and Migration Enhancer. *Int. J. Mol. Sci.* **13** 1762-1777.

[13] Naz S, Jabeen S, Ilyas S, Manzoor F, Aslam F, Ali A. 2010. Antibacterial activity of C longa varieties against different strains of bacteria. *Pakistan Journal of Botany*. **42**, 455 – 462.

[14] Himaja M, Ranjitha A, Ramana M V, Anand M, Asif K. 2010. Phytochemical screening and antioxidant activity of rhizome part of *Curcuma zediaria. International Journal of Research in Ayurveda & Pharmacy*. **1**(2). 414-417

[15] Sasikumar B. 2005. Genetic resources of Curcuma: diversity, characterization and utilization. *Plant Genetic Resources* **3**, 230- 251.

[16] Nasrullah I, Murhandini S, Rahayu WP. 2010. Phytochemical Study from Curcuma aeruginosa Roxb. Rhizome for Standardizing Traditional Medicine Extract. *Journal of International Environmental Application & Science*. **5**, 748-750.

[17] Katsube T, Tabata H, Ohta Y, Yamasaki Y, Anuurad E, Shiwaku K. 2004. Screening for antioxidant activity in edible plant products: comparison of low-density lipoprotein oxidation assay, DPPH radical scavenging assay, and Folin− Ciocalteu assay, *J. Agric. Food Chem.* **52**, 2391-2396

[18] Stoilova I, Krastanov A, Stoyanova A, Denev P, Gargova S. 2007. Antioxidant activity of a ginger extract (Zingiber officinale). *Food Chem.* **102**, 764-770.

[19] Li W, Gao Y, Zhao J, Wang Q. 2007. Phenolic, flavonoid, and lutein ester content and antioxidant activity of 11 cultivars of Chinese marigold. *J. Agric. Food Chem.* **55**, 8478-8484

[20] Gupta A. D, Vipin K. B, Vikash B, Nishi M. 2013. Chemistry, antioxidant and antimicrobial potential of nutmeg (*Myristica fragrans* Houtt). *Journal of Genetic Engineering and Biotechnology*. **11**, 25 -31

Acidity dinamic in low-fat yogurt with the addition of potato starch modified by bacterial amylases

E Nikitina[1] and R A Riyanto[1,2]*

[1]Institute of Food Technology and Biotechnology, Department of Meat and Milk Technologies, Kazan National Research Technological University, K. Marks Str. 68, Kazan 420015, Russian Federation
[2]Department of Food Technology, Faculty of Agriculture, University of Sultan Ageng Tirtayasa, Jl. Raya Jakarta km. 04 Pakupatan, Serang City, Banten Province, Indonesia

*Corresponding author: rifqi.ar@untirta.ac.id

Abstract. Potato starch was modified by two multienzyme bacterial preparation of Amylosubtilin® and α-amylase of *Bacillus licheniformis* at various enzyme concentrations. Eight types of modified starch were obtained and further applied to the mixture of low-fat yoghurt. In this research, we examined the acidity dinamic throughout the fermentation process of low-fat yogurt. It was found that the addition of modified potato starches with the concentration of 2% (w/w) did not affect the acidity dinamic compared to the native starch.

1. Introduction
Yogurt is well known for its various health benefits for human. Commonly the fat content of yogurt somewhere around 3.25% while another types of yogurt are low fat (0.5% to 2% milk fat) and nonfat (less than 0.5% milk fat). Lesser fat content is more preferred due to health concerns [1]. The loss of fat content in yogurt resulted in undesireable characteristics that might be affecting the acceptance of yogurt to the consumer. Those characteristics for instance are lack of firmness, difference in taste, flavor and aroma [2]. One of the solution to overcome the undesireable characteristics of low fat yogurt is by adding the fat replacer. While decreasing the calorific value of food, fat replacer can be used to solve some problems of physical and organoleptic originating from low fat final products [3]. Fat replacers consist of mixtures of lipid-originated fat substitutes, protein- or carbohydrate-originated fat mimetics, or their combinations [4]. In this research we used potato starch modified by bacterial amylases as fat replacer in low fat yogurt. Previous study [5] had revealed the effect of these modified starches in low fat yogurt on textural and sensory characteristic while in this paper we focused on the modified starches addition effect on the acidiy dinamic during fermentation process.

2. Materials and Methods
2.1. Enzyme preparation
Amylosubtilin and α-amylase of *B. licheniformis* preparation and amylase activity detection are described in the earlier studies [6].

2.2. Starch modification
Native potato starch (according to GOST P52791-2007 standard) was used for the research. Modification of these starches were based on the application of amylase of *B. licheniformis* (Bl) and Amylosubtilin® (AM) of different concentrations (Table 1). Modification was carried out in water (30

g/100 ml), pH = 7, 40°C for 1 hour. The hydrolysis reaction was stopped with the addition of concentrated sulfuric acid (pH = 2). The starch was afterwards separated from the liquid by filtration and dried at 40°C.

Table 1. Code of modified starches

Starches	Amylase activity (U/g starch)	Amylosubtilin (g/100 ml reaction mixture)	Amylase *B. licheniformis* (ml/100ml reaction mixture)
AM-0.05	0.415	0.01	-
AM-0.1	0.83	0.0201	-
AM-0.05	4.15	0.1005	-
AM-1	8.3	0.201	-
BI-0.05	0.415	-	0.05
BI-0.1	0.83	-	0.1
BI-0.5	4.15	-	0.5
BI-1	8.3	-	1

2.3. Yogurt preparation
Pasteurized cow's milk (according to GOST 31450-2013 standard) subjected to centrifugation at 3000 g. Analysis of the physicochemical properties of milk was tested on a Klever-2M quality analyzer, according to the method recommended by the manufacturer. *Lactobacillus delbrueckii subsp. bulgaricus* strain 51 (isolated from Gastrofarm®), *Bifidobacterium* BB-12 ™, and *Streptococcus thermophilus* TH-4 (isolated from Bifiform Baby®) were used as the yogurt starter. Overnight yoghurt culture was grown on low-fat milk at 37°C for 16 hours. Milk without starch was used as a control, starch was added to the test samples at a concentration of 2% (w/w), the milk was heated to gelatinize the starch with constant stirring. After cooling the milky-starch mixture, the starting culture of yogurt was added to it in the amount of 5% (v/v). Fermented milk was incubated at 37 ° C.

2.4. Acidity analysis
Measurements of pH were carried out by using a digital pH meter (HI 2216, Hanna Instruments, Germany).

3. Results and Discussion

Table 2. Initial chemical composition of the milk

Chemical composition	Concentration (%)
Fat	1.71
Protein	3.31
Lactose	4.6
Salt	0.74

Chemical composition of the milk used for the fermentation was shown in Table 2. The fat content was 1.71% thus the milk was considered low fat. Also the milk had sufficient amount of protein and lactose as the starting capital for the bacteria to ferment. It is already well known that lactose degradation of LAB resulted in lactic acid primarily and other acids [7].

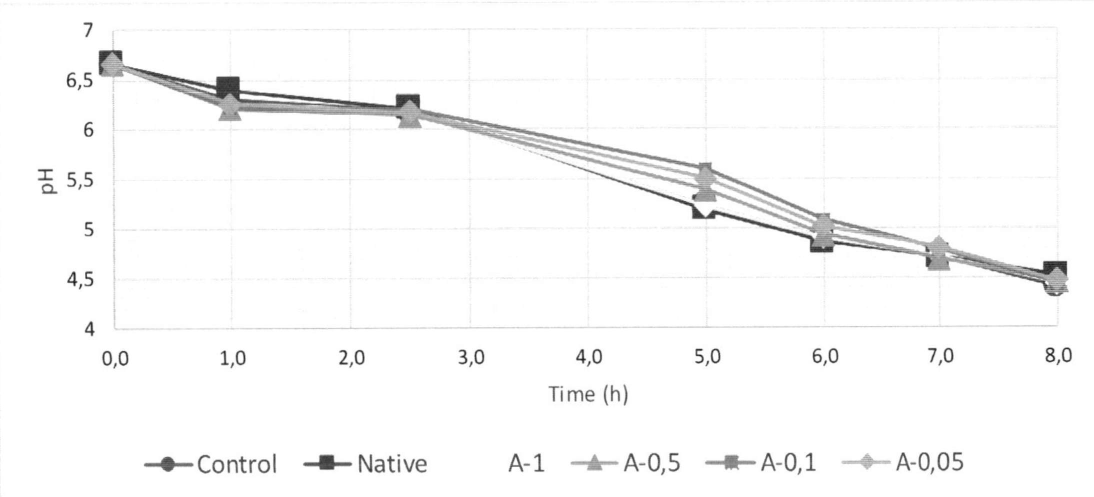

Figure 1. Acidity dinamic of low-fat yogurt with addition of amylosubtilin-modified starches.

All of the treatments' pH were around 6.6 as shown in Figure 1. As the fermentation progressed, the pH decreased for all treatments. In the fifth hour of fermentation there are slight differences in the pH of several treatments, the lowest pH was native starch (5.19) and the highest pH was A-0,1 (5.59). This could be caused by the difference of amylose and amylopectin ratio on the starches based on previous study [6]. Since some LAB produce amylase enzyme [8], the difference of amylose content in the starches might differentiate the time and concentration of acid production. Eventually all the treatments of amylosubtilin-modified starches resulted in pH of 4.5 at the 8th hour of fermentation.

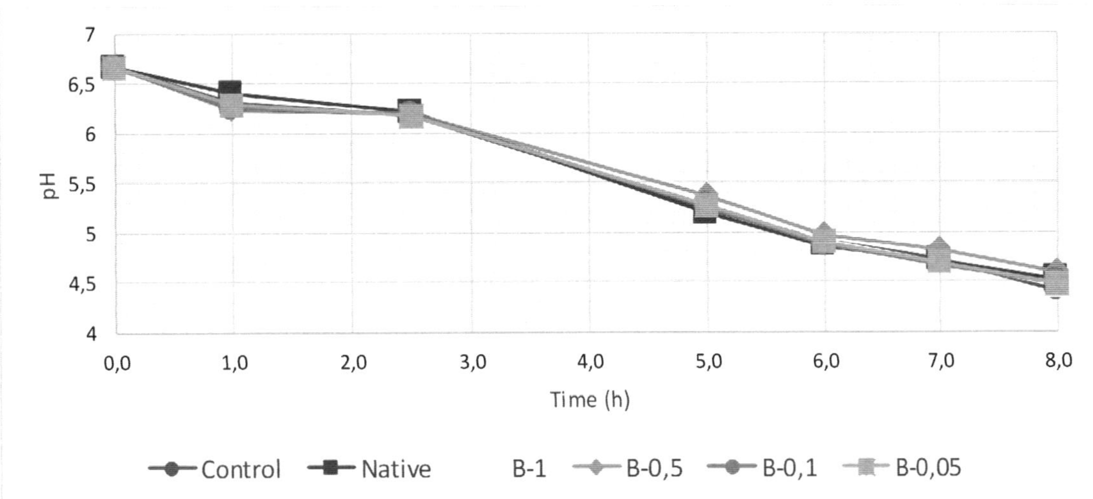

Figure 2. Acidity dinamic of low-fat yogurt with addition of amylase-modified starches.

Compared to previous treatments of amylosubtilin, the treatments of amylase-modified starches shown in figure 2 did not present any significant differences. All of the treatments' final pH are around 4.5 at the 8th hour.

4. Conclusion

The addition of modified potato starches with the concentration of 2% (w/w) did not affect the acidity dinamic during fermentation process of low-fat yogurt compared to the addition of native starch. The fermentation time for the low fat yogurt with the addition of bacterial modified starches is eight hours.

5. References

[1] Routray W and Mishra H N 2011 Scientific and Technical Aspects of Yogurt Aroma and Taste: A Review *Comprehensive Reviews in Food Science and Food Safety* **10** 208-20

[2] Desay N T, Shepard L, and Drake M A 2013 Sensory properties and drivers of liking for Greek yogurts *J. Dairy Sci.* **96** 7454–66

[3] Guven M, Yasar K, Karaca O B, and Hayaloglu A A 2005 The effect of inulin as a fat replacer on the quality of set-type low-fat yogurt manufacture *International Journal of Dairy Technology* **58** (3) 180-4

[4] Huyghebaert A, Dewettinck K and de Greyt W 1996 Fat replacers. In Fat Replacers-Ripening and Quality of Cheese *IDF Bulletin* **317** 10 –5. Brussels : International Dairy Federation

[5] Nikitina E, Riyanto R A, Vafina A, Yurtaeva T, Tsyganov M, and Ezhkova G 2019 Effect of Fermented Modified Potato Starches to Low-fat Yogurt *Journal of Food and Nutrition Research* **7** (7) 549-53

[6] Vafina A, Proskurina V, Vorobiev V, Evtugin V G, Egkova G, and Nikitina E 2018 Physicochemical and Morphological Characterization of Potato Starch Modified by Bacterial Amylases for Food Industry Applications *Journal of Chemistry* **2018**

[7] de Vos W M and Vaughan E E 1994 Genetics of lactose utilization in lactic acid bacteria *FEMS Microbiology Reviews* **15** 217-37

[8] Padmavathi T, Bhargavi R, Priyanka P R, Niranjan N R, Pavitra P V 2018 Screening of potential probiotic lactic acid bacteria and production of amylase and its partial purification *J Genet Eng Biotechnol* **16** (2) 357–62

6. Acknowledgement

This work was supported by Russian Federation Government Scholarship.

Paddy rice farming diversification with *Trigona sp.* in Pandeglang regency, Banten province (case study: diversification income analysis of paddy rice business with bee cultivation)

Elpawati*, U M D A Hudaya, F M Habibie

Department of Agribusiness, Faculty of Science and Technology, Islamic State University of Syarif Hidayatullah Jakarta
Jl. Ir.H. Juanda 95 Ciputat, South Tangerang, Banten Province, Indonesia

*Corresponding author: elpawati@uin.jkt.ac.id

Abstract. The study is aimed to determine the total income and contribution income of Trigona sp. cultivation by paddy rice farmers in Pandeglang District, Banten Province. The study location was purposively selected sub-districts at Pandeglang, Majasari, and Menes. Seven farmers who had at least 30 units of honeycomb colony and managed maximum 0,3 hectare (ha) of paddy field was used as a respondent. Data were collected using questionnaire to obtain the information and analyzed using Microsoft Excel for data analysis. The results showed that bee cultivation diversification with paddy rice business has potentially make a good income. The contribution income of bee cultivation and paddy rice business in 1 year have a rate 59.46% or Rp. 13,291,857,00 and 40.54% or Rp. 8,872,257.00, respectively with rate total income Rp. 22,164,114.00. The highest paddy rice and bee cultivation income were 10,296,000.00 and 18,568,000.00, respectively obtained by same farmer.

Keywords: contribution income, *Trigona sp.* cultivation, diversification, paddy rice business, total income.

1. Introduction

Pandeglang is a regency of Banten province located on the west and south coasts of the island of Java. It has 54.739 ha land of paddies including flooded and dry field with total paddy production. Most famer lived in Pandeglang are small farmer. Therefore, most of them have a low income and very depend to paddy farming, resulted their prosperity are much lower than labor worker. It resulted to decline of farmer household as it seen in table 1 [1]. The diversification with other work, such as labor, fish farming, animal livestock, as alternate to increase their income and prosperity [2].

Table 1. Agriculture population census of small farmer in Banten 2003 and 2013

City/Regency	Census		Growth (%)
	2003	2013	
Pandeglang	111,833	86,000	-23,10
Lebak	104,810	115,758	10,45
Tangerang	199,467	61,928	-68,95
Serang	141,606	85,045	-39,94

City/Regency	Census		Growth (%)
	2003	2013	
Tangerang City	19,006	7,561	-60,22
CilegonCity	14,005	6,072	-56,64
Serang City	22,526	12,841	-42,99
Tangerang Selatan City	21,162	4,683	-77,87
Total	634,415	379,888	-40,12

One of potency to increase their income is diversification through Trigona cultivation. *T. laeviceps* and *T. itama* can be found easily in Banten. These bees have a same type which is can also be found in Sumatra, Bali, and Borneo [3][4][5]. The only difference is their size much smaller and can be found anywhere, even in the house [6].

Trigona sp. products have economic potency due much more profit than other product from other honey bee especially propolis [7][8]. Therefore, it is considered a good candidate to be applied as diversification, especially in Pandeglang because they have much small farmer and also to increase their income as a part to integrate between paddy farming and Trigona cultivation [9].

It is important to study diversification in Pandeglang Regencey to improve prosperity of small farmer. This study is aimed to learn about total income and contribution income from diversification of Trigona cultivation with paddy farming.

2. Method

The research was conducted in at selected districts in Pandeglang Regency, Banten: (1) Pandeglang, (2) Majasari, and (3) Menes from September to December 2018. These location was purposely selected as a sample location because they have biggest beekeeper as a data sources. Research conducted in the form of case study to learn about the diversification by small farmers which cultivated Trigona Sp as another sources of income. This study aimed to know the total income and contribution income cultivation by paddy rice farmers. Data was collected by exploration to obtain an information by questionnaire from respondent and analyzed using Microsoft Excel for data analysis. Data taken from respondent including: (1) time cultivation from start; (2) types of bee product that have been sold in one month; (3) amount of product that have been sold in one month; (4) price of each product (5) percentage of contribution income compared to main income from paddy farming; and (6) tools and utility for bee cultivation.

3. Result and Discussion

Seven farmers who had at least 30 units of honeycomb colony and managed maximum 0,3 hectare (ha) of paddy field was used as a respondent. The diverse of respondent can be seen below (table 2).

Table 2. Farmer respondent

No.	District	Respondent		
		Total	Name	Symbol
1	Menes	3	I.K	Farmer A
			A.S	Farmer B
			S	Farmer C
2	Pandeglang	2	G	Farmer D
			S	Farmer E
3	Majasari	2	M	Farmer F
			S	Farmer G

The data result can be seen at table 3.

Table 3. Contribution income of paddy rice farming diversification with Trigona cultivation in 1 year

No.	Farmer	Paddy income (IDR)	%	Trigona cultivation income (IDR)	%	Total income
1	A	9,282,800	38.40	14,885,000	61.59	24,167,800
2	B	7,374,000	39.77	11,164,000	60.22	18,538,000
3	C	7,098,000	35.79	12,733,000	64.21	19,831,000
4	D	9,094,000	39.87	13,716,000	60.13	22,810,000
5	E	10,296,000	35.67	18,568,000	64.33	28,864,000
6	F	9,828,000	41.74	13,716,000	58.26	23,544,000
7	G	9,133,000	52.51	8,261,000	47.49	17,394,000
	Rate	8,872,257	40.54	13,291,857	59.46	22,164,114

The table result showed that income contribution of diversification Trigona cultivating rice farmer varied each farmer. The average of total income was IDR 22,164,111 with 59.46% come from Trigona cultivation and the rest come from paddy farming. The highest paddy rice and bee cultivation income were IDR 10,296,000 and IDR 18,568,000, respectively, both obtained by farmer E. Farmer C from Menes also obtained high income from Trigona cultivation (64.21%) second only with farmer E, but lose to farmer A, D, and F in term of amount income. The only had contribution income below from 50% is farmer G, as the other farmer have contribution is above 55%.

The Income from Trigona cultivation above 50% is a good contribution. It means farmer has expectancy to increase their income through diversification. Cheap investment and operation affected their income, which resulted high income compared to rice farming. It could be due paddy farming has a high operational cost and can only harvested two times in one year, resulted paddy income much lower than Trigona income, which it can harvested 12 times or more in one year.

As the study result showed that Trigona income much higher than paddy come, diversification can be a good chance to be new alternate to increase their income and prosperity. It was showed that Trigona income reach 59.46% from total income. Another factor that affecting high income is price of honey and propolis. Both can be sold to higher price, about IDR 450,000/liter and IDR 50,000/kg, respectively.

As a comparison, research from Muyana and Hamzah [10] at Musi Rawas, Riau about contribution income from conversion of paddy farming to fish pond for pisciculture. The contribution income showed IDR 22,7333,333,33/0.47 ha in 8 months compared to paddy income 9,441,366,67/0.47 ha for 8 months. Filly [11] studied about the income from bee cultivaton *(apis cerana)* at Batanghari, Lampung showed contribution income of the beekeeper 27.71%. It much lower than other contribution income. It may due the honey price is much cheaper.

4. Conclusion
The contribution income of bee cultivation and paddy rice business in 1 year have a rate 59.46% or IDR 13,291,857,00 and 40.54% or IDR 8,872,257.00, respectively with rate total income IDR 22,164,114.00. The highest paddy rice and bee cultivation income were IDR 10,296,000.00 and IDR 18,568,000.00, respectively obtained by same farmer.

5. References
[1] Wulandari, *et al.* 2016. Peranan Lebah *Trigona laeviceps* (Hymenoptera: Apidae) dalam produksi Biji Kailan *(Brassica oleracea* var. *alboglabra).* Jurnal Agron Indonesia. **45(2)**: 197-204
[2] Budhi, Gelar Satya. 2010. Dilema Kebijakan dan Tantangan Pengembangan Diversifikasi Usahataani Tanaman Pangan. Jurnal Analisa Kebijakan Pertanian. **8(3)**: 241-258
[3] Dinas Lingkungan Hidup dam Kehutanan (DLHK) Kabupaten Pandeglang. 2017. Laporan Tahunan Lingkungan Hidup Kabupaten Pandeglang: Akarsari. DLHK Kabupaten Pandeglang, Pandeglang
[4] Bonawu, Hendri. 2016. Studi Perkembangan Koloni dan Produksi Lebah *Trigon asp.* Dari Posisi Stup yang Berbeda. Skripsi. Fakultas Kehutanan dan Ilmu Lingkungan. Universitas Haluoleo, Kendari

[5] Ramadani, Rosi Fitri. 2016. Keanekaragaman Polen dari Beberapa Spesies Stingless Bee pada Perkebunan Kelapa Sawit dan Karet. Thesis. Bogor.

[6] Salatnaya, Hearty. 2012. Produktivitas Lebah *Trigona* spp. sebagai Penghasil Propolis pada Perkebunan Pala Monokultur dan Polikultur di Jawa Barat. Thesis. Graduate School. Institut Pertanian Bogor. Bogor.

[7] Trubus Majalah. Muhammad Fajar Ramadhani (ed). 2017. Lebah Trigona: Tingkatkan Propolis 35%. Edisi 572/XLVIII. July 2017. p. 24.25.

[8] Reindra, Septiantna Dyah and Krisnawati. 2017. Produksi Propolis Mentah Kebah Madu *Trigona spp.* di Pulau Lombok. Jurnal Hutan Tropis **1(1)**:71-75.

[9] Indrawanto, Chandra, Atman. 2018. Integrasi Tanaman-Ternak Solusi Meningkatkan Pendapatan Petani. Badan Penelitian dan Pengembangan Pertanian. IAARD Press. Jakarta.

[10] Mulyana, E., and Hamzah, M. 2014. Kontribusi Pendapatan Usaha Perikanan Terhadap Pendapatan Rumah Tangga Petani Padi Sawah di Desa Kalibening Kecamatan Tugumulyo Kabupaten Musi Rawas. Proceeding. Seminar Nasional BKS PTN Barat: p. 833-840. Bandar Lampung. August 19-21, 2014.

[11] Filly, N.N. 2018. Kontribusi Usaha Budidaya Lebah Madu Terhadap Pendapatan dan Kesejahteraan Petani Lebah Madu Desa Buana Sakti, Kecamatan Batanghari, Kabupaten Lampung Timur. Undergraduate Thesis. Universitas Lampung. Bandar Lampung

Effect of organic matter of chicken manure with bioactivator application of microbial rhizosphere inoculant to growth and yield of soybean (Glycine max (L.) Merill)

Caskiwan[1], F R Eris[1]*, Nurmayulis[1], Y R Denny[2]

[1]Department of Agroecotechnology, Faculty of Agriculture, University of Sultan Ageng Tirtayasa, Jl. Raya Jakarta KM. 04 Pakupatan, Serang City, Banten-Indonesia
[2]Department of Physics Education, Faculty of Teacher Training and Education, University of Sultan Ageng Tirtayasa, Jl. Raya Jakarta KM. 04 Pakupatan, Serang City, Banten-Indonesia

*Corresponding author: fitria.eris@untirta.ac.id

Abstract. This research was aimed to know effect of organic matter of chicken manure with bioactivator application of microbial rhizosphere inoculant to growth and yield of soybean (*Glycine max* (L.) Merril). The research was conducted from February 2017 to Mei 2017, located at Panimbang District, Pandeglang Regency, Banten Province, Indonesia. This study used a randomized completely block design one factor that is giving of chicken manure matter with IMR bioactivator application consists of six levels. Thus, there are six treatments and each treatment was replicated four times as a group, so there are 24 experimental units. The results showed that the application of organic waste of chicken manure along with bioactivator application gave a significant effect on the number of leaves of soybean plant age of 5 week after planted and no significant effect on soybean plant height aged 5 and 6 week after planted, leaf number 6 week after planted, number of root nodule, number of pods of contents, number of empty pods, dry weight of seeds per plant, weight of 100 seeds, seed weight per plot and seed yield per hectare.

Keywords: bioactivator, chicken manure matter, microbial rhizosphere inoculant (IMR), soybean

1. Introduction

Soybean (*Glycine max* (L.) Merill) is a species of legume that plays an important role as the main vegetable protein source for the Indonesian people. According to [1], as a source of vegetable proteins, soybeans are generally consumed in the form of processed products such as tofu, tempeh, soy sauce, tauco, soy milk and various kinds of snacks. The needs of soybean are increasing, while the efforts to improve the production are still far from expectations which results in imports of soybeans.

In 2014, soybean production in Banten Province reached 6,384 tons with a harvested area reaching 4,815 ha and the average soybean productivity was 13.26 ku/ha [2]. Regarding the relatively low productivity of soybeans, there is a need to the improve cultivation techniques to increase soybean production to meet market needs.

One effort that can be done to overcome the problem of low soybean production is by the use of organic matters and bioactivators. The admission of fresh or undecomposed organic matters can cause nutrients contained in these organic matters cannot be absorbed by plants and also often contain pathogens, eggs and insects that can harm the plants [3].

Chicken manure is an organic fertilizer derived from solid feces and urine of chickens mixed between the leftover food and the henhouse base [4]. The use of organic chicken manure is due to its abundant availability so that it is easily available and also has a higher nutrient content than other livestock manure. Chicken manure contains the highest nutrients, especially for N, because the urine is mixed with the solid feces [5].

One type of biofertilizers resulted from research conducted by BATAN that has not yet been commercialized is the rhizosphere microbial inoculant (IMR) bioactivator which is a consortia inoculant of the irradiated compost-based plant growth enhancing rhizosphere microbes. The irradiated compost-based inoculants with bioactive content in the form of a combination of *Azotobacte*r sp., *Bacillus* sp. and *Trichoderma* sp. This is expected to be a part of efforts to develop microbial inoculants useful to improve health, growth and productivity of plants [6].

The results of the study conducted by [7] showed that the use of rhizosphere microbial inoculant as bioactivator could increase the effectiveness of cocoa pod compost as a nutrient provider for the growth of cacao seedlings and the use of IMR fertilizer as biofertilizers given once a week to better respond to growth of cocoa seedlings compared to without IMR fertilizer. The use of rhizosphere microbial consortia inoculants as *biofertilizer* was able to increase the height, dry weight of plant stover, weight and production of fresh cobs of sweet corn of 23,64%, 63,62%, 29,41% and 29,35% respectively [6].

The purpose of this study was to find out the effect of organic chicken manure with the application of rhizosphere microbial inoculant bioactivator on the growth and yield of soybean plants (*G. max* (L.) Merill).

2. Methods

This research was carried out from February to May 2017 in the Rice Field of Kampung Pematang Gintung, Mekarjaya Village, Panimbang Sub-District, Pandeglang Regency. The tools used in this study were hoes, sickles, sacks, nameplate, handsprayer, buckets, watering cans, scales, measuring cups, spikes, ruler, meters, scissors, plastic ropes, tarps, cameras and stationery. Furthermore, the materials used were Grobogan-variety soybean seeds, Broiler chicken manure, rhizosphere microbial inoculant bioactivator (IMR), sugar, husk, bran, Sidamethrin 50 EC insecticide and water.

This study used the single factor randomized block design. The factor studied was the admission of organic chicken manure with the application of the rhizosphere microbial inoculant bioactivator consisting of 6 levels, namely:

C1: Control
C2: Chicken Manure Bokashi
C3: Chicken Manure Bokashi + IMR watering on plants every week
C4: Chicken Manure Bokashi + IMR watering on plants every 2 weeks
C5: Chicken Manure Bokashi + IMR watering on plants every 3 weeks
C6: Chicken Manure Bokashi + IMR watering on plants every 4 weeks

Each treatment was repeated 4 times so that there were 24 experimental units. Each experimental unit consisted of 35 plants with a spacing of 40 cm x 20 cm, with an area of the experimental plot of 200 cm x 150 cm, the distance between treatment plots of 50 cm and the distance between groups of 100 cm. Each experimental plot was taken 10 plants randomly as sample.

The data obtained from the results of observations were then analyzed using 5% variance. If there is a difference, then a further test is carried out using the Duncan Multiple Range Test (DMRT) at the level of 5%.

The parameters observed in this study included plants aged 5 and 6 weeks, number of leaves aged 5 and 6 weeks, number of nodules, flowering age, number of filled pods, number of empty pods, dry weight of seeds per plant, weight of 100 seeds, seed weight per plot, and seed weight per hectare.

The research processes included land preparation, land processing and plot making, making chicken manure bokashi fertilizer, planting, application of IMR bioactivator, maintenance (planting, thinning, watering, weeding and pest and disease controlling) and harvesting.

3. Results and discussion

3.1. Plant Height

The results showed that at plant height aged 5 and 6 weeks the treatment of organic chicken manure followed by the application of IMR bioactivator in different soybean plants did not have significant effect. The average soybean plant height at the age of 5 and 6 weeks is presented in Table 1.

Table 1. The average of soybean plant height at the age of 5 and 6 weeks after the admission of organic chicken manure and IMR bioactivator

Treatment	Plant Height (cm)	
	5 Weeks	6 Weeks
C1	28,31	39,28
C2	29,30	40,26
C3	29,00	39,39
C4	28,55	38,90
C5	30,44	41,77
C6	29,76	39,26

Table 1 shows that soybean plants treated with organic chicken manure followed by IMR bioactivator watering did not have any effect on plant height. This is presumably because some important nutrients needed by plants are available in the soil. In addition, the nutrient content in bokashi from chicken manure is low and also slow release, so that the effect on soybean plant height is not significant.

Moreover, at the height of plant aged 6 weeks the effects had no significant difference among the treatments. Allegedly, it is because the treated soybean plants have entered the generative phase. According to [9], the reproductive stage of soybean plants starts from the flower formation to pod maturation. The entry to the generative phase results in more carbohydrates produced from photosynthesis for the process of seed forming and filling.

3.2. Number of Leaves

The analysis of number of leaves showed that the admission of organic chicken manure followed by the application of IMR bioactivator had only significant effect on the number of leaves of soybean plants aged 5 weeks and did not have significant effect on the number of leaves of soybean plants aged 6 weeks. The average number of soybean leaves is presented in Table 2.

Based on Table 2, the admission of organic chicken manure followed by IMR bioactivator watering increased the number of leaves of soybean plants aged 5 weeks. The average number of leaves was mostly found in the treatment of chicken manure bokashi by IMR watering in every 3 weeks (C5), but not significantly different from the treatment C2, C3, and C4. This is presumably that the chicken manure bokashi applied in addition to improving soil conditions is also capable of supplying nutrients needed by plants even in small amounts. According to [8], the important function of soil organic matter is to provide nutrients through the process of decomposition and mineralization, boosting the activity of microorganisms and land animals and binding of toxic elements to acid soil.

Table 2. The average of number of leaves of soybean plants at the age of 5 and 6 weeks resulted from the admission of organic chicken manure and IMR bioactivator

Treatments	Number of Leaves	
	5 Weeks	6 Weeks
C1	12,90 a	24,55
C2	14,45 c	29,13
C3	14,48 c	28,90
C4	14,35 c	27,78
C5	14,50 c	28,25
C6	13,90 b	27,43

Information: The numbers followed by the same lowercase letter on the same column showed insignificant difference in the DMRT test of 5%.

3.3. Number of Nodules

The results of analysis of variance of the variable number of nodules showed that the treatment of the admission organic chicken manure followed by the application of the IMR bioactivator on soybean plants has no significant difference. The average number of nodules of soybean plants can be seen in Table 3.

Based on Table 3, the effect of organic matter of chicken manure followed by IMR bioactivator watering on the number of nodules of soybean plants has no significant difference. Legume plants have a symbiotic relationship with bacteria from the genus *Rhizobium*. In their root nodules, bacteria fix the atmospheric nitrogen to ammonia, which plants assimilate into amino acids [10].

Rhizosphere microbial inoculant bioactivator (IMR) is a rhizospheric microbial consortia inoculant which enhances the growth of plants based on irradiated compost with bioactive content in the form of *Azotobacte*r sp., *Bacillus* sp. and *Trichoderma* sp [11]. *Azotobacter* is an N_2 fixing bacteria that is able to produce stimulating substances to grow gibberellins, cytokines and indole acetic acid, so that the utilization can stimulate root growth. Azotobacter are found in almost all soils and grow well in soils with a pH between 6-7, which is mineral rich and lacks of N [10].

3.3. Flowering Age

The results of analysis of variance of variable flowering age showed that the treatment of the admission of organic chicken manure followed by the application of IMR bioactivator on soybean plants was not significantly different. The average age of flowering of soybean plants can be seen in Table 3.

Based on table 3, the administration of bokashi followed by bioactivator watering has no significant effect on the flowering age of soybean plants. This is presumably because the flowering age is more influenced by genetic factors. It is in accordance with the opinion of [12] that the pattern of flowering of soybean plants depends on the variety.

3.4. Number of filled pods

The results of analysis of variance in the number of filled pods showed that the treatment of the admission of organic chicken manure followed by the application of IMR bioactivator in soybean plants has no significant difference. The average number of soybean pods can be seen in Table 3.

Table 3. The average of the number of nodules, flowering age, number of filled and empty pods of soybean plant after the admission of organic chicken manure and IMR bioactivator

Treatments	Number of Nodules	Flowering Age (days)	Number of filled pods	Number of empty pods
C1	46,95	36,25	59,18	8,45
C2	45,90	36,00	63,25	7,38
C3	45,10	36,25	58,48	7,83
C4	48,08	36,25	55,83	5,73
C5	44,45	36,00	58,98	6,90
C6	39,65	36,00	59,98	7,40

Based on Table 3, the effect of organic matter of chicken manure followed by IMR bioactivator watering on the number of nodules of filled pods has no significant difference. This is presumably because the high nutrient content in the soil causes the chicken manure bokashi has not given any significant effect. In addition, the bioactivator watering is also not optimal because of the occurrence of *leaching* due to high rainfall. High rainfall causes inhibition of mineralization (the release of organic nutrients into nutrients available to plants) in the process of decomposition of organic matter which results in disrupted growth and development of plants because they do not get the necessary nutrients [13].

3.5. Number of Empty Pods

The results of analysis of variance in the number of empty pods showed that the treatment of organic chicken manure followed by the application of IMR bioactivator on soybean plants has no significant difference. The average number of soybean pods can be seen in Table 3.

Based on Table 3, the effect of organic matter of chicken manure followed by IMR bioactivator watering on the number of nodules of empty pods has no significant difference. The number of pods per plant that is not followed by sufficient *source* capacity will cause the photosynthetic distribution to fill the pods unevenly so that many empty pods will be produced. In addition, pest and disease attacks can also increase the percentage of empty pods. High rainfall causes greater leaf rust attack. According to [14], rust can reduce photosynthesis so that when the attack is heavy many pods are not fulfilled.

3.6. Dry Weight of Seeds per Plants

The results of analysis of variance of variable dry weight of seeds per plant showed that the treatment of the admission of organic chicken manure followed by the application of IMR bioactivator on soybean plants has no significance difference. The average dry weight of seeds per soybean plant can be seen in Table 4.

Based on Table 4, the treatment of the admission of organic chicken manure followed by the application of IMR bioactivator shows that the different watering intervals of IMR bioactivators tend to make difference in the dry weight of seeds per plants. The lowest average of dry weight of seeds was found in the treatment without chicken manure bokashi and IMR bioactivator watering (C1) which equals to 22,25 g, but the difference is not significant in all treatments. This is presumably because chicken manure bokashi can improve soil conditions and supply nutrients needed by plants even in small amounts. According to [15] chicken manure applied through soil media can help in fulfilling the soil nutrient availability and improving soil structure so that it can be a good growing medium for plants. However, if the soil has enough N, P and K, the plants are no longer responsive to the admission of chicken manure bokashi.

3.7. Weight of 100 Seeds

The results of analysis of variance of variable weight of 100 seeds showed that the treatment of the admission of organic chicken manure followed by the application of IMR bioactivator on soybean plants has no significance difference. The average weight of 100 soybean seeds can be seen in Table 4.

Table 4. The average of dry weight of seeds per plant, weight of 100 seeds, weight of seeds per plot and weight of yield per hectare resulted from the admission of organic chicken manure and IMR bioactivator.

Treatment	Dry weight of seeds per plant (g)	Weight of 100 seeds (g)	Weight of seeds per plot (g)	Weight of yield per hectare (ton/ha)
C1	22,25	16,00	700,00	2,33
C2	24,85	16,25	735,00	2,45
C3	23,05	17,00	700,25	2,34
C4	22,58	16,75	721,25	2,41
C5	23,15	16,00	751,25	2,51
C6	23,73	16,50	687,25	2,30

Based on Table 4, the effect of the treatment of the admission of organic chicken manure followed by the application of IMR bioactivator to the weight of 100 seeds has no significant difference. This is presumably because the initial growth conditions of the soil have been optimal for the growth and development of soybean plants. This is supported by the results of the initial soil analysis which showed that the soil pH has neutral reaction (7,44). According to [14], the value of soil pH that is suitable for soybean crops ranges from 5,8–7,0. Soil pH determines whether the nutrients are absorbed by plants and by plant roots in general at neutral pH around [5].

3.8. Weight of Seeds per Plot

The results of analysis of variance of variable weight of seeds per plot showed that the treatment of the admission of organic chicken manure followed by the application of IMR bioactivator on soybean plants has no significance difference. The average weight of seeds per plot can be seen in Table 4.

Based on Table 4 the effect of treatment of organic chicken manure followed by the application of IMR bioactivator showed that the different watering intervals of IMR bioactivators tend to give different weight yields of seeds per plot. According to [16] the increase in seed weight both dry seed weight per experimental plot and the weight of 100 dried seeds are closely related to photosynthetic matter which is translocated to the pods and seeds. The greater the photosynthate translocated, the higher the filling of seeds. The absence of a significant response to the weight of seed yield per plot is suspected because the land used has very high levels of organic C so that soybean plants are not responsive to the administration of chicken manure bokashi fertilizer and IMR bioactivator. In addition, the land used is also used for rice cultivation which still contains a lot of nutrients N, P and K due to the residue of the fertilizer left before.

3.9. Weight of Yield per Hectare

The results of analysis of variance of variable weight of seeds per hectare showed that the treatment of the admission of organic chicken manure followed by the application of IMR bioactivator on soybean plants has no significant difference. The average weight of seeds per hectare can be seen in Table 4.

Based on Table 8, the effect of organic chicken manure followed by the application of IMR bioactivator on weight of seeds per hectare has no significant difference. It is allegedly because some of the important nutrients needed by plants are already available in the soil and high rainfall results in leaching so that the IMR bioactivator watering is ineffective and inefficient. In addition, the attack of rats in the filling phase until pod maturation caused a considerable decrease in yield.

IOP Conf. Series: Earth and Environmental Science **383** (2019) 012020 doi:10.1088/1755-1315/383/1/012020

4. Conclusion

Based on the results of the research, it can be concluded that the admission of organic chicken manure followed by IMR bioactivator watering every 3 weeks has only good effect on the number of leaves aged 5 weeks (14,50 leaves). Meanwhile, there is no effect found in the plant height is at the age of 5 and 6 weeks, number of leaves aged 6 weeks, number of nodules, number of filled pods, number of empty pods, dry weight of seeds per plant, weight of 100 seeds, weight of seeds per plot, and yield of seeds per hectare.

5. References

[1] Sudaryanto T and Dewa KSS. 2013. Ekonomi Kedelai di Indonesia. Pusat Analisis Sosial Ekonomi dan Kebijakan Pertanian. Bogor.

[2] BPS Provinsi Banten. 2015. Banten Dalam Angka 2015. Badan Pusat Statistik Provinsi Banten. Serang.

[3] Hartatik W and Diah S. 2013. Pemanfaatan Pupuk Organik untuk Meningkatkan Kesuburan Tanah dan Kualitas Tanaman. Prosiding Seminar Nasional Teknologi Pemupukan dan Pemulihan Lahan Terdegradasi. Balai Besar Penelitian dan Pengembangan Sumberdaya Lahan Pertanian. Bogor.

[4] Sutedjo MM. 2008. Pupuk dan Cara Pemupukan. Rineka Cipta. Jakarta.

[5] Hardjowigeno S. 2007. Ilmu Tanah. Akademika Pressindo. Jakarta.

[6] Mulyana N and Dadang S. 2012. Formulasi Inokulan Konsorsia Mikroba *Rhizosfer* berbasis Kompos Teriradiasi. Pusat Aplikasi Teknologi Isotop dan Radiasi-BATAN. Jakarta.

[7] Pratama S. 2015. Respons Pertumbuhan Bibit Kakao (*Theobroma cacao* L.) pada Beberapa Dosis Kompos Kulit Buah Kakao dengan dan Tanpa Aplikasi IMR-51. Skripsi. Jurusan Agroekoteknologi Fakultas Pertanian Universitas Sultan Ageng Tirtayasa. Serang. unpublished.

[8] Handayanto E and Hairiah K. Biologi Tanah. Pustaka Adipura. Yogyakarta.

[9] Adisarwanto. 2005. Kedelai. Penebar Swadaya. Jakarta.

[10] Munawar A. 2011. Kesuburan Tanah dan Nutrisi Tanaman. IPB Press. Bogor.

[11] Sudrajat D, Nana M, and Arief A. 2014. Seleksi Mikroba Rhizosfer Lokal untuk Bahan Bioaktif pada Inokulan Berbasis Kompos Iradiasi. Jurnal Ilmiah Aplikasi Isotop dan Radiasi Vol. 10 No.1 p:23-34. Pusat Aplikasi Isotop dan Radiasi BATAN. Jakarta.

[12] Cahyono B. 2007. Kedelai Teknik Budidaya dan Analisis Usaha Tani. Aneka Ilmu. Semarang.

[13] Nurmayulis, P Utama, dan R Jannah. 2014. Pertumbuhan dan Hasil Tanaman Selada (*Lactuca sativa*) yang Diberi Bahan Organik Kotoran Ayam Ditambah Beberapa Bioaktivator. Jurnal Agrologia, Vol. 3, No. 1, April 2014, p: 44-53. Fakultas Pertanian. Universitas Pattimura. Ambon.

[14] Suprapto. 2004. Bertanam Kedelai. Penebar Swadaya. Jakarta.

[15] Hamzah S. 2014. Pupuk Organik Cair dan Pupuk Kandang Ayam Berpengaruh Kepada Pertumbuhan dan Produksi Kedelai (*Glycine max* L.). Jurnal Agrium, April 2014 Vol. 18 No 3.p: 228-234 Jurusan Agroekoteknologi. Fakultas Pertanian UMSU. Medan.

[16] Fuady, Z., Mawardi dan Melizawati. 2012. Teknik Pengendalian Gulma dan Pengelolaan Tanah terhadap Pertumbuhan dan Produksi Tanaman Kedelai (*Glycine max* L. Merril). Jurnal Ilmiah Sains dan Teknologi Vol. 12 No.3. Fakultas Pertanian Universitas Almuslim. Aceh.

IOP Conf. Series: Earth and Environmental Science **383** (2019) 012021 doi:10.1088/1755-1315/383/1/012021

Development strategies of agricultural sector toward environmental externalities: case study in East Java, Indonesia

G Wirakusuma*, Irham, S Hartono, and J Mulyo

Department of Agricultural Socio-economics, Universitas Gadjah Mada, Yogyakarta, Indonesia

*corresponding author: gilang_wirakusuma@ugm.ac.id

Abstract. This research was conducted to measure the agricultural sector's externalities and to generate a model of proper strategies to encourage sustainable economic development through the agricultural sector. This research employed time-series data of 1987-2016 periods at selected province in Indonesia (East Java). Agricultural sector's externalities were addressed by environmental degradation variables (greenhouse gas emissions) and environmental carrying capacity variable. The 2006 IPCC guidelines for green house gas inventories tier 1 and land balance approach was utilized to calculate those externalities. Furthermore, for generating a model of proper strategies, the second law of thermodynamics was applied through a simultaneous equation model that was analyzed with Two-Stage Least Square method. This research revealed that agricultural income, environmental degradation, and environmental capacity have simultaneously significant interrelationship. Intensifying credit allocation for rural farm household and adding direct investment in the agriculture sector can decrease environmental degradation, escalate environmental carrying capacity, and increase agricultural income.

1. Introduction

Along with developing civilization, the 2nd Thermodynamics Law point of view provided awareness that every effort to fulfill human's needs, by natural resources utilization in industry, agriculture, fishery, services, and so on, would not get 100% of efficiency. It means that every benefit and effect yielded by the utilization is not only perceived by the subjects of the economic activities solely but also come into other parties outside of the activities, such as environment and society who utilize environmental services [1]. According to the opinion, every economic activity does not only yield economic value but also generates externalities that can be a benefit or negative effect on the whole societies.

Growing population demands highly increasing food production in order to ensure food availability and food security. Intensive farm management, which is employed by many agricultural practitioners allows high natural resources exploitation. This fact has been postulated by Kuznets, who identified the relationship between various indicators of environmental degradation and economic growth, well-known as the environmental Kuznets curve (EKC). EKC describes that in the early stages of growing economic process, environmental degradation (pollution) increase. Furthermore, the trend will reverse and indicate high-income economic growth generates environmental improvement [2].

Nowadays, approximately 1 billion people in developing countries (Sub-Saharan Africa and Asia) are undernourished. It was predicted that one person in twenty will still be undernourished by 2050 although agricultural production doubles. This production growth would indicate agriculture remaining vital sector of economic development, environmental services, and rural poverty eradication [3]. Thus, sustaining resources to retain the ability for providing sufficient food is the central part of agricultural development [4].

Agricultural development policies have been successful at emphasizing external inputs as the significant part to increase food production. This has produced high growth in global consumption of inorganic fertilizers, pesticides, tractors, animal feed, and other machinery. These external inputs have replaced natural balance process and resources, delivering them more vulnerable [5]. Moreover, the current universal agriculture system cannot achieve an important objective. That is meeting the needs of the present without compromising the ability of future generations to meet their own needs. Modern agriculture system implemented by most of the farmers drives to lead environmental degradation and turns the earth system beyond the 'safe operating space' for humanity. However, agriculture will still play a pivotal role in maintaining environmental services, which can be yielded by its activities.

Many researchers have examined and measured the value of environmental externalities of the economic activities [6–10]. The critical point to do it was the decision to limiting various externalities to get specific proper variable. Two main negative externalities of agricultural activities are represented by greenhouse gases emissions. In 2007, the Intergovernmental Panel on Climate Change (IPCC) published the fourth report that apparently indicated the primary determinant of global climate warming. Above 90% of global climate change is generated by mankind's activities. Regarding the report, agricultural activities play a role in contributing to greenhouse gases emissions, thus on climate change [11]. Several types of research in the past time revealed that greenhouse gases emissions have a relationship with agricultural economic growth [12, 13].

Beside agricultural activity generates negative externalities, it generates positive externality also, which is formed of environmental carrying capacity. The previous study stated that mankind's need for environmental resources could be declared as the width of the area which is needed for supporting mankind's life. This indicator is defined as an ecological footprint. Moreover, their report explained that determining natural and environmental resources sustainability level could be calculated by comparing finger footprint to the size of the actual production area. Thus, agricultural cultivation could be the leading party for increasing environmental carrying capacity[14].

Indonesia earns a pivotal role in the agricultural sector in its economy. The agricultural sector is the most significant source of employment in which around 40.12 million people are employed, equal to 33% of the whole labor force. Moreover, the agricultural sector still contributes to 14% of the Gross Domestic Product (GDP) in 2014. Although playing a significant role in Indonesia's macroeconomic, it still cannot afford the significant level of poverty alleviation. Poverty in Indonesia is still dominated by the rural and agricultural region, where over 60% of the miserable life [15].

Indonesia has the opportunity to maintain agricultural sector as a vital part in its development agenda since it has 55 million ha of agricultural land and 129 million ha of forest, besides prominent agricultural labor force [16]. Being the fourth of the most populous countries provide a large number of the food market, but this could be challenging while population growth leads to a significant number of land pressure, food demand, and high input exploitation [15].

We selected East Java as representative of this research. East Java is the most populated province and one of the most vital provinces for Indonesia's economic development. Supported by its infrastructure and local resources, economic development in this region has good progress at an economical value side. That economic value is manifested in Regional Gross Domestic Product (RGDP) value and employment. Based on Statistical Agency, in 2013, East Java has yield Rp. 1,136,326.87 billion, and it was equal to 14,99% of Indonesia GDP. At the same time, employment in this province was 19.081.995 employees, and this number is the most significant employment in Indonesia [16]. This significant role in Indonesia's economy, it has shown that East Java has become a pioneer in economic success at the macroeconomic side. One thing that has to be realized, the success of macroeconomic can't be the primary indicator that economic development will sustain in the long term. In fact, the agricultural sector has a significant role in employment and RGDP of East Java. In

2013, 40.41% of all employees worked in this sector. Meanwhile, only 13.83% RGDP of East Java was created by this sector [16].

A significant role of the agricultural sector in the economy has to be a consideration for the farmer and other stakeholders. The agricultural sector creates externalities, which may be negative or positive. The other important thing which has to be a consideration for the long-term economic development is the environmental degradation caused by negative externalities of the agricultural sector. Degradation or decreasing of environment quality is decreasing environment ability to provide environmental services and goods for a human being. According to that definition, if the agricultural sector were managed unwisely now, agricultural sector performance, even the whole regional economic, will be declined in the future.

Based on this background, this research was conducted to provide an appropriate approach to facing environmental externality yielded by the agricultural sector. The appropriate approach could generate raw-model of proper strategies to encourage sustainable economic development through the agricultural sector for many emerging-economic and developing countries which have many challenges in environmental issues and deal with constraints in employing their resources. Moreover, the strategies have to be able to increase agricultural production, increase positive environmental externality, and decrease negative externality.

2. Materials and Method

2.1. Location and Data

The research location was selected by purposive method considering local condition. Employment, Long-Term Development Plan, and agricultural RGDP were the reason for selecting East Java as a research location.

Type of data used in this research was secondary data. Data was collected from agencies that provide suitable data for this research. This research employed time-series data of 1987-2016 periods

2.2. Data analysis method

Environmental externalities in this study are represented by two indicators, namely greenhouse gas emissions and environmental carrying capacity generated by agricultural activities. Greenhouse gas emissions denote negative externalities, while environmental carrying capacity implies positive externalities. Both indicators are combined with the actual economic value variables of the agricultural sector in the analysis model.

2.2.1. Measurement of Green House Gases (GHG) of Agricultural Sector

Measurement of GHG emission in agricultural activity was based on the Intergovernmental Panel on Climate Change (IPCC) 2006 Guidelines. This study accommodates four sources of agricultural greenhouse gas emissions, including enteric fermentation of livestock, livestock manure, application of urea fertilizer, and rice cultivation.

An approximate number of CH_4 emission from livestock enteric fermentation was measured with the formulation as:

$$Es = EF_T \times N_T \times 10^{-3} \tag{1}$$

where:

Es = total methane emissions from livestock enteric fermentation, (ton/year)
$EF_{(T)}$ = emission factor for the defined livestock population, (kg CH_4/head/year)
$N_{(T)}$ = the number of head of livestock species/category T (head)
T = species / category of livestock

An approximate number of CH_4 emission from manure management was measured with the formulation as this:

$$CH_4\,Manure = \sum_T \frac{(EF_{T}*\,N_T)}{10^3} \tag{2}$$

where:
$CH_4\,Manure$ = CH_4 emissions from manure management (ton/year)
EF_T = emission factor for the defined livestock population, (kg CH_4/unit/year)
N_T = the number of head of livestock species / category T (head)
T = Species / category of livestock

The basic equation to estimate CH4 emissions from rice cultivation are estimated by multiplying daily emission factors by cultivation period of rice and annually harvested areas. In its purest form, this equation is implemented using national activity data (i.e., national average cultivation period of rice and area harvested) and a single emission factor.

$$RE_{CH4} = L_t \times H_t \times EF_{rf} \times 10^{-3} \quad (3)$$

where:
RE_{CH4} = methane emissions from rice cultivation (ton/year)
L_t = harvested area of rice (ha)
H_t = cultivation period of rice (200 days).
EF_{rf} = a daily CH_4 emission factor of rice (0,14 kg CH_4/ha/day)

The approximate number of CO_2emission from urea application was measured by the following formula:

$$CO_2UREA = M_{urea} \times EF_{urea} \quad (4)$$

where :
CO_2UREA = annual CO_2 emissions from urea application (ton/year)
M_{Urea} = annual amount of urea fertilization (ton/year)
EF_{Urea} = urea emission factor ton CO_2 per unit of urea. (Default IPCC Apply an overall emission factor (EF) of 0.20 for urea, which is equivalent to the carbon content of urea on an atomic weight basis (20% for $CO(NH_2)$)).

2.2.2. Environmental Carrying Capacity Measurement

Environmental carrying capacity measurement of the agricultural sector was estimated by the Indonesian Ministry of Environment Rule Number 17 2009 approach [17]. This approach utilized the ratio of potential land supply and land demand, which generated by agricultural activities. Furthermore, measurement of environmental carrying capacity was formulated as:

1. Land Supply:

$$S_L = \frac{\Sigma(P_i \times H_i)}{H_b} \times \frac{1}{Ptvb} \quad (5)$$

where:
S_L = land supply (ha)
P_i = actual production of all agricultural commodities
H_i = Price of each agricultural commodity (Rp/unit)
H_b = Price of rice (Rp/kg)
$Ptvb$ = rice productivity in one ha rice field (kg/ha)

2. Land Demand:

$$D_L = N \times KHL_L \quad (6)$$

where:
D_L = land demand (ha)
N = the number of population (head)
KHL_L = Expediency of sufficient ned (assumed equal to 1 ton of rice)

Environmental carrying capacity value (*CC*) is a comparison ratio of land supply (S_L) to land demand (D_L). If $CC > 1$, it means that environmental carrying capacity is categorized in surplus status (high). If $CC < 1$, it means that environmental carrying capacity is categorized in deficit status (low).

IOP Conf. Series: Earth and Environmental Science **383** (2019) 012021 doi:10.1088/1755-1315/383/1/012021

2.2.3. Simultaneous Equation Analysis Model

Based on the theory and a review of several kinds of literature, such as those already reconciled in Chapter 1, we can note the relationship between production activities and environmental externalities. Production activities in the agricultural sector not only produce economic value from the main products but also produce externalities related to environmental services. In the next stage, the performance of agricultural production will be affected by the consequences of the reciprocal process. In this study, we are using simultaneous equations. A study stated that where simultaneity exists, use of ordinary least squares to approximate single equation relationships might give a biased and erroneous estimate [2]. In order to approximate the strong environment-income relationship, it is, therefore, more appropriate to employ a simultaneous equation model. [18], however, noted that because of the difficulty in model specification and getting the needed data, the simultaneous relationship between economic growth and environmental quality had not been employed by any empirical studies. Three structural equations are used in the model in this study by testing three endogenous variables, namely agricultural RGDP, environmental degradation, and environmental carrying capacity in East Java Province. These structural equations are formulated as follows:

1. Structural equation of environmental degradation

$$DG_t = \rho_0 + \rho_1 PRC_t + \rho_2 PJG_t + \rho_3 PUKY_t + \rho_4 PUJ_t + \rho_5 PKD_t + \rho_6 LAND_t + \rho_7 RPOV_t + \gamma_1 YAGR_t + \rho_8 D_1 + U_1 \quad (7)$$

where:

DG	= environmental degradation (ton CO_2)
PPADI	= rice productivity (ton/ha)
PJG	= maize productivity (ton/ha)
PUKY	= cassava productivity (ton/ha)
PUJ	= sweet potatoes productivity (ton/ha)
PKD	= soybean productivity (ton/ha)
YAGR	= agricultural RDGP (IDR)
LAND	= farmland width (Ha)
RPOV	= rural poverty rate (%)
D_1	= *Dummy* Long Term Development Plan of East Java Province 2006-2025 (1 if t ≥ 2006, 0 if < 2006)
U_1	= disturbance
P	= exogenous variable coefficient
Γ	= endogenous variable coefficient
T	= year t

2. Structural equation of environmental carrying capacity

$$CC_t = \delta_0 + \delta_1 DEN_t + \delta_2 LAB_t + \delta_3 LAND_t + \delta_4 D_1 + \gamma_2 YAGR_t + U_2 \quad (8)$$

Where:

CC	= environmental carrying capacity
YAGR	= agricultural RGDP (IDR)
DEN	= population density (head/km^2)
LAB	= the number of agricultural labor (head)
LAND	= farmland width (ha)
D_1	= *Dummy* Long-Term Development Plan of East Java Province 2006-2025 (1 if t ≥ 2006, 0 if < 2006)
U_2	= disturbance
Δ	= exogenous variable coefficient
Γ	= endogenous variable coefficient
T	= year t

3. Structural equation of agricultural RGDP

$$YAGR_t = \beta_0 + \beta_1 CRD_t + \beta_2 LAB_t + \beta_3 INV_t + \beta_4 LAND_t + \beta_5 POP_t + \beta_6 RPOV_t + \beta_7 D_1 + \gamma_4 CC_t + \gamma_5 DG_t + U_3 \quad (9)$$

Where:

YAGR	= agricultural RGDP (IDR)
CC	= environmental carrying capacity
CRD	= credit allocated to the agricultural sector (IDR)
LAB	= the number of agricultural labor (head)
DG	= environmental degradation (ton CO_2)
INV	= investment value of agricultural, chemistry, and timber industry (IDR)
LAND	= farmland width (Ha)
RPOV	= rural poverty rate(%)
POP	= population (head)
D_1	= *Dummy* Long Term Development Plan of East Java Province 2006-2025 (1 if t ≥ 2006, 0 if < 2006)
U_3	= disturbance
Β	= exogenous variable coefficient
Γ	= endogenous variable coefficient
T	= year t

3. Result and Discussion

3.1. Environmental Degradation: GHG Emissions of Agricultural Sector

Table 1 shows that all components of the environmental degradation of the agricultural sector in East Java Province increased in 1987-2016 periods. CH_4 from poultry manure growth rate is the biggest; it's 7.82% per year. It gives the implication that increasing performance in husbandry sub-sector, especially poultry commodity, is followed by increasing GHG emissions. Even though CH_4 from poultry manure growth rate is the biggest; it isn't the most dominant emission component produced by the agricultural sector. Rice cultivation is the most dominant component which produces GHG of the agricultural sector in East Java.

Table. 1. Approximate environmental degradation of the agricultural sector in East Java province.

GHG Emissions	Average Value 1987-2016 (ton/yr)	Growth Rate 1987-2016 (%/yr)
CH_4 from livestock enteric fermentation	187.910,94	0,89
CH_4 from livestock manure	9.725,52	2,24
CH_4 from poultry manure	3.038,65	7,82
CO_2 urea application	216.901,90	0,76
CH_4 rice cultivation	472.070,37	1,69

Source: Data analysis, 2019.

3.2. Simultaneous Analysis Model of Environmental Degradation, Environmental Carrying Capacity, and Agricultural RGDP in East Java Province

We used *E-views 9* software to analyze the simultaneous model by Two-Stage Least Square (TSLS) analysis result of environmental carrying capacity equation is shown in Table 2. TSLS analysis yields F-probability as 0,000002, and this result shows that at α=0,01, simultaneously, agricultural RGDP, population density, farmland width, the number of agricultural labor, dummy East Java Province Long Term Development Plan 2006-2025 are significant determinants of environmental carrying capacity. Based on the partial test (t-test), we know that agricultural RGDP and farmland width are positively significant determinants of environmental carrying capacity. It means that if agricultural RGDP or farmland width increase, *ceteris paribus*, environmental carrying capacity will increase too. This condition is appropriate with the hypotheses expected by this research.

Table 2. TSLS analysis result of environmental carrying capacity

Variable	Coefficients	Probability	F-probability	Adjusted-R^2
C	0,330058	0,6617	0,000002***	0,8614
YAGR	$2{,}69.10^{-14}$	0,0020**		
DEN	-0,002652	0,0824*		
LAND	$3{,}45.10^{-7}$	0,0112**		
LAB	$6{,}49.10^{-8}$	0,2258		
D1	0,059824	0,3931		

= Significant at α=0,1; **= Significant at α=0,05; *= Significant at α=0,01*
Source: Data analysis, 2019.

Two-Stage Least Square (TSLS) analysis result of environmental degradation equation is shown in Table 3. TSLS analysis yields F-probability as 0,077756, and this result shows that at α=0,01, simultaneously, agricultural RGDP, rice productivity, maize productivity, cassava productivity, sweet potatoes productivity, soybean productivity, farmland width, rural poverty rate, and dummy East Java Province Long Term Development Plan 2006-2025 are significant determinants of environmental degradation. Based on the partial test (t-test), we know that maize productivity and sweet potatoes productivity are negatively significant determinants of environmental degradation. It means that if maize productivity or sweet potatoes productivity increase, *ceteris paribus*, environmental degradation will decrease. This condition is contrary to the hypotheses expected by this research.

Table 3. TSLS analysis result of environmental degradation

Variable	Coefficients	Probability	*F-probability*	Adjusted-R^2
C	$-7{,}17.\ 10^{13}$	0,1240	0,077756*	0,4210
YAGR	-0.806848	0,5112		
PJG	$-1{,}63.\ 10^{13}$	0,0950*		
PUKY	$4{,}49.\ 10^{12}$	0,0129**		
PKD	$2{,}72.\ 10^{13}$	0,2832		
PPADI	$1{,}15.\ 10^{13}$	0,1508		
PUJ	$-1{,}96.\ 10^{12}$	0,0057**		
RPOV	$-3{,}93.\ 10^{11}$	0,3111		
LAND	16727814	0,1939		
D1	$4{,}13.\ 10^{12}$	0,4451		

**= Significant at α=0,1; **= Significant at α=0,05*
Source: Secondary Data Analysis, 2019.

Moreover, Two-Stage Least Square (TSLS) analysis result of agricultural RGDP equation is shown in Table 5. TSLS analysis yields F-probability as 0,000001, and this result shows that at α=0,01, simultaneously, environmental degradation, carrying capacity, the number of agricultural labour, farmland width, investment on agricultural industry, the number of population, credit allocated to agricultural sector, rural poverty rate, and dummy East Java Province Long Term Development Plan 2006-2025 are significant determinants of agricultural RGDP.

Based on the partial test (t-test), we know that investment in the agricultural industry, the number of population, and credit allocated to the agricultural sector are positively significant determinants of agricultural RGDP. It means that if the agricultural industry, the number of population, or credit allocated to the agricultural sector, *ceteris paribus*, agricultural RGDP will increase. Meanwhile, environmental degradation is a negatively significant determinant of agricultural RGDP.

Table 4. TSLS analysis result of agricultural RGDP

Variable	Coefficients	Probability	F-probability	Adjusted-R^2
C	$2,16.10^{13}$	0,3076	0,000001***	0,9553
DG	-0,116638	0,0572*		
CC	$6,5.10^{12}$	0,5835		
LAB	-1585568	0,1819		
LAND	-4551353	0,2192		
INV	0,036145	0,0349**		
POP	1409777	0,0213**		
CRD	1510366	0,0086***		
RPOV	$-2,11.10^{11}$	0,1026		
D1	$-1,24.10^{-14}$	0,3656		

*= *Significant at α=0,1*; **= *Significant at α=0,05*; ***= *Significant at α=0,01*
Source: Data analysis, 2019.

3.3. Policy Simulation

We used *E-views 9* software to do a simulation with a simultaneous equation model, which is generated by TSLS analysis. Raising credit allocated to the agricultural sector is a policy alternative to decrease environmental degradation, increase environmental carrying capacity, and increase agricultural RGDP all at once. We tried to raise credit allocated to the agricultural sector for 15,1% (according to its average growth rate (BPS, 2015)).

Table 5. Simulation result of raising credit allocated to agricultural sector for 15,1% rate

Type of Value	Environmental Degradation Value (Rp. 000.000)	Environmental Carrying Capacity	Agricultural RGDP (Rp. 000.000)
Actual Average	23.520.596,60	1,0679	44.770.165,00
Simulated Average	22.961.472,00	1,0865	45.463.138,00
Gap (%)	-2,38%	1,72%	1,55%

Source: Data analysis, 2019.

Raising credit allocated to the agricultural sector as many as 15,1% is proven to decrease environmental degradation 2,38%, increase environmental carrying capacity 1,72% and increase agricultural RGDP 1,55%. Credit is one of the most critical resources to enhance farm management's performance. Accessing credit for agricultural activities increases farmers' ability to multiply their productivity by purchasing more input; thus, the increase of farmers' income. According to the EKC, an increase in income will lead farmers to assign environmental conservation cost for their farm management. This conservation will retain environmental capacity and decrease environmental degradation.

Table 6. Simulation result of raising investment in agricultural, chemistry, and timber industry for 5% rate

Type of Value	Environmental Degradation Value (Rp. 000.000)	Environmental Carrying Capacity	Agricultural RGDP (Rp. 000.000)
Actual Average	23.520.596,60	1,0679	44.770.165,00
Simulated Average	23.486.430,00	1,0690	44.812.508,50
Gap (%)	-0,145%	0,103%	0,095%

Source: Data analysis, 2019.

Raising investment in agricultural, chemistry, and timber industry at 5% level is proven to decrease environmental degradation 0,145%, increase environmental carrying capacity 0,103%, and increase agricultural RGDP 0,095%. Investment in agricultural, chemistry, and timber industry represents many activities for manufacturing various harvest at down-stream of agribusiness. These activities can add more value to agricultural raw commodities and increase income indeed. Based on EKC, this condition could be a source of conserving capital.

4. Conclusion and Policy Implication

Economic development has a trade-off in the form of externalities that affect the environment, especially the agricultural sector. Growing population and food demand have to be emphasized as the challenges of providing sufficient food for the communities. High capital allocation for accessing external input not only can multiply food production but also can degrade the environment. This issue can be more difficult to answer by many developing countries because of their constraint on economic resources and high dependency on the agricultural sector (also environment certainly). This research tried to generate proper strategies to deal with environmental externalities yielded by the agricultural sector through macro-scope analysis. This research employed agricultural income as the principal value of production, while environmental capacity and environmental degradation as externalities. It revealed those three variables have a simultaneous correlation. Raising credit allocated to the agricultural sector and investment in agricultural, chemistry, and timber industry (agricultural manufacturing) is proven to decrease environmental degradation, increase environmental carrying capacity, and increase agricultural production all at once. This research encourages the government and other stakeholders to allocate more credit loan to the rural and agricultural area. The credit loan has to be invested in on-farm activities in order to gain higher productivity and to utilize the eco-friendly farming system. Furthermore, developing rural agroindustry has to be initiated and empowered to generate a better alternative for the rural/farm households in gaining more income. Gaining income from agroindustry could reduce their dependence on agricultural activities and even on exploiting environmental resources.

5. References

[1] Burley P, Foster J. *Economics and Thermodynamics – New Perspectives on Economic Analysis*. Netherland: Springer, 1994.
[2] Stern DI. The Environmental Kuznets Curve. *Encyclopedia of Ecological Economics in June 2003*.
[3] FAO. *The State of Food Insecurity in the World 2010. Addressing food insecurity in protracted crises*. Rome: Food and Agriculture Organization (FAO), 2010.
[4] Apata TG. Effects of global climate change on Nigerian agriculture : An empirical analysis. *CBN J Appl Stat* 2006; **2**: 31–50.
[5] Coan R, Ann W. *Farming for The Future: An Introduction to Low External Inputs and Sustainable Agriculture*. London: Macmillan, 2001.
[6] Sutono S, Tala'ohu SH, Sopandi, et al. Erosi pada berbagai penggunaan lahan di DAS Citarum.

In: *Prosiding Seminar Nasional Multifungsi Dan Konversi Lahan Pertanian*. 2002, pp. 113–133.
[7] Alam S. Globalization , poverty, and environmental degradation: Sustainable development in Pakistan. *J Sustain Dev* 2010; **3**: 103–114.
[8] Hung M, Shaw D. *Economic Growth and the Environmental Kuznets Curve in Taiwan : A Simultaneity Model Analysis*. 2006.
[9] Omri A, Nguyen DK, Rault C. Causal interactions between CO2 emissions, FDI, and economic growth: Evidence from dynamic simultaneous-equation models. *Econ Model* 2014; **42**: 382–389.
[10] Vu QM, Le QB, Frossard E, et al. Socio-economic and biophysical determinants of land degradation in Vietnam: An integrated causal analysis at the national level. *Land use policy* 2014; **36**: 605–617.
[11] Intergovernmental Panel on Climate Change. *2006 IPCC Guidelines for National Greenhouse Gas Inventories Volume 4 Agriculture, Forestry and Other Land Use*. Intergovernmental Panel on Climate Change (IPCC), 2007.
[12] Li Z, Zheng X. Study on relationship between Sichuan agricultural carbon dioxide emissions and agricultural economic growth. *Energy Procedia* 2011; **5**: 1073–1077.
[13] Coderoni S, Esposti R. Is there a long-term relationship between agricultural GHG emissions and productivity growth? A dynamic panel data approach. *Env Resour Econ* 2014; **58**: 273–302.
[14] Lenzen M, Murray S a. The Ecological Footprint – Issues and Trends. *Univ Sydney* 2003; 1–27.
[15] Quincieu E. *Summary of Indonesia's agriculture, natural resources, and environment sector assessment*. 2015.
[16] Badan Pusat Statistik. *Statistik Indonesia 2014*. Jakarta: Badan Pusat Statistik, 2015.
[17] Menteri Negara Lingkungan Hidup. Peraturan Menteri Negara Lingkungan Hidup Nomor 17 Tahun 2009 Tentang Pedoman Penentuan Daya Dukung Lingkungan Hidup Dalam Penataan Ruang Wilayah. Indonesia, 2009.
[18] De Bruyn SM. *Economic Growth and the Environment An Empirical Analysis*. Netherland: Springer, 2000.

The effect of citric acid and sodium bicarbonate concentration on the quality of effervescent of red ginger extract

Giyatmi and D K Lingga

Study Program of Food Technology, Sahid University Jakarta, Jl. Prof. Supomo 84 Tebet, Jakarta, Indonesia

E-mail giyatmi@hotmail.com

Abstract. The effervescent product of red ginger extract was made from a mixture of red ginger extract, citric acid, sodium bicarbonate, dextrose, mannitol, sucralose and polyethylene glycol. This study used a single factor randomized complete design method with the ratio of different concentrations of citric acid and sodium bicarbonate (50:50, 51:49, 52:48, 53:47, 55:45) in three replication. Analysis of Variance (ANOVA) was used to see the effect of each treatment, if there was an effect of each treatment then continued with DMRT (Duncan Multiple Range Test) to see which levels produced differences in quality. The results showed that the addition of different concentrations of citric acid and sodium bicarbonate had significant effect (sig $<\alpha = 0.01$) on the value of soluble and effervescent pH of red ginger extract and significantly affected the level of color preference, taste and effervescent solubility of red ginger extract. The best effervescent product of red ginger extract was selected based on the hedonic test with the ratio of citric acid and sodium bicarbonate 52:48 with the following quality test results: 0.00% ethanol residue; levels of active red ginger 0.65%; and the level of effervescent red ginger extract 0.52%.

1. Introduction

Red ginger (*Zingiber officinale* var. Rubrum) is widely used as a spice, traditional medicine ingredients, sweets and refreshing drinks. People use red ginger more often as medicines because it has the highest volatile oil content, consisting of, among others, turpentine oil, zingiberene, and curcumin. Ginger also contains camphene, farnesene, sesquiphellandrene, linalool, nerolidol, neutral geranial and 1,8-cineol [1]. According to Hapsoh and Julianti [2], the spicy taste in ginger is produced by gingerol and shogaol which are widely found in ginger oleoresin. Red ginger extract is also used to stimulate appetite, alleviate coughs, colds, inflammation, facilitate digestion, reduce nausea and vomiting [3]. Processed ginger is usually enjoyed as a warm drink in cold weather.

Many red ginger preparations on the market have been made in traditional beverage preparations such as *wedang jahe*, instant ginger drinks and *bandrek*. The use of traditional drinks in the form of effervescent is still rare, only limited to products of supplements and vitamins. Whereas this has several advantages such as time efficient and producing a fresh taste because of the presence of carbonates which help improve the taste of the preparation[4].

Effervescent is a product made by mixing active ingredients in the form of acidic sources and alkaline sources such as citric acid or tartrate acid and sodium bicarbonate. When this product is put into the water, a chemical reaction begins between citric acid and sodium bicarbonate forming sodium salt from acid and produces CO_2 [5]. The reaction is quite fast and usually completed in a minute or less. Effervescent produce clear solutions and produce a good taste because of the presence of carbonates that help improve certain flavors [3].

Acid sources and alkaline sources have an important role in making effervescent. Sources of acid commonly used are citric acid, malic acid and fumaric acid and the alkaline source commonly used is sodium bicarbonate. This research refers to previous research [6] regarding the manufacture of effervescent granules. In this study we used only citric acid. Citric acid and sodium bicarbonate have an important role in formulating an effervescent preparation because it is the main source of carbon dioxide that determines the effervescent system produced [7]. Thus, drinks given in the form of effervescent granules will provide a refreshing sensation caused by the release of carbon dioxide [8].

The aim of this study was to determine the quality of effervescent granules of red ginger extract with a ratio of citric acid and sodium bicarbonate 50:50, 51:49, 52:48, 53:47 and 55:45. The parameters tested were flow test, stationary angle, bulk density, dissolution time, moisture content, pH, organoleptic, levels of active ingredients of red ginger extract, ethanol residue, and test of purchase decision.

2. Materials and Methods

2.1. Effervescent ingredients

The ingredients used in making effervescent red ginger extract were citric acid, sodium bicarbonate, dextrose, red ginger extract, sucralose, ethanol, water and PEG 6000, while the chemicals used for analyzing samples were H_3PO_4, acetonitrile and methanol.

2.2. Effervescent granule preparation

The process of making effervescent in this study was by using a wet granulation method [9], which included several stages: weighing raw material, mixing I, wet granulation process, sieving I, drying, sieving II, mixing II, so that wet granule mass was formed.

Dextrose, red ginger extract, mannitol, citric acid, sodium bicarbonate and sucralose are mixed and shaken in the plastic until the mixture was homogeneous. After the mixture of homogeneous ingredients, the material was put into a stainless steel basin container for wet granulation process by adding an ethanol binder solution, then the moist granule was sieved with a 6 mesh sieve in order to made granule homogeneous. Then the granules were dried using an oven at 50 ° C for 1 hour to obtain a moisture content of 0.4-0.7%. The dried granules were sieved again with a 20 mesh sieve to form a granule mass. The mass of the formed granules was mixed with PEG 6000 lubricating material to help increase the flow of effervescent granules. The process of making effervescent granules with a ratio of citric acid and sodium bicarbonate with 50:50, 51:49, 52:48, 53:47 and 55:45 can be seen in figure 1.

2.3. Methods

2.3.1. Flow time

The time needed to flow a number of granules or powder on the tool was measured. The flowing granules or powders is influenced by the shape, surface area, density and humidity of the granules. The inequality and the smaller size of the granule will increase the power of cohesion so that the granules clot and not easily flow. Usually the flow rate $\leq$ 10g / sec is considered good [9].

2.3.2. Angle of repose

Fixed angles that occur between heaps of cone-shaped particles with horizontal fields, if a number of powders or granules were poured into a measuring device. The size of the stationary angle is influenced by the shape, particle size and humidity of the granule. Granules will flow well if they have a stationary angle between 25° to 45° [10]. Angel of repose can be calculated by the formula:

$$Tg\ \beta = \frac{h}{R} \qquad (1)$$

Note :

Tg β	= angle of repose (^{0})
h	= cone height (cm)
R	= radius (cm)

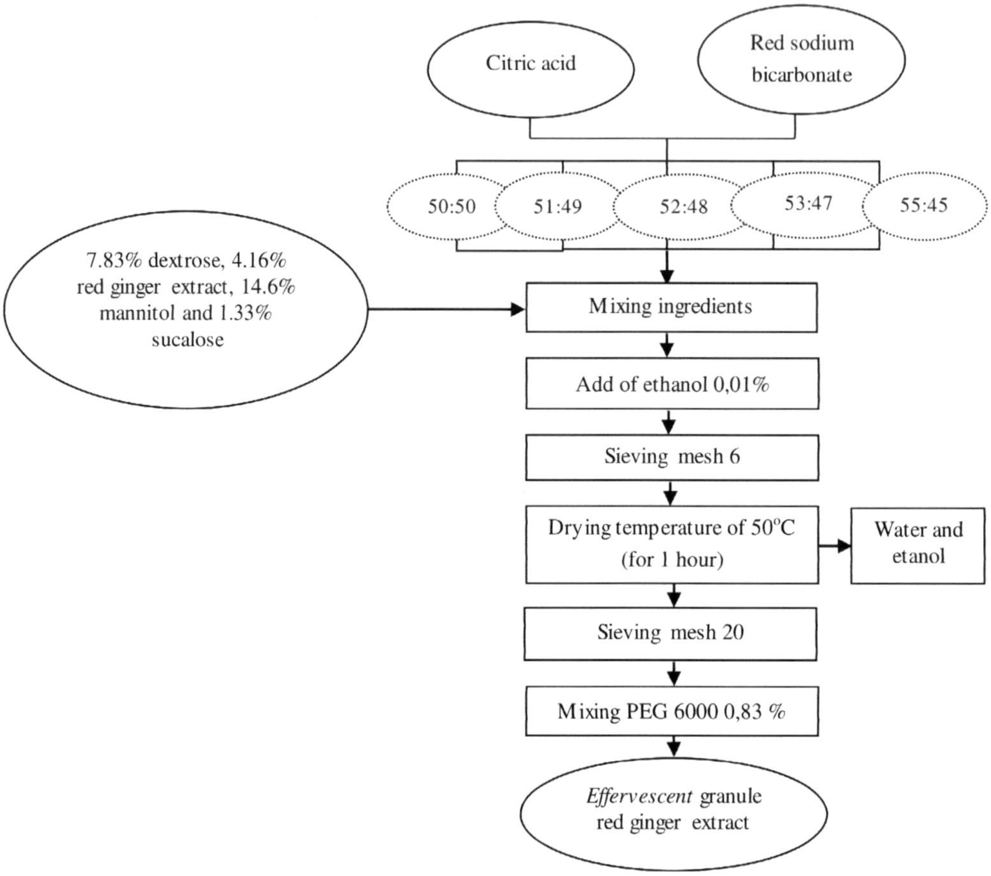

Figure 1. The process of making red ginger extract effervescent granules.

2.3.3. Bulk Density
Decrease in the volume of a number of powder granules due to shock with vibration. The smaller the tapping index (in percent), the better the flow properties. The tapping test was carried out by a volumenometer device which consists of a measuring cup that can move regularly up and down with the help of a driving motor. Granules or powders that have a determination index of less than 20% have good flow properties. This determination process can also be calculated the value of its bulk density with the formula [10]:

$$\rho b = \frac{M}{Vb} \quad (2)$$

Note:
M = particle mass
Vb = final volume after being pressed
ρb = bulk density after being pressed

2.3.4. Solubility time
200 ml of cold water at a temperature of 15°C - 25°C were poured in a 200 ml glass. After that, 6 grams of effervescent granules were added to the water. The requirements for dissolving time are less than 5 minutes, when the effervescent was completely diluted and had an ideal dissolving time ranging from 1-2 minutes. If the effervescent is well dispersed in water and completes the reaction in less than 5 minutes, it shows that the preparation meets the standards of dissolution and the requirements of dissolution time [11].

2.3.5. Moisture content [12]

Determination of water content began with drying of aluminum plates at a temperature of 105°C for 15 minutes in oven, then cooled in a desiccator and weighed. A total of 1-2 grams of sample was put into the aluminum dish and dried in an oven at 105°C for five hours then cooled in a desiccator, and weighed until the dry sample weight was relatively constant.

The moisture content of the granule can affect the flow rate of the granule. Conversely, high moisture content can cause instability of granules which can be seen from changes in physical appearance, color, smell, taste, and texture of the granule [8]. In this study the author used a moisture analyzer to test the water content analysis.

$$\text{Moisture content } (\%\, b/b) = (a - b) \times \frac{100\,\%}{B} \tag{3}$$

Note :
a = initial weight
b = final weight

2.3.6. pH

According to Juita [13], the pH test needs to be done because if an effervescent solution formed too acidic, it might irritate the stomach while if it is too alkaline it might causes a bitter and unpleasant taste. In addition, according to Lestari [4], effervescent has good solubility at pH between 5-7 because effervescent is not stable at pH> 8. While according to Estiasih and Ahmadi [14], effervescent need a pH <6 as a quality standard. Balance of acid base body tissue and human blood must be at pH 7.3 - 7.5 means that the condition of the body is somewhat alkaline or alkaline, in order to stay healthy and function optimally. Above pH 7.8 or below pH 6.8 will cause metabolic disorders, which in turn is also a disruption to health [15].

2.3.7. Organoleptic test

The organoleptic test of taste, smell, color and solubility were carried out by a preference / hedonic test and hedonic quality test. The parameters observed were the taste, smell, color, and solubility of the effervescent granules of red ginger extract produced.

2.3.8. Active compound [16]

Measuring the concentration of gingerol content was carried out using a chromatographic method, the sample was ultrasonically heated by sonication using methanol at room temperature. The extract was then analyzed by HPLC. Oven column temperature 40°C. 280 nm wavelength. and the mobile phase flow rate of 1 ml / minute. The solvent used in the column is acetonitrile (mobile phase).

2.3.9. Ethanol residue

Alcohol / ethanol residues were examined by the method of GC (Gas Chromatography) or HPLC (High Performance Liquid Chromatography). Analysis of ethanol residues using GC with a detection limit of at least 20 ppm. Analysis of alcohol / ethanol levels can be done using GC equipped with flame ionization (FID) detectors and software integrated with computers. The temperature of the FID detector was set at 150°C and the injection port temperature is set at 250°C. Helium was used as a carrier gas and was used with a flow rate of 3°C / minute. The capillary column used is SGE ID-BP20, 30 mm, and a diameter of 0.25 mm. The oven temperature was set at 60°C at the start, up to 80°C and programmed 20°C / minute.

In food products, the limits for receiving ethanol residues are differ by countries and organizations. In general, ethanol residues of 0.5% are allowed in food products. However, the Islamic Food and Nutrition Council of America sets the permissible alcohol level at 0.1% [17].

2.3.10. Purchase decision test

The decision test to buy effervescent products of red ginger extract was done by determining whether the panelist will buy the product or not based on general price and acceptance ratings.

2.4. Data analysis
The data displayed was in the form of mean ± standard deviation. The analysis technique used was Variant Analysis (ANOVA) with five levels with three repetitions using Statistical Package for The Social Science (SPSS) version 20. If there was an influence of the ratio of citric acid and sodium bicarbonate to the quality of red ginger extract effervescent products, then proceed with Test DMRT (Duncan's Multiple Range Test) to find out which treatment level is different. The variable in this study was the ratio of citric acid and sodium bicarbonate (A) which consists of 5 levels, namely A1 (50:50), A2 (51:49), A3 (52:48), A4 (53:47), and A5 (55:45).

3. Results and Discussion

3.1 Physical properties
The quality of effervescent granules of red ginger extract was assessed in the form of physical quality, i.e. flow time, angle of repose, bulk density, and dissolution time. The effervescent value of red ginger extract treated with physical tests can be seen in table 1.

Table 1. Physical test value of effervescent value of red ginger extract.

Physical Test	A1 (50:50)	A2 (51:49)	A3 (52:48)	A4 (53:47)	A5 (55:45)
Flow time	9,35 ± 0,05	9,40 ± 0,00	9,40 ± 0,10	9,4 ± 0,10	9,45 ± 0,05
Angle of repose	42,30 ± 0,00	42,34 ± 0,00	42,36 ± 0,03	42,34 ± 0,04	42,34 ± 0,04
Bulk density	13,33 ± 0,58	13,66 ± 1,15	13,33 ± 0,58	13,00 ± 0,00	13,66 ± 0,58
Solubility time	2,25 ± 0,05^{a}	2,13 ± 0,03^{b}	2,07 ± 0,04^{b}	1,60 ± 0,00^{c}	1,56 ± 0,02^{c}

Note : same codes mean that they are not significantly different between treatment and different codes mean that they are significantly different between treatments.

Flow time is the time needed to drain a number of granules or powder on the tool used. The ease of flowing granules or powders is influenced by the shape, surface area, density and humidity of the granules. The inequality and the smaller size of the granule will increase the power of cohesion so that the granules clot and not easily flow. The value of flow time ranging from 9.35 to 9.45 (s) is in accordance with the quality requirements for effervescent flow time provided that the flow velocity ≤10g / sec is considered good [9]. Based on the results of variance, the ratio of citric acid and sodium bicarbonate did not affect the quality of the effervescent flow time of red ginger extract (α> 0.01 and 0.05).

Angle of repose is a fixed angle that occurs between a pile of cone-shaped particles and a horizontal plane, if a number of powders or granules are poured into a measuring device. The size of this angle is influenced by the shape, particle size and humidity of the granule. Based on the results of variance, the ratio of citric acid and sodium bicarbonate did not affect the quality of the stationary angle (α> 0.01 and 0.05). The value of the stationary angle ranges from 42.30° to 42.36° according to the requirements for the quality of the angle of repose of the effervescent granule provided that the granule will flow well if it has a stationary angle between 25° to 45° [10].

Based on the results of variance, the different ratio of citric acid and sodium bicarbonate did not affect the bulk density value (α> 0.01 and 0.05). Bulk density values ranging from 13.00% to 13.66% according to the requirements of bulk density of effervescent granules with granules or powders that have a bulk density index of less than 20% means having good flow properties [10].

Dissolution time test was carried out by 200 ml of granules put into beaker glass containing aquadest. When it dissolved, it was determined that the granule was inserted into the glass until all the granules were dissolved in the distilled water. Good effervescent has a dissolving time of less than 5 minutes [11]. Soluble time testing was done using a stopwatch. Based on the results of variance, the ratio of citric acid and sodium bicarbonate added influenced effervescent dissolution time (α> 0.01 and 0.05). Based on Duncan's test to find out on which comparison affected the quality of dissolved time in the product, it is known that the higher the ratio of citric acid is added, the faster the solubility time, this is because citric acid has a high solubility in water and very hygroscopic [18].

3.2 Chemical properties
The chemical quality of effervescent granules of red ginger extract was assessed based on water content and pH. Effervescent value of red ginger extract chemical test treatment can be seen in Table 2.

Table 2. Chemical properties of effervescent red ginger extract.

Chemical Test	A1 (50:50)	A2 (51:49)	A3 (52:48)	A4 (53:47)	A5 (55:45)
Moisture content	0,41 ± 0,02	0,43 ± 0,05	0,43 ± 0,02	0,45 ± 0,05	0,47 ± 0,06
pH	5,45 ± 0,02[a]	5,37 ± 0,02[c]	5,31 ± 0,02[c]	5,23 ± 0,04[b]	5,15 ± 0,02[d]

Note : same codes mean that they are not significantly different between treatment and different codes mean that they are significantly different between treatments.

Water content was tested using a moisture content analyzer by weighing 2.5 grams of granules and inserted into a moisture meter. The device was run by setting the temperature to 60°C. The water content value ranged from 0.41% to 0.47%. In accordance with the effervescent granule water quality requirements with the condition that standard granules with moisture content of ≤ 5%, the effervescent moisture content of red ginger extract met the requirements. Based on the results of variance, the comparison of different citric acid and sodium bicarbonate did not affect the effervescent water quality of red ginger extract ($\alpha > 0.01$ and 0.05).

Effervescent granule pH measurements were carried out by 6 grams dissolved in 200 ml of beaker glass. The pH meter used previously was calibrated first using a pH 7 buffer. Then the electrode is cleaned with distilled water and dried with a tissue. Effervescent pH quality according to ranges from ≤ 6 [14]. Based on the results of variance, the comparison of citric acid and sodium bicarbonate added affects the effervescent pH level of red ginger extract ($< \alpha = 0.01$ and 0.05). Duncan test results $\alpha = 0.01$ indicate the average value of effervescent pH of red ginger extract differed very significantly at the level of $\alpha = 0.01$ for each treatment. Based on the results of these observations, the higher the citric acid, the effervescent pH value of red ginger extract will lower the pH. The decrease in pH is caused by an increase in the amount of citric acid added and when reacting with carbonate and water sources can increase the activity of H^+ ions in organic acids in food products [19].

3.3 Organoleptic result
The quality of effervescent granules of red ginger extract was assessed as a hedonic test for the color, smell, taste and effervescent solubility of red ginger extract. The quality test of hedonic effervescent red ginger extract can be seen in Table 3.

Table 3. Hedonic test of effervescent red ginger extract.

Hedonic Test	A1 (50:50)	A2 (51:49)	A3 (52:48)	A4 (53:47)	A5 (55:45)
Color	68,00[bc]	80,00[a]	82,67[a]	79,33[b]	78,00[b]
Smell	68,00	71,00	74,00	71,33	70,00
Taste	67,00[a]	71,00[ab]	79,33[b]	67,33[a]	63,00[a]
Solubility	67,33	68,00	74,00	75,00	77,67

Note : same codes mean that they are not significantly different between treatment and different codes mean that they are significantly different between treatments.

The average results of the panelists' favorite values for the effervescent color of red ginger extract ranged from 68.00% - 82.67%, from dislike to very like. 82.67% of panelists liked the A3 sample with the addition of citric acid and sodium bicarbonate 52:48. Based on the results of variance, the comparison of citric acid and sodium bicarbonate added influenced the hedonic test of the effervescent color of red ginger extract ($< \alpha = 0.01$ and 0.05). The Duncan test results showed a hedonic test of the effervescent color of red ginger extract with a ratio of citric acid and 50:50 sodium bicarbonate was significantly different from the ratio of citric acid and sodium bicarbonate 51:49, 52:48, 53:47 and 55:45. Winarno [20] states that color is influenced by raw materials and processes. Color changes occur in the effervescent dough of red ginger extract after the mixture has undergone a wet granulation process.

Citric acid is added because it affects the solubility of effervescent powder to be faster, so the color looks brighter [18].

The average value of the panelists' preference for the effervescent smell of red ginger extract ranged from 68.00% to 74.00%, from dislike to very like. 74.00% of panelists liked the A3 sample with the ratio of citric acid and sodium bicarbonate 52:48. Based on the results of variance, the ratio of citric acid and sodium bicarbonate added did not affect the hedonic test of the effervescent smell of red ginger extract (α> 0.01 and 0.05).

The average results of panelists' preference for the effervescent taste of red ginger extract ranged from 63.00% -79.33%, from dislike to very like. 79.33% of panelists liked the A3 sample with the ratio of citric acid and sodium bicarbonate 52:48. Based on the results of variance, the comparison of citric acid and sodium bicarbonate added influences the hedonic test of the effervescent taste of red ginger extract (<α = 0.01 and 0.05). The Duncan test results showed a hedonic test of red ginger taste in effervescent with a ratio of citric acid and sodium 52:48 not significantly different from a ratio of 51:49 that it would be different from the citric acid ratio of 55:45, 50:50 and 53:47. This shows the ratio of citric acid and sodium bicarbonate to affect the taste of the panel effervescent red ginger extract liking the ratio of citric acid and sodium bicarbonate 52:48.

The average value of panelists' preference for effervescent solubility of red ginger extract ranged from 67.33% -77.67%, from very insoluble to very soluble. 77.67% of panelists like the P5 sample, which is the ratio of citric acid and sodium bicarbonate 55:45. Based on the results of variance, the comparison of citric acid and sodium bicarbonate added affects the hedonic test of the effervescent solubility of red ginger extract (<α = 0.01 and 0.05). The Duncan test results showed a hedonic test of the effervescent solubility of red ginger extract with a ratio of citric acid and sodium bicarbonate 55:45 not significantly different from the ratio of citric and sodium bicarbonate 52:48 and 53:47 but significantly different from the ratio of citric acid and sodium bicarbonate 50 : 50 and 51:49. Based on observations the higher the ratio of citric acid added, the faster the solubility will be, this is because citric acid has a high solubility in water and is very hygroscopic [18].

3.4 Hedonic quality result

The quality of effervescent granules of red ginger extract was assessed in the form of hedonic quality tests on the color, smell, taste and effervescent solubility of red ginger extract. The quality test of hedonic effervescent red ginger extract can be seen in Table 4.

Table 4. Hedonic quality test of effervescent red ginger extract.

Hedonic Quality Test	A1 (50:50)	A2 (51:49)	A3 (52:48)	A4 (53:47)	A5 (55:45)
Color	$73{,}11^{c}$ Slightly bright	$80{,}00^{b}$ Bright	$81{,}44^{a}$ Bright	$79{,}78^{b}$ Bright	$77{,}00^{bb}$ Bright
Smell	71,44 Slightly ginger smell	72,33 Slightly ginger smell	72,33 Slightly ginger smell	71,89 Slightly ginger smell	68,83 Slightly ginger smell
Taste	$67{,}44^{ac}$ Slightly spicy	$74{,}44^{b}$ Spicy	$78{,}89^{b}$ Spicy	$63{,}89^{ac}$ Slightly spicy	$63{,}66^{a}$ Slightly spicy
Solubility	66,89 Slightly soluble	71,11 Slightly soluble	74,56 Soluble	74,44 Soluble	76,82 Very soluble

Note : same codes mean that they are not significantly different between treatment and different codes mean that they are significantly different between treatments.

Color is the most important quality attribute. The color of the product is considered as a factor that determines quality, also can be used as an indicator of whether or not the standard method of mixing or processing can be characterized by the presence of color [20]. Based on the results of variance, the comparison of the concentration of citric acid and sodium bicarbonate affects the hedonic quality test on the brightness of the effervescent color of red ginger extract (α <0.05). The Duncan test results show that the ratio of citric acid and natrium bicarbonate affects the hedonic quality test. Panelists liked the ratio of citric acid and sodium bicarbonate 52:48.

The smell caused by chemical stimulation which is smelled by olfactory nerves in the nasal cavity [20]. Based on the results of variance, the comparison of citric acid and sodium bicarbonate did not affect the hedonic quality test on the smell of effervescent ginger red ginger extract.

Taste is judged by the presence of chemical stimulation responses by the senses and trigeminal nerves located in the mouth of the tasting mouth, namely the tongue and nose. Sensitivity to taste varies depending on the substance being tested, as well as the panelist and temperature conditions. Based on the results of variance, the comparison of the concentration of citric acid and sodium bicarbonate affects the hedonic quality test of the effervescent taste of red ginger extract (<α = 0.01). The Duncan test results showed a hedonic test of effervescent spicy flavor of red ginger extract with a ratio of citric acid and sodium bicarbonate 52:48 not significantly different from a ratio of 51:49 but was markedly different from the concentration ratio of citric acid and sodium bicarbonate 50:50, 53:47 and 55:45.

Standard effervescent preparations have a solubility time of less than 5 minutes [11]. Based on the results of variance, the comparison of the concentration of citric acid and sodium bicarbonate affects the hedonic quality test of the effervescent solubility of red ginger extract (<α = 0.01 and 0.05). The Duncan test results showed that the comparison of citric acid and sodium bicarbonate in the making of effervescent red ginger extract can affect the effervescent solubility of the resulting red ginger extract, where the higher the concentration of citric acid is added, the faster the solubility of the product [18].

3.5 Supporting analysis

Effervescent products of red ginger extract with the best treatment were carried out supporting tests in the form of red ginger active ingredient test, red ginger extract extract test, ethanol residue and purchase decision based on product prices and general product acceptance. The best product was obtained based on the highest value hedonic test results, the product made using the ratio of citric acid and sodium bicarbonate 52:48 (A3). It is known that the test for the level of active ingredients of red ginger is 0.65%, while the active ingredient in effervescent red ginger extract is 0.52%. Testing of ethanol residue on effervescent red ginger extract is known to be 0.00%.

Supporting analysis was also carried out on the decision to buy effervescent red ginger extract products. The decision to buy a product is the action of the consumer to buy or not buy the product. Of the various factors that influence consumers in purchasing a product or service, usually consumers always consider the quality, price, and products that have been known by the community. The buying decision factor used in this study is the organoleptic quality and product selling price. The price of effervescent products from the research is Rp. 2,000 / 6 gram packaging. The diagram of the test results for the decision to buy effervescent red ginger extract products can be seen in figure 2.

The results of the buying decision test based on product acceptance in general as many as 60% of panelists liked and intented to buy products with a concentration of citric acid and sodium bicarbonate 52:48 and based on the price of products sold Rp.2,000/6 gram packaging as much as 40% panelists liked and intented to buy the product.

4. Conclusion

Based on the physical properties, the difference in the ratio of the concentration of citric acid and sodium bicarbonate to the effervescent red ginger extract only affected the dissolution time, but did not affect the flow time, angle of repose and bulk density. All levels of comparison of the concentration of citric acid and sodium bicarbonate could still be categorized as conformed with the physical requirements of effervescent products.

Based on chemical properties, the difference in the ratio of concentrations of citric acid and sodium bicarbonate to effervescent red ginger extract affected pH, but did not affect water content. All levels of comparison of concentrations of citric acid and sodium bicarbonate could still be categorized as conformed with the chemical requirements of effervescent products.

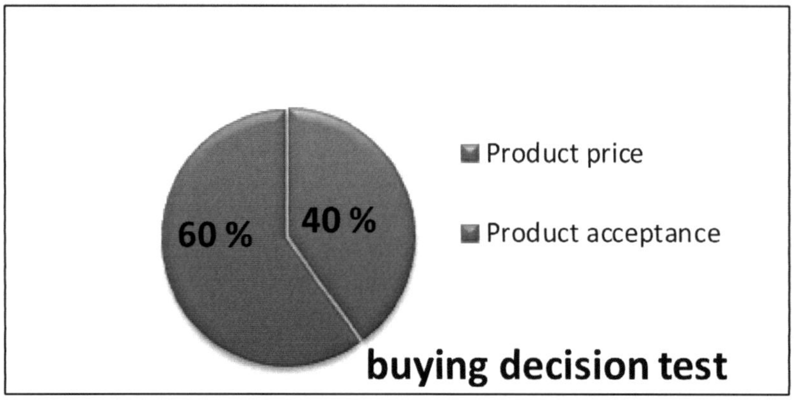

Figure 2. The results of the test diagram for the purchase of red ginger extract effervescent products.

Based on the hedonic test, the difference in the ratio of the concentration of citric acid and sodium bicarbonate to the effervescent red ginger extract only affected color and taste, but did not affect smell and solubility. Comparison of the concentration of citric acid and sodium bicarbonate which had the best level of preference is 52: 48. Based on the hedonic quality test of the best effervescent product of red ginger extract had a bright color, the smell of the slightly ginger smells, tastes spicy, and a good solubility.

The effervescent product of red ginger extract with the best ratio of citric acid and sodium bicarbonate (52:48), the raw material (red ginger) had characteristics of red ginger active content of 0.65%: the level of effervescent red ginger extract 0.52%; 0.00% ethanol residue. In addition, the results of the test of buying decisions were also obtained based on product acceptance, 60% panelist liked and intented to buy and based on product selling prices of 40% panelist liked and intented to buy too.

5. References

[1] Harwati C H T 2009 Khasiat jahe bagi kesehatan tubuh manusia *Jurnal Inovasi Pertanian* **8(1)** 54-61

[2] Hapsoh H Y and Julianti E 2008 Budidaya dan Teknologi Pascapanen Jahe. USU Press *Medan* 1,3

[3] Ervina A T 2010 Formulasi Tablet Effervescent Ekstrak Jahe Merah (*Zingiber officinale* Rosc) Dengan Kombinasi Asam Sitrat dan Asam Malat sebagai sumber asam dan natrium karbonat sebagai sumber basa. Fakultas Farmasi Universitas Muhammadiyah Surakarta

[4] Lestari P M, Radjab N S and Octaviani A 2014 Formulasi dan evaluasi fisik granul effervescent sari buah naga *(Hylocereus undatus).*

[5] Imanuela M, Sulisyawati, Ansori M 2012 Penggunaan asam sitrat dan natrium bikarbonat dalam minuman jeruk nipis berkarbonasi. *J Food and Culinary Education* Univ Negeri Semarang

[6] Rantiasih M D 2007 Optimasi natrium bikarbonat dan campuran asam tartrat-asam fumarat sebagai eksipien dalam pembuatan granul effervescent ekstrak rimpang temulawak [*Curcuma xanthorrhizaroxb*.] secara granulasi basah dengan metode desain faktorial. *Skripsi thesis*, Universitas Sanata Dharma

[7] Kumullah I R 2016 Optimalisasi Formulasi Bahan Pengikat Dan Bahan Penghancur Terhadap Karakteristik Effervescent Ampas Stroberi *(Fragaria Chiloensis L.)* Program Studi Teknologi Pangan fakultas Teknik Universitas Pasundan. Bandung

[8] Ansel H C 2008 Pengantar Buku Sediaan Farmasi UI-Press *Jakarta*

[9] Siregar C J P and Wikarsa S 2010 Teknologi Farmasi Sediaan Tablet :Dasar-Dasar Praktis EGC *Jakarta*

[10] Lachman L, Lieberman H A and Kanig J L 2008 Teori dan Praktek Farmasi Industri Universitas Indonesia Press *Jakarta*

[11] Anshory, Syukri, and Malasari 2007 Formulasi Tablet Effervescent Dari Ekstrak Ginseng Jawa *(Tlinumpaniculatum)* Dengan Variasi Kadar Pemanis Aspartam. *Jurnal Ilmiah Farmasi*
[12] AOAC 2012 Official Method of Analysis of the Assosiation Analytical Chemistry Inc. Washington D.C.
[13] Juita Y 2008 Formulasi Tablet Effervescent Tepung Daging Lidah Buaya. Skripsi FMIPA UI *Depok*
[14] Estiasih T and Ahmadi K 2009 Teknologi Pengolahan Pangan PT Bumi Aksara *Jakarta*
[15] Misbah D 2014 Pengaruh Asam Basa dalam Makanan terhadap Kesehatan https://klinikpengobatanalami.wordpress.com/(*accessed 21 March 2018*)
[16] Lee S, Choo C, Halstead C W, Huynh T, and Bensoussan 2007 Liquid chromatographic determination of 6-, 8-, 10-gingerol, and 6-shogaol in ginger *(Zingiber officinale)* as the raw herb and dried aqueous extract *Journal of AOAC International* **90(5)** pp 1219-1226
[17] Husni P, Norisca A P, and Imam A W 2017 Metode Deteksi Kandungan Babi dan Alkohol dalam Eksipien Farmasi dan Produk Obat untuk Menjamin Kehalalan Sediaan Obat *Majalah Farmasetika* **2 1** pp 1-7
[18] Fung K Y and King K M 2003 Product Centered Processing: Pharmaceutical and Capsules *J AICHE.* **49: 5** pp 1193-1218
[19] Dede Y 2005 Studi Kasus Fisika Pangan Pembuatan Tablet Effervescent Sari Buah Tomat *Skripsi* IPB Bogor
[20] Winarno F G 2004 Kimia Pangan dan Gizi Gamedia Pustaka Utama *Jakarta*

Study of chemical components of liquid smoke from cocoa rind in two variations of moisture content by using GC-MS method

I K Budaraga*, R Aga Salihat

Department of Agricultural Product Technology, Faculty of Agriculture, Ekasakti University, Veteran Dalam Street No. 26 B Padang, Indonesia

*Corresponding author: budaraga1968@gmail.com

Abstract. The purpose of this study was to determine the chemical components contained in cocoa smoke liquid smoke at two different moisture contents by using gas chromatography-mass spectrometry (GC-MS) method. The sample was cocoa rind with a moisture content of 10% and 25% that were hydrolyzed at 200°C. The results showed that the dominant chemical components in liquid smoke at 10% moisture content were: hexadecanoic acid (palmitic acid), butyrolactone, 2,6-dimethoxy-phenol; 5-methyl-5-hexen-3-in-2-ol and 2-methoxyphenol. While the liquid smoke at 25% moisture content contained: hexadecanoic acid (palmitic acid), octadecanoic acid, 2,6-dimethoxy-phenol, (9z)-octadec-9-enoic acid and 2-methoxy phenol. The abundance of beneficial chemical components contained, making the liquid smoke of cocoa rind has a high economic value. However, the phenol compound contained, resulting in the liquid smoke of the cocoa rind requires a further refining process.

Keywords: Cocoa Rind, Liquid Smoke, Palmitic Acid, GC-MS

1. Introduction

Cocoa farmers generally only use cocoa beans during harvesting, while cocoa rind is usually used as animal feed ingredients and the rest is left to become waste around the plantation area. Cocoa peel waste produced in large quantities will be a problem if it is not handled properly because this solid waste production reaches more than 60% of the total fruit production, this will be a great potential to pollute the surrounding environment. Therefore, there is a need for a solution, one of which is the manufacture of liquid smoke [1].

Liquid smoke can be obtained by condensing or condensing biomass from the pyrolysis process. Liquid smoke is a mixture of solutions from wood smoke suspensions in water made by condensing or condensing the results of wood pyrolysis [2]. This pyrolysis process produces gas, solids and liquid smoke. Smoke is produced by incomplete combustion which involves decomposition reactions of polymer constituents into low molecular weight organic compounds due to the influence of heat which includes oxidation, polymerization, and condensation reactions. During combustion, components of wood including cellulose, hemicellulose, and lignin will undergo the pyrolysis process to produce various compounds including phenols and their derivatives, carbonyl (ketones and aldehydes), carboxylic acids, furans, alcohols, lactones, polycyclic aromatic hydrocarbons and so on. The most important components in contributing to the smoking reaction are three compounds, namely phenol, acid and carbonyl [4].

Several studies on the production and utilization of liquid smoke have been carried out, among others: the development of liquid smoke producing devices from rice husks to produce organic insecticides [5], making liquid smoke from rubber shells as latex coagulants [6], chemical components of pyrolysis liquid smoke solid palm oil waste [7], making liquid smoke from coconut fiber with non pyrolysis technique [8], activity test and antibacterial liquid smoke originating from cinnamon bark and peanut shell [9].

Pyrolysis involves various reaction processes, namely decomposition, oxidation, polymerization, and condensation. The process that occurs during pyrolysis according to the increase in temperature is removal of water from wood at a temperature of 120-150°C, pyrolysis of hemicellulose at a temperature of 200-250°C, pyrolysis of cellulose at temperatures of 280-320°C and pyrolysis of lignin at 400°C. Pyrolysis at a temperature of 400°C produces compounds that have high organoleptic quality and at higher temperatures a condensation reaction for the formation of new compounds and oxidation of condensation products followed by linear increases in tar and polycyclic aromatic hydrocarbons [4].

The chemical components of wood which decompose during the pyrolysis process are hemicellulose, cellulose, and lignin. Hemicellulose is the component that undergoes the earliest pyrolysis producing furfural, furan, acetic acid, and homologs. Pyrolysis of pentosan forms furfural, furans and their derivatives along with a long series of carboxylic acids. Together with cellulose pyrolysis hexose forms acetic acid and homologs. Lignin in pyrolysis produces compounds that play a role in the smell of smoke from smoked products. These compounds are phenols and phenolic ethers such as guaiakol (2-methoxyphenol) and their homologs and derivatives. The next process is cellulose pyrolysis producing acetic acid compounds, and carbonyl compounds such as acetaldehyde, glyoxal and akreolin. Lignin pyrolysis will produce phenol compounds, guaiakol, syringol together with homologs and their derivatives.

Liquid smoke has functional properties as an antioxidant, antibacterial and forming colors and distinctive taste. These functional properties are related to the components contained in the liquid smoke. Liquid smoke has the ability to preserve food ingredients because of the presence of acidic compounds, phenol derivatives, and carbonyl [3]. To produce liquid smoke that is of high quality and optimum liquid smoke yield, a pyrolycerator is needed which can increase the quantity and quality of liquid smoke.

In some situations, GC can help in identifying a complex such as liquid smoke. In gas chromatography, the mobile phase is a carrier gas, which is usually a pure gas such as helium or non-reactive, such as nitrogen gas. A stationary phase is a microscopic stage of a liquid layer or polymer that supports pure gas, in a part of glass or metal pipe system called a column. The instrument used to carry out this analysis is called gas chromatography - mass spectroscopy (GC-MS). So far there has been no study to look at the chemical components of cocoa pod liquid smoke. In this research, the chemical components of liquid smoke from cocoa peel with different moisture content than GC-MS were examined.

2. Methodology

This research was carried out by bringing raw materials from brown rind from Padang Pariaman Regency. Furthermore, the production process of brown leather liquid smoke was carried out in the mechanical engineering production laboratory of the Faculty of Engineering, Ekasakti University, and the liquid smoke maker was produced in the welding workshop which was reached in Padang City. The chemical components using the GC-MS method were analyzed in the Center Instrumentation Laboratory of the Faculty of Agricultural Technology Andalas University in May 2019.

The tools used include: thermometer IR temperature 600°C sellery brands, brand value vacuum pumps, San-EI brand water pumps SE 125A model with pump suction power 42 liters / minute, digital scales, tools to make pyrolysis tank (pyrolyzer) namely stainless steel pan volume of 32 liters, diameter of 30 cm, height of 30 cm and thickness of 5 mm equipped with tar size 10 x 12 cm, stainless steel pipe diameter of 0.75 inches, length of 1.5 meters connecting pyrolysis tank with condenser, for condenser pipes using stainless pipes stell 0.75-inch diameter 8 meters long, there are 12 rounds of diameter 30 cm, for operation the win gas brand 0100 gas stove is used, and 2 containers of liquid smoke from plastic and glass. The material used is cocoa fruit rind with 10% and 25% moisture content as measured by the MBI-TM-20 type multi-digital moisture content measuring device. The MS GC Shimadzu brand is equipped with a carrier gas backup (injector), a place for injecting substances, columns, detectors and recording devices displayed on the computer.

The procedure for making liquid smoke starts with preparing 2 kg of raw material for cocoa fruit skin, the water content is 10% and 25%. Each raw material is first cleaned, reduced in size from 2 to 4 cm, dried in the sun, and measured in its moisture content using the MBI-TM-20 type multi-digital moisture meter. Then, the ingredients of the cocoa peel with different moisture content are inserted into the pyrolyzer and closed tightly so that the condition is completely airtight. Pyrolysis pipe joints are checked and are really strong and do not leak, and a vacuum compressor pump can work safely. Cooling water is sure enough and the water circulation pump runs smoothly. The heating source (gas stove) is turned on, while the temperature moves on the temperature gauge according to what is desired and the time of pyrolysis is noticed until the first smoke comes out and liquid smoke comes out.

During the pyrolysis process, the flowing water is sufficient and inundates the condenser spiral pipe so that the condensation process can take place perfectly. The pyrolysis process is considered complete if there is no more liquid smoke coming out from the pyrolysis tank. After the pyrolysis stage is complete, the cooling process is carried out by waiting until the temperature drops (room temperature), after the temperature drops, the lid is opened valve and activated charcoal pyrolysis results are weighed, including the volume of liquid smoke measured, tar separated by opening tar reservoir on the pyrolysis pipe and the weight of charcoal obtained. Documentation of liquid smoke making equipment can be observed in Figure 1.

Figure 1. Construction of the liquid smoke maker: a). Pyrolysis tube b). Smoke pipe, c) Condenser, d). Liquid smoke container.

After liquid smoke with a moisture content of 10% and 20% were completed at a pyrolysis temperature of 200°C, all GC-MS equipment was prepared such as columns, detection devices, temperature, and carrier gas flow. Everything is arranged until conditions such as those printed on each monograph, a solution of a number of substances listed on each monograph are injected at the injection site using a micro-injector. Separation of components is detected and described in chromatography. The X axis in the chromatogram is expressed in retention time (time from sample injection to peak curve on the chromatogram) or retention volume (retention time x carrier gas flow rate) which is constant for each substance under a constant condition. This basis is used for identification. From the area of the curve peak or the height of the curve, the component of the substance can be determined quantitatively.

Usually GC-MS equipment before being used in the sample, there is a calibration treatment with the procedure as follows: one row of solution is made. After that, the solution is injected with the same volume into the injection site. The calibration line is drawn from the chromatogram, with the weight of the substance on the horizontal axis, and the height of the curve peak or area of the peak curve on the vertical axis. Substance solutions are made as shown on each monograph. From the chromatogram obtained with the same conditions such as how to obtain a calibration line, the area of the curve peak area or the height of the curve peak is measured. The amount of substance is calculated using a calibration line. In this way of working, everything must be done with conditions that are truly constant. The GC-MS instrument can be observed in Figure 2.

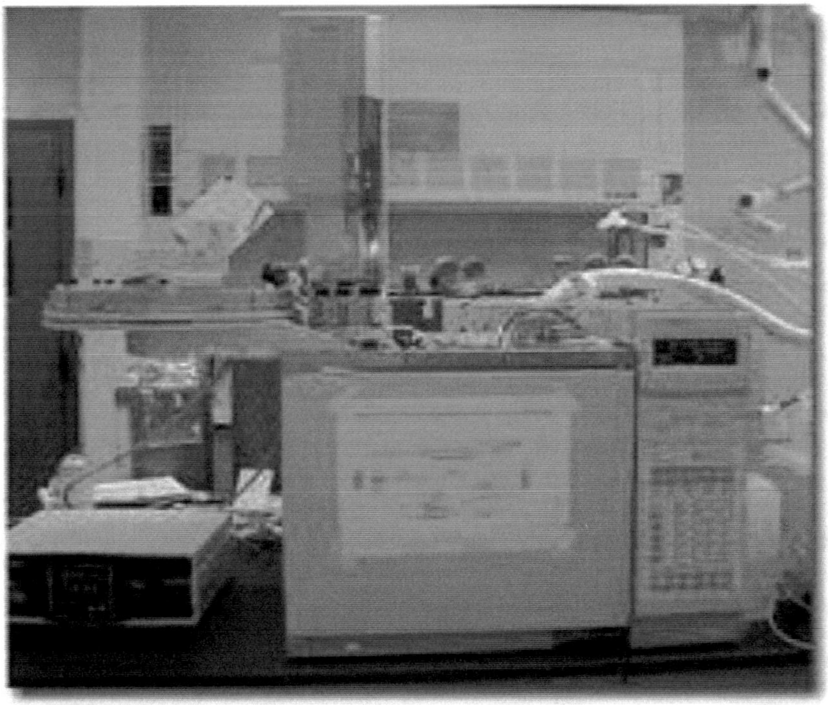

Figure 2. GC-MS instrument for analysis of chemical components of cocoa rind liquid smoke

3. Results and Discussion

Identification of liquid smoke components from cocoa peel with variations in water content was carried out using the Gas Chromatography - Mass Spectroscopy or Gas Chromatography - Mass Spectroscopy (GC-MS) method. This method can be used to identify a compound, both one component and a mixture [10]. This GC-MS method has advantages in identifying molecular structures based on fragmentation [11]. In addition to identifying, the GC-MS method also functions as a separation technique, especially for essential oils [12]. Analysis using GC-MS showed that liquid smoke from cacao rind contained 48 organic compounds for 10% moisture content and 47 organic compounds for 25% moisture content. As many as 22 compounds contained in these two variations have a percentage above 1%. Some of these components are classified as saturated fatty acids, phenolic compounds, lactone compounds, and acetic acid derivatives. In addition, from the results of GC-MS analysis, also found compounds used as food additives, namely 5-methyl-5-hexen-3-in-2-ol at 10% moisture content and 3-ethyl-2-hydroxy-2 -cyclopentene-1-on at 25%

moisture content. Compounds contained in liquid smoke with two variations in water content, can be observed in Table 1 and Table 2.

Table 1. Chemical components with more than 1% in liquid smoke of cocoa rind with a moisture content of 10%

No	Chemical Components	Persentage (%)
1	Hexadecanoic acid ($C_{16}H_{32}O_2$)	17,70
2	Butyrolactone ($C_4H_6O_2$)	7,31
3	2,6-Dimethoxyphenol ($C_8H_{10}O_3$)	6,20
4	5-Methyl-5-hexen-3-in-2-ol ($C_7H_{10}O$)	5,43
5	2-Methoxyphenol ($C_7H_8O_2$)	4,85
6	Propylene carbonate ($C_4H_6O_3$)	4,78
7	Cyclopentanedion, 3-methyl-1,2- ($C_6H_8O_2$)	4,26
8	1,2-Benzenediol ($C_6H_6O_2$)	4,26
9	Acetamide (C_2H_5NO)	3,86
10	Acetamide, N,N-dimethyl- (C_4H_9NO)	3,51
11	Hexanoic acid, 2-methylphenyl ester ($C_{13}H_{18}O_2$)	3,43
12	2H-Pyran-2-one, tetrahydro- ($C_5H_8O_2$)	3,00
13	4-Methyl-1,2-benzenediol ($C_7H_8O_2$)	2,87
14	2,4-Hexadiene, 2,3-dimethyl- (C_8H_{14})	2,58
15	2-Cyclopenten-1-one, 3-ethyl-2-hydroxy- ($C_7H_{10}O_2$)	2,05
16	2-Methyl-2,3-divinyloxirane ($C_7H_{10}O$)	1,90
17	2-Propanone, 1-(4-hydroxy-3-methoxyphenyl)- ($C_{10}H_{12}O_3$)	1,69
18	1,2-Benzenediol, 3-methyl- ($C_7H_8O_2$)	1,37
19	1,4-Benzenediol, 2-methyl- ($C_7H_8O_2$)	1,33
20	1,2-Benzenediol, 3-metoxy- ($C_7H_8O_3$)	1,18
21	2-Pentanone, 1-(2,4,6-trihydroxyphenyl)- ($C_{11}H_{14}O_4$)	1,16
22	Phenol, 2-methoxy-4-methyl- ($C_8H_{10}O_2$)	1,01

Table 2. Chemical components with more than 1% in liquid smoke of cacao rind with 25% moisture content

No	Chemical Components	Persentage (%)
1	Hexadecanoic acid ($C_{16}H_{32}O_2$)	12,30
2	Octadecanoic acid ($C_{18}H_{36}O_2$)	8,84
3	Phenol, 2,6-dimethoxy- ($C_8H_{10}O_3$)	5,78
4	(9Z)-Octadec-9-enoic acid ($C_{18}H_{34}O_2$)	4,90
5	Phenol, 2-methoxy- ($C_7H_8O_2$)	4,50
6	3,5-Heptadien-2-one ($C_7H_{10}O$)	3,06
7	2-Propanone, 1-(4-hydroxy-3-methoxyphenyl)- ($C_{10}H_{12}O_3$)	3,04
8	Butyrolactone ($C_4H_6O_2$)	2,87
9	3-Methyl-1,2-cyclopentanedione ($C_6H_8O_2$)	2,69
10	Benzene, 1,2,3-trimethoxy-5-methyl- ($C_{10}H_{14}O_3$)	2,25
11	1,2-Benzenediol ($C_6H_6O_2$)	2,24
12	Acetamide (C_2H_5NO)	2,23
13	Cyclopropil carbinol (C_4H_8O)	2,13
14	3-Isopropyl-5-(phenoxymethyl)-2-oxazolidinone ($C_{13}H_{17}NO_3$)	1,81
15	Phenol (C_6H_5OH)	1,69

16	1,2-Benzenediol, 3-methyl- ($C_7H_8O_2$)	1,68
17	Phenol, 4-ethyl-2-methoxy- ($C_9H_{12}O_2$)	1,63
18	2-Cyclopenten-1-one, 2,3-dimethyl- ($C_7H_{10}O$)	1,22
19	Phenol, 2-methoxy-4-methyl- ($C_8H_{10}O_2$)	1,11
20	2,5-Dimethylhydroxyquinone ($C_8H_{10}O_2$)	1,11
21	Tetradecanoic acid ($C_{14}H_{28}O_2$)	1,11
22	2-Cyclopenten-1-one, 3-ethyl-2-hydroxy ($C_7H_{10}O_2$)	1,03

The 5 dominant compounds contained in the liquid smoke of cocoa rind with 10% moisture content were hexadecanoic acid; butyrolactone; 2,6-dimethoxyphenol; 5-methyl-5-hexen-3-in-2-ol and 2-methoxyphenol with successive retention times: 18,865 minutes; 2,829 minutes; 9,206 minutes; 2,275 minutes, and 4,551 minutes, respectively. While for liquid smoke of cocoa rind with 25% moisture content, five dominant compounds are: hexadecanoic acid; octadecanoic acid; 2,6-dimethoxyphenol; (9Z)-octadec-9-enoic acid and 2-methoxyphenol with successive retention times: 18,766 minutes; 13,791 minutes; 9,013 minutes; 13,016 minutes and 4,599 minutes. Retention time is the time needed by a compound to move through the column towards the detector. Retention time was measured based on the time at which the sample is injected until the sample shows the maximum peak height on the chromatogram of the compound. Retention time is the most important function of gas chromatography due to as a key parameter for separating, identifying and determining the number of compounds contained in a mixture [13].

The most dominant compound found in the two variations of water content was hexadecanoic acid ($C_{16}H_{32}O_2$), or often known as palmitic acid. The difference between the two variations is the percentage of the compound. For liquid smoke with a moisture content of 10%, palmitic acid has a percentage of 17.70%. While for liquid smoke with 25% moisture content, palmitic acid has a percentage of 12.30%. The percentage of palmitic acid in liquid smoke of cocoa rind with a moisture content of 10% is greater than the moisture content of 25%. This is caused by palmitic acid and water has the opposite polarity. Palmitic acid is an example of saturated fatty acids which are widely found in nature [14]. In industry, palmitic acid is widely used in the field of cosmetics and coloring. In terms of nutrition, palmitic acid is an important source of calories but has a low antioxidant power. The structural formula of palmitic acid can be observed in Figure 3. Characteristics of palmitic acid can be observed in table 3.

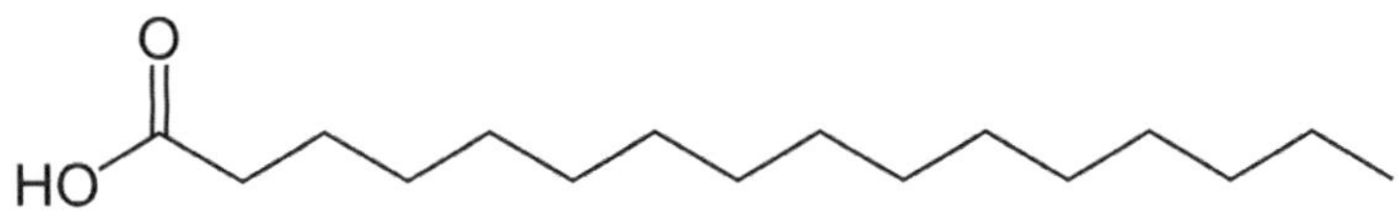

Figure 3. Formula structure of hexadecanoic acid / palmitic acid.

Table 3. Chemical and physical properties of palmitic acid.

Characteristics palmitat acid	
Chemical formula	$C_{16}H_{32}O_2$
Molar mass	256,42 g/mol
Appearance	White crystal
Density	0,853 g/mL pada 62 °C
Melting point	62,9 °C
Boiling point	351-352 °C

Another dominant compound found in liquid smoke of cocoa rind with a 10% moisture content is 2,6-dimethoxyphenol or better known as oleic acid which was a saturated fatty acid and 5-methyl-5-hexen-3-in-2-ol which often used as a food additive or food flavoring. The structure formula for 5-methyl-5-hexen-3-in-2-ol can be observed in Figure 4 and its characteristics in table 4.

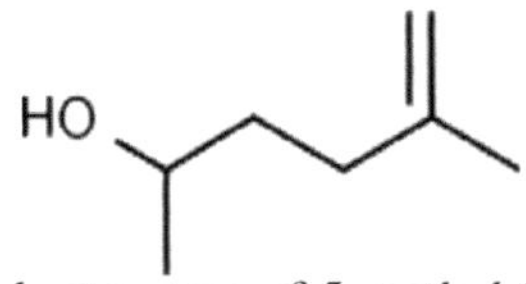

Figure 4. Formula structure of 5-methyl-5-hexen-3-in-2-ol.

Table 4. Chemical and physical properties of 5-methyl-5-hexen-3-in-2-ol.

Characteristics of 5-methyl-5-hexen-3-in-2-ol	
Chemical formula	$C_7H_{14}O$
Molar mass	110,156 g/mol
Appearance	Pale yellow liquid
Density	0,8±0,1 g/mL
Boiling point	163,2±9 °C

Whereas the other dominant compound found in liquid smoke of cocoa peel with 25% moisture content as octadecanoic acid or better known as stearic acid. This compound also includes saturated fatty acids. Stearic acid is used in making medicines. Recently, stearic acid has also been used in the development of drug delivery systems because they are considered inert, inexpensive and biocompatible, and have low toxicity [15]. In addition, stearic acid is also used as a system for carrying cyclosporine-A drugs and to cover the bitter taste of compound drugs [16][17]. The structural formula for stearic acid compounds can be observed in Figure 5 and its characteristics in table 5.

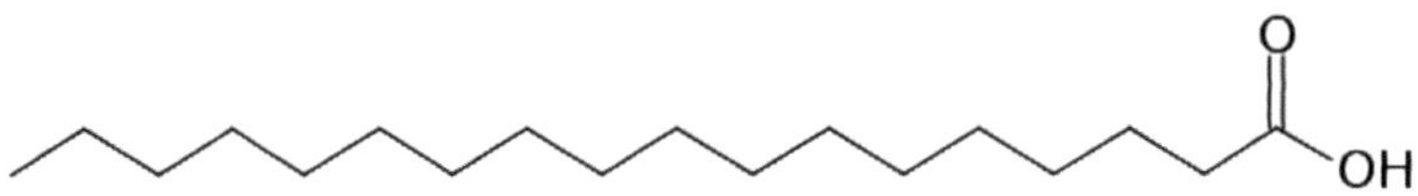

Figure 5. The structure formula of octadecanoic / stearic acid.

Table 5. Chemical and physical properties of stearic acid.

Characteristics of stearic acid	
Chemical formula	$C_{18}H_{36}O_2$
Molar mass	284,484 g/mol
Appearance	White crystal
Density	0,847 g/mL pada 70 °C
Melting point	69,3 °C
Boiling point	361 °C

Phenol compound was also contained in two variations in water content, namely: 0.84% for water content of 10% and 1.69% for water content of 25%. Phenol are corrosive compound that can cause chemical combustion in exposed skin. The percentage of phenol in liquid smoke of cocoa rind with a moisture content of 10% is smaller than 25% moisture content.

Based on the type, the compounds in the liquid smoke of cocoa rind, both with 10% and 25% moisture content, are divided into four groups, namely: aromatic compounds, carboxylic acids and esters. Classification of compounds contained in liquid smoke with variations in water content can be observed in table 6, table 7 and table 8.

Table 6 shows that the liquid smoke of cocoa rind with variations in water content contains benzene aromatic compounds and their derivatives, such as: phenols, phenolic compounds and quinones. Phenol is a compound with an OH group attached to the benzene ring. Meanwhile, phenolic compounds and quinones are phenol derivative compounds [18]. For groups of aromatic compounds, both variations in water content are dominated by phenolic compounds that can be used as medicinal ingredients.

Table 6. Aromatic compounds contained in the liquid smoke of cocoa rind with variations in water content

Cocoa rind liquid smoke	
Water content 10%	Water content 25%
2,6-Dimethoxyphenol	Phenol, 2,6-dimethoxy-
2-Methoxyphenol	Phenol, 2-methoxy-
1,2-Benzenediol	Benzene, 1,2,3-trimethoxy-5-methyl-
4-Methyl-1,2-benzenediol	1,2-Benzenediol
1,2-Benzenediol, 3-methyl-	Phenol
1,4-Benzenediol, 2-methyl-	1,2-Benzenediol, 3-methyl-
1,2-Benzenediol, 3-methoxy-	Phenol, 4-ethyl-2-methoxy-
Phenol, 2-methoxy-4-methyl-	Phenol, 2-methoxy-4-methyl-
Phenol	2,5-Dimethylhydroxyquinone

Table 7 shows compounds classified to carboxylic acids and their derivatives. Carboxylic acid compounds that dominate liquid smoke with two variations of water content were saturated fatty acids, namely: hexadecanoic acid (palmitic acid), octadecanoic acid (stearic acid), (9Z)-octadec-9-enoic acid (oleic acid) and tetradecanoic acid (myristic acid). In addition, the two variations also contain acetamide which is a derivative of acetic acid. The liquid smoke of cocoa rind with 25% moisture content has more saturated fatty acids than liquid smoke from cocoa rind with a moisture content of 10%.

Table 7. Carboxylic acid compounds contained in the liquid smoke of cocoa rind with variations in water content

Cocoa rind liquid smoke	
Water content 10%	Water content 25%
Hexadecanoic acid	Hexadecanoic acid
Acetamide	Octadecanoat acid
Acetamide, N,N-dimethyl-	(9Z)-octadec-9-enoic acid
	Acetamide
	Tetradecanoic acid

The group of ester compounds contained in liquid smoke with two variations in water content can be observed in table 8. The group of ester compounds contained in the two variations were lactone compounds and fatty acid esters. Lactone compounds are cyclic esters which are condensation products of alcohol and carboxylic acids in the same compound. Lactone compounds contained in liquid smoke with two variations of water content are butyrolactone. However, liquid smoke with a water content of 25% contains other lactone compounds, namely 3-Methyl-1,2-cyclopentanedione or better known as maple lactone. Lactone compounds are commonly used in the food industry as the aroma of various foods and fruits [19]. In addition, lactone compounds are also used in other fields, for example, as precursors in the production of biodegradable polyesters such as polycaprolactone (PCL) because of their suitable properties [20].

Table 8. Esters contained in the liquid smoke of cocoa peel with variations in water content

Cocoa rind liquid smoke	
Water content 10%	Water content 25%
Butyrolactone	Butyrolactone
Propylene carbonat	3-Methyl-1,2-cyclopentanedione
Hexanoic acid, 2-methylphenyl ester	Hexadecanoic acid, methyl ester
Butiric acid, 2,2-dimetil-, vinyl ester	
Hexadekanoic acid, methyl ester	

The fatty acid ester compounds contained in liquid smoke with two variations of water content are hexadecanoic acid, methyl ester or better known as methyl palmitate. Methyl palmitate belongs to fatty acid methyl esters (FAME) compounds. Methyl palmitate is one of the main components of biodiesel fuel [21]. Biodiesel is a fuel consisting of monoalkyl esters of long chain fatty acids derived from vegetable oils, animal fats, or mixtures thereof. Biodiesel is produced by transesterification of triglycerides with short chain alcohols, usually methanol or ethanol, in the presence of a catalyst, which leads to the formation of a mixture of fatty acid methyl ester (FAME) or fatty acid ethyl esters (FAEE) [22][23][24]. In addition to methyl palmitate, liquid smoke with a 10% moisture content also contains other fatty acid esters, such as: hexanoic acid, 2-methylphenyl ester (methyl hexanoic) and butyric acid, 2,2-dimethyl-, vinyl ester (vinyl 2,2-dimethylbutanoic).

4. Conclusion

The biggest component in the liquid smoke of cocoa peel both with 10% and 25% moisture content is hexadecanoic acid or better known as palmitic acid which is widely used in the field of cosmetics and dying. In addition there are also components that can be used as food additives, both in liquid smoke with 10% moisture content and 25% moisture content. The phenolic compounds contained can act as scents that are typical of food and drink, as well as antioxidants. Because of many useful chemical components contained, the liquid smoke of cocoa rind has a high economic value. However, the phenol compound contained in it, result in the liquid smoke from the cocoa rind requires further refining processes.

5. References

[1] Harsini and Susilowati Pemanfaatan kulit buah kakao dari limbah perkebunan kakao sebagai bahan baku pulp dengan proses organosolv 2010 *J. Ilmu. Tek.. Lingkungan* **2**(2) 80–89
[2] Darmadji P 1996 Aktivitas antibakteri asap cair yang diproduksi dari bermacam-macam limbah pertanian Laporan Penelitian Mandiri, DPP-UGM **16** 19–22
[3] Darmadji P 2002 *Optimasi pemurnian asap cair dengan metode redistilasi* Proseding Seminar Nasional PATPI
[4] Girrard J P 1992 *Technology of meat and meat product* (New York: Ellis Hardwork)
[5] Putri, R E, Mislaini and Lisa S N Pengembangan alat penghasil asap cair dari sekam padi untuk menghasilkan insektisida organik 2015 J. Tek. Pertanian Andalas **19**(2) 30–36
[6] Prasetyowati, Hermanto M and Salman Farizy S Pembuatan asap cair dari cangkang buah karet sebagai koagulan lateks 2014 *J. Tek. Kim.* **20**(4) 14–21
[7] Haji A G Komponen kimia asap cair hasil pirolisis limbah padat kelapa sawit 2013 *J. Rekayasa Kim. dan Lingkungan* **9**(3) 109–116
[8] Feti F Komposisi dan aktivitas antibakteri asap cair sabut kelapa yang dibuat dengan teknik pembakaran non pirolisis 2011 *Agritech* **31**(4): 305–311
[9] Yefrida, Aprilina F, Leone I T, Refilda and Salim M Uji aktivitas anti bakteri asap cair yang berasal dari batang kayu manis dan kulit kacang tanah 2009 *J. Ris. Kim.* **2**(2) 190–194
[10] Sastrohamidjoyo H 1985 *Spektroskopi* (Yogyakarta: Liberty)

[11] Harbone J B 1987 *Metode Fitokimia. Penuntun Cara Modern Menganalisis Tumbuhan 2* (Bandung: ITB)
[12] Marriot P J, Shellie R and Cornwell C Review: gas chromatographic technologies for the analysis of essential oils 2001 *J. of Chromatography A* **936**(1) 1–22
[13] Etxebarria N, Zuloaga O, Olivares M, Bartolomé L J and Navarro P Review: retention-time locked methods in gas chromatography 2009 *J. of Chromatography A* **1216**(10) 1624–29
[14] Winarno F G 2004 *Kimia Pangan dan Gizi* (Jakarta: PT Gramedia Pustaka Utama)
[15] Killen B U and Corrigan O I Factors influencing drug release from stearic acid based compacts 2001 *Int. J. of Pharm.* **228**(1-2) 189–198
[16] Zhang Q, Yie G, Li Y, Yang Q and Nagai T Studies on the cyclosporine A loaded stearic acid nanoparticles 2000 *Int. J. of Pharm.* **200**(2): 153–159
[17] Robson H J, Craig D Q M and Deutsch D. An investigation into the release of cefuroxime axetil from taste-masked stearic acid microspheres. Part I: The influence of the dissolution medium on the drug release profile and the physical integrity of the microspheres 1991 *Int. J. of Pharm.* **190**(2) 183–192
[18] Fessenden R J and Fessenden J S 1992 *Kimia Organik Jilid 1 Edisi Ketiga* (Jakarta: Erlangga)
[19] Berger R G 2007 *Flavours and Fragrances Chemistry, Bioprocessing and Sustainability* (Berlin: Springer)
[20] Ravotti R, Fellmann O, Lardon N, Fischer L J, Stamatiou A and Worlitschek J Investigation of lactones as innovative bio-sourced phase change materials for latent heat storage 2019 *Molecules* **24**(7) 1300
[21] Knothe G "Designer" Biodiesel: Optimizing fatty ester composition to improve fuel properties 2008 *Energy Fuels* **22**(2) 1358–64
[22] Knothe G, Gerpen J V and Krahl J 2005 *The Biodiesel Handbook* (Champaign: AOCS Press)
[23] Pandey A 2009 *Handbook of Plant-based Biofuels* (Boca Raton: CRC Press Taylor & Francis Group)
[24] Pratas M J, Freitas S, Oliveira M B, Monteiro S C, Lima A S and Coutinho J A P Densities and Viscosities of Fatty Acid Methyl and Ethyl Esters 2010 *J. of Chem. & Engineering Data* **55**(9) 3983–90

6. Aknowledgment

We would like to thank the Directorate of Research and Community Service, the Directorate General of Research Strengthening and, Development of the Ministry of Research, Technology and Higher Education according to the 2019 Research Year contract number 012 / K10 / KM / Contract-Research-J / 2019 March 28 2019 and number 005 / LPPM-UNES / Research Contract-J / 2019 on 8 April 2019. Chancellor of Ekasakti University, Chair of the Institute for Research and Service to the Community of Ekasakti University, Dean of the Faculty of Agriculture, Ekasakti University, team and students who assisted this activity

IOP Conf. Series: Earth and Environmental Science **383** (2019) 012024 doi:10.1088/1755-1315/383/1/012024

Impact of forest fire in peat land on land properties in Pelalawan district region

Ibrahim*, Harlen, Sukendi, Y I Siregar

Doctoral Program in Environmental Sciences. University of Riau, Indonesia
Jl. Pattimura 9, Gobah. Pekanbaru , Riau. Postal Code 28 125

*Corresponding author: kamaruddineddiwan@gmail.com

Abstract. The physical, chemical, and biological properties of peat soil in several levels of fire in Kampar regency. The research was conducted from March to April 2017 in Pelalawan Regency, Riau Province, and Soil Science Laboratory, Faculty of Agriculture, University of Riau. The research was conducted by a survey in the field, determining the sample point using Purposive Sampling Method. The samples were taken from 3 locations ie. the land with heavy, medium and light fire level and control (non-burning/natural forest) 12 units of the experiment. Soil samples were analyzed by the physical and chemical characteristics soil. The results of the physical and chemical characteristics of soil undergoing changes and categorized as broken.

1. Introduction

Riau Province has the charm of peatlands with a million benefits in the ecosystem structure. At present, the privilege is threatened by land degradation that goes hand in hand with the development of plantations, agriculture, and industry. Peat soil island that can be used as an energy source. Peat soil is one of the most abundant lands in the world by originating from weathering or decay from the remnants of plants that are half-rotten and in the process takes a very long time.

Even though it has a myriad of features, peat soil also has deficiencies, such as a very wide range of fires. Even if peatland is on fire, it has a high degree of difficulty to extinguish it. Fire is a serious problem that can have an impact if viewed from various aspects. One of the impacts of forest fires that can be assessed is the quality of the land including the physical, chemical and biological properties of the soil.

The impact of changes in the physical properties of soil due to combustion is seen by the loss of soil organic matter by the heat of the fire destroys the structure of the soil, increases the weight of the soil, reduces soil porosity. This damage will result in decreasing the speed of infiltration and increasing the speed of runoff water [1]. Where, it appears that the chemical properties of peat soil due to combustion are changes in fuel to ash containing various nutrients such as N, P, S, and C / N will be lost by the evaporation process during the process of forest fires. In addition, combustion increases the pH of peat soil. The chemical properties of soil due to fire will not be able to improve soil fertility in the long term but only temporary.

The impact of changes in the biological properties of peat soil due to combustion is the death of flora, fauna and soil microorganisms. This loss will affect the chemical properties of peat soil. This is caused by the secretion and excretion of fauna and soil microorganisms that can indirectly improve the chemical properties of the soil.

2. Materials and Methods

This research was conducted in 4 locations, which consisted of a non-burning area (K1) in Pangkalan Gondai village, (coordinates 00 05 '31 "N 1010 49' 14" E), (K2) slightly burned area in Kerinci Barat village (coordinates 00 23 '39 "N 1010 49' 56" E), Pangkalan Kerinci Subdistrict, (K3) Moderately burned area in Langgam Village (coordinates 00 06 '32 "N 1010 39' 41" E), Langgam District, and (K4) heavily burned area in Kerinci Timur village (coordinates 00 24 '42 "N 1010 52' 40" E), Pelalawan Regency, Riau Province. The location of this study has experienced a fire incident for 1 year; this land is now dominated by peatland with a maturity level between milk and cloth and has a depth of 4-7 meters (Figure 1).

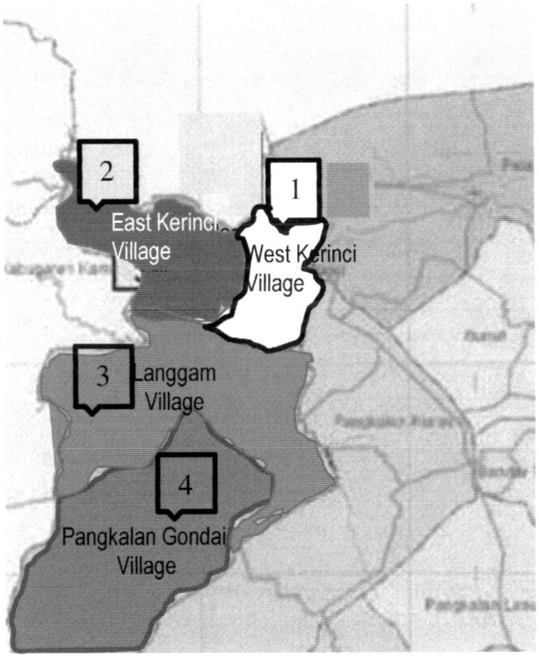

The research location as:

1. East Kerinci Village
2. West Kerinci Village
3. Langgam Village
4. Pangkalan Gondai Village

No fire areas (1)

Low fire areas (2)

Medium fire areas (3)

High fire areas (4)

Figure 1. Map of research location in Pelalawan Regency

2.1 Analysis of the physical properties of peat soil

The tools used to measure the activity of physical properties of soil are soil samples in the form of rings, hoes, peat drills, ovens, analytical scales, porcelain glasses, measuring cups, plastic, soil. Analysis of soil samples was carried out at the Soil Laboratory of the Faculty of Agriculture, Riau University, Pekanbaru, Riau. This study was conducted for 1 month starting from March to April 2018. In analyzing nutrient levels in a solid sample. Nutrients in the soil were analyzed by the gravimetric method in the form of nutrient deposits in the soil. The constant price of nutrient deposition solubility in soil = 1.8 x 10−10.

The implementation is first weighed 2 grams of sample, then dissolved in a little HCl in a small beaker. Then transferred and diluted in a 250 mL measuring flask. Next, pipette from this solution 25 mL into a 400 mL beaker, add 50 mL aquadest and 10 mL 1: 1 HCl. Add 2 mL HNO 3 concentrated and boil carefully until the solution sample is yellow. Dilute to 200 mL. Heat until boiling. Then, slowly add 1: 1 ammonia to it until all the nutrients (III) are deposited (the steam on it smells of ammonia). Boil for one minute and leave the precipitate down to the bottom of the beaker. After most of the sediment has gone down, the solution is decanted through filter paper that has been prepared in a filter funnel, but it is tried so that most of it remains in the beaker.

Added a 1% hot solution of ammonium nitrate into a beaker, stirred briefly, and left the precipitate down to the bottom of the glass, then repeated this decantation work several times, then transferred most of the sediment to filter paper, cleaned the glass wall with policeman and transferred the remaining sediment to filter paper with hot water spray aid. Wash these deposits several times with a hot solution of ammonium nitrate until the filtrate is practically chloride free. While this work took place, it was spawned with a clean porcelain cup with a lid. After cooling in the desiccator, the weight is determined. Next, it is wrapped in sediment in ash-free filter paper or Whatman 41 filter paper and placed in a cup. Heated on low heat until all the water evaporates, then paperized and scraped the paper insufficient airflow. Finally, the cup is spread and the sediment above the Bunsen burner for 15 minutes, cooled in the air and in the exicator before determining the weight. Repeat this work until the weight is constant.

Determination of Gravimetric Nutrient Levels

1. Carefully weighed 2 grams of a soil sample, dissolved in 100 mL beaker with 25 mL of water. Moved quantitatively into a 100 ml measuring flask and diluted to the boundary mark.
2. Piped 25 mL of this solution into a 400 mL beaker, added a little distilled water, and 5 mL of 1: 1 HCl solution, diluted to 200 mL.
3. Heated to around 70oC, added 25 mL Dimethyl Glioksim solution, stirred evenly.
4. Carefully add Ammonium Hydroxide solution until Ni-DMG red precipitate is formed, a little alkaline solution is added until it smells ammonia.
5. Put the beaker and the contents above the water bath and left for 20 to 30 minutes, checking whether all nutrients have settled completely.
6. The solution is cooled for a while and then filtered through a maser glass cup which has known the weight.
7. Washed with cold water, and then dried at 110°C to 120°C. After being cooled in the exicator, it is weighed until the weight is constant, and done in duplicate.

So, in one liter of a solution, there is 0.48 mg of nutrient deposits in the soil which is not deposited. In accordance with the laws of similar ions, the equilibrium reaction will shift towards the formation of deposits. The impact of changes in the chemical properties of peat soils due to forest combustion is that there is a change in organic material to ash, which contains many nutrients such as N, P, S. Then C or N will disappear along with the evaporation process during the process of forest fires. In addition, forest fires can increase the pH level of peat soils, and warm up by the fire. So that all this can destroy the soil structure, increase soil weight, and ultimately reduce soil porosity. Damage to soil structure can result in a decrease in the speed of infiltration and an increase in runoff water speed [1]. Based on the chemical properties of the soil, due to these forest fires, it turns out that it will not be able to increase soil fertility in the long term but is more temporary.

Furthermore, the impact of changing the biological properties of peat soils due to forest fires is the death of flora, fauna and soil microorganisms. The death of flora and fauna will affect the chemical properties of peat soil. This is caused by the results of secretion and excretion of soil fauna and microorganisms, which indirectly can improve the chemical properties of the soil itself.

The tools used to measure the activity of physical properties of soil are soil samples of rings, hoes, peat drills, ovens, analytical scales, porcelain glasses, measuring cups, plastic, soil. Analysis of soil samples was carried out at the Soil Laboratory of the Faculty of Agriculture, Riau University, Pekanbaru, Riau. This research was conducted for 1 month starting March - April 2018.

The impact of changes in the chemical properties of peat soils due to forest combustion is that there is a change in organic material to ash, which contains many nutrients such as N, P, S. Then C or N will disappear along with the evaporation process during the process of forest fires. In addition, forest fires can increase the pH level of peat soils, and warm up by fire. So that all this can destroy the soil structure, increase soil weight, and ultimately reduce soil porosity. Damage to soil structure can result in a decrease in the speed of infiltration and an increase in runoff water speed [1]. Based on the chemical properties of the soil, due to these forest fires, it turns out that it will not be able to increase soil fertility in the long term but is more temporary.

Furthermore, the impact of changing the biological properties of peat soils due to forest fires is the death of flora, fauna and soil microorganisms. The death of flora and fauna will affect the chemical properties of peat soil. This is caused by the results of secretion and excretion of soil fauna and microorganisms, which indirectly can improve the chemical properties of the soil itself.

2.2 Analysis of the physical characteristics of peat soil

The material used in this study is peat soil on burned land and natural forests. Isolates, GPS, cameras, meters, stationery, and tools used during the analysis in the soil laboratory. This research was conducted using a survey method. Determination of sampling points using the Purposive Sampling Method, which is choosing locations according to the purpose of the research conducted in four locations. The four selected locations are as follows:

Pangkalan Kerinci District includes:

- Point K1, land that does not experience a fire in the village of Pangkalan Gondai, (coordinates 00 05 '31" N 1010 49 '14" E),
- Point K2, the area that burned a little in the village of Kerinci Barat (coordinates 00 23 '39" N 1010 49 '56" E),

Langgam District includes:

- K3 point, a moderate burning area in Desa Langgam (coordinates 00 06 '32" N 1010 39 '41" E), and
- Point K4, heavily burned area in Kerinci Timur village (coordinates 00 24 '42" N 1010 52 '40" E),

The 4 points are in the area of Pelalawan Regency, Riau Province. The 3 areas of sampling area under conditions:

- Land with a mild fire level (K1).
- Land with moderate fire (k2).
- Land with a heavy fire level (K3).
- Zero fire level (K4)

The fourth location is the land from the land area where there has never been an event of land burning (natural forest), which in this study was used as a control. Each location is given 3 replications so that all 12 units of observation sample points are obtained.

3. Results and Discussions

3.1 Analysis of soil physical properties

3.1.1 Soil water content (%)

The results of the analysis show that the more severe the level of land fires, the lower the moisture content at the point of observation. The level of soil water in various levels of peatland fires is presented in Figure 2. Figure 2 shows that there has been a decrease in peat water content after peat soil has burned with different levels of fire. The heavier the level of peat fires, the lower the water content in peat soil. The highest water content value occurs in non-combustible peat soils which are used as control treatments.

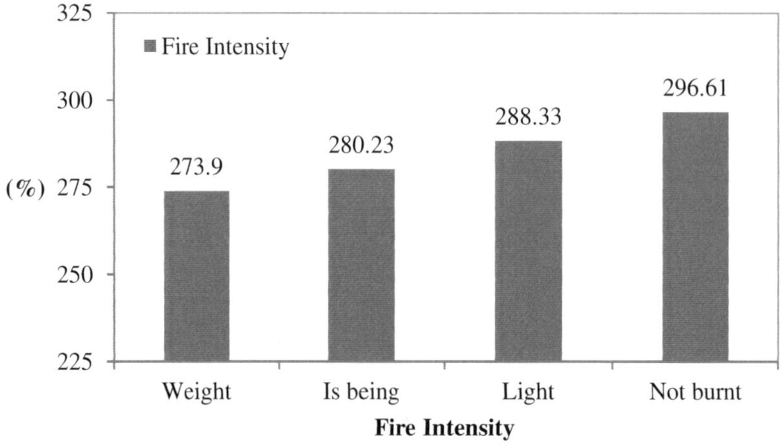

Figure 2. Histogram of water content in various levels of peatland fires

Based on the results of observations, it can be seen that the water content with the highest percentage is found in unburned soil samples 296, 61 %. While with increasing soil temperature the peat evaporation process is faster so that the soil water content decreases. This is consistent with the statement of [2] that fires can increase soil temperature and cause soil water content to decrease.

Fires also produce minerals (ash) which can reduce the moisture content of peat soil, this is due to the ash would fill soil pore spaces are empty, so the pore space for storage of groundwater is reduced. This is consistent with the statement of [2] that ash from burning organic matter due to burning peatland causes clogged soil pores and impacts on disturbing the storage process of groundwater, resulting in reduced peat water content.

3.1.2 Water binding power (%)

The results of the analysis show that the heavier the level of land fires, the lower the value of the binding power of water at the point of observation. Power to bind groundwater at various levels peatland fires are shown in Figure 3.

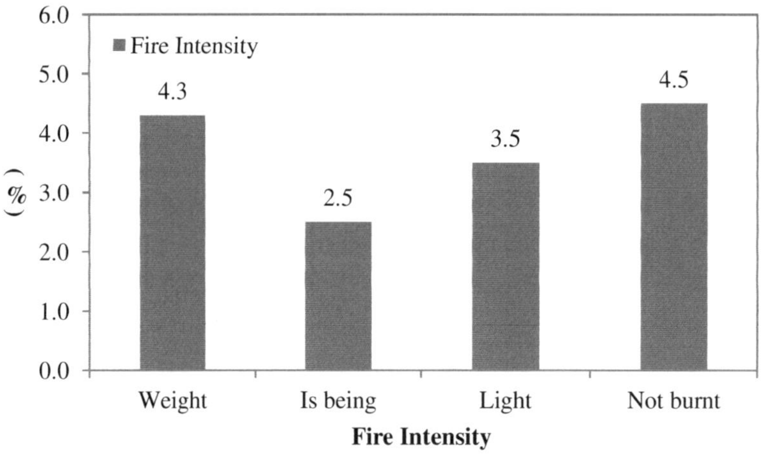

Figure 3. Histogram of power binding to water in various levels of peatland fires

Figure 3 shows that there has been a decrease in the binding power of peat water after burning with different levels of fire. The value of the binding power of water at the level of fire is lower than 164, 76 %. compared with the value of the binding power of water at the moderate fire level of 177.13%, lightly 187.34%, and those that did not burn higher with a value of 208.65%. The presence of fires shows that a clear impact on the physical properties of the soil will affect soil temperature, soil structure, and the ability of the soil to absorb water. Large fires will cause the bulk density to increase, which in turn will reduce the porosity and rate of soil infiltration which in turn has an impact on the ability to bind water. As a result, the surface flow will increase so that the soil becomes sensitive to factors that increase erosion and flooding [3].

In addition to the low power after a fire, because the water-binding fires on peatland produces ash, the ash will move to the bottom layer so that it fills the peat pores [2]. Proof of movement Fire ash, seen by increasing BD and pH (Table 3) at different fire levels. The low ability of peat to bind water after burning with various levels of fires is increasingly heavy due to the destruction of functional groups of water-binding peat material (hydrophilic) that is carboxylic and OH-phenolics because the temperature is too high. This is in line with the statement of [4] that the decreasing ability of peat to absorb water after drying is closely related to the reduced availability of carboxylic and OH-phenolics. Both the organic component is a hydrophilic compound that helps the absorption of water availability, but if the peat is in a dry state, then both the organic compound to be not working [5].

3.1.3 Load weight (bulk density)

The results of the analysis show that the heavier the level of fires in peatlands, the bulk density increases at the point of observation. The weight of the soil in various levels of peatland fires is shown in Figure 4. The graph in Figure 4 shows that there has been an increase in BD (Cultivation) peat after burning with different levels of fire. The BD value at the level of heavy fires was higher at 0.23 g / cm 3 compared to the BD value at the moderate fire level of 0.22 g / cm 3, the light at 0.21 g / cm 3 and those that did not burn as a control treatment 0.2 g / cm3. In addition, there was an increase in BD from control to after a fire from a mild to severe fire.

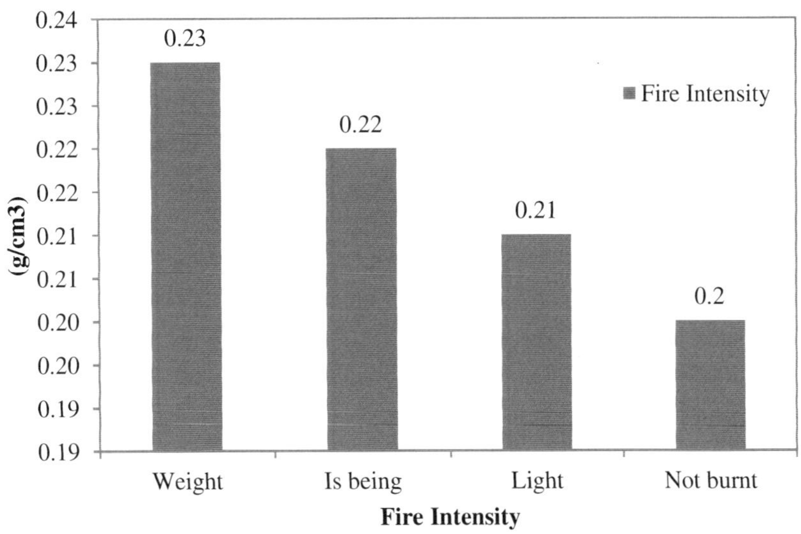

Figure 4. Histogram of bulk density in various levels of peat fires

Based on the results of observations it can be seen that the sample of land with a heavy fire level has an average content weight highest land, that is 0.23 g / cm3. Think this is due to precipitation from the occurrence of ignition due to land fires. Fires will cause a process of ignition on the surface of the burning ground, the ash produced from combustion will enter into the pores of the soil with an intermediary of rainfall, so that the soil becomes dense and BD increases. BD enhancement in the peat layer is accompanied by increasing pH, ash content, and base saturation. Increases in pH, ash content and base saturation at a depth of 10-20 cm indicate the presence of combustion ash, fires provide input of minerals (ash and charcoal) in the soil so as to increase soil pH.

The burning peatland causes loss of organic matter in the top layer of peat and the soil becomes dense. Compaction occurs because of the addition of soil mass in the form of ash. Ash comes from the burning of peat organic matter. The ash fills the pores of the peat so that the peat becomes dense and has increased BD. This is in accordance with the statement of [2] that fires cause peat to become dense, caused by an increase in the weight of contents by the ash produced by combustion. In addition, Wasis (2003) explained that burned land will experience the destruction of soil particles that cause compaction.

3.1.4 Particle density

The results of the analysis show that the heavier the level of land fires, the particle density increases at the point of observation. The density of soil particles in various levels of peatland fires is shown in Figure 5.

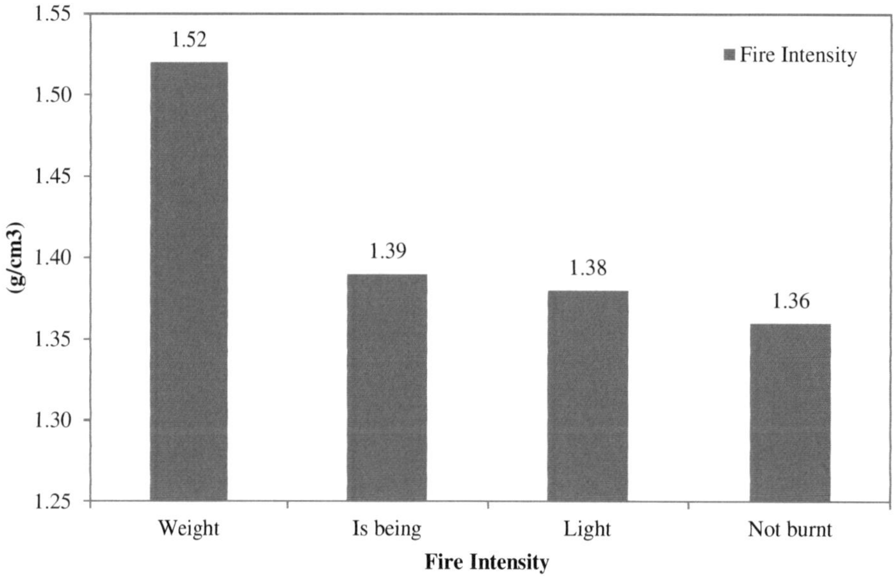

Figure 5. Particle density histories in various levels of peatland fires

Figure 5 shows that there has been an increase in peat PD after burning with different levels of fire. The PD value at the fire level is higher than the PD value at the moderate, mild and non-combustible fire levels as a control treatment. In addition, there was an increase in PD from control until after a fire from a mild to severe fire. Based on the observations obtained data that the particle density in the control treatment (non-combustible soil) has the highest value (1,52 cm^3). Based on PD observations, the results showed that the highest BD was found on land with a heavy fire level. The increase in PD that occurs in burned land is affected by the ash from combustion that fills the pore space peat. PD is strongly influenced by the presence of combustion ash because ash is a mineral material that will mix with peat material so that it can increase the density of solids.

3.1.5 Permeability

The results of the analysis show that the heavier the level of land fires, the permeability value decreases at the point of observation. Soil permeability in various levels of peatland fires can be seen in Table 1.

Increasing bulk density and decreasing the total value of the soil pore space, especially the macropore space will be able to pass large amounts of water per unit time. However, other research found that soil is micropores [6]. Table 2 indicates the value which varies from observing permeability. The highest value of permeability is control (not burning) with a value of 19.43. Furthermore, followed by permeability in mildly burned land (8.13), moderate (4.1) and heavy (2.4). This shows a decrease in permeability ability of each soil sample due to fire.

Table 1. Permeability in various levels of peat fires

No	Fire Level	Permeability (cm/hour)
1	Weight	2.4
2	Is being	4.1
3	Light	8.13
4	Not burnt	19.43

The highest value from the control sample (non-combustible land) makes the land classified as a good category with the highest permeability. This is because the control still has a good soil structure and pore so that it can pass water well when compared to burning land which is much lower in value.

3.1.6 The permeability value of burned land is caused by several factors.

Factors that influence permeability include texture, porosity, soil pore size distribution, aggregate stability, structure, and soil organic matter content. Heating produced by fires will reduce soil pores so that the soil becomes dense and the ability of the soil to hold groundwater that decreases bulk density and increases total soil pore space resulting in increased soil permeability rates.

3.1.7 Total pore space (TRP)

The results of the analysis show that the more severe the level of land fires the more the total pore space (TRP) decreases at the point of observation. The total soil pore space in various levels of peatland fires is presented in Figure 6.

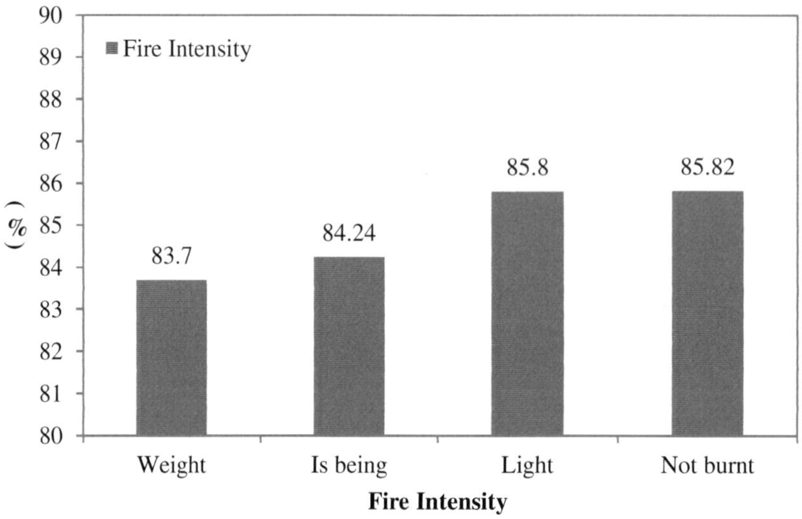

Figure 6. Total pore space histogram (TRP) in various levels of peat fires.

Figure 5 shows that there has been a decrease in peat TRP after burning with different levels of fire. The TRP value at the fire level is lower than the TRP value at the moderate, mild and non-combustible fire levels as a control treatment. In addition, there was a decrease in TRP from control until after a fire from a mild to severe fire.

The highest value of total pore space (TRP) is in the control sample with a value of 85,82. Furthermore, the highest value followed by the value of mild fire land (85,80) and moderate fire land (84.24) and light indicates that the increase in temperature due to fire causes damage to the structure on the ground surface with reduced soil pore space which affects the increase in weight (83.07). This shows that the total pore space value (TRP) is classified as not significantly different and shows the result that the total pore space value (TRP), however, decreases.

The total soil pore space and bulk density are important in the assessment of soil density or density. The soils with increased bulk density mean that their porous space is getting lower and the soil is getting denser [7]. Soil that has a large bulk density, the denser the soil, which means it is difficult to pass water or penetrate the roots of plants.

The results of [8] in the burning of Acacia mangium in Parungkuda, Sukabumi showed that the increase in temperature due to fires caused structural damage to the soil surface with reduced soil pore space which had an effect on increasing soil weight (Figure 6). Fires make the soil open with loss of litter, understorey, and canopy which increases the temperature and rate of evaporation while causing loss of organic matter which decreases the available water content.

3.2 Analysis of Chemical Properties of Soil

3.2.1 Degree of acidity

The results of the analysis show that the heavier the level of land fires, the higher the soil pH value but not significant at the point of observation. Soil pH in various levels of peatland fires can be seen in Table 2. The number of pH in various levels of peat fires Table 2 shows the results that observing soil pH from each soil sample showed results that were not significantly different. It can be seen that the highest pH is in samples with heavy fires with a value of 4.37 and the lowest pH that is in the control (no fire) with a value of 4.11. This shows that the level of fire that has occurred both mild, moderate and severe and control (non-combustible forest) does not significantly affect changes and differences in forest soil pH.

Table 2. Fire level and soil pH

No	Fire Level	Soil pH
1	Weight	4.37
2	Is being	4.34
3	Light	4.23
4	Not burnt	4.11

However, thus, occurrence forest fires can slightly increase the value of forest soil pH when viewed from the value of control and observation of treatment. This shows that with the incidence of forest fires, soil pH will be increased so that the availability of certain nutrients needed for plants becomes available.

3.2.2 Redox
The results of the analysis show that the more severe the level of land fires, the lower the soil redox value at the point of observation. Redox soil in various levels of peatland fires can be seen in Table 3. Table 3 shows the results that observations of soil redox from each soil sample show results different from each treatment. It can be seen that the highest redox is in the sample with control (no fire) with the lowest value of 195.6 and redox, which is in the soil sample with heavy fires with a value of 165. The existence of redox changes due to changes in elements and chemical reactions due to fire. This can also be seen from the explanation of changes in the chemical properties of peat soil. The impact of fires on soil chemistry is explained by [9] that fire will change the chemical properties of the soil in three ways, namely minerals released from the combustion process that is left behind by ash, changes in microclimates after a fire, and decomposition of minerals to see the simplification of organic structures into inorganic materials. The contribution of soil nutrients due to fire does not last long and is limited. If fires occur repeatedly, land degradation will increase and the nutrient impoverishment process will take place.

Table 3. Redox in various levels of peatland fires

No	Fire Level	Redox
1	Weight	165
2	Is being	167
3	Light	175,6
4	Not burnt	195,6

3.2.3 Electrical Conductivity
The results of the analysis show that the heavier the level of fires in peatland, the higher the conductivity of soil electricity at the point of observation. Redox soil in various levels of peatland fires can be seen in Table 4. Table 4 shows that there are differences in the value of the electrical conductivity of each soil sample. The greatest value of electrical conductivity is in heavy fires with a value of 118,6 and followed by a moderate fire level sample (109.6), mild fire (106.1) and non-burning peatland or control (86.9). The greatest value from soil samples of heavy fires is possible to change the physical properties of the soil so that electricity is easier to deliver.

Table 4. Electrical conductivity in various levels of peat fires

No	Fire Level	Electrical Conductivity
1	Weight	118,6
2	Is being	109,6
3	Light	106,1
4	Not burnt	86,9

However, moderate and light fires do not differ significantly. This is possible due to changes in the nature of the soil from heavy fires that facilitate the transmission of electricity so that the value is quite large. Whereas there is no fire in the soil due to the natural state of the soil so that the electrical conductivity remains the same size. However, land the electrical conductivity of fires that are not included in the criteria for chemical properties is damaged (except mild fire land) based on the general criteria for damage to peat soil PP No. 4 of 2001 relating to forest fires and/or land. This is because PP No. 4 of 2001 states that electrical conductivity is included in chemical properties and is classified as damaged if it has a rising value compared to non-combustible soil.

Analysis Of The Biological Properties Of Soil

- Total Soil Microbes (CFU / ml)

The results of the analysis show that the more severe the level of peat fires, the lower the total value of soil microbes at the observation point. The total soil microbes in various levels of peatland fires can be seen in Table 5.

Table 5. Total soil microbes in various levels of peat fires

No	Fire Level	Total Microbes (CFE / ml)
1	Weight	1.86×10^8
2	Is being	3.63×10^8
3	Light	8.53×10^8
4	Not burnt	12.64×10^8

Table 5 shows that there are differences in the total microbial values of each soil sample. The largest total microbial value was in non-combustible soils 12.64 x 108 and followed by samples after mild, moderate and severe fires. The greatest value of non-combustible soil samples is caused by no change in the physical properties of the soil and the availability of organic matter in peat soil.

Physically or visually, the most striking difference in non-burning peat soils is that they become more compact or harden the soil by accumulating organic matter (plant litter), and the appearance of white on peat soil which indicates the presence of microorganisms that grow in it. This is because microorganisms that function as decomposition agents or cellulase decomposers and fibers contained in peat soil. In this study used control variables and the comparison of the number of microorganisms contained in the soil before the fire and after the fire. The table shows that the total microbial amount in unburnt peat soil causes more changes in the amount. Increasingly increasing TPC.

4. Conclusion

Based on the results of the research and discussion described in the previous section it can be concluded that at various levels peatland fires can increase the value of fill weight, particle density, water binding power, soil pH in various levels of peatland fires, total pore space, permeability, expected to increase the activity of decomposition or decomposition of organic matter from peat soil. Where, soil chemistry such as pH is also one of the factors that support microbial life in the soil, as well as macrofauna and soil mesofauna that the soil pH supports the life of microbes in the soil. Usually, soil microbes commonly grow at a soil pH of about 7, but soil microbes can grow at pH 2 - 10 if available organic material in total soil microbes does not increase compared to peat not burn out. water content, redox, electrical conductivity, and total soil microbes do not experience an increase compared to peat not burning.

5. References

[1] Sianturi F 2006 Changes in the physical and chemical properties of soil in burned areas in Puspa stands Department Thesis Cultivation Forest, Faculty Forestry Bogor Agricultural University Not published.

[2] Depari EK and Adinugroho WC 2009 Impact of forest fires to function hydrology Major Paper for Tropical Silviculture School of Post-graduate IPB Bogor.

[3] Sugato IS 2005 Changes in soil physical and chemical properties after 1, 2 and 3 years of combustion in secondary forests, Jasinga Bogor Thesis Department of Forest Cultivation, Faculty of Forestry Bogor Agricultural Institute Not published.

[4] Sabiham S 2000 Water content , criticism of Central Kalimantan peat in relation to non-return dry events *Tropical Land Journal* **11** 21-30.

[5] Stevenson FJ 1994 Humus Chemistry: Genesis, composition and reaction 2nd Edition John Willey & Sons Inc New York.

[6] Khoiri Al 2013 Changes in the physical properties of various types of soil under oil palm stands (Elaeis guineensis jacq.) Which are applied to oil palm empty bunches (tksks) at PT. Salim Ivomas Pratama [thesis] Agrotechnology Study Program, Faculty of Agriculture University of Riau Pekanbaru.

[7] Simatupang,YMA 1999 Effect of bokashi chicken manure and bokashi grass on some physical properties of red yellow gajrug podzolic soil and growth of soybean (Glycine max L. Merr) varieties of Willis Bogor Agricultural Institute.

[8] Prakoso Y 2005. Impact of forest fires on the physical properties of soil in secondary plantations of Acacia mangium in the Village Langensari, Parungkuda District, Sukabumi, Java West. Essay Faculty of Forestry IPB Bogor.

[9] Chandler C, Cheney P, Trabaud L, Williams D 1983 Fire in Forest Vol I Forest Fire Behavior and Effects Jhon Wiley and Sons Inc Canada USA.

The exploration of phosphate solubilizing bacteria in mangrove forest at Teluk Naga, Banten

I J Sari* and S G S Fitri

Department of Biology Education, Faculty of Teacher Training and Education, University of Sultan Ageng Tirtayasa
Jl. Raya Jakarta km. 04 Pakupatan, Serang City, Banten Province, Indonesia

*Coresponding author: indah.juwitasari@untirta.ac.id

Abstract. The aim of this study was to explore phosphate solubilizing bacteria in mangrove forest at Teluk Naga, Tangerang district, Banten. To collect samples, random sampling was applied. Bacteria were isolated by Pikovskaya media, using pour plate method. Nineteen isolates collected from soil demonstrated a potency to solubilize phosphate. Six isolates showed high phosphate solubilization activity from calculation of dissolution index. They were identified morpho-physiologically and biochemically as *Pseudomonas* (isolate R22 and A17), *Klebsiella* (A11, A14, A18, and A38), *Agromyces* (A13), *Bacillus* (A33 and A37), *Azomonas* (A15), and *Enterobacter* (A31 and A32). These bacteria were found in the rhizosphere of mangrove forests which consist of *Avicennia* and *Rhizophora* mangroves. These bacteria are known to have the ability to increase the availability of phosphate in the soil. This ability is one of the capabilities of the potential growth of promoting bacteria (PGPR or Plant Growth Promoting Rhizobacteria).

Key Words: Mangrove forest, Rhizobacteria, Phosphate solubilization bacteria, PGPR.

1. Introduction

Teluk Naga mangrove ecosystem is located in the Banten Bay region, Muara Village, Teluk Naga District, Tangerang Regency, Banten. Teluk Naga mangrove forests is one of mangrove ecosystems in Banten which have many potentials in various fields, such as eco-tourism, fisheries and agriculture. The potential of mangrove forests in agriculture is inseparable from bacteria that support plant growth. [1] and [2], explored epyphitic cyanobacteria isolated from pneumatophore of *Avicennia marina* and planktonic cyanobacteria isolated from brackish water in mangrove forest of Cagar Alam Pulau Dua, Serang, Banten. Some of these cyanobacteria known to have an activity in solubilizing phosphate. Meanwhile, no research has been reported on the exploration of phosphate solubilizing bacteria in the Teluk Naga mangrove forest.

[3] conducted a research about phosphate solubilizing microorganisms associated with Chollangi Mangrove Soil in East Coast of India. They isolated 23 bacterial strains from the soil of Chollangi mangrove environment. Seven isolates showed phosphate solubilization activity. Two isolates out of these bacteria were identified morphophysiologically and biochemically as *Bacillus subtilis*, three

isolates as fluorescent *Pseudomonas* and two isolates as *Azotobacter sp*. Phosphate solubilizing bacteria play an important role in soil enrichment, because of their capability of dissolving phosphate by excreting a number of low molecular weight organic acids such as oxalate, succinate, fumarate, and malate. These organic acids will react with phosphate binders, such as Al^{3+}, Fe^{3+}, Ca^{2+}, or Mg^{2+} to form stable organic chelates so that they can free phosphate ions bound and be utilized by plants [4].

In this study, we aimed to explore rhizobacteria from Teluk Naga mangrove forest which have an ability to solubilize phosphate. The bacterial isolates found can be further applied as biofertilizer that utilize to promote plant growth in agricultural fields.

2. Methods

The research was conducted in two main locations, namely Teluk Naga Mangrove Forest, Tangerang as the location of rhizobacteria sampling and Biology Education Laboratory of Sultan Ageng Tirtayasa University for isolation, selection, and morphological characterization of rhizobacteria. Teluk Naga Mangrove Forest is located in Tanjung Pasir Bay, Muara Village, Teluk Naga District, Tangerang Regency, Banten. Total area of Teluk Naga Mangrove Forest is 270 Ha.

The method sampling used in this study was random sampling. Each transect was made by pulling a 300 m transect line from the sea to the land direction perpendicular to the coastline. The transect was determined based on the zone of mangrove vegetation. Rhizosphere soil samples were taken at low tide. The soil was taken using pipes as deep as 10-20 cm and the roots of the plants were cut 3 cm long. Then stored in a sterile plastic.

As much as 1 g of soil was aseptically weighed and diluted in 0.85% NaCl and put into a sterile container containing 99 ml sterile solution. Soil samples were diluted to 10^{-4} and poured in sterile petri dishes. Serial dilution methods were carried out following [5]. One milliliter of each dilution was spread on Pikovskaya media [6], then incubated at 28oC for 7 days. Colonies that formed clear zones (halo zones) were purified and subcultured on pikovskaya media. The characteristics of each colony were observed morphologically and physiologically following [7]. Isolation of phosphate solubilizing bacteria was carried out using Pikovskaya media, with a composition (in g / L): 5 g $CaCO_3$, 10 g Glucose, 0.5 g $(NH_4)_2SO_4$, 0.2 g NaCl, 0.2 g $MgSO_4$, 0.5 g yeast extract, 0.004 g $MnSO_4 \cdot 4H_2O$, 0.002 g $FeSO_4 \cdot 7H_2O$, and 15 g agar (media pH 7.5) [8].

Qualitative measurements of phosphate solvents were carried out using the dot method using puncture needles on the Pikovskaya medium and incubated at room temperature for 7 days. Solubility index (SI) is measured by the formula of clear zone diameter minus the colony diameter and the results are divided into colony diameter [9]. Bacterial identification is carried out through characterization of bacterial morphology which includes observation of colonies (color, shape, edges, and elevation), gram staining, observation of cell morphology (shape), motility, catalase test and other physiological tests. Gram staining was done to find out whether bacterial cells belong to the Gram Positive or Gram Negative group. The overall observations were adjusted to the Bergey's Manual of Determinative Identification Book.

3. Results and Disscusion

There were 19 isolates that have an ability to dissolve phosphate. Selection of phosphate solubilizing bacteria was carried out by growing the isolates on Pikovskaya media. The ability of bacteria to dissolve phosphate is measured qualitatively by measuring the phosphate dissolution index (IP). This index value is obtained by calculating the ratio between the clear zone diameter difference (DZ) minus the colony diameter (DK) and the diameter of the colony ((DZ-DK) / DK). Clear zones are seen when isolates are grown in Pikovskaya media (Figure 1).

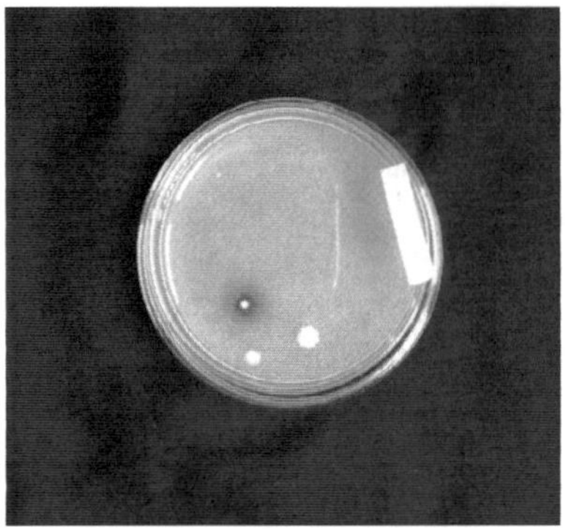

Figure 1. A clear zone on Pikovskaya media resulted by isolate A18

It is assumed that the higher index value, the higher the ability of bacteria to dissolve phosphate. In Table 1, it is seen that isolates which have the highest IP values are A18 and A38. These rhizobacteria were isolated from *Avicennia* mangrove forest, respectively at stations 1 and 3.

Table 1. The phosphate dissolubilizing indexproduced by rhizobacteria at mangrove forest Teluk Naga

Isolat	Diameter		Indeks Pelarutan Fosfat ((DZ-DK)/DK)
	zona bening (DZ)	koloni (DK)	
A18	1.10	0.10	10.000
A38	2.00	0.20	9.000
A14	0.10	0.02	4.000
A13	0.10	0.02	4.000
A15	0.90	0.20	3.500
A37	0.40	0.10	3.000
A12	0.60	0.20	2.000
R21	0.20	0.10	1.000
R22	0.20	0.10	1.000
A33	0.60	0.30	1.000
A34	0.80	0.30	1.670
A31	0.90	0.50	0.800
A32	0.60	0.40	0.500
A36	0.70	0.50	0.400
R24	0.40	0.30	0.330
A11	0.05	0.04	0.250
A17	0.50	0.40	0.250
A35	0.50	0.40	0.250
A16	0.90	0.80	0.125

Based on the calculation of dissolution index, six isolates showed high phosphate solubilizing activity. According to *Bergey's Manual of Systematic Bacteriology, t*hey were identified morpho-physiologically and biochemically as a member of genus *Klebsiella, Bacillus, Pseudomonas,* and *agromyces, and azomonas.*

1. Klebsiella
Isolate A11, A14, A18, and isolate A38 showed a rounded colony, flat elevation, shiny colony surface, flat colony edge and cream coloured. Based on microscopic observations, isolates A11 and A38 have basil cell forms and are Gram negative (Figure 2). Both isolates demonstrated positive reaction on the catalase test, motile, and aerobic for A38 and anaerobic for A11. According to Bergey's Manual of Determinative Bacteriology 7th edition, the characteristics of isolate A11 and A38 mostly are owned by *Klebsiella,* which are bacillus, Gram negative, long or short shaped, and facultative anaerobes. However, *Klebsiella* are non motile. These bacteria measure 0.5-1.5 × 1-2 microns and have a sheath which is 2-3 times the size of germs [10]. They can produce nitrite from nitrate, grows well on ordinary culture media. These organisms can be isolated from various animals and soils, and are often found in the respiratory, intestinal and urogenital tracts of human. *Klebsiella* are also a soil bacteria which known to produce IAA hormones and have a potency to be PGPR [11].

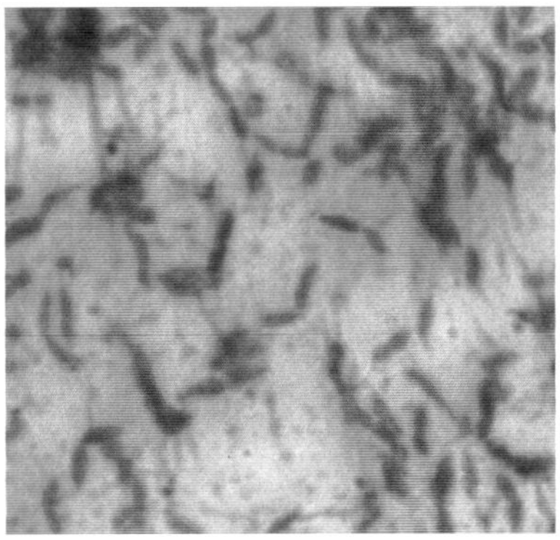

Figure 2. Bacterial isolate A11 (100x magnification)

2. Bacillus
Colonies characteristics of round, flat elevation, shiny surface, flat edge, and creamy white were shown on isolates A33 and A37. These isolates have basil cell form and are Gram positive (Figure 3). Characteristic observation on the biochemical activity test showed that the isolates could produce catalase enzyme (positive reaction). They are motile and aerobic. The results obtained showed that isolates A33 and A37 belong to the *Bacillus* genus [10]. *Bacillus* have trunk-shaped characteristics, and can be found in soil and water including sea water. Some of the bacteria produce extracellular enzymes that hydrolyze proteins and complex polysaccharides. *Bacillus* form an endospore, Gram positive, move by erythricus flagellum, catalase positive, and either aerobic or facultative anaerobic [12].

Bacillus are a true marine occupant that have been investigated by experts in marine research and proven to have several capabilities. Some of these bacteria able to produce antibiotics that can fight pathogenic bacteria *Vibrio cholerae. Bacillus circulans* known as oil breaking bacteria, produce glucan-breaking enzymes which can decompose crude oil and other hydrocarbons. Enzymes produced by *Bacillus* have been produced on an industrial scale, including alanine and formicate, α-amylase, isoamilase, β-amylase, glucoamylase, chitinase, and cholesterol oxidase [13]. The *Bacillus* are widely used as a biocontrol agent, producing antimicrobial substances in the form of bacteriocin, which is an antimicrobial polypeptide or protein produced by bactericidal microorganisms. Bacteriocin kills its target cells by entering the target membrane and causing cell membrane's function to become unstable, causing cell lysis [14]. [15] found that *Bacillus* are also able to dissolve phosphate in soil samples from the cultivation of cayenne.

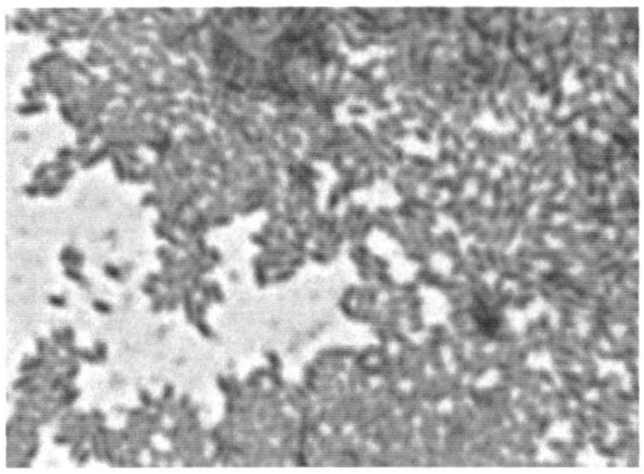

Figure 3. Bacterial isolate A33 (100x magnification)

3. Pseudomonas

Based on macroscopic observations, isolates R22 and A17 showed irregular colonies, flat elevations, dull or non-glossy colonies on R22 isolates but shiny on A17 isolates, the edge of the colony was wavy with milky white. Microscopic observations showed that isolates R22 and A17 have basil cell forms and are Gram negative (Figure 4). Characteristic observations on biochemical activity tests showed that isolate R22 and A17 have a positive reaction on catalase test, motile, and aerotolerant. According to their characteristics, it is assumed that isolate R22 and A17 belong to *Pseudomonas* group. [10] explain that *Pseudomonas* are Gram negative, have long or short shaped, and aerobic. These bacteria measure 0.5-1.0 × 1.5 - 5.0 microns, and do not have a resting phase. In some cases, *Pseudomonas* can use nitrates as alternative electron receptors. Some are oxydase positive, but some are negative. Most of them are spread in nature, some pathogenic species in humans, animals and or plants. According [16], *Pseudomonas* has high activity and production of enzyme phosphatase than *Basilus.*

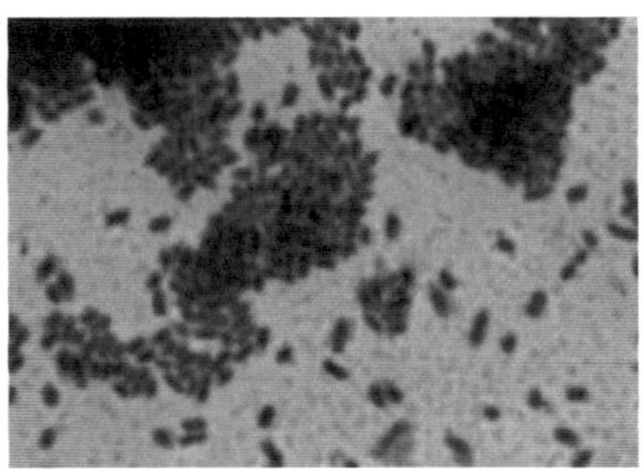

Figure 4. Bacterial isolate A17 (100x magnification)

4. Agromyces

Isolate A13 is assumed to be *Agromyces*. The isolate showed irregular colonies, thick or convex elevations, shiny colony surfaces, flat colonies with white color colonies. Under microscope, isolate A13 have filament cell forms and Gram positive (Figure 5). Their biochemical characteristics are catalase negative, non motile, and facultative anaerobes. According to Bergey's Manual of Determinative Bacteriology 7th edition, *Agromyces* are Gram positive, cells are composed of branched, slender filamenttous element 0.3-1.0 microns and irregular cells in older culture. Widely distributed in soil, where it is very numerous [10]. Strengthened by [17] *Agromyces soli* sp. nov., isolated from farm soil.

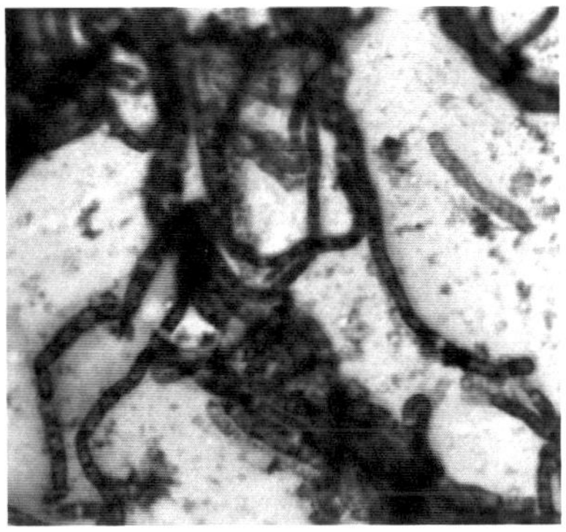

Figure 5. Bacterial isolate A13 (100x magnification)

5. Azomonas

Isolate A15 showed irregular colony, convex elevation, shiny colony surface, flat colony edge and cream coloured. Based on microscopic observations, isolates A14 have coccus cell forms and are Gram negative (Figure 6). The isolate demonstrated positive reaction on the catalase test, motile. According to Bergey's Manual of Determinative Bacteriology 7th edition, the characteristics of isolate A15 mostly is owned by *Azomonas* because of A15 have many characteristics same with *Azomonas*. The characteristics of Azomonas are cell generally stain Gramn negative, motility occurs by peritrichous or polar flagella, aerobic, but can also grow under decreased oxygen tensions. Nitrogen fixers, and catalase positive. They occur in soil and water [10]. Based on the research of [18] *Azomonas agilis* released large amounts of organic acids to solubilize phosphate to increase soil fertility and enhance plant production.

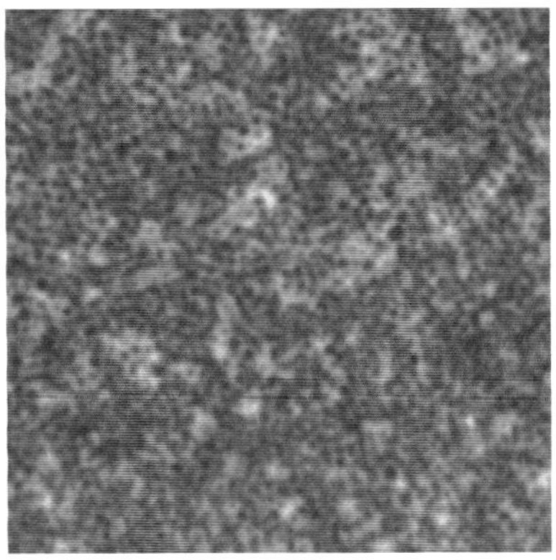

Figure 6. Bacterial isolate A15 (100x magnification)

6. Enterobacter

Based on macroscopic observations, A31 and A32 isolates showed round colonies, thick or convex elevations, shiny colony surfaces, flat colonies with yellow color in A32 isolates and pink color colonies in A31 isolates. Microscopic observations, isolates A31 and A32 have long basil cell forms and are Gram negative. Characteristic observations on biochemical activity tests of A31 and A32 isolates showed that the catalase test was positive, motile, and facultative anaerobes. The results obtained were the possibility of A31 and A32 isolates belonging to *Enterobacter* Genus as identified in the Bergey's Manual of Determinative Bacteriology 7th edition. *Enterobacter* genus has features including Gram negative, long or short shaped facultative anaerobes. This bacterium measures 0.6-1.0 × 1.2-3.0 microns. Pair or line. Most are spread in nature; in fresh water, soil, swamps, plants, vegetables, animals and human faeces. According to [19], *Enterobacter* has been isolated in Sediments from Shallow Eutrophic Lake and a Wetland; also were identified molecularly as bacteria which were phosphorus release determination ability.

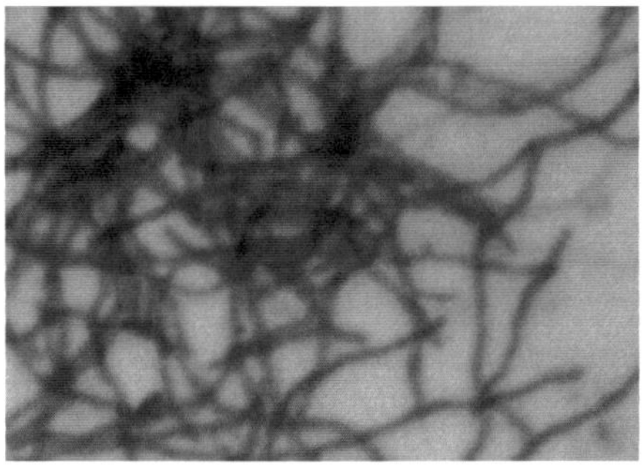

Figure 7 . The isolate bacteria A31 (100x magnification)

Phosphorus (P) is one of the major constituents in energy metabolism and biosynthesis of nucleic acids and cell membranes with an important role in regulation of a number of enzymes. Soil phosphorous is an important macronutrient for plant growth. Macronutrient means that the element needed in large quantities. Total soil P occurs in either organic or in organic form. The major form of organic P soil is not readily available to plant as a source of P [20].

Phosphorus is absorbed by the roots and leads to young leaves first and then transferred to older leaves. Many P elements are found in phloem tissue so they have a function for plant nutrient translocation. The P element is absorbed by plants in the form of ions, both positive and negative ions, for example $H_2PO_4^-$. After an H_2PO_4 ion enter into plant, it will be directly esterified through C chain hydroxy groups to high-energy phosphates, such as ATP [21].

In soil, some microorganisms are known to be able to convert non-soluble forms of phosphorus to dissolved form through acidification, secretion of organic acids or protons [22]. Phosphate solvent bacteria have been reported to be found more in mangrove samples than in the sea and brackish water. However, the abundance of phosphate solvent bacteria is correlated with the level (content) of phosphorus in the soil. This is possible due to the fact that the active bacteria converts insoluble phosphorus compounds into dissolved forms, which are easily transferred from the soil to the waters or have been used by plants and microbes. The mechanism of phosphate dissolution is a cumulative action that is influenced by many factors, including pH, oxygen concentration, wind speed, waves, water flow, the content of organic compounds, and carbohydrate levels [8].

4. Conclusion

Mangrove forest at Teluk Naga, Banten is a potential ecosystem, which vary in rhizobacteria. The result of this study found 19 bacterial isolates demontrated phosphate solubilization activity. Six isolates with high phosphate dissolution index (IP) were identified morpho-physiologically and biochemically, as *Pseudomonas* (isolate A14), *Klebsiella* (A11, A14, A18 and A38), *Agromyces*

(A13), *Bacillus* (A37), *Azomonas* (A15), and *Enterobacter* (A31 and A32). For further study, these potential rhizobacteria can be applied as biofertilizer which promote plant growth in agricultural field.

5. References

[1] Fatimahsari,T.K., S. G. S. Fitri, R. O. Khastini. 2014. Epyphitic Cyanobacteria on *Avicennia marina* pneumatophore in Mangrove Ecosystem of Cagar Alam Pulau Dua (CAPD), Serang, Banten. *Proceeding of International Conference On Research, Implementation And Education Of Mathematics And Sciences 2014, Yogyakarta State University, 18-20 May 2014.],* pp. B177-B182.

[2] Rohmah, W.S. 2014. Eksplorasi sianobakteria pada perairan payau di ekosistem mangrove Cagar Alam Pulau Dua Serang Banten. *Prosiding Semirata 2014 Bidang MIPA. ISBN* 978-602-70491-0-9: pp. 324-331.

[3] Audipudi, A.V., N.P. Kumar, and A. Sudhir. 2012. Phosphate Solubilizing Microorganisms Associated with Chollangi Mangrove Soil in East Coast of India. *International Journal of Scientific & Engineering Research,* **3 (11)**: 1-9.

[4] Simanungkalit, R. D. M., R. Saraswati, R. D. Hastuti, & E. Husen. 2009. Bakteri penambat nitrogen. 28 hlm.

[5] Cappuccino, J. G. dan Sherman, N. 2013. *Manual Laboratorium Mikrobiologi*. Edisi VIII. Jakarta: EGC.

[6] Nautiyal, C.S. 1999. An efficient microbiological growth medium for screening phosphorus solubilizing microorganisms. *FEMS Microbiol. Lett.* 170, 2017-2021.

[7] Holt. J.G., et al. 2000. Bergey's *Manual Determinative Bacteriology*. Baltimore: Williamn and Wilkins Baltimore.

[8] Selvam, M.M. and Kathiresan, K. 2010. Beneficial bacteria from soil of a tropical mangroves. *Asian Jr. of Microbiol. Biotech. Env. Sc.,* **12 (1)**: 1-8.

[9] Chibuogwu, O. J. and E. H. Nmesoma,. 2011. Batch culture studies of Phosphate solubilisation by Micrococcus sp PSB 7 isolated from rhizospheric soil. *American-Eurasian J. Agric and Environ. Sci,,* **10 (4)**: 667-674.

[10] Holt, J.G et al,. 1994. Bergey's Manual of Determinative Bacteriology. Ninth Ed. A Wolters Kluwer Company. Philadelphia.

[11] Schadev, D.P., H. G. Chaudhari, V. M. Kasture., D. D. Dhavale, & B. A. Chopade. 2009. Isolation and Characterization on Indole acetic acid (IAA) producing Klebsiella Pneumoniae strains from rhizosphere of wheat (Triticum aestivum) and their effect on plant growth. *Indian Journal of Experimental Biology*. **47**: 993-1000.

[12] PelczarJr, M J and Reid R. 1986. *Microbiology*. New york. McGraw-Hill.

[13] Hatmanti. 2000. Pengenalan *Bacillus spp. Oseana.* **25(1)**: 31-41.

[14] Compant, S., C. Clement, and A. Sessitsch. 2010. Plant Growth promoting bacteria in the rhizo and endosphere of plant: their role, colonization, mechanisminvolved and prospects for utilization. *Soil Biology and Biochemistry*. **42**: 669-678.

[15] Jayadi, M., Baharuddin and B. Ibrahim. 2013. In vitro selection of rock phosphate solubility by microorganism from Ultisols in South Sulawesi, Indonesia. *American Journal of Agriculture and Forestry* 2013; **1(4)**: 68-73.

[16] Setiawati, M.R., Suryatmana, P., Hindersah, R., Fitriatin, B.N. dan Herdiyantoro, D. 2014. Karakterisasi Isolat Bakteri Pelarut Fosfat untuk Meningkatkan Ketersedian P pada Media Kultur Cair Tanaman Jagung (Zea mays L.). *Bionatura-Jurnal Ilmu-ilmu Hayati dan Fisik*: **16 (1)**: 30 – 34.

[17] Lee, M., L N Ten, S G Wool and J. Park. 2011. Agromyces soli sp. nov., isolated from farm soil. *International Journal of Systematic and Evolutionary Microbiology*. **61**: 1286–1292.

[18] Sunantapongsuk, V., P. Nakapraves, S. Piriyaprin and L. Manoch. 2006. Protease Production and Phosphate Solubilization from Potential Biological Control Agents Trichoderma Viride and Azomonas Agilis from Vetiver Rhizosphere. *International Workshop on Sustained Management of the Soil-Rhizosphere System for Efficient Crop Production and Fertilizer Use*. Land Development Department, Bangkok: Thailand. p. 1-4.

[19] Qian, Y., J Shi , Y Chen, L Lou, X Cui, R Cao, P Li and J Tang. 2010. Characterization of Phosphate Solubilizing Bacteria in Sediments from a Shallow Eutrophic Lake and a Wetland: Isolation, Molecular Identification and Phosphorus Release Ability Determination. *Molecules*. **2010 (15)**: 8518-8533.

[20] Singh, B. and T. Satyanarayana. 2011. Microbial phytases in phosphorus acquisition and plant growth promotion. *Physiol Mol Biol Plants* (April–June 2011), **17(2)**:93–103. *DOI 10.1007/s12298-011-0062-x.*

[21] Šarapatka, B. 2003. Phosphatase activities (ACP, ALP) in agroecosystem soils. *Thesis*. Swedish University of Agricultural Sciences. p. 113.

[22] Bhattacharyya, F.N. and Jha, D.K. 2012. Plant growth-promoting rhizobacteria (PGPR): Emergence in agriculture. *World J Microbiol Biotechnol*, **28**: 1327-1350.

Physical properties on Indonesian local rice varieties

I Utami[1], R Rusmana[1], F R Eris[1], S Sjaifuddin[2], and Susiyanti[1]*

[1]Department of Agroecotechnology, Faculty of Agriculture, University of Sultan Ageng Tirtayasa. Jl. Raya Jakarta km. 04 Pakupatan, Serang City, Banten Province, Indonesia
[2]Department of Science Education, University of Sultan Agung Tirtayasa, Serang City, Banten Province, Indonesia

*Corresponding author: susiyanti@untirta.ac.id

Abstract. Rice is the main staple food for the people of Indonesia. It consists of some variation as its diversity. Each rice variety has a different physical shape and size in the grain. This study aims to determine the physical properties of several local Indonesian rice varieties. Research on the physical properties of grain was carried out on nine Indonesian local rice varieties from Banten, West Sumatra, and East Java provinces. Physical properties parameters observed in grain include length, width, shape, 1000 grain weight, bulk density, and chalkiness percentage of rice. The results showed that the grain shape of Tunggul Hideung and Genjah Nganjuk was slender and the other varieties were oval. While the type of Kewal Bulu Hideung includes round rice, the Kewal Bulu Hideung is slim in shape, and other varieties are oval. The highest 1000 grain weight in grain and rice was possessed by Kewal Gudril (28.23 g and 22.44 g), and the lowest was in Kuriak Kusuik (20.62 g and 14.09 g). The highest density of unhulled bulk density is Kuriak Kusuik (0.62 g..mL-1), and the lowest is Kewal Bulu Hideung (0.3 g.mL-1), while the highest rice bulk density is Susu Putih (0.89 g.mL-1) and the lowest one is the Genjah Nganjuk (0.79 g.mL-1). Based on the percentage of calcareous rice, Kewal Benur and Genjah Nganjuk including premium quality rice, Kewal Bulu Hideung, Tunggul Hideung and Kuriak Kusuik rice including medium quality class 1 rice, and other varieties do not meet the quality requirements set by SNI (> 5%).

Keywords: rice, grain, local rice, physical properties

1. Introduction

Rice is one of the primary food crops of almost half the world's population [1]. About 90% of the world's rice is grown and consumed in Asia [2]. Indonesian poeple has become the largest nation that consumes rice as a staple food and makes it a preeminent source of carbohydrates because it is easy to obtain, tastes good and can be combined with other food ingredients. According to [3], rice as a staple food contributes around 40-80% of energy and 45-55% of the protein in the average menu of the Indonesian people.

Rice *(Oryza sativa)* is also cinsidered the most important cereal crop in the developing world. It is generally considered a semi-aquatic annual grass plant. About 20 species of the genus Oryza are recognized, but nearly all cultivated rice is *O. sativa* L. A small amount of *Oryza glaberrima*, a perennial species, is grown in Africa. So-called "wild rice" *(Zizania aquatica)*, grown in the Great Lakes region of the United States, is more closely related to rice than to rice [4].

Since 2011 to 2017 rice production has continued to increase at 65.75 million tons in 2011 and 81.38 million tons in 2017. The achievements in 2017 have exceeded the rice production target set at 79 million tons [5]. National rice production in the period 2010-2017 continues to increase. In 2017 rice production is estimated to reach 81,382,451 tons, growing 2.56 percent from the previous year [6].

Indonesia has a vast genetic diversity of rice because the archipelago was formerly united with the Asian continent, which is the center of origin (center of origin) rice [7]. Rice germplasm in the form of varieties is very diverse, ranging from national varieties and local varieties. [8] reported that some rice-producing farmers in several regions in the Indonesian province still planted local varieties hereditary. According to [9], differences in varieties show significant differences in morphology, physical-chemistry, and cooking properties.

The marketing values of rice as an agricultural product depend on its physical qualities after the harvesting. Information on nutritional content and physical properties of rice are generally only found in superior varieties of rice. While rice on the market is very diverse in types and varieties, there is still little information related to the physicochemical nature of Indonesia's local rice germplasm. So it is deemed necessary to obtain primary data regarding the physical properties of local Indonesian rice for various purposes, especially for plant breeding activities. Based on this, it was necessary to research the physical properties of Indonesian local rice (*Oryza sativa* L.) varieties.

2. Method

This type of research is descriptive quantitative research. The research has been carried out in the biotechnology laboratory of Faculty of Agriculture, University of Sultan Ageng Tirtayasa and analized in laboratory of Faculty of Agriculture, University of Jember from October 2017 until February 2018.

The research began with the collection and identification of local rice variety from several regions in Indonesia. The local rice varieties used were Kewal Bulu Hideung, Kewal Gudril, Kewal Benur, Kewal Sampai Putih and Tunggul Hideung from the Banten region; Kusuik Kuriak from West Sumatra and MS Pendek, Susu Putih and Genjah Nganjuk from East Java. After collecting and searching for data on local rice varieties used including the date of harvest and the method of storing the local rice, the physical characteristics were analyzed, namely length, width, size, shape, the weight of 1,000 grains, density, and chalkiness..

3. Results and Discussion

3.1. Length, width and shape on grain and rice

Physical properties that are closely related to the quality of grain and rice and the level of consumer acceptance include grain size (length and width) and length and width ratio that reflects the shape of grain and rice. Classification of grain size and shape criteria is based on the SNI scale [10], and rice is based on the FAO scale [11].

Table 1. Average of length (L), width (W), the ratio of L/W, size, and shape of grain

Varieties	Length (mm)	Width (mm)	Ratio L/W	Size Grain	Form Grain
Kewal Bulu Hideung	8.01	3.38	2.37	Very long	Oval
Kewal Gudril	8.42	3.22	2.62	Very long	Oval
Kewal Benur	8.56	3.00	2.85	Very long	Oval
Kewal Sampai Putih	8,12	3.37	2.41	Very long	Oval
Tunggul Hideung	9.68	2.60	3.72	Very long	Slim
Kuriak Kusuik	8.05	2.75	2.93	Very long	Oval
MS Pendek	8.44	2.92	2.89	Very long	Oval

Varieties	Length (mm)	Width (mm)	Ratio L/W	Size Grain	Form Grain
Susu Putih	9.44	3.18	2.97	Very long	Oval
Genjah Nganjuk	9.54	2.88	3.31	Very long	Slim

All rice varieties studied have very long grain size with a length of more than 7.5 mm (Table 1). The layer of the husk primarily determines the length and shape of the grain. The longer and the larger size of the grain show the thicker layer of the husk that wraps the rice grains on the rice.

Table 2. Average of length (L), width (W), the ratio of L/W, size, and shape of rice

Varieties	Long (mm)	Wide (mm)	Ratio L/W	Size Rice	Form Rice
Kewal Bulu Hideung	5.61	2.81	1.99	Medium	Round
Kewal Gudril	5.67	2.74	2.07	Medium	Oval
Kewal Benur	6.08	2.45	2.48	Medium	Oval
Kewal Sampai Putih	5.76	2.70	2.13	Medium	Oval
Tunggul Hideung	6.50	2.10	3.10	Medium	Slim
Kuriak Kusuik	5.57	2.07	2.69	Medium	Oval
MS Pendek	5.71	2.21	2.58	Medium	Oval
Susu Putih	6.12	2.42	2.53	Medium	Oval
Genjah Nganjuk	6.79	2.33	2.91	Long	Oval

The size of the local rice studied was obtained in length ranging from 5.57 to 6.79 mm (Table 2). In general, rice in Indonesia is of medium to long size. Long sized rice is preferred on the international market compared to medium and short sized rice. However, in certain countries, short-sized rice is preferred in Japan, Korea, and Taiwan.

The width of rice obtained ranged from 2.07 to 2.81 mm. The widest rice is Kewal Bulu Hideung rice from Banten, while the rice with the smallest width is Kuriak Kusuik rice from West Sumatra. The length and width ratio of rice obtained ranged from 1.99 - 3.10. Based on the ratio of length/width, the shape of ground rice, according to FAO, is divided into three types, namely slim, oval, and round. Kewal Bulu Hideung Variety from Banten has a round shape with a ratio of 1.99 between its length and width, and Tunggul Hideung variety has a slim shape with a ratio of 3.10 between its length and width, while other varieties have an oval shape with a ratio between 2.00-3.00 between its length and width. According to [12], the size and shape of rice is the dominant character derived from the genetic characteristics of the parent and can be used as a parameter for determining the purity of a variety. The length and shape of rice are influenced by genetic factors, agroecosystems, and land fertility.

All rice varieties studied have a variety of sizes, shapes, and colors, as well as the appearance of grain and rice, as seen in Figure 1 and Figure 2. The grain color is generally yellow, but the yellow Kuriak Kusuik grain is mixed with brown. The variety of Kewal Bulu Hideung and Kewal Sampai Putih also have fur on their grain. The color of rice is not only white, but also yellowish as on the Kewal Gudril variety, red on the MS Pendek variety and pale white color on the Susu Putih glutinous rice variety.

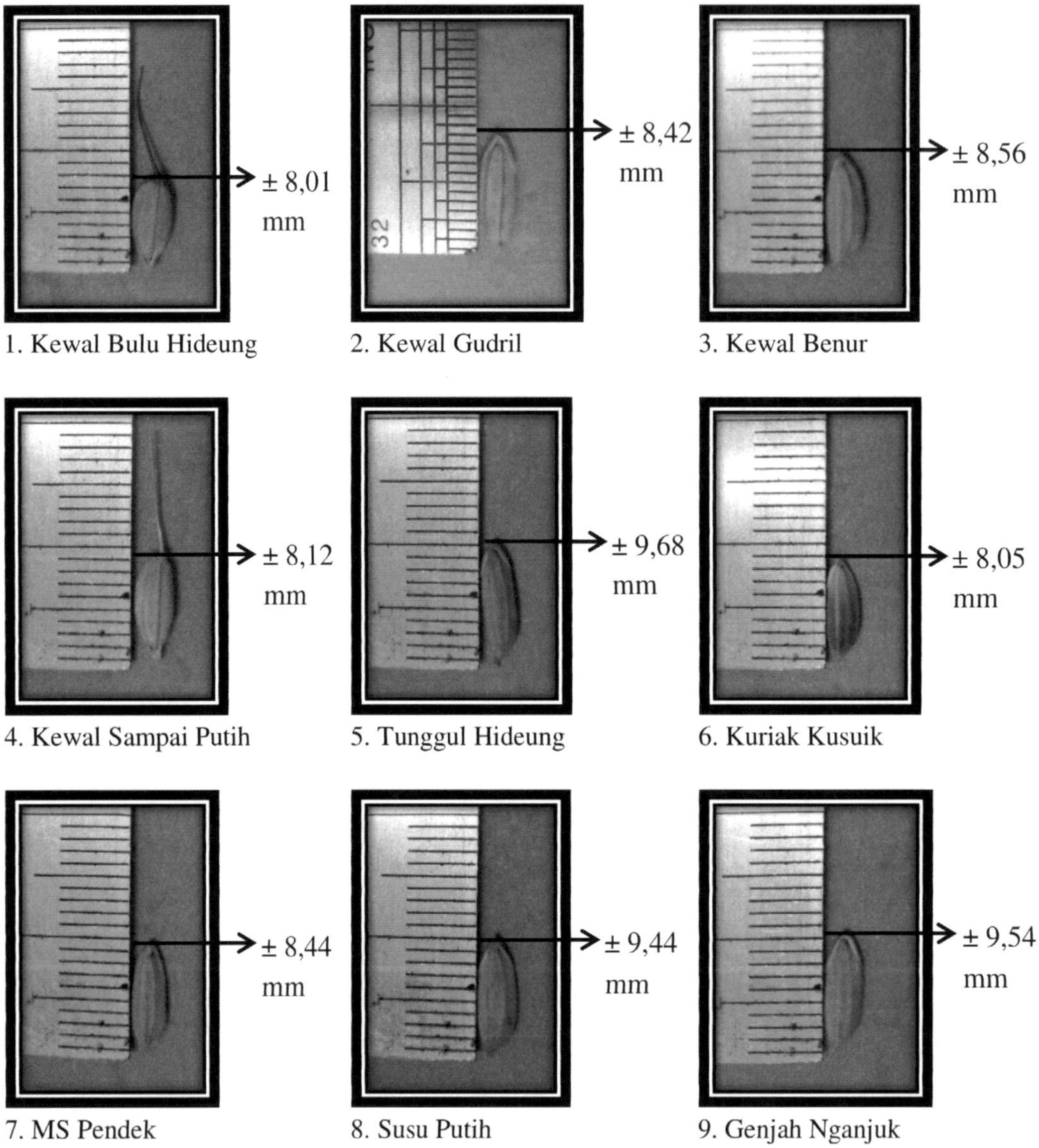

Figure 1. Diversity of shape, size and color of grain in 9 local Indonesian rice varieties

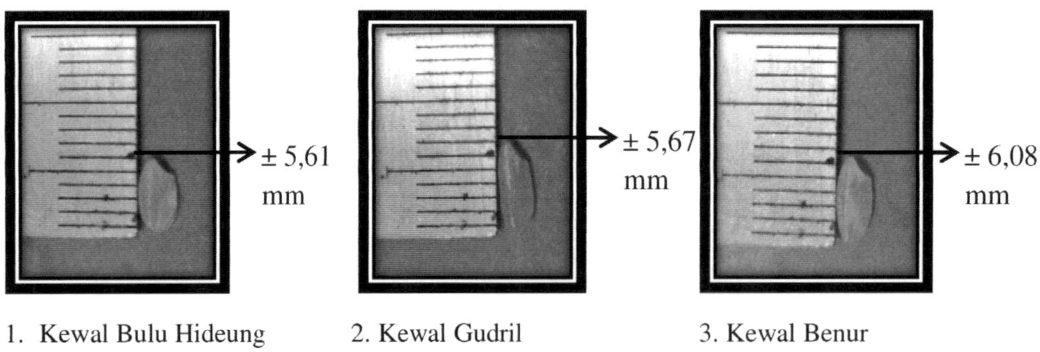

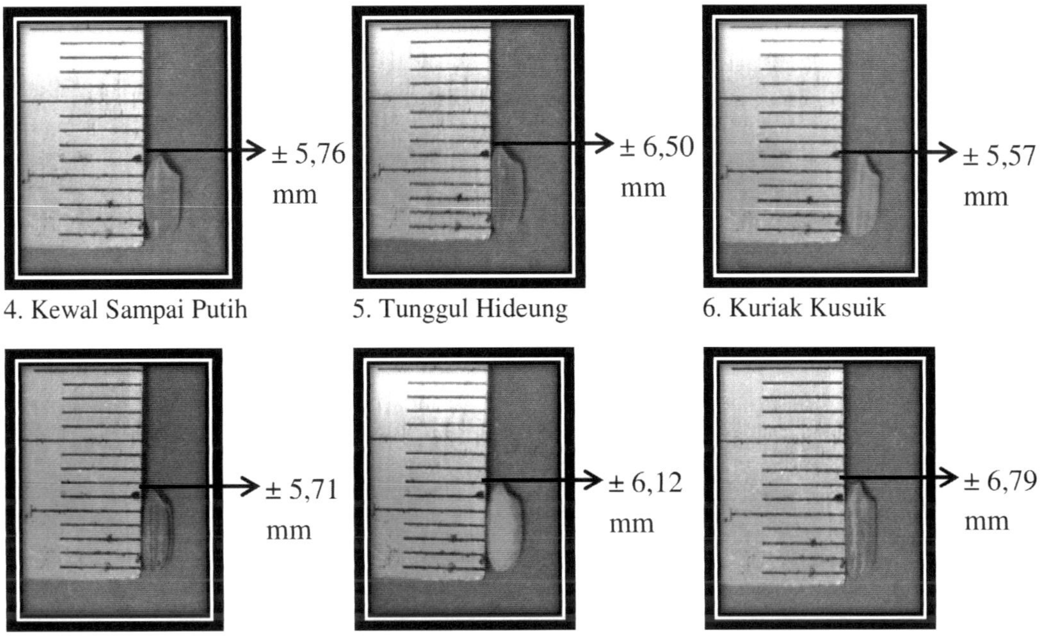

7. MS Pendek 8. Susu Putih 9. Genjah Nganjuk

Figure 2. Diversity of shape, size and color of rice in 9 local Indonesian rice varieties

3.2. 1000 grain weight in grain and rice

The weight of 1000 points indicates the weight of each grain which determines the yield of production. 1000 grain weight grain also determines the yield of milled rice. This value can be used to determine whether there is a mixture in a rice sample on the market. It also can be used to determine the purity of a rice variety. The results of an analysis of the measurement of 1000 grain weights on the grain and rice studied are presented in Figure 3.

The weight value of one thousand grains analyzed ranged between 20.62 - 28.23 grams, while the weight value of one thousand grains of rice ranged between 14.09-22.44 grams. According to [13], one thousand grains are affected by the availability of nutrients in the soil during rice planting. Nutrient deficiencies at the time of planting will cause in the weight of a thousand grains produced lower than its optimum potency.

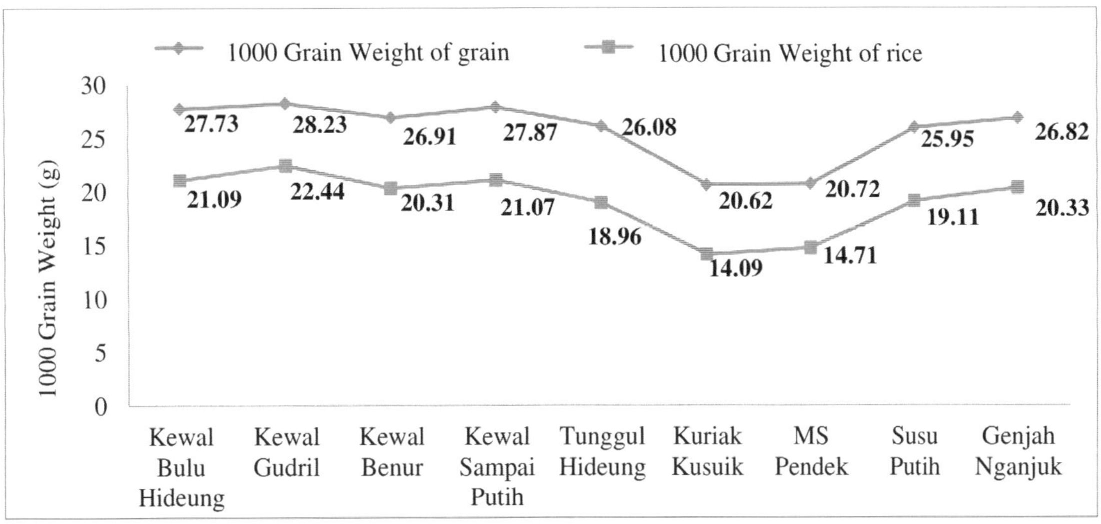

Figure 3. 1000 weight of grain and rice

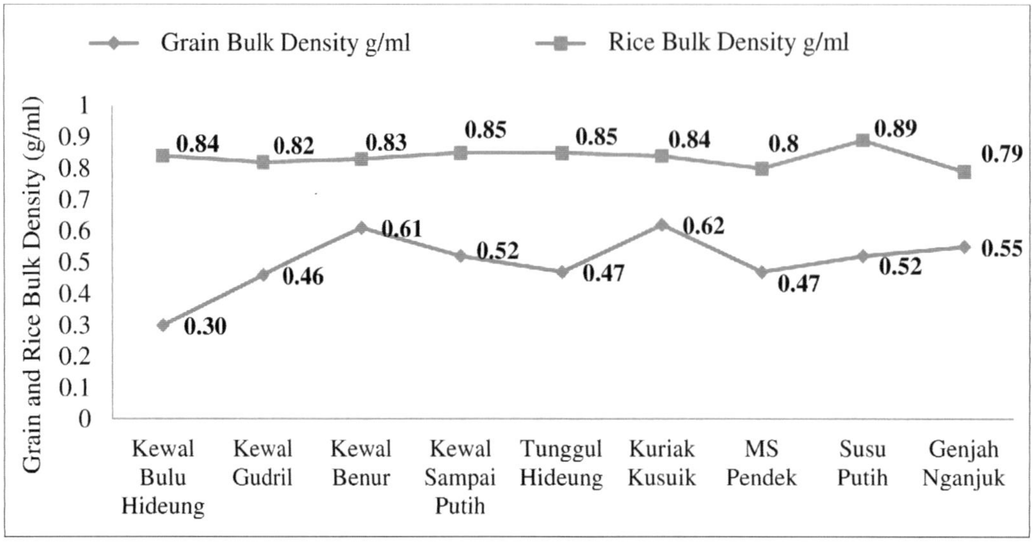

Figure 4. Grain and rice bulk density

Table 3. Percentage of chalkiness rice

Varieties	*White-based Rice* (%)	*White-belly Rice* (%)	*White-core Rice* (%)	*Milky-white Rice* (%)	Rice Total Whitewashing (%)*	*Perfect Rice* (%)
Kewal Bulu Hideung	1,4	44,4	5,2	1,0	52,0	48,0
Kewal Gudril	0,0	73,4	4,6	5,4	83,4	16,6
Kewal Benur	0,0	59,6	9,6	0,0	69,2	30,8
Kewal Sampai Putih	0,0	83,0	9,2	5,8	98,0	2,0
Tunggul Hideung	0,0	4,6	9,2	1,2	15,0	85,0
Kuriak Kusuik	0,6	1,4	32,2	1,8	36,0	64,0
MS Pendek	0,8	24,4	23,0	7,0	55,2	44,8
Susu Putih	-	-	-	-	-	-
Genjah Nganjuk	1,4	0,4	9,0	0,0	10,8	89,2

Note:
*Total Whitewashing Rice (%) = *White-based Rice* + *White-belly Rice* + *White-core Rice* + *Milky-white Rice*
Total observed rice (%) = total whitewashing rice+ *Perfect Rice* = 100%

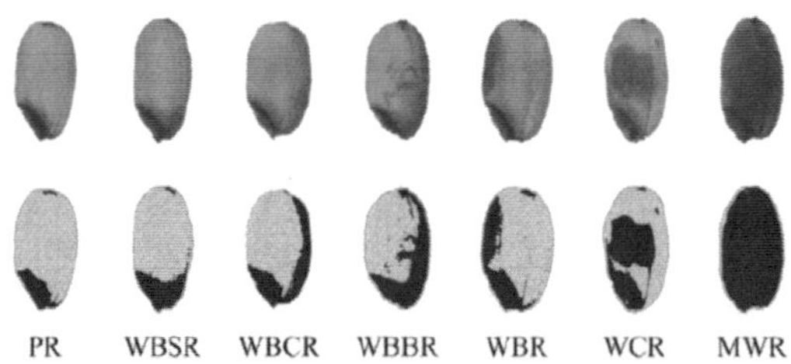

Figure 5. Grouping chalkiness: perfect rice (PR), white-based rice (WBSR), white-back rice (WBCR), white-back and -based rice (WBBR), white- belly rice (WBR), white-core rice (WCR) and milky-white rice (MWR) (Yoshioka, 2007).

3.3. Grain and rice bulk density

The density of grain or rice is a measure that describes the weight of grain or rice per unit volume expressed in units of g/ml. 1000 grain weight grain also determines the yield of milled rice. The size and shape of the grain and rice also directly affect the density. The rough rice of each variety show shows the least value of actual density, which agrees with [14], who stated that the specific gravity of paddy is lower than that of brown rice. Also, [15] reported similar results in the case of the actual density of rice by processing.

The bulk density of rice varieties in Indonesia ranges from 0,45 - 0.58 g/mL. Thus all grain samples are within the bulk density range, except for the Kewal Bulu Hideung grain variety which has a bulk density below the range of only 0.30 g/mL. The cause of this deference is the Kewal Bulu Hideung variety has fur at the tip of its grain, thereby reducing its density in storage. The Kuriak Kusuik has the highest bulk density of rice, and the lowest is possessed by the Kewal Bulu Hideung grain. It means that the Kewal Bulu Hideung grain needs two times more storage space than the Kuriak Kusuik grain, whereas the densest rice is possessed by Susu Putih variety. That is, for the same weight of Genjah Nganjuk rice requires 1.13 times more storage space than Susu Putih rice.

3.4. Percentage of number of chalking rice (chalkiness)

Whitewashing grains are white broken rice skin such as chalk and soft texture caused by physiological factors [16]. This white area is the result of very little starch development in the peripheral area of the endosperm. This is caused by insufficient carbohydrate transport during rice ripening [17].

When referring to the SNI standard for rice (2015) which requires a maximum content of limestone grains of 5%, then the level of whitewashing rice in varieties of Kewal Gudril (5,4 %), Kewal Sampai Putih (5.8%) and MS Pendek (7.0%) does not meet these requirements. Whereas the Susu Putih variety is glutinous rice so it cannot be observed its whitewashing grain because the color of the rice is indeed all murky white like glutinous rice in general. *Chalkiness* grouping based on Figure 5. below.

4. Conclusions

The grain size indicates that all grain is very long. The grain form of the Tunggul Hideung and Genjah Nganjuk varieties is slim, and the other varieties are oval-shaped. Whereas in the size of rice, the Genjah Nganjuk variety is long and the other varieties are medium. Rice forms of Kewal Bulu Hideung are round, Tunggul Hideung is slim and other varieties are oval-shaped.

The weight value of one thousand grains analyzed ranged between 20.62 - 28.23 grams, while the weight value of one thousand grains of rice ranged from 14.09 - 22.44 grams. The grain samples studied have densities in the range of 0.3 - 0.55 g / mL. Kewal Bulu Hideung variety has the smallest density (0.3 g / ml) because there are long black feathers in each grain of grain. The level of whitewashing rice in the studied rice fulfills SNI requirements (maximum 5%), except varieties of Kewal Gudril (5.4%), Kewal Sampai Putih (5.8%) and MS Pendek (7.0%).

Rice varieties on the market are very diverse. Therefore, the physical characteristics of them need to be known by consumers in order to obtain consumers preferences. As a result, we can find out the characteristics of rice they preferred and can arrange appropriate steps to gain those rice.

5. References

[1] Childs, NW 2004. Production and Utilization of Rice. In Elaine T. Champagne (ed). Rice: Chemistry and Technology. American Association of Cereal Chemists, Inc. Minnesota, pp.

[2] Tyagi, AK, Khurana JP, Khurana P., Raghuvanshi S., Gaur A., Cretaceous A., Gupta V., Kumar D., Ravi V., Vij S. and Sharma S. 2004. Structural and Functional analysis of Rice Genome. Journal Genetic. **Vol 83**: 79-99.

[3] Widjayanti, E. 2004. Potentials and Prospects of Indonesia's Indigenous Functional Food: Potential, Regulation, Security, Efficacy, and Market Opportunities. Bandung, 6-7 October 2004.

[4] Juliano, Bienvenido O. 1993. Rice in Human Nutrition. Published With The Collaboration Of The International Rice Research Institute Of Food And Agriculture Organization Of The United Nations. Rome, Paris.

[5] Ministry Agriculture of Republic of Indonesia. 2018. Rice Production 2018. Indonesia. Jakarta

[6] BPS. Statistic Central Agency. 2018. National Rice Production 2010-2017. Indonesia. Jakarta.

[7] Abdullah, B. 2006. The Potential of Wild Rice as a Genetic Source in Rice Breeding. Food Crop Science and Technology Bulletin **1(2)**: 143-152.

[8] Ariani, M., A. Saryoko, Kardiyono, and Z. Yursak. 2011. Appropriate Engineering Technology Engineering Study in Supporting Food Security: Potential and Opportunities for Local Rice Development and Local Tubers. Final Report of Research Cooperation BPTP Banten - Balitbangda Banten Province.

[9] Yadav R B, Khatkar B S, Yadav B S. 2007. Morphological, physicochemical and cooking properties of some Indian rice (*Oryza sativa* L.) cultivars. J Agric Technol, **3(2)**: 203–210.

[10] Indonesian National Standards. 1987. Grain, Quality Standards. National Standardization Agency.

[11] FAO. 1980. Dimensions and shapes (of grain, brown rice, or milled rice). In: Descriptors for Rice (L.) by IBPGR-IRRI Advisory committee. The International Rice Research Institute. Manila, Philippines.

[12] Setyono, A and Wibowo P. 2008. Rice Quality Selection and Its Relationship with the Characteristics of Several Inbred and Hybrid Strains. National Rice Seminar 2008.

[13] Hernawan. 2016. Analysis of the Physical Characteristics of White Rice, Red Rice and Black Rice (*Oryza sativa* L., *Oryza navira* and *Oryza sativa* L. indica). Bakti Tunas Husada Health Journal, Siliwangi University. **15(1)**: 79-91.

[14] Araullo EV, De Padua DB, and Graham M., 1976. Rice: Postharvest Technology. Int. Devel Res. Center Press, Ottawa, Canada.

[15] Correa PC, Schwanz da Silva F., Jaren C., Afonso Junior P.C., and Arana I., 2007. Physical and mechanical properties in rice processing. *J. Food Eng*., **79.** 137-142.

[16] Indonesian National Standards. 2015. Rice. National Standardization Agency.

[17] Juliano. 1973. The Chemical Base of Rice Grain Quality. Proceedings of the Workshop on: Chemical Aspects of Rice Grain Quality. IRRI. Los Banos, Philippines.

Growth rate of *Pseudomonas fluorescens* in liquid fertilizer from brown seaweed (*Sargassum* sp.) extracts

J Basmal[1], M E Aribowo[2], Nurhayati[1] and R Kusumawati[1]

[1] Research Center for Marine and Fisheries Product Processing and Biotechnology, Indonesia
[2] University of the Brawijaya, Malang, East Java, Indonesia
E-mail jamalbasmal24@gmail.com

Abstract. The purpose of this study was to observe the growth rate of *Pseudomonas fluorescens* bacteria as biological agents to stimulate plant growth and inhibit the growth of pathogenic bacteria. The method used were aimed to extract the micro, macro and growth stimulating substances contained in the thallus of *Sargassum* seaweed which had been fermented for 5 days using 0.3% KOH solution and then heated at 70°C for 2 hours. The filtrate obtained was formulated by adding silage and molasses which were then sterilized at 121°C for 15 minutes. The *P. fluorescence* bacteria were inoculated into the formulation solution and observed for 20 days with a span of 5 days against pH values, thickness, total nitrogen, $C_{organic}$, electric conductivity (EC), total dissolved of solid (TDS) and growth rate of *P. flourescens*. The results showed that the pH value did not show significant differences but the viscosity value, total nitrogen, $C_{organic}$, EC, TDS, and TPC were significantly different. The best results were found on the 10th day of inoculation of *P. fluorescence* in a non-sterile formulation solution with a pH value of 3.76, thickness of 2.95 cPs, total N-0.62%, $C_{organic}$ 0.1%, EC 9,814 µS / cm, TDS 4,602 ppm and Total Plate Count of 10.6 log CFU/ml.

1. Introduction

It is known that seaweed besides containing hydrocolloid also contains vitamins, proteins, minerals and growth substances. Basmal *et al.* [2] found that seaweed contained growth-promoting substances such as auxin, cytokinin and gibberellin. The mineral content and growth booster substances are mostly contained in the liquid seaweed. The mineral content and growth stimulating substances that exist in the seaweed thallus are very useful for increasing crop production. According to Serrani *et al.* [15] the content of growth booster substances can increase tomato production.

Although seaweed contains N-P-K but the amount does not meet the requirements set by Indonesian Minister of Agriculture which says that N-P-K in organic fertilizer must contain a minimum of 4%. To overcome this, seaweed or sap liquid can be added to other organic matarial such as fish meal or fish silage which is rich in minerals, vitamins and complete amino acids. Besides that, fermented manure can also be added which is also rich in organic matter. Other materials that can be used are molasses which have high $C_{organic}$ and are very much needed by plants.

The combination of seaweed, fish silage and molasses is not enough if applied as organic fertilizer as required by the Minister of Agriculture which states that single or multiple organic fertilizers must contain a minimum of 15% $C_{organic}$ and a minimum of 4% N-P-K content. This is impossible to achieve by organic fertilizers unless there are additional inorganic ingredients. Another alternative combination of sap liquid with other organic materials such as fish silage and molasses can be used as biological fertilizers by adding microbes as a N fixator from air, Other nutrients, P and K, are sufficiently available in the soil but are still mixed with other ingredients, therefore decomposers are needed to release P and K from other materials.

Elekhtyar [5] reported genera such as *Pseudomonas, Bacillus, Azospirillium, Agrobacterium, Arthrobacter, Alcaligenes, Serratia, Rhizobium, Enterobacter, Burkholderia, Klebsiella, Clostridium, Vario–vovax, Xanthomonas* and *Phyllobacterium* can be used to realease N-P and K also release plant growth promoting Rhizobacteria (PGPR). PGPR is a secondary metabolite (regulator of plant growth, fitohormones and biologically active substances) produced by microbes which facilitates the availability and absorption of certain nutrients from the root environment and inhibits plant pathogenic organisms in the rhizosphere. The present of the PGPR will increase the root exudation component e.g. sugars, amino acids, organic acids, vitamins, enzymes and organic or inorganic ions. PGPR are secondary metabolite, isolated from the rhizosphere, which when applied to seeds or crops, enhanced the growth of the plant or reduce the damage from soilborne plant pathogens. Suyono and Salahudin [16] said that many *Pseudomonas* found on land, plants and water such as *Pseudomonas aeruginosa*, *Pseudomonas* sp, *Pseudomonas putida, Pseudomonas fluorescens*, *Pseudomonas syringae, Pseudomonas stutzeri* dan lain-lain.

Seed germination was significantly improved with inoculated rice seeds by PGPR *Pseudomonas fluorescens* (PGPR*Pf*) as inoculation [5] while leaf mustard plants inoculated with *Pseudomonas fluorescens* and *Bacillus subtilis* had longer roots than those without the inoculation of *Pseudomonas fluorescens* and *Bacillus subtilis* [4]. Research [1,6] showed that PGPR (Plant Growth Promoting Rhizobacteria) could be used as a biological control agent by inducing resistance in plants. PGPR also serves to increase nitrogen capture, synthesis of phytohormones, dissolving minerals such as phosphorus and siderophores to availability of iron in plant roots. Luttge *et al.* [8] stated that phytochrome and phytohormon are signal system defense of plants. This can be seen with the increasing of resistance compound such as peroxide that occurs in the plant tissues. Park *et al.* [20] said that *Pseudomonas fluorescens* SS101 can produce volatile organic compounds (VOC) as a driver of plant growth and play an important role in modulating plant growth and inducing systemic resistance (ISR) to pathogens.

By paying attention to the amount of environmental damage due to the use of inorganic fertilizers, the modern agricultural system of crop fertilization will shift from the use of non-environmentally friendly inorganic fertilizers to organic fertilizers or biological fertilizers.

The purpose of this study was to look at the growth rate of *Pseodomonas fluorescens* in biological fertilizer formulations.

2. Materials and Methods

The type of seaweed used was *Sargassum* seaweed obtained from Binuangeun beach - Banten. Seaweed was collected from the collector washed with seawater, dried and then taken to the Center for Marine and Fisheries Product and Biotechnology Research for further treatment. Fish to be produced into fish silage was obtained from Kronjo-Tenggerang Fish Landing Place while molasses was obtained from sugar factories in the Bantul area of Central Java. The species of microbes used was *Pseudomonas fluorescens* obtained from the collection of the Research Center for Marine and Fisheries Biotechnology and Product Processing. The chemicals used were technical grade such as phosphoric acid and potassium hydroxide.

Sargassum washed and then soaked in a solution of 0.3% (w/v) KOH with a ratio of 1 part *Sargassum* : 10 for well water for 5 days. Then cooked at 70°C for 2 hours. The filtrate then separated from the seaweed. The resulting seaweed liquid was then separated from thallus. The filtrate obtained was then mixed with fish silage and molasses with a ratio: 72%: 21%: 7%. Then divided into two to get sterilization treatment at 121°C for 15 minutes and non-sterilization. After reached room temperature, the mixed medium was then inoculated with *P. fluorescens* with a ratio of 1: 10. The process of observing the growth rate of *P. fluorescens* was carried out for 20 days from a range of 0, 5, 10, 15 and 20 days. Observations made were the growth rate of *P. fluorescens*, N (Metode Kjeldal Foss), viscosity using Brookfield, P, K, & C organic used method from Balai Penelitian Tanah, EC & TDS used EC-Meter.

3. Results and discussion

3.1. pH value

The pH value of the liquid biofertilizer before the sterilization process was 3.52 and after the sterilization process the pH value dropped to 3.28. During the process of fermentation the pH value in the sterilization treatment in ranged from 3.28 to 3.41 while the treatment of non-sterile pH values during the fermentation process ranged from 3.52 to 3.33 (figure 1). In the treatment of the sterilization process there was a decrease in pH value from 3.52 to 3.28. This decrease is the possibility of a reaction between the sap liquid and silage during the sterilization process. The observation of the pH value of fish silage was 2.7. While the pH value of sap liquid was 9.64, the mixing of sap liquid with fish silage and molasses had resulted in a pH mixture of 3.52. The sterilization process had caused a reaction between fish silage and sap liquid to reduce the pH value. Differences in pH values between sterilization and non-sterilization treatments have different pH values up to the tenth day of the fermentation process, then in the fermentation process from day 15 to 20 days no significant differences occur. The highest pH value for the first 10 days of fermentation was found in non-sterilizing treatments. The pH value has an effect on microbial growth. Suyono and Salahudin [16] said that *Pseodomonas* spp will grow at pH 5-9 and arobic condition. Based on the pH value produced in sterile and non-sterile treatments, the highest change in pH value was found in the non-sterile treatment, this may be in the non-sterile treatment there were other microbes that develop and most likely come from fish silage. Handajani [7] reported that lactic acid bacteri (LAB) was found in fish silage to produced acidic pH conditions.

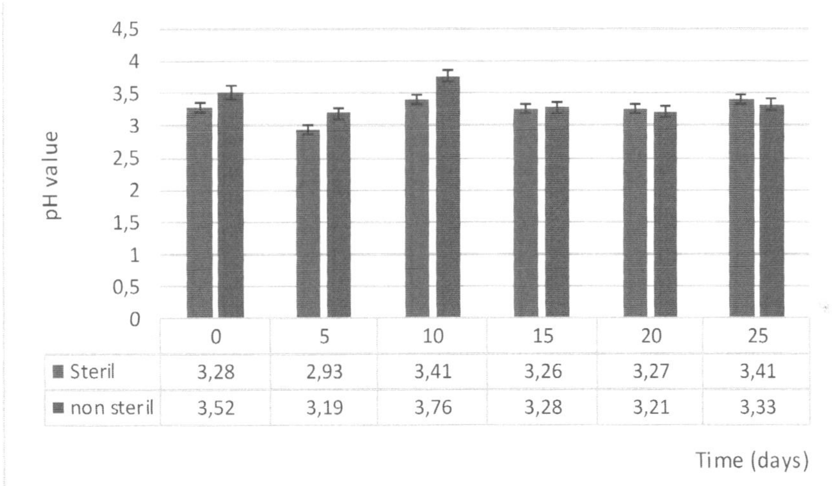

Figure 1. pH value on sterile and non-sterile treatments during fermentation.

3.2. Viscosity

The viscosity value in the sterilization treatment of *P. fluorescens* growing medium ranged from 2.12 - 2.75 cPs during the 20 days fermentation process, whereas in the non-sterilization treatment the medium of *P. fluorescens* ranged between 2.275 - 2.835. The pH value tended to increase in the sterilization treatment while the non-sterilization tended to decrease. At the end of the fermentation process the highest viscosity was found in the sterilization treatment. Viscosity values could be derived from sap liquid, fish silage and molasses, but increased viscosity values during the microbial inoculation process might be caused by microbial activity during the fermentation process. Based on figure 2 changes in viscosity values have no significant effect and tended to return to the initial pH value like mentioned by Patricio [11] the sudden slope change occurs during fermentation, and the viscosity value will return to a value close to the initial value.

Basmal *et al.* [2] found that viscosity value in the sap liquid of *Sargassum* in range of 2.7 - 3.1 cPs while the results of this research on the treatment of sterilized and non-sterilized liquid fertilizers ranged from 2.12 - 2.96 cPS. The low viscosity value of sterilized and non-sterilized liquid fertilizer treatment after *P. fluorescens* inoculation might probably caused by the presence of microbial activity of *P. fluorescens* which makes the resulting low viscosity of biological fertilizers. The results also found the pH value of sterilization and non-sterilization treatments ranged from 2.93 - 3.76. The decrease in pH value had caused alginate in the biological fertilizer solution to be converted to alginic acid. As a result, the thickness of the resulting biological fertilizer was low.

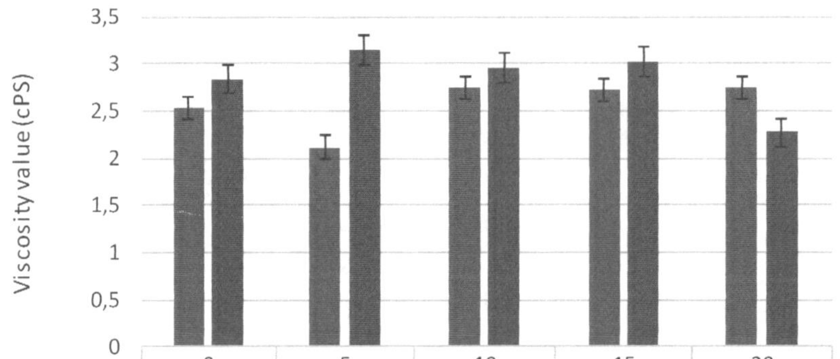

	0	5	10	15	20
Steril	2,535	2,12	2,745	2,73	2,75
Non steril	2,835	3,15	2,955	3,02	2,275

Figure 2. Viscosity value on sterile and non-sterile treatments during fermentation.

3.3. Total Nitrogen

The total N value in the sterilization treatment ranged between 0.66 - 0.65%, whereas in the non-sterile treatment between 0.60 - 0.61% during the inoculation process 20 days at room temperature. In figure 3, the fluctuating total N curve shows that in the 10th day of inoculation there was a decrease and then there was an increase in the 20th day inoculation. In the sterilization treatment, the total N reduction was 0.01% while the sterilization treatment took a 0.01% increase at the end of the 20th day inoculation. The total N in the sterilization treatment was higher than the non-sterilization treatment during the 20th day inoculation process (figure 4). Overall, the total N sterilization process on the growth medium of *P. fluorescens* was higher than the non-sterilization treatment.

It is well known that *P. fluorescens* is a group of N fixation bacteria that can take N from the air. In the sterilization treatment, the growth medium of *P. fluorescens* had no direct contact with the outside air while in the non-sterilization treatment the medium grew *P. fluorescens* was left open so that the air could be in contact with the growing medium. The total N content in the *P. fluorescens* growing medium non-sterilization could attract N from the air. In the sterilization process, there is no contact with air, as a result, *P. fluorescens* utilizes the total N present in the growing medium *P. fluorescens* as a result of the total N decreasing (figure 3). Mantelin, and Touraine [9] said that relatively large amounts of atmospheric N reach the plant as ammonia released by the bacteroids

3.4. Organic carbon value

Organic C (organic material) is a part of the land which is a complex and dynamic system, which is sourced from plants and/or animals remaining in the soil that is constantly changing shape because it is influenced by biological, physical and chemical factors. As a source of nutrients for organic C plants, it can improve environmental characteristics, sustainable highland productivity, improve soil structure, while improving the consistency and stability of soil aggregates, improving soil quality, energy sources and C sources for microbodies (prokaryotes), can reduce micronutrient poisoning (eg Al on acid soils), increases the ability of the soil to hold water available to plants and improves the living system of soil bodies, especially bacteria so that all microbiological processes in the soil run more perfectly.

The results showed that organic C in the treatment of sterilized and tightly closed *P. fluorescens* growth medium without contact with outside air ranged from 0.09 - 0.10%, while the treatment of non-sterilizing growth medium and left contact with air was found to be constant at 0.1% for 20 days *P. fluorescens* inoculation. Clark and Burki [3] said that *Pseudomonas* and *Achromobacter* required oxygen in the growth.

C organic is obtained from the process of composting or weathering organic matter both by the influence of the changing environment (temperature, pressure, humidity, open air or anaerobic conditions), enzyme and microbial activity. In the digestive process by microorganisms, a combustion reaction occurs between elements of carbon and oxygen into calories and carbon dioxide (CO_2). This carbon dioxide is released into gas, then the decomposed nitrogen element is captured by microorganisms to build up its body. When these microorganisms die, the nitrogen element will stay with compost and become a source of nutrition for plants. Carbon is the main food source for anaerobic bacteria so that the optimum growth of bacteria is strongly influenced by this element, where carbon is needed to supply energy. Total carbon measures all inorganic and organic carbon.

.

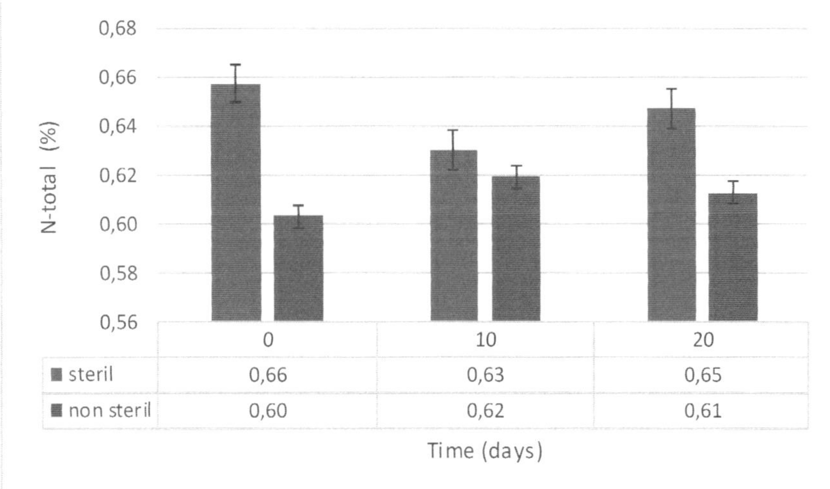

Figure 3. Total nitrogen value on sterile and non-sterile treatments during fermentation.

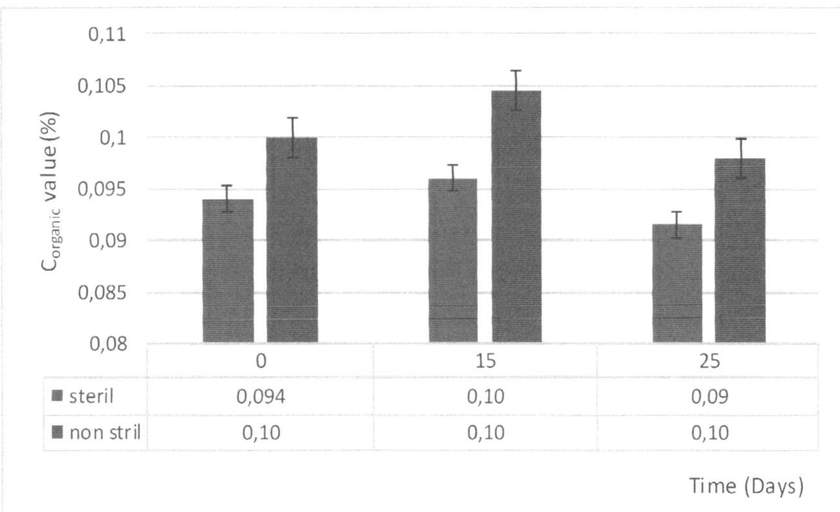

Figure 4. $C_{organic}$ value on sterile and non-sterile treatments during fermentation

3.5. Electrical conductivity (EC)

The value of Electrical Conductivity (EC) and pH of the solution have been used to determine the quality of liquid fertilizer. The higher the concentration of the solution means the more concentrated the salt content in the solution, so the ability of the solution to deliver electric current is higher which is indicated by high EC values as well. The density of the nutrient solution is influenced by the total salt content and the accumulation of ions present in nuts. Electrical conductivity in solutions affects plant metabolism, namely in terms of the speed of photosynthesis, enzyme activity and the potential for absorption of ions by roots. The concentration of nutrient solutions will also determine the duration of the use of nutrient solutions in hydroponic systems.

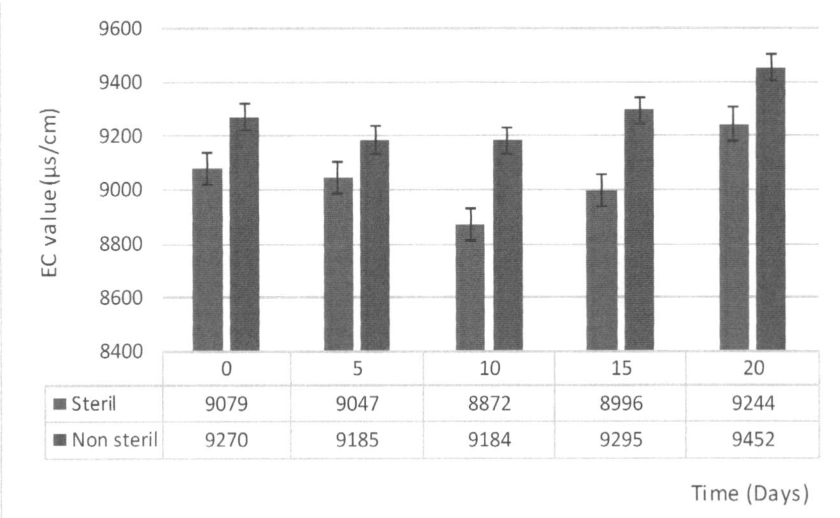

Figure 5. EC value on sterile and non-sterile treatments during fermentation.

The results showed that the liquid fertilizer sterilization inoculated with *P. fluorescens* had a smaller value compared to the liquid fertilizer which was sterilized during the inoculation process. In the sterilization treatment the EC value was between 8872 - 9244 µS / cm (micro Siemens / cm) while the EC sterilization value was between 9184 - 9452 µS / cm. Overall the EC value during the 20-day inoculation process showed a tendency to rise both in sterile treatment and in non-sterile medium growing *P. fluorescens*. The high value of EC in the non-sterile treatment was probably not only by *P. fluorescens* which works to decipher mineral materials in liquid fertilizer medium but also assisted by other microbes. The presence of air contact between liquid fertilizer and air medium was also one of the causes of high EC because *P. fluorescens* in multiplying itself requires oxygen. Basmal *et al.* [2] reported that the EC value in liquid *Sargassum* sap ranged from 2.230 to 3.520 micro Siemens/cm, while the results of the study found that the EC value of sterilization and non-sterilization treatments ranged from 8,871 - 9,452 micro Siemens/cm which meant inoculation into liquid fertilizer can increase the EC value of biological fertilizers.

3.6. Total dissolved solute (TDS)

The TDS value of the sterilized liquid fertilizer treatment ranged from 4427 - 4636 ppm, while the treatment of non-sterilized liquid fertilizer ranged from 4600 - 4757 ppm for 20 days in the *P. fluorescence* inoculation process (figure 6). In the treatment of TDS, value sterilization tended to increase while the treatment of non-sterile TDS values tended to decrease. TDS values indicate the number of minerals dissolved in the liquid fertilizer due to *P. fluorescens* activity during the 20-day inoculation process. The more dissolved minerals, the higher the TDS value. The decrease and increase in TDS value was likely due to microbial activity in the fertilizer medium. In the treatment of medium that had been sterilized the TDS value was lower than the treatment of non-sterilized medium. The high TDS value in the treatment of non-sterilized medium due to large amount of fertilizer composting process by various types of microbes present in the medium of fertilizer that is not sterilized, while the fertilizer medium which is sterilized is only carried out by *P. fluorescens* bacteria.

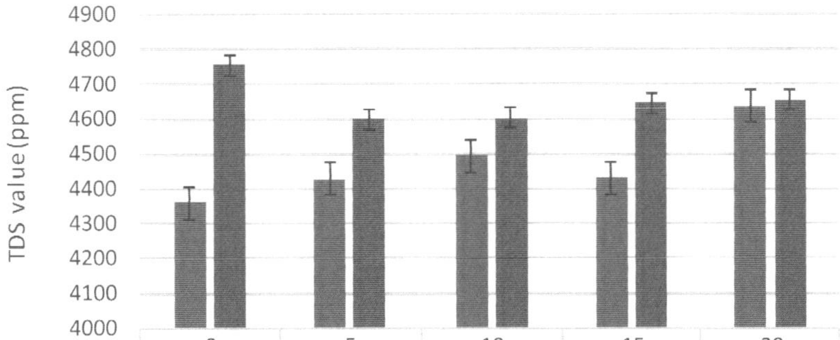

	0	5	10	15	20
■ Steril	4360	4427	4494	4430	4636
■ Non steril	4757	4600	4602	4645	4654

Figure 6. Total dissolved solute value on sterile and non-sterile treatments during fermentation.

3.7. Total Plate Count (TPC)

The experimental results showed that the sterilization and non-sterilization treatment had an effect on the growth rate of *P. fluorescens*. The highest growth rate was found on the 10th day of inoculation of *P. fluorescens* bacteria with the highest value added to the treatment of non-sterilizing growth medium of 10.6 log CFU / ml in the log phase while the sterilization treatment of the medium was only reached 4.7 log CFU / ml in the log phase. After the tenth day, on the 15th and 20th day, the treatment of growth medium both sterilized and non-sterilized had decreased but the highest value was still in the treatment of non-sterilizing growth medium. The occurrence of differences in the amount of CFU / ml was possible in the sterilization process because there are some nutrients lost. While the non-sterilized did not occur loss of nutrients, for example, some vitamins needed by bacteria would be lost due to heat treatment. Several factors that influence the rate of microbial growth are the availability of nutrients, conditions of growing medium and environmental conditions which include: temperature, oxygen, pH and humidity levels. There are several treatments that can damage the nutrition of growing medium because some of the vitamins and enzymes that are present in the medium grow if given high heat treatment and high pressure it is likely that some of the nutrients given will disappear. Missing some nutrients given will have an impact on the rate of microbial growth.

P. fluorescens is a bacterium that belongs to the rhizosphere group which can produce plant growth promoting rhizobacteria (PGPR) which is very influential to increases root surface area, increases nutrient uptake and improve plant production, inhibits the growth of other bacteria [9]. These bacteria can dominate the rhizosphere and develop rapidly, are gram-negative, motile, aerobic/facultative anaerobes [13] and are able to degrade and use large amounts of organic and inorganic compounds, interact with plants and associate in the rhizosphere that benefits agriculture [12]. *Pseudomonas* groups can produce large amounts of phytohormones especially IAA to stimulate growth [19] besides that it also produces cytokines, isopentenyl adenosine, and ribose zeatin [14]. IAA is a growth hormone auxin group which is useful for stimulating plant growth. Auxin is useful for increasing stem cell growth, inhibits the process of leaf abortion, stimulates fruit formation, and stimulates cambium growth, and inhibits the growth of armpit shoots[17]. *Pseudomonas* sp. also known to produce silicic acid which is capable of controlling tobacco necrosis virus in tobacco [10] and able to enhance crop yield by direct and indirect mechanisms [18].

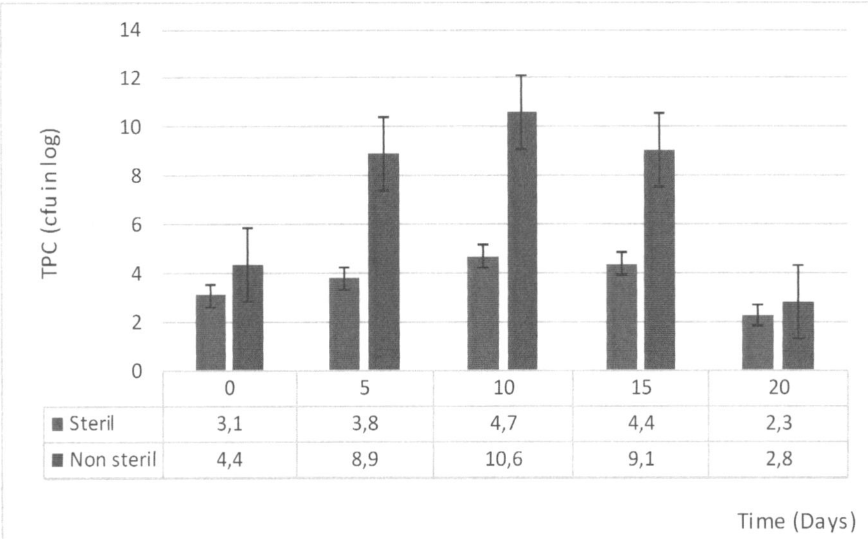

Figure 7. TPC value *P. flourescen* on sterile and non-sterile treatments during fermentation.

4. Conclusion

The growth rate of *P. fluorescens* has been influenced by the physical properties of growing medium and its environmental factors. In this experiment which uses two different growing medium, it is proven that there was a real effect on *P. fluorescens* growth rate, total N, organic C, EC, TDS and viscosity value of the resulting biological fertilizer. The highest *P. fluorescens* growth rate was found in the treatment of non-sterile growing medium namely: the highest stationer phase was achieved on the 10th day of *P. fluorescens* inoculation which was 10.6 log CFU/ml, with a pH value of 3.76, thickness of 2.95 cPs, total N-0.62%, Corganic 0.1%, EC 9,814 µS / cm, TDS 4,602 ppm.

5. References

[1] Ashrafuzzaman M, Hossen F A, Ismail M R, Hoque M A, Islam M Z, Shahidullah S M, and Meon S 2009 Efficiency of plant growth-promoting rhizobacteria (PGPR) for the enhancement of rice growth *African Journal of Biotechnology* **8(7)** pp 1247-52

[2] Basmal J, Saputra R, Karnila R and Leksono T 2019 effects of extraction temperature and time on the total extracted macro-nutrients from seaweed *Sargassum* sp JPPBKP *Jakarta*

[3] Clark D S and Burki T 1971 Oxygen requirements of strains of *Pseudomonas* and *Achromobacter CANADIAN JOURNAL OF MICROBIOLOGY* **18**

[4] Diyansah B, Aini L Q and Hadiastono T 2013 The effect of PGPR (Plant Growth Promoting Rhizobacteria) *Pseudomonas fluorescens* and *Bacillus subtilis* On Leaf Mustard Plant (*Brassica juncea* L.) Infected by TuMV (Turnip Mosaic Virus) *J. Trop. Plant Prot.* **1 (1)** pp 30- 38

[5] Elekhtyar N M 2015 Efficiency of *Pseudomonas fluorescens* as Plant Growth-Promoting Rhizobacteria (PGPR) for the Enhancement of Seedling Vigor, Nitrogen Uptake, Yield and Its Attributes of Rice (*Oryza sativa L.*) *International Journal of Scientific Research in Agricultural Sciences* **2** (Proceedings) pp 057-67

[6] Hai-Bi L, Singh R K, Singh P, Song Q, Xing Y, Yang L and Li Y 2017 Genetic Diversity of Nitrogen-Fixing and Plant Growth Promoting Pseudomonas Species Isolated from Sugarcane Rhizosphere *Frontier in Mycobiology* **8** p 20

[7] Handajani H 2014 Quality Improvement Silage Fish Biological Waste Using Lactic Acid Bacteria *JURNAL GAMMA* **9(2)** pp 31-9

[8] Luttge U and Higinbotham N 1979 Transport in Plants Springer-Verlag *New York* Heidelberg Berlin pp 235-54

[9] Mantelin S and Touraine B 2004 Plant growth-promoting bacteria and nitrate availability impacts on root development and nitrate uptake *J Exp Bot* **55** pp 27–34

[10] Maurhofer M, Hase C, Meuwly D, Metraux J P and Defago G 1994 Introduction of systemic resistance of tobacco to *tobacco necrosis virus* by root-colonizing *Pseudomonas fluorescens* strain CHAO: Influence of the gacA gene and of pyoverdine production *Phytopathology* **84** pp 139-46

[11] Patricio P, Almeida P I, Portela R, Sobral R G, Grilo I R, Cidade T and Leal C R 2014 Living bacteria rheology: population growth, aggregation patterns and cooperative behaviour under different shear flows 9 https://arxiv.org/pdf/1403.1405

[12] Palleroni J N and Moore E R B 2004 Taxonomy of pseudomonas : experimental approaches *Pseudomonas* Vol 1 *New York* Plenum Publishers

[13] Pelczar M J and Chan E S C 1986 Dasar-Dasar Mikrobiologi I Hadioetomo HS, Imas T, Angka S L Translated from *Element of Microbiology* Jakarta UI Press

[14] Salamone I E G, Heynes R K and Nelson L M 2001 Cytokinin production by plant growth promoting rhizobacteria and selected mutants *Can J Microbiol* **47** pp 404-11

[15] Serrani J C, Fos M, Atare´s A and Garcı´a-Martı´nez J L 2007 Effect of gibberellin and auxin on parthenocarpic fruit growth induction in the cv Micro-Tom of tomato. *J Plant Growth Regul* **26** pp 211-21

[16] Suyono Y and Salahudin F 2011 IDENTIFIKASI DAN KARAKTERISASI BAKTERI *PSEUDOMONAS* PADA TANAH YANG TERINDIKASI TERKONTAMINASI LOGAM. *Jurnal BIOPROPAL INDUSTRI.* **8***:***02** No. 01

[17] Tjondronegoro P D, Natasaputra M, Gumawan A W, Djaelani M and Suwanto A 1989 Botani Umum *Bogor* PAU Ilmu Hayati Institut Pertanian Bogor

[18] Walsh U F, Morrissey J P and O'Gara F 2001 Pseudomonas for biocontrol of phytopathogens: from functional genomics to commercial exploitation *Curr Opin Biotechnol* **12** pp 289–95

[19] Watanabe I, So R, Ladha J K, Katayama-Fujimura Y and Kuraishi H 1987 A new Nitrogen-fixing species of Pseudomonad: *Pseudomonas diazotrophichus,* nov. Isolated from rice *Can J Microbiol* **33** 670-78

[20] Yong-Soon P, Dutta S, Ann M, Raaijmakers J M and Park K 2015 Promotion of plant growth by *Pseudomonas fluorescens* strain SS101 via novel volatile organic compounds. *Biochemical and Biophysical Research Communications* pp 1-5

Response growth and yield of okra (*Abelmoschus esculentus* l.) to giving dosage levels of chicken manure and variation of plants spacing

Kartina A M *, P Utama, and I Dimyati

Department of Agroecotechnology, Faculty of Agriculture, University of Sultan Ageng Tirtayasa. Jl. Raya Jakarta KM.4 Campus Pakupatan, Serang City, Banten.

*Corresponding author: kartina_plg@yahoo.com

Abstract. This research was aimed to the response of dosage of chicken manure and variation of plants spacing on the growth and yield of Okra (Abelmoschus esculentus L.). This research was conducted from December 2018 until March 2019 in the Experimental Garden of Indonesia Vegetables Research Institute, Situ Gadung, Tangerang, Banten. This research used a Randomized Completely Block Design (RCBD) with two factors, the first factor was a dosage of chicken manure consisting of three levels P1 (25 ton/ha), P2 (30 ton/ha), and P3 (35 ton/ha) and the second factor was variation of plants spacing consisting of three levels J1 (40 cm x 50 cm), J2 (50 cm x 50 cm), and J3 (60 cm x 50 cm). The results showed that treatment of 35 ton/ha chicken manure (P3) had the best effect at plant height (24.89 cm) and number of leaves (7.59 leaf) on 30 days after planting, number of flowers (10.48 flowers) and number of fruits per plant (10.00 fruits) on 80 days after planting, number of fruits per plot (154.44 fruits) and mass of fruits per plot (4.02 ton/ha) on 90 days after planting, and mass of fruits per plant (4.64 ton/ha) on 100 days after planting. The variation of plants spacing had the best effect to number of fruits per plot (114.67 fruits) on 80 days after planting and mass of fruits per plot (1.9 ton/ha) on 70 days after planting. There was interaction between dosage of chicken manure and variation of plants spacing at number of leaves on 30 days after planting, number of flowers on 70, 80, and 100 days after planting, number of fruits per plant on 50, 70, 80, and 100 days after planting, number of fruits per plot on 50, 60, 70, 80, and 90 days after planting.

Keywords: okra, chicken manure, plant spacing, growth, crop yields.

1. Introduction

Okra is a plant that belongs to the *Malvaceae* family and comes from the tropical part of Africa. Okra fruit can be used as a vegetable, in 100 g okra fruit contains 88% water, 2.1% protein, 0.2% fat, 8% carbohydrate, 1.7% fiber, and 0.2% ash [1]. The importance of the nutrients contained in the okra fruit makes the plant widely produced commercially. However, in some tropical countries the optimum yield of okra (2-3 ton/ha) and high quality cannot be achieved, due to the continued decline in soil fertility [2]. According to [3], the use of organic materials such as chicken manure has an important role in improving the quality and nature of the soil, among others, increasing the binding capacity of sandy soil (improving sandy soil structure), improving clayy soil structure so that the soil that was originally heavy will be light , increase the ability of the soil to collect water so that the soil can provide more water for plants, improve drainage and soil air system (especially heavy soil) so that the water content is sufficient, enhance the binding capacity of the soil so that the soil becomes more resistant, not easily dissolved by irrigation water. According to [4], fertilizer dosage of chicken manure with a dose of 30 ton/ha, gives the best effect on the parameters of stem diameter, wet weight and dry weight of stems of okra plants. Based on the results of research by [6], the spacing of 50 cm x 50 cm gives the best results on the parameter number of fruits per plot (321.25 fruits). This research aims to determine the effect of dosing of chicken manure fertilizer and variation of planting distance on the growth and yield of okra (*Abelmoschus esculentus* L.).

2. Research Methods

This research was conducted from December 2018 until March 2019 took place at the Experimental Garden of the Situ Gadung Vegetable Research Institute, Pagedangan District, Tangerang Regency, Banten Province. The materials used in this research were the seeds of green okra varieties of Naila IPB, fertilizer for chicken manure, pesticides, soil and water.The tools used in this study are paper, buckets, scales, hoes, roll meters, stationery, scissors, embrats, stakes, rapia ropes, shovels, machetes, tractors, seed tray and cameras.

This research used a Randomized Completely Block Design (RCBD) with two factors, the first factor was a dosage of chicken manure consisting of three levels P1 (25 ton/ha), P2 (30 ton/ha), and P3 (35 ton/ha) and the second factor was variation of plants spacing consisting of three levels J1 (40 cm x 50 cm), J2 (50 cm x 50 cm), and J3 (60 cm x 50 cm). There were 9 treatment combinations, each of which was repeated 3 times so that there were 27 experimental units and 3 plants were used as samples. The parameters observed in this study are as follows: Plant Height (cm), Number of Leaves (strands), Number of Flowers (flowers), Number of Fruits Per Plant (fruit), Number of Fruits Per Plot (fruit), Weight of Fruits Per Plant (g), and Fruit Weight Per Plot (g).

In this study, Anova analysis of variance was used using DSAASTAT 1.101 software. If the results of the variance have a significant effect (*) or very real (**), then further tests are conducted. In this study, Duncan Multiple Range Test 5% was tested further.

3. Results and Discussion

3.1. Plant Height (cm)

Based on the variance, the treatment of fertilizer levels of chicken manure gave a very significant influence on the growth of okra plant height at the age of 10 DAP and 30 DAP. Table 1 depict the dosage of chicken manure fertilizer dose of 35 ton/ha giave the best effect on the height of okra plants at 30 DAP (24.89 cm).

Table 1. Average height of okra (*Abelmoschus esculentus* L.) plants in the administration of chicken manure dose doses and variations in plants spacing

Plant Age (DAP)	Manure Feces Chicken Animals (P) (Ton/ha)	Plant Height			Average
		Plants Spacing (J)			
		40 cm x 50 cm (J1)	50 cm x 50 cm (J2)	60 cm x 50 cm (J3)	
10	25 (P1)	10.27	9.78	9.55	9.87 a
	30 (P2)	10.61	1.,83	9.66	10.37 a
	35 (P3)	10.66	11.66	11.72	11.35 b
	Average	10.51	10.76	10.31	10,53
20	25 (P1)	13.44	13.50	13.61	13.52
	30 (P2)	16.06	14.50	13.83	14.80
	35 (P3)	14.78	15.83	15.17	15.26
	Average	14.76	14.61	14.20	14.52
30	25 (P1)	20.55	20.11	21.78	20.81 a
	30 (P2)	23.50	21.67	22.50	22.56 b
	35 (P3)	24.89	26.22	23.56	24.89 c
	Average	22.98	22.67	22.61	22.75

Note: numbers followed by the same letters in the same column show no significant difference according to the DMRT test of 5% level.

This is presumed that the application of fertilizer at a dose of 35 ton/ha is able to meet the needs of okra plant nutrients, one of them is nutrient N which can meet the needs of okra plants. This is also in accordance with the opinion of [6], the biggest contribution of chicken manure fertilizer was the nutrient content of N. This nutrient can affect the vegetative growth of plants which ultimately determines the reproductive phase and crop yield.

3.2. Number of Leaves (strands)

The treatment of chicken manure fertilizer dosage has a very significant influence on the growth of the number of leaves of okra at 30 DAP, and there is an interaction between treatments at 30 DAP. This is in accordance with the opinion of [7], the availability of sufficient N can increase the growth of plant organs, one of which is the process of leaf formation.

Table 2. average number of leaves of okra (*Abelmoschus esculentus* L.) leaves on dosing of chicken manure feces and variations in plants spacing.

Plant Age (DAP)	Manure Feces Chicken Animals (P) (ton/ha)	Number of Leaves Plants Spacing (J)			Average
		40 cm x 50 cm (J1)	50 cm x 50 cm (J2)	60 cm x 50 cm (J3)	
10	25 (P1)	4.33	3.67	4.22	4.04
	30 (P2)	4.56	3.67	4.00	4.07
	35 (P3)	4.00	4.44	5.00	4.48
	Average	4.26	3.93	4.41	4.20
20	25 (P1)	3.89	4.78	5,00	4.56
	30 (P2)	5.44	4.33	5.11	4.96
	35 (P3)	5.89	5.56	4.44	5.30
	Average	5.07	4.89	4.85	4.94
30	25 (P1)	5.00 a	5.78 b	6.11 b	5.63 a
	30 (P2)	7.22 c	6.22 b	7.11 c	6.85 b
	35 (P3)	8.11 d	8.89 e	5.78 b	7.59 c
	Average	6.78	6.96	6.33	6.69

Note: numbers followed by the same letter in the same row or column show no significant difference according to the DMRT test of 5% level.

Based on Table 2, it shows that the dosage of chicken manure 35 ton/ha gives the best number of leaves at 80 DAP (7.59 strands) based on these data it is suspected that in chicken manure 35 ton/ha is able to provide better nutrient needs compared to other animal manure fertilizer doses.

There was a treatment interaction at the age of 30 DAP, a combination of 35 ton/ha chicken manure fertilizer treatment with a spacing of 50 cm x 50 cm gave the best number of leaves (8.89 strands). This is because the combination of nutrients available is able to meet the needs of plants and a spacing of 50 cm x 50 cm is able to provide more efficient space for plants to grow.

IOP Conf. Series: Earth and Environmental Science **383** (2019) 012028 doi:10.1088/1755-1315/383/1/012028

3.3. Number of flower (flower)

The treatment of chicken manure fertilizer dosage has a significant effect on the number of flowers of okra plants at the age of 70 DAP and 80 DAP, very significant effect on 100 DAP. And there are interactions between treatments at the age of 70 DAP, 80 DAP and 100 DAP. Based on Table 3, it shows that the fertilizer dosage of chicken manure 35 ton/ha gives the best amount of flowers at 80 DAP (10.48 flowers).

There was an interaction between treatments at the age of 80 DAP, a combination of 35 ton/ha fertilizer dosage with a spacing of 50 cm x 50 cm. gives the best amount of interest (14.78 flowers). This is presumed at a dose of 35 ton/ha provides a better nutrient P compared to other dung fertilizer doses. And the spacing of 50 cm x 50 cm gives room to grow better for okra plants compared to other planting distances.

This is consistent with the opinion of [8], that the nutrient P plays an important role in plant reproduction, the availability of sufficient P directs higher crop production. The more flowers will produce more and more fruit.

Table 3. The average amount of flowers of okra (*Abelmoschus esculentus* L.) in the dosing of chicken manure feces and variations in plants spacing

Plant Age (DAP)	Manure Feces Chicken Animals (P) (ton/ha)	Number of flower Plants Spacing (J)			Average
		40 cm x 50 cm (J1)	50 cm x 50 cm (J2)	60 cm x 50 cm (J3)	
60	25 (P1)	1.22	2.00	2.78	2.00
	30 (P2)	2.67	1.89	3.11	2.55
	35 (P3)	3.22	3.22	2.44	2.96
	Average	2.37	2.37	2.78	2.51
70	25 (P1)	3.00 a	6.89b c	7.56b c	5.81 a
	30 (P2)	6.33 b	4.00 a	8.78 d	6.37 a
	35 (P3)	8.44 cd	12.11 e	6.44 b	9.00 b
	Average	5.93	7.67	7.59	7.06
80	25 (P1)	4.89 a	7.22 b	9.44 c	7.19 a
	30 (P2)	844 bc	5.78 a	11.78 d	8.67 b
	35 (P3)	9.56 c	14.78 e	7.11 b	10.48 c
	Average	7.63	9.26	9.44	8.78
90	25 (P1)	5.11	9.56	9.22	7.96
	30 (P2)	10.11	6.78	8.33	8.41
	35 (P3)	10.67	12.11	7.33	10.04
	Average	8.63	9.48	8.30	8.80
100	25 (P1)	300 a	5.11 c	5.44 c	4.52 a
	30 (P2)	5.78 c	4.00 b	10.67 e	6.81 b
	35 (P3)	7.78 d	7.33 d	5.11 c	2.67 b
	Average	5.52	5.48	7.07	6.02

Note: numbers followed by the same letter in the same row or column show no significant difference according to the DMRT test of 5% level.

3.4. Number of Fruits Per Plant (fruit)

The treatment of chicken manure fertilizer dosage has a significant effect on the number of fruits per plant at the age of 70 DAP, 80 DAP, and 100 DAP. And there are interactions between treatments at the age of 50 DAP, 70 DAP, 80 DAp, and 100 DAp.

Based on Table 4, shows the fertilizer dosage of chicken manure 35 ton/ha gives the best number of fruits per plant at 80 DAP (10.00 fruits).

There was an interaction between treatments at the age of 80 DAP, a combination of 35 ton/ha fertilizer dosage with a spacing of 50 cm x 50 cm. gives the best number of fruits per plant (13.78 fruits). This is presumably at a dosage of 35 ton/ha of chicken manure provides N elements that are able to meet the needs of okra plants.

This is in accordance with the opinion of [9], that nitrogen is one of the photosynthetic materials which directly affects plant production. Nitrogen plays an important role in producing assimilates needed to produce fruit, N deficiency can reduce fruit production.

Table 4. Average number of fruits per plant okra (*Abelmoschus esculentus* L.) plant in the dosing of chicken manure feces and variations in plants spacing

Plant Age (DAP)	Manure Feces Chicken Animals (P) (ton/ha)	Number of Fruits Per plant			Average
		Plants Spacing (J)			
		40 cm x 50 cm (J1)	50 cm x 50 cm (J2)	60 cm x 50 cm (J3)	
60	25 (P1)	1.22	2.00	2.78	2.00
	30 (P2)	2.56	1.89	3.00	2.48
	35 (P3)	3.00	3.22	2.44	2.89
	Average	2.26	2.37	2.74	2.46
70	25 (P1)	2.78 a	6.11 cd	7.67 cde	5.52 a
	30 (P2)	6.33 cd	3.89 b	8.56 f	6.26 a
	35 (P3)	8.44 ef	11.89 g	6.44 cd	8.93 b
	Average	5.85	7.30	7.56	6.90
80	25 (P1)	4.33 a	6.45 bc	9.44 d	6.78 a
	30 (P2)	7.78 c	5.56 b	11.33 e	8.22 b
	35 (P3)	9.11 d	13.78 f	7.11 c	10.00 c
	Average	7.07	8.63	9.30	8.33
90	25 (P1)	4.89	8.78	9.00	7.56
	30 (P2)	8.89	6.67	8.33	7.96
	35 (P3)	10.67	12.11	6.89	9.89
	Average	8.15	9.19	8.07	8.47
100	25 (P1)	3.00 a	4.89 c	5.44 cd	4.44 a
	30 (P2)	5.78 d	4.00 b	10.00 f	6.59 b
	35 (P3)	7.67 e	7.33 e	5.44 cd	6.81 b
	Average	5.48	5.41	6.96	5.95

Note: numbers followed by the same letter in the same row or column show no significant difference according to the DMRT test of 5% level.

And at a distance of 50 cm x 50 cm okra plants can grow better because at the planting distance reduces the risk of competition between plants both nutrient competition and growing space. This is in accordance with the research of [5], that the use of spacing of 50cm x 50 cm can produce optimum okra production.

3.5. Number of Fruits Per plot (fruit)

The treatment of chicken manure fertilizer dosage has a significant effect on the number of fruits per plot at the age of 50-70 DAP, and 90-100 DAP. The treatment of variations in plant spacing has a significant effect on the age of 80 DAP. And there are interactions between treatments at the age of 50-90 DAP.

Based on Table 5, shows the dosage of chicken manure 35 ton/ha gives the best number of fruits per plot at 90 DAP (154.44 fruits). And the variation of plant spacing of 50 cm x 50 cm gives the best number of fruits per plot (122.78 fruits). The combination of 35 ton/ha fertilizer dosage with spacing of 50 cm x 50 cm gives the best number of fruits per plot (184.67 fruits).

Table 5. Average number of fruits per plot of okra (*Abelmoschus esculentus* L.) plant in the dosing of chicken manure feces and variations in plants spacing

Plant Age (DAP)	Manure Feces Chicken Animals (P) (ton/ha)	Number of Fruits Per Plot plants spacing (J)			Average
		40 cm x 50 cm (J1)	50 cm x 50 cm (J2)	60 cm x 50 cm (J3)	
60	25 (P1)	21.67 a	36.00 b	38.67 b	32.11 a
	30 (P2)	47.67 cd	22.67 a	40.00 bc	36.78 a
	35 (P3)	54.33 d	70.33 e	32.33 b	52.33 b
	Average	41.22	43.00	37.00	40.41
70	25 (P1)	53.33 a	74.33 bc	76.33 bc	68.00 a
	30 (P2)	89.00 c	59.33 a	86.33 c	78.22 a
	35 (P3)	123.33 d	160.00 e	65.00 ab	116.11 b
	Average	88.56	97.89	75.89	87.44
80	25 (P1)	60.33 a	114.00 e	89.33 bcd	87.89
	30 (P2)	95.67 d	74.33 ab	90.00 bcd	86.67
	35 (P3)	92.00 cd	155.67 f	78.33 bc	108.67
	Average	82.67 a	114.67 b	85.89 a	94.41
90	25 (P1)	81.00 a	98.00 bc	104.00 cd	94.33 a
	30 (P2)	116.33 d	85.67 ab	120.00 d	107.33 a
	35 (P3)	160.67 e	184.67 f	118.00 d	154.44 b
	Average	119.33	122.78	114.00	118.70
100	25 (P1)	66.33	83.67	91.33	80.44 a
	30 (P2)	100.67	72.67	90.67	88.00 a
	35 (P3)	148.67	134.00	83.67	122.11 b
	Average	105.22	96.78	88.56	96.85

Note: numbers followed by the same letter in the same row or column show no significant difference according to the DMRT test of 5% level.

It is assumed that at a distance of 50 cm x 50 cm, there is a quite large population, but the level of competition between plants is not as high as other plant spacing, and the fertilizer dosage of 35 ton/ha is able to meet the nutrient needs of okra plants so that it can produce more fruit. This is in accordance with the opinion of [5] in his research, that the provision of fertilizer with a higher dose causes plants to absorb more nutrients so that it affects the increase in vegetative growth of plants which ultimately also affects the increasing number of okra fruit.

3.6. Fruit Weight Per Plant (g)

The treatment of chicken manure fertilizer dosage has a significant effect on fruit weight per plant at 100 DAP. And there are interactions between treatments at the age of 50 DAP, 80 DAP, 90 DAP and 100 DAP. Based on Table 6, it shows that the treatment of chicken manure 35 ton/ha gives the best fruit weight per plant at the age of 90 DAP (131.59 g). There is an interaction between treatments, a combination of 35 ton/ha fertilizer dosage with a spacing of 50 cm x 50 cm gives the best fruit weight per plant (165.83 g).

This is presumably at a dose of chicken manure 35 ton/ha is able to provide nutrients for plants, and it is estimated that the spacing of 50 cm x 50 cm is a more favorable planting distance for the growth of okra plants. This is in accordance with the opinion of [10] growing spacing can reduce competition between plants, so that the process of photosynthesis can be maximized. The resulting photosynthate can be used for plant growth and reproduction. And in the opinion of [11] that competition between plants will not occur as long as plant population density has not reached the threshold, where the resources needed by plants are limited. Wide planting distance causes plants not to be shaded with each other so that the absorption of sunlight can be maximized, so that the process of photosynthesis in the leaves also runs optimally and the high photosynthate produced can be used in plant growth.

Tabel 6. Average weight of fruits per plant okra (*Abelmoschus esculentus* L.) plant in the dosing of chicken manure feces and variations in plants spacing

Plant Age (DAP)	Manure Feces Chicken Animals (P) (ton/ha)	Weight of Fruits Per Plant			Average
		Plants Spacing (J)			
		40 cm x 50 cm (J1)	50 cm x 50 cm (J2)	60 cm x 50 cm (J3)	
60	25 (P1)	12.23	17.98	37.84	22.69
	30 (P2)	26.25	16.24	34.41	25.63
	35 (P3)	39.82	42.83	22.67	35.11
	Average	26.10	25.69	31.64	27.81
70	25 (P1)	40.62	57.87	59.07	52.52
	30 (P2)	54.04	31.25	60.76	48.68
	35 (P3)	89.72	107.83	43.95	80.50
	Average	61.46	65.65	54.59	60.57
80	25 (P1)	39.05 a	48.01 a	85.23 cd	57.43
	30 (P2)	73.59 bc	40.97 a	90.33 d	68.30
	35 (P3)	65.49 b	115.65 e	61.81 b	80.98
	Average	59.37	68.21	79.13	68.90
90	25 (P1)	55.48 a	104.31 c	130.73 cd	96.84
	30 (P2)	106.93 c	71.36 a	136.79 d	105.03
	35 (P3)	140.96 d	165.83 e	87.98 b	131.59
	Average	101.12	113.83	118.50	111.15
100	25 (P1)	47.33 a	71.58 b	91.41 c	70.11 a
	30 (P2)	91.97 c	56.52 ab	140.98 e	96.49 b
	35 (P3)	116.68 d	125.51 de	92.51 c	111.57 c
	Average	85.33	84.53	108.30	92.72

Note: numbers followed by the same letter in the same row or column show no significant difference according to the DMRT test of 5% level.

3.7. Fruit Weight per Plot (g)

The treatment of chicken manure fertilizer dosage has a significant effect on the number of fruits per plot at the age of 50-70 DAP, and 90 DAP. The treatment of variations in plant spacing has a significant effect on the age of 80 DAP. And there are interactions between treatments at the age of 50-90 DAP. Based on Table 7 shows that the dosage of chicken manure 35 ton/ha gives the best fruit weight per plot (2413.90 g). And the treatment of spacing of 50 cm x 50 cm gives the best average fruit weight per plot (1924.17 g). The combination of 35 ton/ha fertilizer dosage with spacing of 50 cm x 50 cm gives the best fruit weight per plot (3152.82 g).

Table 7. Average Weight of Fruits Per Plot of Okra (*Abelmoschus esculentus* L.) Plant in the Dosing of Chicken Manure Feces and Variations in Plants Spacing

Plant Age (DAP)	Manure Feces Chicken Animals (P) (ton/ha)	Weight of Fruits Per Plot			Average
		Plants Spacing (J)			
		40 cm x 50 cm (J1)	50 cm x 50 cm (J2)	60 cm x 50 cm (J3)	
60	25 (P1)	213.62 a	354.48 bc	446.09 c	338.06 a
	30 (P2)	559.44 d	264.17 ab	406.13 c	409.91 a
	35 (P3)	815.16 e	877.96 e	370.88 bc	688.00 b

Plant Age (DAP)	Manure Feces Chicken Animals (P) (ton/ha)	Weight of Fruits Per Plot			Average
		Plants Spacing (J)			
		40 cm x 50 cm (J1)	50 cm x 50 cm (J2)	60 cm x 50 cm (J3)	
	Average	529.41	498.87	407.70	478.66
70	25 (P1)	508.41 ab	833.64 de	702.39 cd	681.48 a
	30 (P2)	969.35 e	590.47 bc	820.20 de	793.34 a
	35 (P3)	1320.56 f	2023.55 g	375.49 a	1239.87 b
	Average	932.77 b	1149.22 c	632.69 a	904.90
80	25 (P1)	617.05 a	849.64 bc	1073.63 cd	846.77
	30 (P2)	1113.36 d	790.95 ab	1053.96 cd	986.09
	35 (P3)	626.14 a	1907.19 e	1031.94 cd	1188.43
	Average	785.52	1182.60	1053.18	1007.10
90	25 (P1)	971.61 a	1589.05 b	1617.19 b	1392.62 a
	30 (P2)	1663.13 b	1030.64 a	1681.77 b	1458.51 a
	35 (P3)	2326.19 c	3152.82 d	1762.69 b	2413.90 b
	Average	1653.64	1924.17	1687.22	1755.01
100	25 (P1)	924.14	1167.28	2763.48	1618.30
	30 (P2)	1858.62	1087.52	1284.74	1410.29
	35 (P3)	2397.35	2261.50	1334.77	1997.87
	Average	1726.70	1505.43	1794.33	1675.49

Note: numbers followed by the same letter in the same row or column show no significant difference according to the DMRT test of 5% level.

[12] states that the increase in pod yields at certain plant spacing and N doses is due to the efficient utilization of resources and leads to optimum morphological growth character to support the increase in yield of okra fruit.

4. Conclusion and Recomendation

Based on the results of research that has been done, it can be concluded as follows:

(1). Chicken manure 35 ton/ha (P3) showed the best effect at plant height (24.89 cm) and number of leaves (7.59 leaf) on 30 days after planting, number of flowers (10.48 flowers) and number of fruits per plant (10.00 fruits) on 80 days after planting, number of fruits per plot (154.44 fruits) and mass of fruits per plot (4.02 ton/ha) on 90 days after planting, and mass of fruits per plant (4.64 ton/ha) on 100 days after planting.

(2). The variation of plants spacing 50 cm x 50 cm showed the best effect to number of fruits per plot (114.67 fruits) on 80 days after planting and mass of fruits per plot (1.9 ton/ha) on 70 days after planting.

(3). There was interaction between dosage of chicken manure and variation of plants spacing to parameters number of leaves on 30 days after planting, number of flowers on 70, 80, and 100 days after planting, number of fruits per plant on 50, 70, 80, and 100 days after planting, number of fruits per plot on 50, 60, 70, 80, and 90 days after planting.

Therefore, the cultivation of okra plants can be used dosage of chicken manure fertilizer 35 ton/ha with a spacing of 50 cm x 50 cm. Further research is needed to investigate impact of other types of animal manure fertilizers and spacing planting variation to the growth and yield of okra plants.

5. References

[1] Akanbi, W.B., A.O. Togun., J.A. Adediran., dan E.A.O. Ilupeju. 2010. Growth, Dry Matter and Fruit Yields Components of Okra under Organic and Inorganic Sources of Nutrients. American-Eurasian Journal of Sustainable Agriculture. **4(1)**: 1-13.

[2] El-Kader, A.A.A., S.M. Shaaban, dan M.S. Abd El-Fattah. 2010. Effect of Irrigation Levels and Organic Compost on Okra Plants (Abelmoschus esculentus L.) Grown in Sandy Calcareous Soil. Agriculture and Biology Journal of North America. **1(3)**: 255-231.

[3] Murbandono, H. S. 2008. Compost Making. Jakarta: Swadaya.

[4] Martini, Riska Amalia. 2018. Effectiveness Test Application of 6.5 L Chicken and Atonic Manure on Growth and Yield of Okra (Abelmoschus esculentus L.) Plants. Mataram. University of Mataram.

[5] Raditya, J., E.D. Purbajanti, and W. Slamet. 2017. Growth and Production of Okra (Abelmoschus esculentus L.) at Different Nitrogen Fertilization Levels and Spacing. J. Agro Complex **1(2)**: 49-56.

[6] Barus, R.A.A., Chairani H, and Rosita S. 2018. Response of Growth and Production of Two Okra Varieties (Abelmoschus esculentus L) to Provision of Various Types of Organic Fertilizers. Journal of Agroecotechnology. **6(2)**: 253-258.

[7] Maghfoer, M.D., R. Soelistyono, and N. Herlina. 2013. Response of eggplant (Solanum melongena L.) to combination of inorganic-organic N and EM4. Agrivita. **35(3)**: 296 – 303.

[8] Pranata I., D.R. Lukiwati, and W. slamet. 2017. Growth and Production of Okra (Abelmoschus esculentus L.) with Various Organic Fertilizer enriched with Phosphate Rock. J. Agro Complex 1 (2): 65-71.

[9] Ayissa, T. dan F. Kebede. 2011. Effect of Nitrogenous Fertilizer on the Growth and Yield of Cotton (Gossypium hirsutum L.) Varieties in Middle Awash, Ethiopia. J. of the Drylands 4 (1): 248 – 258.

[10] Arifah, S.H., Murti A and Yulia E.S. 2019. Effectiveness of Manure and Plant Spacing on Okra (Abelmaschus esulentus L. Moench) Crops. Journal of Tropical and Subtropical Agricultural Sciences. 4 (1): 38-42.

[11] Maurya, R. P., J. A. Bailey, and J. S. A. Chandler. 2013. Impact of Plant Spacing and Picking Interval on The Growth, Fruit Quality and Yield of Okra (Abelmoschus esculentus L.) Moench). American Journal of Agriculture and Forestry 1(4): 48-54.

[12] Onyegbule U. N., S. O. Ngbede, E. I. Nwanguma, dan A. Ohaneze. 2012. Evaluation of Okra Performance to Different Rates of Poultry Manure and Plant Population Density. Continental. J. Agric. Sci. 6(3): 56-63.

Total bacteria and milk nutrition using ZnO doped N thin film with visible light to maintain milk quality

M Z Asror[1], M I Ilsatoham[2], N I Pratiwi[2] and H Susanto[2]
[1] Department of Animal Science, Faculty of Animal and Agricultural Sciences, Diponegoro University, Indonesia
[2] Department of Physics, Faculty of Science and Mathematic, Diponegoro University Indonesia
email zainulasror28@gmail.com

Abstract. Milk is susceptible to microbial contamination due to its high nutrient contents. The presence of microbial contamination can damage the nutrients in milk such as protein, fat and lactose. Therefore, there is a need to keep milk quality. ZnO doped N thin film can kill pathogenic bacteria. The aim of this research is to investigate the number of bacteria and milk nutritions (protein, lactose and fat) due to ZnO doped N thin film. Milk samples were obtained from smallholder dairy farm in Central Java. Fresh milk was poured in a beaker glass, then ZnO doped with N thin films and illuminated with visible light. The testing scheme was carried out for 0, 60, 120 and 180 minutes in the refrigerator. Total bacteria was calculated using Total Plate Count (TPC) method. Milk nutrition received using lactoscan (MMC, Bulgaria). The conclusion of this research is ZnO doped N thin film with visible light could reduce total bacteria by 95.2%, *Coliform* by 66%, *S. aureus* by 94.4% and milk nutritions in accordance with Indonesian National Standards of fresh milk. Suggestions from this study, it is necessary to apply ZnO doped N thin films in food technology as an effort to maintain milk quality.

1. Introduction

Cow milk is a food that has very important nutrients for a human such as protein, lactose, fat, and mineral [1]. The high level of nutrient content in cow milk gives a chance for the growth of pathogenic bacteria. Generally, dairy farmers in Indonesia are folk farmers who are still traditional. The milk quality produced by folk farms has not been in line with the standard as the total bacteria 5 x 10^6 CFU/mL and the total contamination of coliform 229,76 MPN/mL [2,3].

The high level of bacterial contamination can cause the degradation of milk quality (protein, lactose, and fat). This is due to the non-hygiene handling process of milking and post-milking. Sources of bacterial contamination in milk are affected by non-sterile equipment, sanitation during milking and post-milking. Milk has a standard based on SNI (Indonesian National Standard) that is the maximum bacterial total of 1x10^6 CFU/mL, protein 2.8%, fat 3%, lactose 4 %, and density 1.024 [4]. On the other hand, the milk quality that is not convenient with SNI will be rejected by the dairy industry. This can cause losses economically. Therefore, that needs to solve in reducing bacterial contamination in order to maintain the milk quality.

Today nanotechnology has grown such as photocatalysts that are interesting because of the efficiency and the cost relatively low [5]. The photocatalyst is an efficient and cheap method of bacterial degradation. ZnO is one of the most popular photocatalysts because it has high, inexpensive and non-toxic photoactivity [6]. ZnO can be active under UV light. Therefore ZnO must be doped with nitrogen so that it can be active under visible light. ZnO doped N under visible light will produce reactive oxygen species (ROS) such as OH^-, $^1O_2$, O_2H, H_2O_2 which can exist as antibacterial [7]. In previous studies, photocatalysts can reduce the contamination of pathogens bacteria such as *Staphylococcus aureus* and *E. coli* [8]. Thus, this study aimed to know the effect of ZnO doped N thin film on bacterial contamination in milk. The results of this study could provide a method for maintaining the quality of milk and to ensure food safety from bacterial contamination.

2. Material and Method

2.1 Materials

All chemicals were reagent grade and used without futher purification. Zinc acetate dehydrate ($Zn(COOCH_3)_2.2H_2O$ as precursor of ZnO, isopropanol ($(CH_3)_2CHOH$), monoethaolamine (C_2H_7NO) as a solvent and stabilizer, urea (CH_4N_2O) as doping and fresh milk.

2.2 Fabrication of N doped ZnO thin film

Synthesis of doped ZnO N nanoparticles was carried out by sol-gel method with Zn $(CH_3COO_2)_2.2H_2O$ precursors as a source of Zn and isopropanol (IPA: (CH_3) 2CHOH), monoethanolamine (MEA: $HOCH_2CH_2NH_2$), and urea (CH_4N_2O) as a nitrogen source. Making ZnO nanoparticles by dissolving Zinc Asetate into propanol and monoethanolamine solutions at room temperature with a concentration of 0.3 M Zinc Asetate with a 1: 1 molar ratio and then sterile for 1 hour. The result was a clear white solution. The next process was coating on a glass substrate with a temperature of 450^0C. Before deposition, glass substrates were cleaned by the Radio Corporation of America (RCA) method. In this method, glass substrates were cleaned by acetone and methanol for 10 min. An ultrasonic cleaner was used to wash dust and oil out. Then, we immersed the glass slide into the double-distilled water for 8 min and dried the substrates off using a compressor gun. The obtained solutions were sprayed onto glass substrates for 1 hour at 450^0C. Furthermore, the films were annealed at 450 ^{0}C for 1 hour.

2.3 Testing process

This research was carried out on a laboratory scale. Milk sampels were obtained from smallholder dairy farm in Central Java. The testing process was carried out on a lab scale. 500 ml fresh milk were poured in a beaker glass. Then ZnO doped with N thin films placed on top and illuminated with visible light. The testing scheme was carried out for 0, 60, 120 and 180 minutes in the refrigerator. The total bacteria data was taken at 0, 60, 120 and 180 minutes while milk quality was taken at 0 and 180 minutes.

2.4 Total plate count analysis

The test of total bacteria uses the TPC method (Total Plate Count) based on Indonesian National Standar (SNI) [9]. The sample of milk is dissolved with BPW 0,1 %. Then it is diluted with 10^{-3}, 10^{-4}, 10^{-5} levels dilution. 1 mL of suspension was taken at 10^{-5} level dilution and inserted into a petri dish. Then add Plate Count Agar (PCA) then incubate at 37° for 24 hours [10]. Calculate the number of colonies with the following formula

$$\text{Total Bacteria} = \text{Number of Colonies}\ \frac{1}{\text{Dilution level factor}}$$

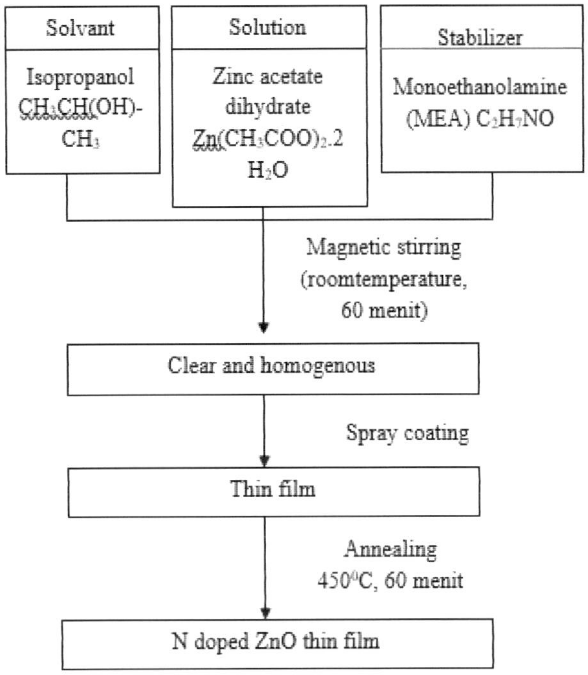

Figure 1. Schematic of fabrication N doped ZnO thin film.

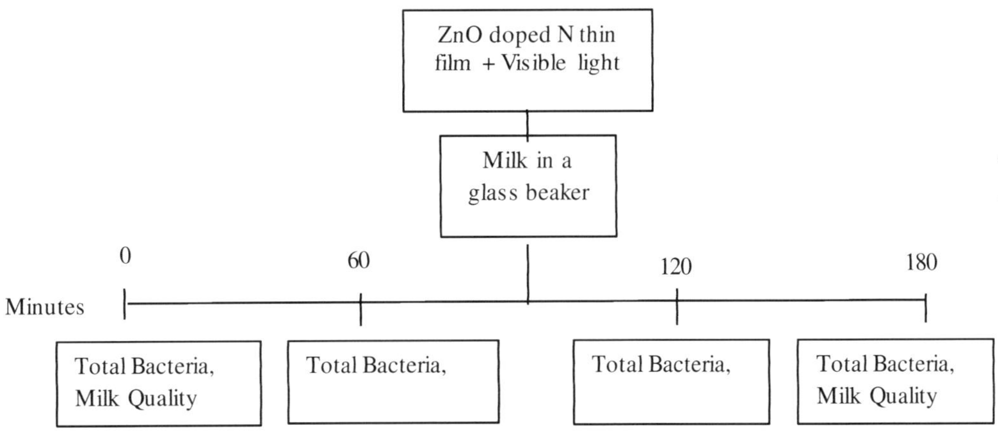

Figure 2. Schematic diagram of testing process.

2.5 Milk quality analysis

The testing of dairy components in this study was performed with Lactoscan (MCC, Bulgaria). The testing includes protein, fat, lactose and density of milk. The pH of milk was measured with pH meter 5011 (Ezdo Gondo, Chinese Taipei). The milk sample was poured 50 ml into glassware, then the pH meter was dipped into milk and the result of pH recorded. Each sample was duplicated and the result calculated on average. The test of somatic cells in milk using the breed method [11].

3. Results and discussion

3.1 Bacterial analysis

The total bacteria in milk before treatment was 2.5 x 10^6 CFU/mL. It shows that the milk samples from the dairy farmers have high bacterial contamination. The high levels of bacterial contamination in milk are closely related to hygiene and sanitation during the handling of milking. Based on Hassan *et al.* [12] sources of bacterial contamination in milk are influenced by non-sterile equipment, sanitation during milking and post-milking treatment.

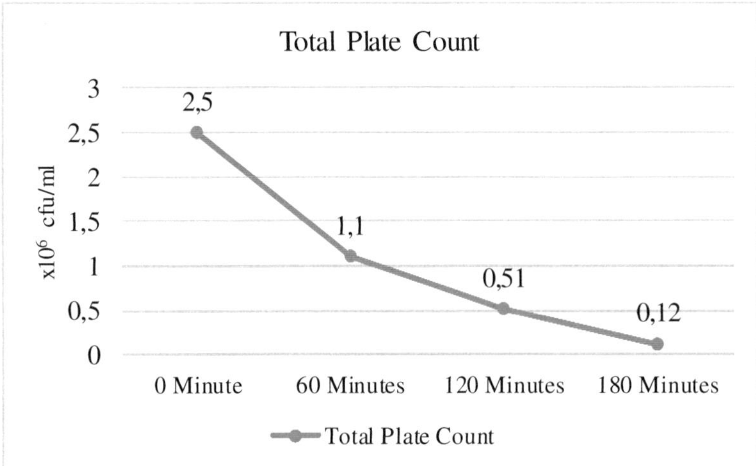

Figure 3. Schematic diagram of reaction mechanism of ZnO doped N thin film with visible light for 0, 60, 120 180 minutes.

Based on figure 3 shows the effect of the thin film of ZnO doped N on the total bacteria in milk. The degradation of total bacteria increased in time to 0, 60, 120 and 180 minutes. At 0 minute the total bacteria was 2,5x10^6 CFU/mL, at 60 minutes was 1,1x10^6 CFU/ml, at 120 minutes was 0,5x10^6 CFU/ml and at 180 minutes was 0,16x10^6 CFU/ml or the percentage of amount degradation of 120 minutes was 95,2%. Furthermore, based on figure 3 shows the effect after ZnO doped N thin film treatment for 180 minutes, *Coliform* bacteria decreased by 66% and *S. aureus* was 94,4%. It shows that the treatment of ZnO doped N thin film for 120 minutes could affect the decrease in the number of bacteria in milk. This is because of the existing of photocatalysts that can decrease total bacteria. Photocatalysts of ZnO doped N are capable of producing hydroxyl radicals (OH-) under visible light that can kill bacteria. According to Prasanna and Vijayaraghavan [13] photocatalyst N doped ZnO has antibacterial activity so it can decrease bacteria. The result of total bacteria was according to SNI of fresh milk. Total fresh milk bacteria maximum limit is 1x10^6 CFU/ml [4].

The photocatalyst mechanism of killing bacteria, when ZnO is photo-induces by solar litght with photonic energy (h_v) equal or greater than excitation energy (E_g) e^- from the filled valence band (VB) are promoted to an empty conduction band (CB). This photoinduced process produces electron-hole (e^-/h^+) pairs. The photoexcited e^-_{CB} could be reacted with surface adsorbed oxygen to produce superoxide radical anions which subsequently converted to hidroxyl radicals [14, 15, 16].

$$ZnO\text{-}N \rightarrow h^+_{VB} + e^-_{CB} \quad (1)$$

$$e^-_{CB} + O_2 \rightarrow .\ O^-_2 \quad (2)$$

$$O^-_{2\,ads} + H_2O \rightarrow .OH + OH^- \quad (3)$$

$$H_2O \rightarrow OH^-_{ads} + H^+ \quad (4)$$

$$OH^-_{ads} + h^+_{VB} \rightarrow .OH \quad (5)$$

$$OH \rightarrow \text{organic contaminants} \rightarrow \text{degradation products} + CO_2 + H_2O \quad (6)$$

The h^+_{VB} that react with the adsorbed hydroxyl anion will form hydroxyl radicals. The hidroxyl radicals as a result of the transfer of an electron from the valance band (VB) to the conduction band (CB) will be capable of damaging the cell wall and can decompose the cellular materials [8]. Less oxidative species such as superoxide anions and singlet oxygen ($^1O_2$) are believed to be responsible for the antibacterial action during the visible light induced photocatalysis [17, 18].

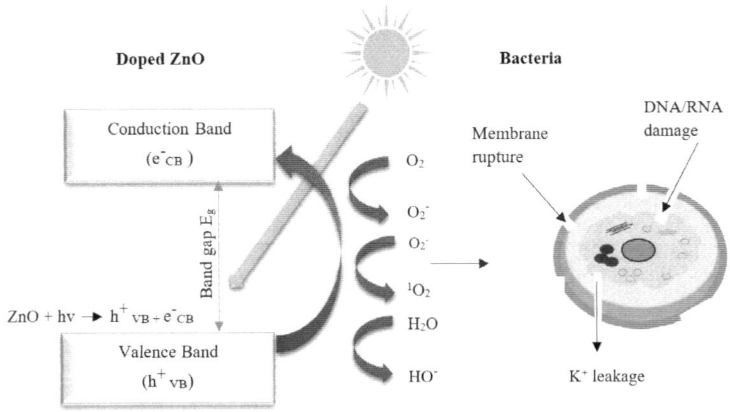

Figure 4. Schematic diagram of reaction mechanism of ZnO doped N thin film with visible light kills bacteria [8].

Table 1. Effect of ZnO doped N thin film with visible light on bacteria before and after treatment.

Test	Standard Quality [4]	Before Storage	After Storage	Percentage Decrease
Total Bacteria cfu/ml	1x10^6	2,5x10^6	1,2x10^5	95,2%
Coliform MPN/ml	20	24	8	66%
S. Aureus cfu/ml	1x10^2	3,06x10^3	0,17x10^3	94,4 %

3.2 Milk Quality Analysis

Milk quality on ZnO doped N thin film treatment showed in table 2. The data showed that milk quality in accordance with Indonesian national standard of fresh milk.

The protein of milk before treatment was about 2.90% while after treatment was 2.90%. The lactose of milk before treatment was 5.03% dropped to 4.0%. The fat of milk before treatment was 3.90% become 3.80 and the density of milk did not change. The test results of protein, lactose, fat and milk density was still in accordance with SNI. According to SNI [4], standards for fresh milk are protein 2.8%, lactose 4%, fat 3%, and density1.024 g / mL. It shows that photocatalysts do not damage the milk component. This is because ZnO doped N photocatalysts can decrease bacteria that become a cause of damage to the milk component. The result of the pH test before treatment was 6.8 and after treatment becomes 6.4. The thin-film of ZnO doped N could kill bacteria while not changing the pH of the milk. Based on the table 2, the result of somatic cells test before treatment was 530.100 cells/ml and after treatment was 321.590 cells/ml. There were decrease in somatic cells by 95.2%. The results showed that the milk produced was better than SNI. According to SNI [4] standard fresh milk for somatic cells is at least 400.000 cells/ml. During the storage process, somatic cell decline was due to photocatalyst N-ZnO that able to kill cells. According to Sirelkhatim *et al.* [19] ZnO photocatalysts produce radical hydroxyl (OH-) that can kill cells and damage DNA.

Table 2. Result of milk quality analysis.

Test	Standard Quality [4]	Before Storage	After Storage	Unit
Protein	Min 2,8	2,90	2,90	%
Lactose	4	5,03	4,08	%
Fat	Min 3	3,90	3,80	%
Density	1,024	1,026	1,026	g/ml
Somatic Cell	400.000	530.100	321.590	cell/ml
pH	6,3 – 6,8	6,8	6,4	-

4. Conclusion

ZnO doped N thin film with visible light for 180 minutes could reduce bacteria in milk such as total bacteria by 95.2%, *coliform* by 66% and *S. aureus* by 94.4%. Futhermore, ZnO doped N thin film with visible light produced milk with quality parameters such as protein, lactose, fat, density, somatic cell and milk pH which were in accordance with Indonesian National Standards for milk. Thus, ZnO doped N is one of the effort to maintain milk quality and food safety..

5. References

[1] Remnant J G, Green M J, Huxley J N and Hudson C D 2015 Variation in the interservice intervals of dairy cows in the United Kingdom *J of Dairy Science* **98 (2)** pp 889-97

[2] Mukson, Ekowati T, Handayani M and Harjanti D W 2009 Faktor-faktor yang mempengaruhi kinerja usaha ternak sapi perah rakyat di kecamatan getasan kabupaten semarang *Seminar Nasional Kebangkitan Peternakan* Semarang

[3] Suhendar G E, Sambodho P W and Harjanti D W 2017 Pengaruh ekstrak daun belimbing wuluh (*Averrhoa bilimbi* Linn.) sebagai bahan *teat dipping* puting terhadap jumlah coliform dan pH susu *J Sains Peternakan Indonesia* **12 (3)** pp 265-76

[4] Badan Standardisasi Nasional Indonesia 2011 Susu Segar SNI 01-3141-2011 Jakarta: Badan Standardisasi Nasional

[5] Gaya U I, Abdullah A H, Zainal Z and Hussein M Z 2009 Photocatalytic treatment of 4-chlorophenol in aqueous ZnO suspensions: intermediates, influence of dosage and inorganic anions *J of Hazardous Materials* **168 (1)** pp 57-63

[6] Qin H, Li W, Xia Y and He T 2011 Photocatalytic activity of heterostructures based on ZnO and N-Doped ZnO *ACS Appl Mater Interfaces* **3 (8)** pp 3152-6

[7] Liu L, Zhang G, Wang L, Huang T and Qin L 2011 Highly active modified ZnFe2O4 heterogeneous catalyst and its photo-fenton behaviour under UV-visible irradiation *Ind Eng Chem Res* **50** pp 7219– 27

[8] Carroll J P, Myles A, Quilty B, McCormack D E, Fagan R., Hinder S J and Pillai S C 2017 Antibacterial properties of F-doped ZnO visible light photocatalyst *J of hazardous*

materials **324** pp 39-47

[9] Standard Nasional Indonesia (SNI 2897 2008) Metode Pengujian Cemaran Mikroba dalam Daging, Telur, dan Susu, serta Hasil Olahannya Badan Standardisasi Nasional (BSN)

[10] Acharya A, Bimali N K, Shrestha S and Lekhak B 2017 Bacterial analysis of different types of milk (pasteurized, unpasteurized and raw milk) consumed in Kathamandu Valley. *Tribh. Univ. J. of Microb.* **4 (1)** pp 32-8

[11] Prescott S C and Breed R S 1910 The determination of the number of body cells in milk by a direct method *Am J Public Hygiene* **20** pp 663–4

[12] Hassan G M, Meshref A M S and Gomaa S M 2015 Microbiological quality and safety of fluid milk marketed in Cairo and Giza Governorates *Current Research in Dairy Sciences* **7 (1)** pp 18-25

[13] Prasanna V L and Vijayaraghavan R 2017 Chemical manipulation of oxygen vacancy and antibacterial activity in ZnO *Materials science and engineering* **77** pp 1027–34

[14] Nagaveni K, Hegde M S, Ravishankar N, Subbanna G N and Madras G 2004 Synthesis and structure of nanocrystalline TiO2 with lower band gap showing high photocatalytic activity *Langmuir* **20 (7)** pp 2900-7

[15] Dong F, Zhao W, Wu Z and Guo S 2009 Band structure and visible light photocatalytic activity of multi-type nitrogen doped TiO2 nanoparticles prepared by thermal decomposition *J of hazardous materials* **162 (2-3)** pp 763-70

[16] Chu F H, Huang C W, Hsin C L, Wang C W, Yu S Y, Yeh P H, and Wu W W 2012 Well-aligned ZnO nanowires with excellent field emission and photocatalytic properties *Nanoscale* **4 (5)** pp 1471-5

[17] Etacheri V, Valentin C D, Schneider J, Bahnemann D and Pillai S C 2015 Visible-light activation of TiO2 photocatalysts: Advances in theory and experiments *J of Photochemistry and Photobiolog* **25** pp 1-29

[18] Keane D A, McGuigan K G, Ibáñez P F, Polo-López I M, Byrne J A, Dunlop P S and Pillai S C 2014 Solar photocatalysis for water disinfection: materials and reactor design *Catalysis Science and Technology* **4 (5)** pp 1211-26

[19] Sirelkhatim A, Mahmud S, Seeni A, Kaus N H M, Ann L C, Bakhori S K M, Hasan H and Mohammad D 2015 Review on zinc oxide nanoparticles: antibacterial activity and toxicity mechanism *Nano-Micro Lett* **7 (3)** pp 219-42

6. Acknowledgement

The research was funded by The Ministry of Research, Technology and Higher Education, the Republic of Indonesia through 2018's Student Creativity Program.

Application of piper betel leaf (*piper betle* linn) extract to control fish pathogenic bacteria in-vitro

M M Bond[1*] and B Senggagau[1]

Station for Investigation of Fish Health and Environment (SIFHE) of Serang
Jl. Raya Carita, Cinangka, Serang Districts, Banten Province, Indonesia 42167

*Corresponding author: meyky78@yahoo.com

Abstract. Betel leaf extracts have some natural antimicrobial. This extracts has been used as antimicrobial agent in some aqua farmer in Indonesia for an alternative method to control fish diseases. This study aims to determine the effects of betel leaf extract on fish pathogenic bacteria in-vitro. Three kinds of solvents were used in betel leaf extracts. There were ethanol, n-hexane and ethyl-acetate. About 100 g of betel leaf simplicia were diluted by 1000 ml of solvents in stainless container cap for 24 hours maceration. This extracts were filtered with 0.42 µm paper filter and then evaporated at 50°C during 30 minutes in 50 rpm with vacuum rotary evaporator. This concentrated extracts were used to antimicrobial testing. The pathogenic bacteria for in-vitro testing were from Gram-negative bacteria such as *Aeromonas hydrophila*, *Vibrio alginolyticus* and *Edwardsiella tarda*. The in-vitro test of herbal extract from betel leaf againts Gram-negative pathogenic bacteria indicated that the betel leaf extract could inhibited the growth of pathogenic bacteria with diameter mean of inhibition zone were more than 14 mm. The solvents variations were not significantly different ($p > 0.05$) to diameter inhibition zone of *V. alginolyticus* and *A. hydrophila* but significantly different ($p < 0.05$) to *E. tarda*.

Keywords : Betel leaf, pathogenic bacteria, A. hydrophila, V. alginolyticus, E. tarda

1. Introduction

In fish culture, many fish farmers complain of a decrease in productivity due to fish disease attacks by pathogenic bacteria and it was very detrimental. Vibriosis and Motile Aeromonas Septicemia (MAS) occurs in many cases of fish disease. *Edwardsiellosis/Emphysematous Putrefactive Disease* of *Catfish* (EPDC) or *Edwardsiella Septicaemia* (ES) was caused by *Edwarsiella tarda* [1]. These Gram-negative bacteria are widely spread into the water in any condition. *Aeromonas hydrophila* can affect fish farming and often lead to outbreaks of disease with mortalities occasionally approaching 80-100 percent in a short period of 1-2 weeks [2]. Outbreaks of *E. tarda* have occurred only in fish larger than 450 g during the period of July through October. Disease outbreaks are most numerous when water temperatures are above 30°C and when high levels of organic fertility are present [3]. Vibriosis is caused by bacteria belonging to the genus Vibrio. Some species are pathogenic and can generate serious epizootic disease, but some other species are only opportunistic pathogens which impact disease if the fish was physically injured, parasitic injuries and stress [4].

Handling of fish diseases caused by pathogenic bacteria using antibiotics such as oxytetracycline, erythromycin, and kanamycin has been widely studied, but the results were still unsatisfactory. The use of antibiotics or medicine to control bacterial diseases can cause new problems. It could affect or kill non-target organisms, the emergence of pathogens that are resistant to antibiotics, residues in fish meat, affect growth and breeding and cause environmental pollution [5]. It might be safer to use natural antibiotics or antimicrobial to prevent fish disease in aquatic culture.

Indonesia as a tropical country has a wealth of plants that have the potential to become medicinal plants. Many types of plants compounds are antimicrobial. A number of plants compounds are bactericidal (bacteria killers), and bacteriostatic (inhibitors of bacterial growth). Betel leaves are known to be antioxidant, antiseptic, bactericidal and fungicidal. This extract has been used as an antimicrobial agent in some aqua farmer in Indonesia for an alternative method to control fish diseases**.** The use of betel leaves to control fish diseases include for African catfish [6-7], carp [8], tetra ornamental fish [9], tiger grouper fish [10], disinfecting giant prawn larvae [11], catfish [5,12] and tilapia GIFT [13].

The ability of betel leaf as a medicinal plant is due to the many active compounds contained in it, such as essential oils, fatty acids and fat esters [14], while the identified betel extract contains alkaloids flavonoids, tannins, steroids, and triterpenoid which are antibacterial [15-16]. Based on the function and application of betel leaf extract, a technical study is needed to determine the effect of betel leaf extract on fish pathogenic bacteria in vitro.

2. Material and methods

2.1. Sterilization of Tools

Equipment such as needles of use before used first was sprayed alcohol 70% then conducted burning directly. Petri dish was sterilized by autoclave at a temperature of 121 °C with pressure 1 atm for 10-15 minutes.

2.2. Preparation of Gram-negative bacteria isolation

The Gram-negative bacterial isolates were from the Laboratory of Microbiology of SIFHE Serang. The types of bacteria isolates were *Aeromonas hydrophila*, *Edwardsiella tarda,* and *Vibrio alginolyticus*. Bacterial isolates that have been frozen in a deep freezer (-80° C) are thawed at room temperature inside the Biosafety Cabinet (BSC). When the temperature of bacterial isolates was the same as the temperature of the isolation room (25°C), the bacterial isolates were cultured into Tryptic Soy Agar (TSA) and saline TSA media. Bacterial isolates that have been replanted in TSA media were incubated in an incubator at 37°C for 18-24 hours. The Gram-negative bacterial isolates were examined for bacterial biochemical tests to ensure their purity.

2.3. Preparation of betel leaf extract

Betel leaf extract was made by dissolving 100 g of betel leaf simplicia, respectively, into 1000 ml of several solvents types (ethanol, n-hexane, and ethyl-acetate). The extraction was conducted by maceration method for 24 hours in a closed container. It was filtered with 0.42 µm filter paper and then concentrated using vacuum rotary evaporator at 50°C and 50 rpm for 30 minutes. The concentrated extract was stored in a 3 mL microtube at room temperature. Pieces of filter paper with a diameter of 5-6 mm were used to be paper disc put into a solution of betel leaf extract during 2 x 24 hours.

2.4. Bacterial sensitivity testing

Gram-negative bacteria isolates from TSA media were re-isolated into Mueller Hinton media with a scratch method on all parts of the petri dish evenly. Then, the isolates were tested respectively with paper discs filled with betel leaf extract and incubated at 37°C for 18-24 hours. The experimental design of this sensitivity testing in three replications for each type of extract treatment. After 24 hours, the inhibition diameter zones were measured if the inhibition zone around the disc has been formed for each treatment.

2.5. Data analysis
All data from diameter inhibition zone were analyzed with one-way ANOVA. The hypothesis of this study was to find out the differentiation of betel leaf extract with various types of solvents to the size of the diameter of the inhibition zone formed.

3. Result and Discussion
The in-vitro test results of betel leaf extracts to the diameter of the inhibition zone in Gram-negative bacteria isolates were shown as follows:

Table 1. Diameter of the inhibition zone (mm) of *V. alginolyticus*

Solvents	Repetition			Mean
	1	2	3	
A	19.00	19.00	21.00	19.67 ± 1.15^{a}
B	16.25	13.25	22.00	17.17 ± 4.45^{a}
C	14.00	18.75	14.25	15.67 ± 2.67^{a}

($p > 0.05$), the same letter on superscript is not significant

Table 2. Diameter of the inhibition zone (mm) of *E. tarda*

Solvents	Repetition			Mean
	1	2	3	
A	34.00	32.75	34.25	33.67 ± 0.80^{a}
B	18.25	16.50	25.75	20.17 ± 4.91^{c}
C	33.25	24.25	27.50	28.33 ± 4.56^{b}

($p < 0.05$), the different letter on superscript is significant

Table 3. Diameter of the inhibition zone (mm) of *A. hydrophila*

Solvents	Repetition			Mean
	1	2	3	
A	23.00	27.00	23.00	24.33 ± 2.31^{a}
B	12.50	15.75	26.00	18.08 ± 7.04^{a}
C	15.25	23.75	16.75	18.58 ± 4.54^{a}

($p > 0.05$), the same letter on superscript is not significant

Solvent A gave the largest diameter of the inhibition zone than solvent B and C for each type of Gram-negative bacteria tested (Table 1, Table 2, Table 3). Solvent A was an herbal extract made from extracting betel leaves using ethanol solvents. While B and C were using n-hexane and ethyl-acetate. This could have happened that the ethanol is a polar solvent which can dissolve the antibacterial compounds contained in betel leaves higher than other organic solvents.

However, the *p-value* showed that the differentiation of solvent extraction for betel leaf was not significant to the diameter of the inhibition zone for *V. alginolyticus* and *A. hydrophila*. Therefore the diameter of the inhibition zone of *E. tarda* bacteria could be influenced by differences solvents extracted from herbal medicines given.

Diameter of the inhibition zone formed in the colonies of Gram-negative bacteria tested with herbal extracts showed that antibacterial active compounds in herbal medicine extracts. The constituent components of betel leaf essential oil consisted of 82.8% phenol compounds and 18.2% compounds not phenols [17]. Phenol compounds which are the main components of essential oils are thought to act as anti-microbes of betel leaves [18]. The slow growth of Gram-negative bacterial diameter in the

IOP Conf. Series: Earth and Environmental Science **383** (2019) 012030 doi:10.1088/1755-1315/383/1/012030

treatment of betel leaf extract was suspected that there has been a reaction between antibacterial compounds from betel leaf extract to the three types of bacteria.

The phenol content will be higher as the concentration of betel leaf extract increases so that the reaction will be stronger. Antioxidant activity will levitate as more as phenol compounds increased [19]. The presence of phenol as anti-microbial agent found in betel leaf extract has damage bacterial cell walls, causing slow growth of bacteria so that the inhibition zone of *A. hydrophila*, *V. alginolyticus* and *E. tarda* were formed. Phenol compounds are able to break the cross linkage of peptidoglycan in an attempt to break through the bacterial cell wall [20].

Betel leaf can be used as a general antimicrobial such as bacteria, fungi and protozoa [21]. Antibacterial activity can be effective in Gram positive and negative bacteria [21]. The diameter of the inhibition zone of bacterial mix by 20% betel leaf extract was better than 3% hydrogen peroxyde [22]. About 25% of betel leaf extract could inhibited *A. hydrophila* growth [5]. Arifin reported that betel leaf extract had antibacterial effect on *Haemophilus influenzas*, *Staphylococcus aureus* and *Streptococcus haemoliticus* [23]. On the antimicrobial activity of 10 medicinal plants used in traditional medicine in Colombia showed that betel and water extracts respond to antimicrobial activity against *Staphylococcus aureus*, but were not effective for hemolitic β *Streptococcus* and *Pseudomonas aeruginosa* [24].

Betel leaf can be used as a medicine that functions as antibacterial, anti protozoa and antifungal [25]. In addition, it can be used to cure several diseases such as abscesses, rheumatism, abrasion, hysteria, mastoiditis and so on [25]. This antibacterial activity occurs because betel leaves contain essential oils in which there are phenolic compounds that are bactericidal. Phenol compounds when interacting with cell walls of microorganisms will cause protein denaturation and increase the permeability of microorganisms [25].

Interaction between microorganisms results in changes in the balance of charge in protein molecules and causes coagulation. Proteins that are denatured and coagulated will lose physiological activity so they cannot function properly. Changes in the structure of proteins in the bacterial cell wall will increase cell permeability so that cell growth will be inhibited and then cells become damaged.

In addition, the kavikol compound (giving a distinctive odor to betel leaf) has a bacterial killing power five times greater than phenol. Charyophilic compounds are antiseptic and local anesthetics, while eugenol compounds are antiseptic and topical analgesics [25].

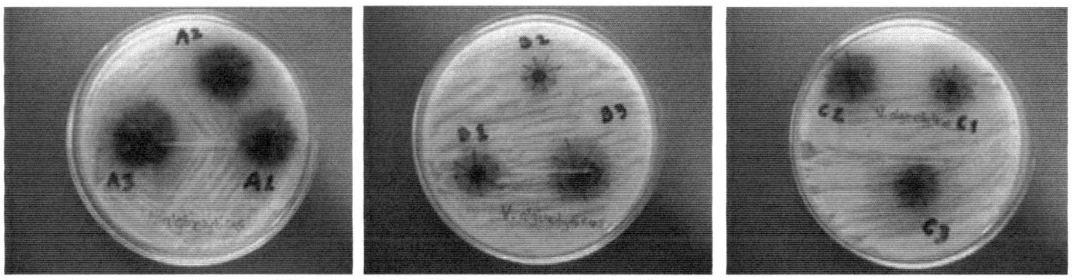

Figure 1. *V. alginolyticus* inhibition zone

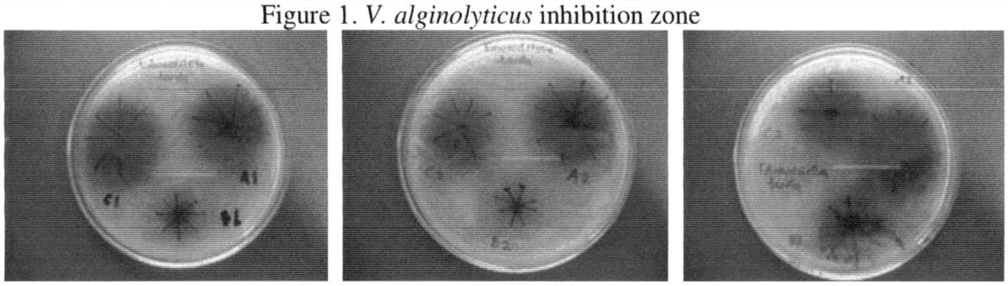

Figure 2. *E. tarda* inhibition zone

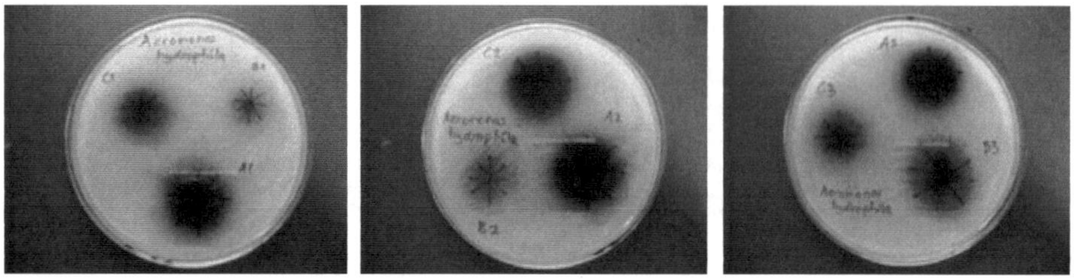

Figure 3. *A. hydrophila* inhibition zone

4. Conclusion

Betel leaf extract is effective in inhibiting the growth of pathogenic bacteria *Aeromonas hydrophila*, *Vibrio alginolyticus* and *Edwardsiella tarda* in vitro. The average diameter of the inhibition zone of betel extract formed more than 14 mm for each type of test pathogenic bacteria. The type of solvent has significantly difference to the diameter of the inhibitory zone of the bacteria *E. tarda* bacteria in-vitro, except to *A. hydrophila* and *V. alginolyticus*.

5. References

[1] Amanu S, Imanudin K, Narwiyani S 2017 Identification of *Edwardsiella tarda* in Indonesia *International Invention Journal of Medicine and Medical Sciences* **4**(3) 24-27

[2] Lukistyowati I, Kurniasih 2012 Pelacakan Gen Aerolysin dari *Aeromonas hydrophila* pada Ikan Mas yang Diberi Pakan Ekstrak Bawang Putih *Jurnal Veteriner* **13** (1) 43-50

[3] Meyer FP, Bullock GL 1973 *Edwardsiella tarda*, a New Pathogen of Channel Catfish (*Ictalurus punctatus*) *Applied Microbiology* **25**(1) 155-156

[4] Zafran D, Roza I, Koesharyani, Johnny F 1998 Panduan untuk diagnosis penyakit ikan dan krustase laut diIndonesia JICA dan Loka Penelitian Perikanan Pantai Gondol p 44

[5] Aisiah S, Muhammad, Anita 2011 Penggunaan Ekstrak Daun Sirih (*Piper betle* Linn) untuk Menghambat Bakteri *Aeromonas hydrophila* dan Toksisitasnya pada Ikan Patin (*Pangasius hypophthalmus*) *Fish Scientiae* **1** (2) 190-201

[6] Sutama IKJ 2002 Efektivitas Ekstrak daun jambu Biji (*Psidium guajava* L.), Sambiloto (*Andrographis paniculata* Nees) dan Daun Sirih (*Piper betle* L.) terhadap Infeksi Bakteri *Aeromonas hydrophila* L_{31} pada Ikan Lele Dumbo (*Clarias* sp) Skripsi Institut Pertanian Bogor Bogor p 68

[7] Wahjuningrum D, Tarono, Angka SL 2007 Efektivitas Rebusan Campuran Sambiloto (*Andrographis paniculata* (Burm.f.) Ness), Daun Jambu Biji (*Psidium guajava* L.) dan Daun sirih (*Piper betle* L.) untuk Pencegahan Penyakit *MAS* (*Motil Aeromonad Septicemia*) pada Ikan Lele Dumbo (*Clarias* sp) *Jurnal Akuakultur Indonesia* **6**(2) 127-133

[8] Maryani, Rosita 2006 Efektivitas Ekstrak Daun Jambu Biji (*Psidium guajava* L.), Daun Sambiloto (*Andrographis paniculata*), dan daun Sirih (*Piper betle* l.) dalam Menanggulangi Infeksi bakteri *Aeromonas hydrophila* pada Ikan Mas (*Cyprinus carpio* L.) *Journal of Tropical Fisheries* **1**(2) 32-139.

[9] Herawati VE 2009 Pemanfaatan Daun Sirih (*Piper betle*) untuk Menanggulangi Ektoparasit pada Ikan Hias Tetra *PENA Akuatika* **1** (1) 8-13.

[10] Dewi J, Marnis H, Tusihadi T 2010 Penggunaan Serbuk Daun Sirih (*Piper betle* L.) sebagai Pengendali Penyakit Bakterial Pada Pemeliharaan Ikan Kerapu Macan (*Epinephelus fuscoguttatus*). *Prosiding Indonesian Aquaculture*. **VI** 183-189

[11] Khasani I, Sopian A 2012 Efektivitas Ekstrak Daun Sirih Sebagai Bahan Desinfektan Larva Udang Galah (*Macrobrachium rosenbergii*) *Prosiding Forum Inovasi Teknologi Akuakultur* p 977-981

[12] Mulia DS, Husin A 2012 Efektivitas Ekstrak Daun Sirih dalam Menanggulangi Ikan Patin yang Terinfeksi *Aeromonas hydrophila* Laporan Penelitian FKIP Universitas Muhammadiyah Purwokerto p 12
[13] Yuniarani N 2013 Pengaruh Penambahan Ekstrak Daun Sirih (*Piper betle* L) pada Pakan terhadap Tingkat Kesehatan Ikan Nila Gift (*Oreochromis niloticus*) Skripsi Universitas Padjadjaran
[14] Nalina T, Rahim ZH 2007 The Crude Aqueous Extract of *Piper betle* L. and Its Antibacterial Effect Towards *Streptococcus* mutans *American Journal of Biotechnology and Biochemistry* **3**(1) 10-15
[15] Ningrum CR 2009 Uji Aktivitas Antioksidan Ekstrak Air dan Ekstrak Etanol 70% Daun Sirih Hijau (*Piper betle* Linn) dan Daun Sirih Merah (*Piper cf. fragile benth.*) dengan Metode Perendaman radikal Bebas FU Jakarta p 17
[16] Bond MM, Senggagau B, Penataseputro T 2015 Pengendalian Penyakit Bakterial dengan Aplikasi Ekstrak Herbal. Laporan Tahunan Loka Pemeriksaan Penyakit Ikan dan Lingkungan Serang Kementerian Kelautan dan Perikanan
[17] Koesmiati S 1996 Daun sirih (*Piper betle* Linn) sebagai desinfektan Skripsi Institut Teknologi Bandung Bandung p 65
[18] Pelczar MJ, Reid RD 1979 Microbiology MC Graw Hill Book Co New York
[19] Andarwulan N 2000 Phenolic synthesis in selected root cultures, and seeds Food Science Study Program Post Graduated Program Bogor Agricultural University Bogor p70
[20] Ingram LO 1981 Mechanism of lysis of *E. coli* by ethanol and other chaotropic agents *Journal of Bacteriology*. **146** (1) 331-335
[21] Cowan MM 1999 Plant products as anti-microbial agents *Clinical Microbiology Reviews* **12** (4) 564-582
[22] Dian AW 2007 Perbedaan khasiat antibakteri bahan irigrasi antara hidrogen peroksida 3% dan infusum daun sirih 20% terhadap bakteri mix Surabaya Fakultas Kedokteran Gigi Universitas Airlangga
[23] Arifin H 1993 Evaluasi aktivitas antibatuk ekstrak air daun sirih *Jurnal Penelitian Tanaman Obat di Beberapa Perguruan Tinggi di Indonesi*.
[24] Rojas JJ, Ochoa VJ, Ocampo SA, Munoz JF 2006 Screening for antimicrobial activity of ten medical plants used in Colombian folkloric medicine: A possible alternative in the treatment of non-nosocomial infections *Biomedical Center Complementary and Alternative Medicine*
[25] Guha OD 2006 Betel leaf: The neglected green gold of India *J Hum Ecol* **19** (2) 87-93

Feeding appropriate formulated diet for improving gonad maturation and spawning of brooder of some native fishes of indonesia

M T D Sunarno[1]*, A Saputra[1], and M B Syamsunarno[2]

[1)]Research Institute for Freshwater Aquaculture and Fisheries Extension, Jl. Sempur No 1 Bogor Indonesia 16154.
[2)]Department of Fisheries, Agriculture Faculty, University of Sultan Ageng Tirtayasa, Jl. Raya Jakarta Km 04 Pakupatan, Serang Indonesia 42124.

*Correspondence address: mastrimm@yahoo.co.id

Abstract. More than 1200 native Indonesian fish species inhabiting Indonesian inland waters have been used as human daily fish consumption. It brings to decline their population and even its species by year. Efforts of spawning some native fish species have been conducted, but the successful spawning limited during the rainy season. This paper, therefore, reviews the results of research related to using appropriate formulated diets for gonadal maturation and spawning some economically important native fishes. The tested fishes used were hoven's carp (*Leptobarbus hoevenii*), knife-fish (*Chitala lopis*), Asian red-tail catfish (*Hemibagrus nemurus*), kissing gourami (*Helostoma temmincki*), snake-head (*Channa striata*), and hard-lipped barb (*Osteochilus hasselti*), This study showed that in general, formulated diet containing about 35% protein fed on these species resulted in gonad maturation whole year and successfully spawned under controlled hatchery. The larvae survived at a high rate after feeding a natural diet up to 30 day-age. Mass production of the juvenile could support the development of its culture and be used as restocking in inland open-waters as well.

Keywords: native fishes, formulated diet, gonad maturation, spawning

1. Introduction

Waters cover more than 38 million Ha of the land area of Indonesia, consisting of the river and its branch, lake, human-made lake and swampy area. This inland-waters inhibits more than 1200 fish species, mostly spreading in Jawa, Sumatera, and Kalimantan. The fish resources are used to be utilized as daily fish consumption, generating income of the family, district, and national as well. Even, some fishes are used for special food menu and/or a part of material of particular food such knife fish (*Chitala* sp) in South Sumatera [1, 2], climbing perch (Anabas testudineus) in South Kalimantan [3], for biomedical material of albumin of snakehead (*Channa striata*) [4] and for ornamental or exotic fish of clown loach *(Chromobotia macracanthus)* [5]

Fish population and even species, particularly having high economic importance tends to decline by a year because of many ecological pressure [6] such as industrial waste [7, 8], mining [9, 10] and reclamation of swampy area [11] as well as incredible increasing fish catch. Many efforts to recover fish resources in inland waters have been made [12] i.e. establishing fisheries reservart [13, 14, 15, 16, 17], integrated management of rivers [18, 19], and fish restocking [20]. These efforts, however, were not impact on increasing fish production in nature. Restocking of introuced fish species of nile tilapia

(*Oreochromis* sp) is bringing to the negative impact such a case of decreasing knife fish in Man-made of Riam Kanan, South Kalimantan [1]. Development of native fish culture for satisfying market demand and decreasing fish catch as well in the next future, therefore, is necessary.

Sustainable development of fish culture has to be supported by the availability of seed in quality and quantity as well as of commercial feed having economical price. A traditional native fish culture such as Hoeven's carp [21] and snake-head [22] using seed collected from nature have been done, but the availability of the seed is seasonal. The seed-producing from a hatchery in a mass is required.

The success of mass production of native fish species needs a domesticated process of the fish to environmental culture (waters, stocking density, and artificial diet given). It is usually started from the brooder size of fish. Artificial diet plays an essential role in fish growth (somatic and gonad). It should contain complete and balance nutrients required by fish. Up to date, commercial artificial feed for brood fish is less available in the market. Fish farmers used to grower feed for any size and species as well of fish in their own business fish culture and sometimes added animal protein sources in wet basis for growing candidate broodstock of fish. This treatment influences gonadal development of fish to achieve at maturation level. The effort of native breeding fish have been conducted since 1980 used commercial feed; therefore, it is unpredictable. If it succeeds, it may occur during the rainy season.

Gonado development of native fish is commonly required relatively high protein containing in feed, besides micro vitamins particularly vitamin E and C. Information on nutrient requirements for brood fish needs basic research taking relatively long time. Besides, native brood fish is difficult to catch in nature, should be raised in captive from juvenile size. Basic data on fish habitat, food and feeding habit of native fish is valuable for predicting requirements of culture enviroment and protein diet as well. Since commercial feed containing high protein level is less available, feed for brood native fish should be made by breeders. Pellet machines having been distributing in fish culture centers [23, 24], particularly since 2015 through Program Gerakan Pakan Mandiri (GERPARI) initiated by the Government of Indonesia is hopefully able to produce local feed for native brooder fish. This paper, therefore, reviewed a series of research on formulated feed for increasing performances of reproduction and spawning of some native brooder fish in Indonesia. Juvenile production of the native fish under controlled condition in a hatchery could hopefully be used for needs of their culture and restocking in nature as well. This would impact on the development of native fish culture, reducing thier exploitation and inclining their productin in nature.

2. Methodology

This research used the desk study. Data and information was collected from previous study conducted by authors and/or others. The data needed was related to habitate, food habit, nutrient requirement, formulated diet application for native fish and feeding strategy for growing larvae up to juvenile size. This study focused on native fishes such as hoeven's carp (*Leptobarbus hoevenii*), knife-fish (*Chitala lopis*), Asian red-tail catfish (*Hemibagrus nemurus*), kissing gourami (*Helostoma temmincki*), snake-head (*Channa striata*), and hard-lipped barb (*Osteochilus hasselti*).

3. Results and Discussion

3.1. *Native fish habitate*

In general, various wild fishes live along river flow from up-stream to down having water salinity range of 3-5 ppt. Middle part of river has high fish production and water productivity, inhabited by many fish species which could be grouped to black fish i.e. kissing gouramy and snake-head, and white fish i.e. hoeven's carp, knife-fish, Asian red-tail catfish, and hard-lipped barb. During rainy season, wild fishes live in deep part of waters (called lebung) for their gonad maturation. They will leave this area for spawning propose during rainy season where the water flood surounding river area. Newly flooded area is rich in natural food for growing larvae of many species to juvenile size.

3.2. Food habit and nutrient requirments of fish

In nature, each fish species is looking for food and feed them on specifically diet according to complitiviness of their digestive organs, enzyme activities, and physiological condition. Less active fishes such as snake-head and kissing gourami tend to live aroung a substrate as their shelter and getting a food. Active fishes such as hoeven's carp, asian red-tail catfish, and hard-lipped barb are always moveable to look for food. Based on food habit, wild fishes could be grouped in to three, namely: herbivourous fish relying on plant materials as a nutrient source, carnivorous fishes mostly feeding animal protein, and omnivorous fishes being able to feed both plant and animal proteins. Snake-head and knife-fish are true carnivorous, and the rest are omnivorous tending to herbivorous.

In captive, food habit of all fishes basically could be changed. These fishes could be trained gradually their food habit to grasp an artifcial diet. Their growth performance will depend on the quality of diet and quantity as well. The quaity of diet affects on balance and complete nutrient in diet, relates to protein-essential amino acids (EAA), fat-essential fatty acids (AFA), charbohydrates, and varius vitamins and minerals as well. Protein and EAA should be considered in diet formulation. Brooder fishes in general require relatively high protein of 35-40% to support protein body and gonad as well. Maintaining energy requiments for daily activity of the brooders should be provided in diet, and its energy source is avaoided from protein, but from fats and charbohydrates. Ratio of energy and protein in diet ranges 8-10 kcal per g protein.

Fat and AFA are required for gonado development, and reproduction proccess of the brooder [25], and for embriological development of larvae. Freshwater brooder fishes generally need linoleat fatty acid and linolenat in an enaugh amount and a balance for gonado development, but each fish varies in number [26]. Brooders of catfish-pangaside (*Pangasius* sp) and Asian red-tail catfish require a diet containing fat at a respective level of 4% [27] and 12 - 14% [28]. According to [27], lack and/or over ratio of fatty acid of n-3 and n-6 in yolk eggs of fish larvae will affect on inhibitation of embriology development of the larva.

3.3. Diet formulation for brooder native fish species

Artificial feed is formulated using various ingredients to fill requirements of energy and essential nutrients (EAA, EFA, vitamins and minerals) of brooder of native fishes. These nutrients are bounded in various ingredients. Feed ingredients and their compostion in feed must considered capability of fish to digest them and availabiliy of essential nutrients. In diet formulation, ingredients containing essential nutrient should be considered suh as fish meal and soybean cake meal as sources of protein and EAA, oils of fish and plant as source of fat and EFA, vitamins (E and C) and minerals. Adding attractance and yeast in the diet formulation is recommended [29]. Besides, considering anti nutrient factors at minimum level, fiber and ash at a maximum level in diet formaltion to increase feed digestion is necessary. The SNI-for fish feed could be used as a refference. Feed price is not a problem for formulating the brooder feed.

Jelawat jouvenile was brought to Bogor, from East Kalimantan, raised in a concrete pond and fed on commercial feed (protein 26-28%). After the next ten years, the fish grew up to 3 kg in individual weigh and successfully spawned artificially using hormone treatment during the rainy season [30]. In 1980, candidate of brooder of Hoeven's carp was about 1.5 kg in individual weight raised in floating wooden cage. The cage was 5x2x1.5 m in dimension size with 50 fish per cage in oxbow-lake Mudung, Jambi and fed on a commercial diet (protein 26-28%) for a year did not develop their gonad at maturation level [31]. After feeding commercial feed changed to formulated feed having protein of 35% and added fresh Ipomea sp weekly, the gonadal brooder was matured [32] and the fish could be spawned artificially at any time using hormone injection [33,34,32,35]. Up to now, the technology of juvenile production of Hoeven's carp has widely been developing in West Kalimantan [36,37] and other provinces of Indonesia.

Knife fish. This carnivorous fish plays an significant role in Sumatera and Kalimantan as a material for using traditional human food [2] and exotic fish [38]. Brooder of knife fish was raised in pen set in swampy area in South Sumatera and fed on natural food for gonado maturation [39]. Also, the knife fish could be spawed naturally during rainy season. According to [40], the big size of knife fish preferred live food such common carp, nile talapia and prawn in small size. Application of feeding these live food in pond in swampy area indicated that the knife fish could be spawned naturally and/or semi-naturally using hormone treatment at any time kolam [41]. The brooder pond was covered by water hyacinth (*Eichormia crassipes*) at much as 50% of total pond area.

Asian red-tail catfish. This species was available in Sumatera, Kalimantan, and Java. Domestication program of this species has been conducted in Reseach Institute for Inland water Fisheries, Palembang since 1981 and continued since 2000 in Bogor. Feeding an artificial diet containing 30-40% protein resulted in gonadal development at a maturation level during the rainy season. Since 2016, feed formulated using local ingredients to have protein of 35% and the ratio of n-3 and n-6 at 1:1 and given to the brooder produced matured gonad at any time and could be spawned every 20-30 days [42]. Value of survival activity index (SAI) was up to day-10. This formulated diet have been applied in West Java (Districts of Cianjur, Bogor, and Majalengka).

Kissing gouramy, this species inhabited in the swampy area and tended to grasp plankton [43]. Spawning this species was during the rainy season. In the pond, kissing gouramy was raised traditionally, relying on natural food and rice-brain as supplementary diet. In the captive tank, this species could be spawned in semi-natural using hormone treatment at any time after feeding a formulated diet containing protein of 37-42%. Relative fecundity of this species was about 26.000 eggs per 100g fish weight.

Hard-lipped barb. This species is a herbivore. Spawning this species is easy using the natural method. In culture condition, feeding formulated diet containing protein 30% produced gonadal development at a mature level monthly [44].

Snake-head. Beside for human consumption, snake-head have been used as biomedical material of albumin [4]. This situation brings to increase market demand of the snake-head. In culture, the snake-head accepted commercial feed after being adapted in those feed that produced gonadal development at a maturation level. The brooder of snake-head could be spawned using natural method and semi-natural using hormone treatment at the rainy season [45].

3.4. Artificial diet for larvae and juvenile

A critical phase in the life cycle of fish is on the larvae stage. Yolk-egg of larvae is a primary source of nutrient required by the larvae, but before yolk sac empty, the nutrient source for larvae was gradually replaced by natural food and then changed artificial diet after 30 day-olds of the larvae. Feeding strategy on larval rearing is necessary i.e., Hoeven's carp [46,47]. Plankton collected from greenish pond or chicken yolk eggs emulsion could be fed on the Hoeven's carp larva as an initial feed for 14 days od larvae rearing [48], changed by the combination of Tubifex sp live and artificial feed at 1:1 in a ratio. After 40 days old, the juvenile required a diet containing protein of 38-40% [49,50] that fed as much as 10% of biomass a day [51], divided at five times a day[52].

Larvae of any fish species are carnivorous habit, and if lack of diet, they tend to prey each other, including snakehead [53,54,55] resulted in mortality rate of up to 70 % [56,57]. Feeding live food of Moina sp on snake-head larvae raised in greed media containing live of Chlorella sp at twice a day increased its survival up to 93.42% [58]. Transition feeding period of the snake-head of 30-day old using a combination of artificial diet (protein 40%) and Tubifex sp dried at 1:1 could increase its survival and growth performance [59]. Juvenile of snake-head required protein of 35% in the diet [60]. Live food of Artemia sp is used to feed on larvae of any fish in hatchery. The use of Artemia sp could be reduced by direct stocking of larvae at seven days old in the fertilized pond [61]. This ponds could be fertilized using urea dan TSP at a respective rate of 1,5 kg N dan 0,75 P per m2 per day for reducing protein diet from 30% up to 25% [50], and in line to increase stocking rate at an optimum level (Sunarno and Yakupitiyage 2002).

4. Conclusion

Artificial feed containing protein at 35-40% could be fed on native fish species of Hoeven's carp (Leptobarbus hoevenii), Asian redtail catfish (Hemibagrus numerous), and kissing gourami (Helostoma temmincki), and snakehead (Channa striata), but at 30% for hard-lipped barb (Osteochilus hasselti). Knife-fish (Chitala lopis) still relied on live food of fish at a small size. These feed could increase gonad maturation at any time. At larvae rearing under controlled condition, live food of Moina sp enriched by Chlorella sp, Artemia sp, or Tubifex sp could be feed as an initial diet for the larvae. Reducing this natural diet could be executed with direct stocking larvae at seven days old to fertilized ponds.

5. References

[1] Makmur S, Sunarno MTD, Wibowo A, Subagdja, Kumari K. 2006. Perikanan belida di Waduk Riam Kanan Kalimantan Selatan. Prosiding Forum Perairan Umum Indonesia III 2006. Jakarta: Pusat Riset Perikanan Tangkap. 146-156.

[2] Makmur S, Wibowo A, Sunarno MTD, editor. 2008. *Belida*. Palembang: Balai Riset Perikanan Perairan Umum.

[3] Mustakim M, Sunarno MTD, Affandi R, Kamal MM. 2009. Pertumbuhan ikan betok (*Anabas testudineus* Bloch) di berbagai habitat di lingkungan Danau Melintang-Kalimantan Timur. *Jurnal Penelitian Perikanan Indonesia.* 15 (2): 113 – 121

[4] Wahab SZ, Kadir AB, Hussein NH, Omar J, Yunus R, Baie S, Noor NM, Hassan II, Mahmood WH, Razak AB, Yusoff WZ. 2015. The effect of (*Channa striatus* Haruan) extracts on pain and wound healing of post-lower segment caesarean section women. *Evidance-Based Complementary and Alternative Medicine.* 2015:1-6

[5] Sunarno MTD. 1992. Prospek pengembangan budidaya ikan hias botia (*Botia macracanthas* Blkr.). *Warta Litbang Pertanian.* XIV (1): 11-12

[6] Wibowo A, Hidayah T, Sunarno MTD. 2005. Berbagai bentuk tekanan ekologis perairan sungai di Indonesia. *Prosiding Forum Perairan Umum I.* Pusat Riset Perikanan Tangkap, Jakarta.

[7] Oktaviani D. Sunarno MTD. 2001. Kontribusi limbah cair efluen pabrik karet terhadap perubahan kualitas air Sungai Musi, Sumatera Selatan. *Lokakarya Pengelolaan Sungai Terpadu.* Pusatlitbang Limnologi LIPI; Jakarta

[8] Purnomo K, Kartamihardja ES, Sunarno MTD. 2004. Pengaruh Pencemaran DAS Way Seputih di Lampung Tengah Terhadap Sumberdaya Perikanan (Status Kasus Pentingnya Pengelolaan Secara Terpadu). *Jurnal Ilmu-Ilmu Perikanan dan Budidaya Perairan.* 2 (1): 13-22.

[9] Kasim K, Sunarno MTD. 2009. Concentration of nickel (Ni) and chromium (Cr) in Sediment and Suso Snail *(Tylomelania patriarchalis)* at Bottom Layer of Matano Lake, South Sulawesi. *Indonesian Fisheries Research Journal.* 15 (1): 1-7

[10] Makmur S, Wibowo A, Sunarno MTD, Subagja. 2009. Kualitas lingkungan perairan Sungai Batanghari Jambi sebagai media budidaya ikan dalam keramba. *Prosiding Forum Perairan Umum Indonesia VI.* Balai Riset Perikanan Perairan Umum, Palembang.

[11] Kartamihardja ES. 2002. Pembukaan lahan gambut di Kalimantan Tengah: Mega proyek pemusnahan sumber daya perikanan? *Makalah Falsafah Sains (PPs 702)*, Program Pascasarjana S3 IPB; Bogor.

[12] Kartamihardja ES, Purnomo K, Tjahjo DWH, Umar C, Sunarno MTD, Koeshendrajana S. 2008. *Petunjuk Teknis Pemulihan Sumberdaya Ikan di Perairan Umum Daratan Indonesia*. Jakarta: Pusat Riset Perikanan Tangkap.

[13] Utomo AD, Sunarno MTD, Susilo A. 2005. Teknik peningkatan produksi perikanan perairan umum di rawa banjiran melalui penyediaan suaka perikanan. *Dalam* NN. Wiadnyana dkk, editor. *Prosiding Forum Perairan Umum I.* Jakarta: Pusat Reset Perikanan Tangkap.

[14] Rupawan, Sunarno MTD. 2006. Karakteristik habitat dan keragaman jenis ikan di suaka perikanan Propinsi Kalimantan Selatan. *Dalam* A.K. Gaffar dkk, editor. Prosiding Forum Perairan Umum Indonesia III. Jakarta: Pusat Riset Perikanan Tangkap.

[15] Satria H, Sunarno MTD. 2009. Karakteristik habitat calon suaka perikanan di Danau Sentani. *BAWAL Widya Riset Perikanan Tangkap.* 2 (4): 163-169.

[16] Krismono ASN, Nuriarini A, Kartamihardja ES, Sunarno MTD. 2009. Penilaian kesesuaian lokasi calon suaka perikanan di Waduk Koto Panjang. *Bawal Widya Riset Perikanan Tangkap.* 2 (5): 193-202.

[17] Nasution Z, Sunarno MTD. 2005. Pengelolaan perairan umum sungai dan rawa banjiran secara terpadu dan berkelanjutan. *Prosiding Forum Perairan Umum I.* Jakarta: Pusat Riset Perikanan Tangkap.

[18] Nasution Z, Sunarno MTD. 2009. Pengembangan model pengelolaan suaka perikanan di perairan umum daratan berbasis ko-manajemen. *Jurnal Kebijakan Perikanan Indonesia.* 1 (1): 17-29.

[19] Musthapia I, Sunarno MTD. 2007. Pengelolaan sungai secara terpadu. *Prosiding Forum Perairan Umum Indonesia IV.* Jakarta: Pusat Reset Perikanan Tangkap.

[20] Sunarno MTD. 2009. Beberapa jenis ikan asli perairan umum daratan sebagai kandidat komoditas restoking dan budidaya. *Prosiding Forum Nasional Pemacuan Sumberdaya Ikan II.* Jakarta: Pusat Riset Perikanan Tangkap

[21] Christensen MS. 1989. *Techniques and Economics of Intensive Cultivation of Jelawat and Lampam in Floating Cages: A Handbook for Extension Workers and Farmers.* Jakarta: Percetakan Persada Utama

[22] Reksalegora O. 1979. Fish cage culture in the town of Jambi, Indonesia. International Workshop on Pen and Cage Culture of Fish, 11-12 February 1979. IDRC-SEAFDEC, Philippines, p: 51-53

[23] Sunarno MTD, Oktaviani D. 2001. Industri rumah tangga pakan ikan. Prosiding Seminar Nasional Hasil Penelitian dan Pengkajian Teknologi Pertanian Spesifik Lokasi Sumatera Selatan. Bogor: Pusat Penelitian dan Pengembangan Sosial Ekonomi Pertanian.

[24] Sunarno MTD. 2002. Effect of different protein levels on the growth and feed efficiency of tawes (*Barbodes gonionotus*) in fertilized ponds. *Indonesian Fisheries Research Journal.* 8 (1): 35–40.

[25] Izquierdo MS, Fenandez-Palacios H, Tacon AGJ. 2001. Effect of broodstock nutrition on reproductive performance of fish. Aquaculture, 197: 25-42.

[26] Furuichi, M. 1988. Dietary Activity of Carbohydrates. In Fish nutition and Mariculture. Watanabe, T. Departement of Aquatic Biosciences. Tokyo University of Fishes, Pp.1-77. Tokyo.

[27] Mokoginta I, Jusadi D, Setiawati M, Takeuchi T, Supriyadi A. 2000. The effect of different level of dietary n-3 fatty acid on the egg quality of catfish (*Pangasius hypophthalmus*). JSPS DGHE. *International symp: Sustainable Fisheries in Asia in the new millenium*, p. 252-256.

[28] Utiah A. 2006. Penampilan reproduksi ikan baung (Hemibagrus nemurus Blkr) dengan pemberian pakan buatan yang ditambahkan asam lemak n-6 dan n-3 dan dengan implantasi estradiol-17â dan tiroksin. *Disertasi*. Program Pasca Sarjana. IPB-Bogor.

[29] Sunarno MTD, Sulhi M, Samsudin R, Heptarina D, editor. 2012. Teknologi Pakan Ikan Ekonomis dan Efisien Berbasis Bahan Baku Lokal. Bogor: PT. Penerbit IPB Press.

[30] Hardjamulia A, Atmawinata S, Suseno D. 1978. Pemijahan ikan domestik jelawat, *Leptobarbus hoeveni* Blkr, dengan teknik hipofisasi. *Bulletin Penelelitian Perikinan Darat.* 1: 43.

[31] Sunarno MTD, Reksalegora O. 1982. Pematangan calon induk ikan jelawat (*Leptobarbus hoeveni* Blkr.) di Danau Mudung, Jambi. *Pewarta BPPD.* 3 (1): 30 – 31.

[32] Sunarno MTD. 1989a. Pengamatan fekunditas ikan jelawat (*Leptobarbus hoeveni* Blkr). *Bulletin Penelitian Perikanan Darat.* 8: 26 – 30

[33] Ondara, Sunarno MTD. 1987. Pemijahan ikan jelawat (*Leptobarbus hoeveni* Blkr.) dengan suntikan hormon dalam sangkar terapung di Danau Teluk, Jambi. *Bulletin Penelitian Perikanan Darat.* 6 (1): 21 -28.

[34] Sunarno MTD, Reksalegora O, Nurdawaty S. 1988. Penelitian pembenihan ikan jelawat (*Leptobarbus hoeveni* Blkr.). *Bulletin Penelitian Perikanan Darat.* 2: 102 – 108.

[35] Pathmasothy S. 1985. The effect of three diets with variable protein levels on ovary development and fecundity in *Leptobarbus hoevenii.* In C.Y. Cho *et al*, editors. Finfish Nutrtion in Asia: Methodological Approaches to Research and Development. Ottawa Canada: International Development Research Centre.

[36] Sunarno MTD. 1991. Pengembangan domestikasi ikan jelawat (*Leptobarbus hoeveni* Blkr.) di Kalimantan Barat. *Prosiding Puslitbangkan No. 20. TKI PLHP/91*. Jakarta: Pusat Penelitian dan Pengembangan Perikanan.

[37] Kristanto AH, Suhenda N, Wahyudi E, Hardjamulia A, Sunarno TD, Sukadi F. 1992. Penelitian pengembangan teknologi budidaya ikan jelawat (*Leptobarbus hoeveni*) di Kalimantan Barat. *Prosiding Seminar Hasil Penelitian Perikanan Air Tawar 1991/1992*. Jakarta: Pusat Penelitian dan Pengembangan Perikanan.

[38] Wibowo A, Sunarno MTD, Makmur S. 2006. Ikan belida *(Chitala lopis)*: Potensinya sebagai ikan hias unggulan dan berbagai hambatan pengembangannya. Jakarta: Pusat Riset Perikanan Budidaya. Ikan Hias Nusantara: 143-151.

[39] Mennakern S. 1986. Breeding of *Notopterus borneonsis* (Blkr) carried out in South Sumatera Indonesia. Directorate General of Fisheries-USAID.

[40] Sunarno MTD. 2012a. Respon ikan belida (*Chitala lopis*) terhadap berbagai jenis pakan hidup dalam wadah terkontrol. *Bulletin Litbang Budidaya Air Tawar*. 1 (2): 113-120.

[41] Sunarno MTD, Syamsunarno MB. 2015. Pengaruh naungan terhadap pematangan gonad dan pemijahan ikan belida (*Chitala lopis*) di kolam rawa. *Jurnal Ilmu Pertanian dan Perikanan*. 4 (1): 71 – 77.

[42] Sunarno MTD, Jusadi D, Samsudin R. 2016a. Suplementasi asam lemak pada pakan induk ikan air tawar. Makalah disajikan pada Seminar Hasil Penelitian BPPBAT 2016. Bogor: Balai Penelitian dan Pengembangan Budidaya Air Tawar.

[43] Handayani. 2003. Kebiasaan makanan ikan tambakang (*Helostoma temmincki* CV) dan keterkaitannya dengan ketersediaan plankton di Danau Sebuah. *Tesis*. Sekolah Pascasarjana, Institut Pertanian Bogor.

[44] Sunarno MTD, Jusadi D, Samsudin R. 2016b. Kebutuhan protein induk ikan nilem. *Makalah disajikan pada Seminar Hasil Penelitian BPPBAT 2016*. Bogor: Balai Penelitian dan Pengembangan Budidaya Air Tawar.

[45] Saputra A. 2018. Teknologi Produksi Benih Ikan Gabus ***Channa striata*** melalui Pendekatan Fisiologi, Manajemen Pakan, dan Rekayasa Lingkungan. *Disertasi*. Bogor: Sekolah Pascasarjana Institut Pertanian Bogor.

[46] Sunarno MTD. 1997. Penyediaan pakan larva ikan jelawat. *Warta Litbang Pertanian*. XIX (5 dan 6): 11-12.

[47] Sunarno MTD. 2001. Kebutuhan nutrisi benih ikan jelawat (*Leptobarbus hoeveni*). *Warta Litbang Pertanian*. 7 (3): 6-9.

[48] Sunarno MTD, Syamsunarno MB. 2016. Performa pertumbuhan larva ikan jelawat (*Leptobarbus hoeveni*) pada berbagai kombinasi pakan alami dan buatan. *Jurnal Ilmu Pertanian dan Perikanan*. 5 (1): 7 - 13.

[49] Sunarno MTD. 1989b. Kebutuhan protein benih ikan jelawat (*Leptobarbus hoeveni* Blkr). *Prosiding Puslitbangkan No. 17/89*. Jakarta: Pusat Penelitian dan Pengembangan Perikanan.

[50] Sunarno MTD. 2002. Growth and nutrient digestibility of jelawat (*Leptobarbus hoeveni*) fry fed various dietary protein levels. *Indonesian Fisheries Research Journal*. 8 (1): 19–26.

[51] Sunarno MTD. 1998. Ransum harian untuk benih jelawat. *Warta Litbang Pertanian*. XX (2): 5-6.

[52] Sunarno MTD. 1989c. Pertumbuhan benih jelawat (*Leptobarbus hoeveni* Blkr.) pada frekuensi pemberian pakan yang berbeda. *Bulletin Penelitian Perikanan Darat*. 10 (2): 76 - 80.

[53] War M, Altaff K. 2014. Preliminary studies on the effect of prey length on growth, survival and cannibalism of larval Snakehead, (*Channa striatus* Bloch, 1793). *Pakistan Journal of Zoology*. 46(1):9-15.

[54] Qin J, Fast AW. 1996. Size and feed dependent cannibalism with juvenile snakehead (*Channa striatus*). *Aquaculture*. 144(4):313-320.

[55] Amornsakun T, Sriwatana W, Promkaew P. 2011. Feeding behaviour of snakehead fish, *(Channa striatus)* larvae. *Songklanakrin Journal of Science and Technology*. 33(6):665-670.

[56] Haiwen B, Shaoyu H, Lwin U, Swe UT, Qiufen D, Song Z, Yong Y. 2014. The snakehead fish: a success in Myanmar. *Fish Culture : AQUA Culture Asia Pacific Magazine*. p 20-23.

[57] Extrada E, Ferdinand HT, Yulisman. 2013. Kelangsungan hidup dan pertumbuhan benih ikan gabus (*Channa striata*) pada berbagai tingkat ketinggian air media pemeliharaan. *Jurnal Akuakultur Rawa Indonesia.* 1(1):103-114.

[58] Saputra A, Jusadi D, Suprayudi MA, Spriyono E, Sunarno MTD. 2018. Pengaruh frekuensi pemberian *Moina* sp. sebagai pakan awal pada pemeliharaan larva ikan gabus *Channa striata* dengan sistem air hijau. Jurnal Riset Akuakultur, 13 (3): 239-249.

[59] Sunarno MTD. 2012b. Pertumbuhan, sintasan dan konversi pakan ikan gabus (*Channa striatus*) yang diberi berbagai kombinasi pakan *Tubifex sp* kering dan pakan komersial. *Bulletin Litbang Budidaya Air Tawar*. 2 (1): 229-240

[60] Sulhi M, Sunarno MTD, Samsudin R. Jusadi D. 2016. Optimasi pakan dengan tingkat protein berbeda pada pemeliharaan benih ikan gabus. *Prosiding Forum Inovasi Teknologi Akuakultur*. Jakarta: Pusat Penelitian dan Pengembangan Perikanan; 2016

[61] Lamin S, Sunarno MTD. 2004. Kelangsungan hidup dan pertumbuhan larva ikan jambal siam (*Pangasius hypopthalmus*) melalui berbagai perlakuan umur larva dalam kolam pemeliharaan. *Prosiding Seminar Lokakarya Nasional Hasil Litkaji Teknologi Pertanian Spesifik Lokasi*; 2004 Juni 28 – 29. Bogor: Pusat Penelitian dan Pengembangan Sosial Ekonomi Pertanian.

[62] Suarno, MTD, and A. Yakupitiyage. 2002. Effect of Stocking Density on the Culture of Silver. *Indonesian Fisherles Research Journal* 8(1): 27–34.

IOP Conf. Series: Earth and Environmental Science **383** (2019) 012032 doi:10.1088/1755-1315/383/1/012032

The local farming system based on custom and tradition to achieve sustainable agriculture in Baduy indigenous community

Mirajiani*, Sulaeni, T Sutisna

Department of Agribusiness, Faculty of Agriculture, University of Sultan Ageng Tirtayasa, Jl. Raya Jakarta km. 04 Pakupatan, Serang City, Banten, Indonesia

*Corresponding author: mirajiani@yahoo.com

Abstract. The Baduy Indigenous Community are in the Lebak Regency of Banten Province. The Baduy depends entirely on the agricultural sector, where the availability of agricultural land becomes very important. Baduy people have their customs related to the existence of agricultural land and have customary rights to the area and agricultural land. To ensure sustainable agriculture, the regulation of the local agricultural system, especially land use and management of the agricultural system is essential. This research explores how customs and traditions contribute to the achievement of sustainable agriculture in the indigenous Baduy community. The study used the qualitative methodology approach with research techniques observation, expert depth interview, people depth interview, literature study, and focus group discussion (FGD). The results of the study show that customs and traditions regulate local agricultural activities, both land tenure, and land-use arrangements and local agricultural activities. The existence of customary arrangements becomes the basic system of local Baduy farming that has an economic, social, and environmental perspective that is the basis for achieving sustainable agriculture in Baduy.

Keywords: Baduy, indigenous community, customs, local agricultural system, sustainable agriculture.

1. Introduction

Baduy tribe is an indigenous community inhabits in Kanekes Village, Leuwidamar subdistrict, Lebak Regency, Banten Province. The existence of the Baduy community is recognized legally by the existence of regional regulations in the Regency of Lebak. In Indonesian national law, the existence of indigenous peoples who have customs, habits, and ways of life that are different from the general community, are acknowledged and protected by law. Baduy indigenous communities have different customs and habits from other communities in the province of Banten so that the differences that make their lives unique. Their main livelihood is farming. The farming activities have their own unique rules based on the customary rules. The Baduy community consists of two groups, namely Inner Baduy and Outer Baduy. Both are two groups of people who obey the traditions that have been inherited by the past generations [1].

Baduy Community agricultural activities are regulated by the customary and have to be carried out, without anyone being violated. With the existence of regulations that have been traditionally and hereditary inherited, the application of developing agricultural technology is a taboo. They have to apply agricultural techniques that are fixed, without any innovations that come from the development of modern agricultural technology. However, between preserving their customs and improving their lives as a human being has been discussed as a dilemma for years [2]. In terms of tenure and use of agricultural land, Baduy customs have regulated a permanent and fundamental way. Then this regulation has to be implemented by farmers in the community. This study will explore how the customary and traditional arrangements governing the local agricultural system are related to the achievement of sustainable agricultural goals in Baduy. Are agricultural activities based on the application of customs and traditions able to achieve three essential aspects in sustainable agriculture, namely the fulfillment of social, economic, and environmental aspects?. Is the existing local farming system capable of supporting the sustainability of community livelihoods in the future?.

2. Methode

This study used a qualitative approach using several research techniques that are deepening the community under study. Researchers conducted qualitative research techniques emically, tried to stick with the community to obtain valid and accurate data so that they can understand the actual condition of the community. In-depth interview techniques, observations, and discussions were also applied to achieve the research objectives.

3. Result and Discussion

3.1. Baduy indigenous land management and utilization

The Baduy community has its territory based on communal rights. In the Lebak Regency regional sheet Number 65 of 2001 Series C contains the Lebak Regency Regulation Number 32 of 2001 concerning the protection of the customary rights of the Baduy people. Customary rights are authority according to customary law owned by specific customary law communities over certain areas which are the environment of their citizens to take advantage of natural resources, including land in the area called Ulayat land. Ulayat Land is the land upon which there are customary rights of a particular adat community. Customary land is intended for survival and life, arising from outward and inner relations between generations and uninterrupted between the customary law community and the area concerned. Their customary legal arrangements bind Baduy people as indigenous peoples. They are also citizens together with a legal alliance that recognizes and applies the provisions of its legal alliance in daily life, has customary territory and has a relationship with the territory. Baduy people in conducting relations with their territory are regulated and limited to their traditional land, so it needs to be protected. Protection is a consequence of the existing regulations, by the Regional Government and the community must protect the order of the Baduy community from interference originating from outside the Baduy community.

Three kinds of land utility exist in Baduy region, namely for dry agricultural land, residential land, and protected forest. The highest religious leader, Puun govern rules for the use of dry land for agriculture. The Puun regulates and divides the land to residents who are married — not allowed to work on land that is not designated. The custom determines the time to start farming activities, likewise for settlement arrangements. Customs through an arrangement decided by the religious leader (Puun) to determine which areas are used as residential land. Every married Baduy will get land to build houses and fields for farming. However, this ownership status is not owned land. Existing land remains customary property. For protected forests, divided into two areas, the first area is a protected forest area that can be entered by all Baduy residents. Baduy residents can take forest products with the conditions set by adat. In certain areas, it is justified to develop mixed agroforestry system that has become a culture for the Baduy community in utilizing their land [3]. In utilizing forests, Baduy residents are encouraged to use them wisely and pay attention to forest sustainability. In protected forest areas no land is allowed for agriculture. They have developed systems from their longstanding experience as their local wisdom than can sustainably adapt to their environment [4].

The custom in Baduy determines land for agriculture. There is an agricultural area that can only be cultivated by the religious leaders, which is called Huma Puun and Huma Girang Seurat. A specific field for rice considered sacred and located in Baduy is known as Huma Serang. Other, which is cultivated by all Baduy people is called Huma Tuladan. Farming activities in this area are carried out jointly, and the yield is also for the common good. Usually, the products are to be served on a sacred ceremony dedicated to the rice goddess, Dewi Sri. Apart from these three types of areas, other agricultural areas can be cultivated privately by each family head and their members. The distribution of agricultural land has a difference in for two disparate groups of Baduy communities, the Inner Baduy community, and the Outer Baduy community. The Agricultural land distributed for Inner Baduy is a type of prohibited land. The Outer Baduy community can not enter this prohibited land. The area is more extensive than agricultural land for Outer Baduy. Therefore, due to lack of arable land, some Outer Baduy residents work on agricultural land owned by communities outside Baduy with rental status or profit-sharing. The existence of farming activities outside the Baduy tribe territorial makes the outer Baduy mobility more dynamic than the Inner Baduy.

Baduy people cultivate the land with communal ownership status due to the lands belong to the local custom, which is arranged by the religious leader. The lands are no privately owned and may not be traded. Available status of land ownership is only loans from communal rights. However, recently, Outer Baduy people tend to buy agriculture land owned by Non-Baduy communities. Therefore, that farm is privately owned by outer Baduy residents. The aim of those purchasing is to fulfill food needs since Outer Baduy people receive a less small number of communal land than Inner Baduy people.

In recent time, the problem related to the use of communal land, primarily agricultural land, arise. The supply of agricultural land area is fixed, while the population of Baduy continues to grow. So, the carrying capacity of agricultural land decrease continuously to meet the needs of communal food demand. On the other hand, the custom regulation has set land sustainability as the main focus in utilizing the communal land. The existence of customary rules prohibit exploitation of protected forests, exploitation of agricultural land, and changing the land contour to sustain the condition of the land. This rule is also supported by the application of a typical Baduy local farming system which also negates the negative impact of agricultural activities on land conditions.

3.2. Application of agricultural systems based on local custom

The Baduy community implements a local farming system that is different from the agricultural systems around it. These differences come from the application of traditional ways that have been inherited for generations. This difference is based on customary regulations and is a typical farming system. These characteristics include the tools and technology used, cultivation techniques, and the belief accompanied farming activities that are the spiritual aspects of these activities. These three things have an impact on aspects of production, social aspects, and environmental aspects of society.

The Baduy community recognizes dryland agriculture or cultivation only. Customary rules do not justify farming wetlands or rice fields. Generally, they grow rice and secondary crops in their fields. Farms use simple equipment, in the form of wooden sticks, small hoes, cleaver (a metal which is forged to be very thin and used to cut tree branches and clean the broth) and machetes. The custom does not allow the use of modern agricultural facilities and infrastructure. Not only modern equipment, but agricultural equipment and facilities that are used by the surrounding community alone also should not be used, such as hoes and hand tractors. They only use equipment that will not damage or change nature.

The farming is conducted in accordance with the land use cycle customary determined. The land is used only once a year, with a dry land paddy age of six months. Land that is tilled will move after only two harvests to four harvests, meaning that the land use cycle is two to four years, then they will work on new land. After two to four years of arable land, they will return to the previously arable land or open new land.

Rice seeds applied in dryland agriculture are local seeds which have cultivated for generations. Baduy people rely on rainfed farming because irrigation has not existed and prohibited in the agricultural system. Before starting farming activities, the religious leader will calculate the planting schedule based on astrology and their inspiration or spiritual message they accept from their ancestors (wangsit). The religious leader will instruct all residents to start clearing land and set a time for planting. All residents will obey this instruction. If there are residents who do not comply will be warned and get custom sanctions.

Several stages of farming activities, particularly for rice commodities in the local Baduy cultivation system include:

(1). *Narawas* (selecting land), this activity takes the form of selecting for *huma* (dry field paddy) as well as weeding. They apply the principle of shifting cultivation with the aim of resting the cultivated fields to restore land fertility. The custom also determined these land selection arrangements and governed by *Puun* (the highest religious leader). The regulation also sets the boundaries of arable land, to ensure fair and equitable distribution of arable land to all community members. If the selection of arable land has been determined jointly, then it will be followed by a sacred ceremony, begging *Sang Hyang* (God) for the cultivation activities to be carried out to fruition.

(2). *Nyacar* (land clearing), is the activity of cutting down grass, shrubs, and pruning small branches and collect them to dry. The farmers do not justify cutting down trees when clearing land. They only prune and cut tree branches so as not to block the sunlight needed in the process of planting rice. The land clearing ways of Badul people is known as pruning, not burnt cutting. According to tradition, the tools that can be used in this activity are the only cleaver. The cleaver is thin and even easy to bend. With tools like this, they cannot cut down trees, only small branches, twigs, shrubs, and grass that can be cut and then burned

(3). *Nukuh* means drying grass, pieces of twigs, shrubs, and other slash products by the sun then collected into several piles for burning.

(4). *Ngaduruk* (burning) is an activity to burn garbage that has been collected in the time of *Nukuh.*

(5). Ngaseuk is making a small hole for planting rice seeds. Planting seeds into the hole is called by *muuhan.* Men usually works for *ngaseuk* followed by women who works for *muuhan.* The ngaseuk ceremony is only carried out in the field of *huma Serang* and *huma Tuladan.* The ceremonial activity has been started with bring down seeds *(nurunkeun binih)* from the granary which was carried out by women. In performing this ritual the participant have to wear a white shawl, white belt, and bun. At the time of ceremony, participan were forbidden from talking.

(6). *Ngaseuk* is making a small hole for planting rice seeds. Planting seeds into the hole is called by *muuhan.* Men usually work for *ngaseuk* followed by women who work for *muuhan.* This ceremony is only carried out in the field of *huma Serang* and *huma Tuladan.* The ceremonial activity is started with bring down seeds *(nurunkeun binih)* from the granary which was carried out by women. In performing this ritual, the participant has to wear a white shawl, white belt, and bun. At the time of the ceremony, participants were forbidden from talking

(7). *Ngirab Sawan* (cleaning), namely the activity of removing garbage or remaining twigs from the sidelines of the rice groves.

(8). *Ngored* is an activity to clear land between rice groves using a small hoe. The small hoe is called *kored,* which has a short stem and is used explicitly by women.

(9). *Made* (harvesting) is carried out when the rice has yellowed and full filled or about six months after planting using ani-ani as a harvester. At this time, Baduy people present the largest traditional and religious ceremony, called *Kawalu.* During this ceremony, the whole Baduy community area was declared closed to all guests without exception, and angklung buhun is performed and played.

Baduy farming system does not use chemical pesticide to eradicate the pest. The Baduy people only use specific methods inherited from tradition and custom, such as pounding certain leaves to be applied to plants affected by the disease. Another way is asking the religious leaders to pray and recite mantras for plants affected by the disease. They believe that plant pests are an incarnation of a spirit disorder that can be eradicated by spells and prayers. They also cannot use chemical fertilizers. The fertility of plants is very dependent on the fertility of the land, which is maintained by shifting cultivation techniques.

Conducting farming activities for the Baduy community is not merely a means of earning a living, but also has an intrinsic meaning part of their beliefs. Farming is part of the implementation of customs and traditions that they hereditary maintain so that economic activities are attached to the custom activities and traditions. Farming is not only seen from the results of agricultural production, but also the meaning of blessing, which has been bestowed upon carrying out customary obligations. Therefore, in their agricultural activities, it is closely related to the implementation of sacred ceremonies related to agriculture which is an expression of gratitude that the Almighty still gives residents life and agricultural products.

Traditional ceremonies related to agricultural activities include:

(1). *Kawalu Mitumbeyen* is held after harvesting by holding a salvage at the traditional hall in each of the customary institutions *(Kepuunan)*. Three customary institutions exist, namely *Kepuunan Cikesik, Kepuunan Kertawana,* and *Kepuunan Cibeo* located in Inner Baduy. Participants of the ceremony came from the inhabitants of the Baduy tribe and outside Baduy residents. Provided dishes in this ceremony was taken from the Huma Serang granary (the results of past years of *Huma Serang*), and the innate ceremonial participants gathered and then distributed again after the ceremony was finished. The local religious leader of the Kepuunan where it took place presided over the ceremony and assisted by the assistants appointed by the religious leader. This ceremony is a gesture to the Baduy residents to harvest paddy plants in their private land. The harvested rice is tied and sun-dried to the floor (laid on bamboo provided) and covered by thatched roof along with the bamboo for a month.

(2). *Kawalu Panengah.* After a month from the time of *Kawalu Mitambeyen* held, *Kawalu Panengah* led by *Girang Seurat (Huma Serang)* and by *Kokolot* (private paddy field), namely to select rice seeds for the planting in next season. Then, it was continued with the man flocked bring crops to the Huma Serang barn *(ngunjal)*, while the women brought some rice to a place for pounding *(saung lisung)*. This ceremony was conducted for a week.

(3). *Kawalu Tutug*, was a ceremony held at the traditional hall which was centered in the traditional hall of *Kepuunan Cikeusik*. This ceremony is a peak event of the series of the previous ceremony because it is the first event to taste this year harvested rice, in a manner determined by the religious leader *(Puun)*. The event ended after sunset with a torch *(bunar)* led by *Girang Seurat* to return to their respective homes. On the following day, all Baduy residents were allowed to consume rice this year harvested.

(4). Pilgrimage Ceremony to the *Domas* Area. After *Kawalu Tutug*, then a particular event is held namely Pilgrimage to the *Domas* Statue. This ceremony is kept a secret, so that most Baduy residents do not know the time, place, and who is leaving. However, usually, this pilgrimage is carried out after *Kawalu Tutug* as a thank giving to God *(Sanghiang Tunggal)* and the ancestors for the harvest they produced.

(5). *Seba* (visiting the Government Officials). *Seba* is a traditional ceremony held after the pilgrimage to the *Domas* Area which has become a tradition, even according to the Baduy tribe, it is an obligation since their ancestors had to carry it out when they have been completed *Kawula Tutug*. Seba was a mass visit with the intent and purpose of submitting a report on behalf of the religious leader about Baduy's condition to the *Bapak Gede* (Lebak Regent, Serang Regent, and Banten Governor) for one year. *Seba* is as a sign of friendship, recognition, and request for protection from the government official. The ceremony is also for surrender yields from Baduy people to the government to taste each other's harvests this year and usually reciprocated by the government by dividing food that is not available in Baduy area and giving other supports. In performing *Seba*, the Baduy people walked in gathering from the Baduy area to get the government office. The *Seba* ceremony is the culmination of a series of religious ritual conducting every year to express gratitude to God and the government for the prosperity they have gained during the year [5].

3.3. Achievement of Sustainable Agriculture

Achievement of sustainable agriculture is determined by the accomplished of three factors: social factors, economic factors, and environmental factors. This point means that the achievement of sustainable agriculture in the Baduy community will depend on the Baduy farming system recognize these three factors. The social achievement has the meaning of how agricultural activities and utilization of agricultural resources to achieve welfare and provide a sense of social security for the community. Achievement economically means how the economic needs of the community related to the achievement of agricultural production at a certain level that can meet the needs of the community. Achievement in the environment means how agricultural activities can maintain and preserve natural resources, especially agricultural resources and the utilization of agricultural resources while taking into account environmental aspects of the present and the future.

Baduy agricultural activities have a very high social value, because farming is considered not solely to earn a living, but also in order to find a living blessing and carry out obligations that have been hereditary regulated by the customs. Agricultural activities are considered as sacred activities which are part of worship that is obligatory for all citizens. The Baduy people interpret the hard work of cultivating agriculture as an effort to fulfill their obligations to God.

Baduy farming activities also have significant economic value. Whole farm product is consumed to meet family needs. The costumery mandates a ban on selling agricultural products in the form of rice commodities. After harvesting, the rice will be stored in a granary and become a family food reserve. Rice fields are the most important economic sector for all inhabitant. All married people are required to have their rice barn called *Leuit*. The leuit have not to be empty. Therefore they have to work hard in farming to keep getting good agricultural products then fill the *leu*it. The leuit which is never empty is a reflection of the availability of family food that is guaranteed. Baduy people are hard workers who work as a manifestation of their obedience to the customs and traditions that have been determined.

The environmental perspective can be analyzed from eco-friendly local wisdom accompanying each economic activity of the community, including the agricultural sector. From the economic view, besides cultivating dry land with the primary commodity of rice, they also take forest products to meet the economic needs of the family. They took some products from the protected forest recognized as customary land with a total area of around three thousand hectares. This forest can only be entered and took its product by Baduy residents. Forest products that can be collected and economically valuable include fruits, high-value wood such as *gaharu*, and honey. The customary rules noticed that inhabitant could only take them as much they need, not excessive. Taking forest products is also considered a sacred economic activity. Usually, residents ask permission from the religious leaders *(Puun)*, then the *Puun* gives his blessing by praying together. Commonly, the customary leaders also get a share of forest products taken based on the sincerity of the people. There is a belief of inhabitant if they are greedy and excessive taking of forest products, they will get a punishment from the spirits of forest watchman. Picking forest products is justified, but cutting down wood in the forest without permission from the customary leaders is considered a taboo. Local people have been internalized to always take care of the forests that they acknowledged to be entrusted in their life.

In the field of agriculture, the Baduy protect land resources in environmentally friendly ways including using pest control of plant diseases using natural materials, not using pesticides because they think they will damage the environment, water sources, and the soil. To maintain soil fertility, they apply a shift cycle. They implement a local agricultural system based on local wisdom that pays attention to environmental sustainability and natural balance. The Baduy do not use modern agricultural tools and facilities because they are considered to damage agricultural resources in the long run.

6. Conclusions

The existence of customary arrangements and the belief of the inhabitant to persistently obey it brings them possible to achieve sustainable agriculture in the Baduy community. Sustainable agriculture in Baduy reflects the existence of an agricultural system that prioritizes the use of natural resources that can maintain thc carrying capacity of the environment. Agricultural activities are not only intended for economic benefits but also seen how agricultural activities can be socially and environmentally valuable. It makes the local agricultural system in Baduy can avoid problems of damage to agricultural resources and can maintain natural conditions in the long run.

References

[1] Batubara, Syukri. 2010. Perlindungan terhadap Masyarakat Hukum Adat: Studi Mengenai Pelaksanaan Hak Ulayat dan Sumberdaya Alam Suku Baduy. Universitas Islam Indonesia. Yogyakarta.

[2] Hakim, Nurul. 2006. Cultural and Spiritual Values of Forests In Baduy Region, Banten, Indonesia. Master Thesis. Forest and Nature Conservation Policy. Wageningen University.

IOP Conf. Series: Earth and Environmental Science **383** (2019) 012032 doi:10.1088/1755-1315/383/1/012032

[3] Senoaji, Gunggung. "Perilaku masyarakat Baduy dalam mengelola hutan, lahan, dan lingkungan di Banten Selatan". *Jurnal Humaniora*, 2011, **23.1**: 1-15.

[4] Widyarti, M., & Arifin, H. S. (2012). Evaluasi keberlanjutan masyarakat Baduy dalam berdasarkan community sustainability assessment. *Jurnal Lanskap Indonesia*, **4(1)**.

[5] Rusnandar, Nandang. Seba: Puncak Ritual Masyarakat Baduy di Kabupaten Lebak, Provinsi Banten. *Patanjala: Jurnal Penelitian Sejarah dan Budaya*, 2013, **5.1**: 82-98.

Water quality at silvoaquaculture pond in indramayu regency

Muarif*, Y Wahyudin, D Merdekawati

Department of Fisheries, Faculty of Agriculture, University of Djuanda, Jl. Tol Ciawi no 1, Bogor, Jawa Barat Province, Indonesia

*Corresponding author: muarif.faperta@unida.ac.id

Abstract. Silvoaquaculture is one of environmentally friendly aquaculture, which is integration between fish ponds (brackish water aquaculture) and mangrove tree in the coastal area. In Indramayu Regency, silvo-aquacultures are located in Karangsong, Brondong and Tegur village. Milk fish (*Chanos chanos*) was cultured at silvo-aquaculture in Indramayu Regency. The objective of this study is to know water quality at silvo-aquaculture ponds in Indramayu Regency. The water quality has been measured includes temperature, salinity, pH, brightness, DO, nitrate, phosphate, and ammonia. The ranges of water quality values at silvo-aquaculture ponds in Indramayu Regency are Brondong Village {temperature (29.8-33.0 °C), salinity (26-30 ppt), pH (7.27-7.80), brightness (26.5-91 cm), DO (3.1-10.7 mgL^{-1}) , nitrates (12.5 mgL^{-1}), phosphates (0.03 mgL^{-1}), and ammonia (0.25 mgL^{-1})}; Karangsong Village {temperature (29.5-31.0 °C), salinity (17-25 ppt), pH (7.29-7.51), brightness (39.5-47 cm), DO (2.2-4.4 mgL^{-1}) , nitrates (12.5 mgL^{-1}), phosphates (0.03 mgL^{-1}), and ammonia (0.25 mgL^{-1})}; and Tegur Village {temperature (30.4-31.8 °C), salinity (15-19 ppt), pH (7.14-8.18), brightness (32.25-48.25 cm), DO (8.8-9.5 mgL^{-1}) , nitrates (12.5 mgL^{-1}), phosphates (0.03 mgL^{-1}), and ammonia (0.25 mgL^{-1})}. Analysis of variance showed temperature, pH, brightness, phosphate, nitrate, and ammonia were not significantly different between locations, but salinity and DO were significantly different between locations. These results indicate all water-quality parameters are suitable for fish culture. The lower salinity presented in Tegur Village was affected by rainy. Interestingly, DO concentrations in the Karangsong silvo-aquaculture pond are lower than optimal value (<5.00 mg/l), but they are still suitable for milkfish life (> 2 mg/l).

1. Introduction

Pond cultivation has long been one of the main livelihoods of the community in the coastal area. The development of ponds has degraded the existence of mangrove ecosystems. In Indramayu, in 2000-2014 it was predicted that around 689 ha of mangroves had been lost and one of them was caused by the development of ponds [1]. The sustainability of mangroves must be maintained because damage to the mangrove ecosystem also affects the sustainability of fish farming [2]. Cultivation design that integrates mangrove and aquaculture will be able to prevent mangrove damage and maintain community welfare.

In the Indramayu coastal area, milkfish (*Chanos chanos*) culture has developed using silvo-aquaculture techniques. Silvo-aquaculture is an environmentally friendly cultivation technique that integrates between pond cultivation and mangrove forest. The integration between the conservation of mangrove and ponds in the form of silvo-aquaculture [3] is a mutually beneficial solution between mangrove conservation and aquaculture [4].

Water quality greatly determines the production of milkfish cultivation in ponds. The presence of mangroves in silvo-aquaculture ponds will be able to maintain the quality of pond water because mangroves are able to act as water pollution biofilter [5] so that the productivity of the pond remains high [6]. This research examines the characteristics of silvo-aquaculture pond water quality in the three villages (Karangsong, Brondong, and Tegur) in Indramayu Regency. The coastal environment conditions of each different villages are expected to affect the quality of pond water for each village can be different.

2. Research Methods

This research was conducted in Indramayu Regency (Karangsong, Brondong, and Tegur Villages). Water quality measurements were carried out in June 2019. The number of water quality observation stations was 21 stations, where each village had 7 observation stations. Water quality measurements are carried out in pond waters, inlets, and outlets. Water quality data measured include temperature, salinity, pH, brightness, dissolved oxygen (DO), phosphate, nitrate, and ammonia. All water quality was measured in situ using TDS-meter, refractometer, seccidisc, DO-meter, and water quality kit.

Water quality data were analyzed qualitatively using criteria for water quality criteria for fish farming [7]. Analysis of water quality data was also carried out quantitatively using an analysis of variance of Kruskal-Wallis test and the results of different types of analysis carried out using the Mann-Whitney test.

3. Results and Discussion

The distribution of silvo-aquaculture pond water quality in the three villages in Indramayu District is presented on Table 1. Water quality data which includes temperature, salinity, pH, brightness, dissolved oxyigen (DO), phosphate, nitrate, and ammonia in the three villages are classified as suitable for milkfish cultivation.

Temperatures range from 29.5-33.0 ^{o}C, with the lowest temperature found in the village of Karangsong and the highest temperature found in the village of Brondong. Fish can live in a temperature range of 15-35 ^{o}C and live well at temperatures of 20-30 ^{o}C [7]. Based on these criteria, the water temperature in all observation stations is appropriate for the life and growth of milkfish. Water temperature will affect the fish's metabolic rate and physiological processes of fish which will affect fish growth [8-9]. The ideal temperature will support the high production of milkfish.

The temperature difference at each station is caused by differences in observation times, where morning observations produce low temperatures and daytime observations produce high temperatures. The results of the variance analysis showed that the quality of silvo-aquaculture pond water in the three villages was not significantly different (Table 2). These temperature characteristics indicate that silvo-aquaculture area in Indramayu Regency is suitable for milkfish cultivation.

Table 1. Silvoaquaculture pond water quality in Karangsong, Brondong, and Tegur villages

Parameter	Unit	Water quality			Water Quality Criteria [7]		
		Karangsong	Brondong	Tegur	Acceptable range	Desirable range	Stress
Temperature	^{o}C	29.5-31.0	29.8-33.0	30.4-31.8	15-35	20-30	<12, >35
Salinity	ppt	17-25	26 - 30	15-19	0-60	10-25	-
Brightness	cm	39.5-47	26.5-91.0	32.25-48.25	-	30-80	<12,>80
pH		7.29-7.51	7.27-7.80	7.14-8.18	7.0-9.5	6.5-9.0	<4, >11
DO	mgL^{-1}	2.2-4.4	3.1-10.7	8.8-9.5	3-5	5-8	<5, >8
Phosphate	mgL^{-1}	0.03	0.03	0.03	0.03-2	0.01-3	>3
Nitrate	mgL^{-1}	12.5	12.5	12.5	<100	0.1-4.5	>100, <0.01
Ammonia	mgL^{-1}	0.25	0.25	0.25	<0.05	<0.025	>0.3

Salinity shows the salt content of the water. Different salinity levels affected on fish growth [10]. Milkfish are classified as euryhaline fish that can live in a wide range salinities. The salinity range at the observation site ranges from 15-30 ppt. The distribution of salinity in all observation stations is good because milkfish live well at salinity of 10-25 ppt [7]. Salinity will affect the fish osmoregulation process [11]. Appropriate salinity for fish conditions good osmoregulation so that fish growth will also be good.

The results of the variance analysis showed that salinity in the three locations was different (Table 2). Mann-Whitney's test results (Table 3) provided information that between silvo-aquaculture ponds in Brondong, Karangsong, and Tegur villages had significantly different salinity. The lowest salinity is found in the Tegur village as a result of rain and freshwater flow from the Cimanuk river.

Brightness is an indicator of the total intensity of light in the water body [12]. The brightness of silvoaquaculture pond water in each village us not significantly different (Table 2). The brightness ranges from 26.5-91.0 cm. The brightness distribution is quite good because fish is feasible at a brightness value of 30-80 cm. At station 3 in the Brondong village has a high brightness (> 80 cm) which can affect fish stress. The existence of mangrove trees will provide shade for the waters so that the negative effects of sunlight on aquaculture fish can be reduced.

pH indicates the acidity [13] of pond water' and It is one of the critical factors that limit the survival rate of aquatic organisms [14]. The ideal pH for fish ranges from 6.5-9.0 [7]. The water quality of silvo-aquaculture ponds showed a range between 7.14-8.18 which indicates that the pH of water is ideal for milkfish cultivation. The results of variance analysis (Table 2) showed that each location had a relatively similar pH so that the pH values in all locations of silvo-aquaculture ponds were ideal for the survival and growth of milkfish.

Table 2. The results of the variance analysis of water quality between locations based on the Kruskal-Wallis analysis

Water Quality	Asymp.Sig.	Information
Temperature	0.505	not significantly different
Salinity	0.000	significantly different
Brightness	0.953	not significantly different
pH	0.100	not significantly different
DO	0.040	significantly different
Phosphate	1.000	not significantly different
Nitrate	0.591	not significantly different
Ammonia	1.000	not significantly different

Dissolved oxygen (DO) is the water quality which plays an important role [15] in the process of fish respiration [16]. In waters that are low in oxygen content, fish will be weak and can experience death [16-17]. The results of the analysis of variance of each village had a different DO (Table 2), and the results of the Mann-Whitney test showed the oxygen content of the waters of each village was different, except between the village of Brondong and the village of Tegur (Table 3).

Dissolved oxygen in a silvo-aquaculture pond in Karangsong village ranged from 2.2-4.4 mgL^{-1}. This dissolved oxygen level was lower than DO in silvo-aquaculture pond in the village of Brondong and the village of Tegur. The location of silvo-aquaculture ponds in Karangsong village is close to the ship's traffic flow and the fishing port of Indramayu so that the waters in this area receive a higher pollution impact compared to other villages. Pollution sources in Karangsong come from the activities of fishing port, agriculture, shipbuilding industries, tourism, and fish auction sites [18]. This pollution is thought to cause dissolved oxygen in silvo-aquaculture ponds in the Karangsong village to be lower. Fish are still able to live at oxygen levels below 1.5 mgL^{-1}, but the growth is not optimal [19]. Dissolved oxygen content 5-8 mgL^{-1} is the ideal range for fish farming [7]. Oxygen levels in silvo-aquaculture ponds in the Karangsong village range from 2.2-4.4 mgL^{-1} classified as suitable for the life of milkfish, but not good for milkfish growth (below the optimal value). Dissolved oxygen levels in silvo-aquaculture ponds in the Brondong village (3.1-10.7 mgL^{-1}) were included in the criteria to be feasible until optimal, while the dissolved oxygen levels in silvo-aquaculture ponds in the Tegur village (8.8-9.5) were classified as optimal for the growth of milkfish.

Table 3. The results of Mann-Whitney test at salinity and dissolved oxygen

Parameter	Location 1	Location 2	Sig.	Information
Salinity	Karangsong	Brondong	0.002	significantly different
	Karangsong	Tegur	0.008	significantly different
	Brondong	Tegur	0.002	significantly different
DO	Karangsong	Brondong	0.047	significantly different
	Karangsong	Tegur	0.021	significantly different
	Brondong	Tegur	0.654	not significantly different

Ammonia is a dissolved gas in the waters that are toxic to fish [20-21]. The effect of ammonia toxicity dependent on temperature and pH [21]. Ammonia content that is good for fish farming is a maximum of 0.025 mgL^{-1}[7]. Total ammonia levels in silvo-aquaculture pond water in all research locations were 0.25 mgL^{-1}. This ammonia content is higher than the optimal value for fish growth but is lower than the fish stress limit (ammonia levels >0.03 mgL^{-1}), so that this ammonia level is still safe for milkfish survival.

Ammonia in silvo-aquaculture pond waters come from the biological decomposition [21] of the remaining feed and the rest of the metabolism of fish [22]. About 75 % of the feed is not eaten by fish [10] and dissolved in water as organic waste which has the potential to produce ammonia. The frequency of replacement of pond water which follows caused organic matter and ammonia to be formed that are not quickly wasted out of the pond.

Phosphate and nitrate are water quality associated with aquatic nutrients [23,24,25]. The presence of phosphate and nitrate in the waters is needed for the growth of plankton [24, 25]. The presence of nitrate and phosphate in silvo-aquaculture pond waters is needed to grow plankton as natural food for milkfish seeds. Phosphate and nitrate are not toxic, even so at phosphate levels above 3 mgL^{-1} and nitrate above 100 mgL^{-1} can cause fish stress [7,26]. Nitrate and phosphate content in silvo-aquaculture ponds in all research locations are uniform, that is nitrate of 12.5 mgL^{-1} and phosphate of 0.03 mgL^{-1}, so that is was considered good for milkfish cultivation.

4. Conclusion

Silvo-aquaculture pond water quality in Indramayu Regency is in the proper and optimal range for milkfish cultivation. Almost all parameters of silvo-aquaculture pond water quality in each village are relatively the same, except for salinity and DO.

5. References

[1] Yulianto G 2017 *Desain kebijakan pengelolaan terpadu mangrove dan perikanan (studi kasus di Kabupaten Indramayu Jawa Barat)* Sekolah Pascasarjana Institut Pertanian Bogor p 259

[2] Muliawan I, Zamroni A, Priyatna FN 2016 Kajian keberlanjutan pengelolaan budidaya ikan bandeng di Gresik *J. Kebijak. Sosek KP* **6** 11.

[3] Fitzgerald WJ 2002 Silvofisheries: Integrated Mangrove Forest Aquaculture Systems Ecological *aquaculture: the evolution of the llue revolution* Oxford Blackwell Science Ltd p 100.

[4] Bosma RH, Nguyen TH, Siahainenia AJ, Tran HTP, Tran HN 2016 Shrimp-based livelihoods in mangrove silvo-aquaculture farming systems *Rev. Aquac.* **8** 18.

[5] Kariada N and Irsadi A 2014 Peranan mangrove sebagai biofilter pencemaran air wilayah tambak bandeng Tapak Semarang *J. Mns. dan Lingkung.* **21** 7.

[6] Muntalif BS, Hasian O, Sembiring E 2013 Mangrove di kecamatan Muara Gembong *J. Tek.* Lingkung. **19** 9.

[7] Bhatnagar A, Devi P 2013 Water quality guidelines for the management of pond fish culture *Int. J. Environ. Sci.* **3** 30.

[8] Walberg E 2011 Effect of Increased Water Temperature on Warm Water Fish Feeding Behavior and Habitat Use *J. Undergrad. Res. Minnesota State Univ.* **11** 13.

[9] Singh SP, Sharma JG, Ahmad T, Chakrabarti R 2013 Effect of Water Temperature on the

Physiological Responses of Asian Catfish Clarias batrachus (Linnaeus 1758) *Asian Fish. Sci.* **26** 13.

[10] Hamed SS, Jiddawi NS, Poj B 2016 Effect of salinity levels on growth , feed utilization , body composition and digestive enzymes activities of juvenile silver pompano Trachinotus blochii *Int. J. Fish. Aquat. Stud.* **4** 4.

[11] Lisboa V, Barcarolli IF, Sampaio LA 2015 Effect of salinity on survival , growth and biochemical parameters in juvenile Lebranch mullet Mugil liza (Perciformes : Mugilidae) *Neotrop. Ichthyol.* **13** 5.

[12] Cajas RFC, Selz OM, Ripmeester EAP, Seehausen O, Maan ME 2012 Species-Specific Relationships between Water Transparency and Male Coloration within and between Two Closely Related Lake Victoria Cichlid Species *Int. J. Evol. Biol.* **2012** 12.

[13] Jellyman PG, Harding JS 2014 Variable survival across low pH gradients in freshwater fish *J. Fish Biol.* **85** 6.

[14] Reynalte-tataje DA, Baldisserotto B, Zaniboni-filho E 2015 The effect of water pH on the incubation and larviculture of curimbatá Prochilodus lineatus (Valenciennes, 1837) (Characiformes : Prochilodontidae) **13** 7.

[15] Ren Q, Zhang L, Wei Y, Li D 2018 Original papers A method for predicting dissolved oxygen in aquaculture water in an aquaponics system *Comput. Electron. Agric.* **151** 7.

[16] Null SE, Mouzon NR Elmore LR 2017 Dissolved oxygen , stream temperature , and fish habitat response to environmental water purchases *J. Environ. Manage.* **197** 12.

[17] Franklin PA 2014 Dissolved oxygen criteria for freshwater fish in New Zealand : a revised approach *New Zeal. J. Mar. Freshw. Res.* **48** 15.

[18] Natadisastra GG, Hasan Z, Sriati, Lili W 2018 Kemampuan penyerapan logam berat tembaga (Cu) pada akar Avicennia marina di perairan Karangsong , Kabupaten Indramayu *J. Perikan. dan Kelautan.* **IX** 7.

[19] Boyd CE 2015 *Water Quality* New York Springer p 374.

[20] Randall DJ and Tsui TKN 2002 Ammonia toxicity in fish **45** 6.

[21] Eddy B 2005 Ammonia in estuaries and effect on fish Ammonia in estuaries and effects on fish *J. Fish Biol.* **67** 19.

[22] Babatunde A, Ajadi A, Tola-fabunmi AS 2019 Waste production in aquaculture Sources, components and managements in different culture systems Aquac. Fish. 4 7.

[23] Milhazes-cunha H, Otero A 2017 Valorisation of aquaculture ef fl uents with microalgae The Integrated Multi-Trophic Aquaculture concept Algal Res. 24 9.

[24] Adesuyi A A, Nnodu VC, Njoku KL, Anuoluwapo J 2015 Nitrate and Phosphate Pollution in Surface Water of Int. J. Geol. Agric. Environ. Sci. 14 7.

[25] Santi I, Tsiola A, Dimitriou PD, Fodelianakis S, Kasapidis P, Papageorgiou N, Daffonchio D and Pitta P 2019 Prokaryotic and eukaryotic microbial community responses to N and P nutrient addition in oligotrophic Mediterranean coastal waters : Novel insights from DNA metabarcoding and network analysis *Mar. Environ. Res.* **150** 13.

[26] Davidson J, Good C, Welsh C, Summerfelt ST 2014 Aquacultural Engineering Comparing the effects of high vs . low nitrate on the health , performance , and welfare of juvenile rainbow trout Oncorhynchus mykiss within water recirculating aquaculture systems *Aquac. Eng.* **59** 10.

6. Acknowledgments

Authors would like to thank KEMENRISTEKDIKTI for funding this research (PTUPT Research Programme), to LPPM and Department of Fisheries University of Djuanda for support project.

Effects of neem plant extract (*Azadirachta indica* A. Juss) and bio-surfactant diethanolamide olien from palm oil to the mortality of cacao moth pest (*Conopomorpha cramerella*)

Nurmayulis[1]*, F R Eris[1], D Hastuti[1], R T Utami[1], Y R Denny[2], T Firmansyah[3]

[1]Department of Agroecotechnology, Agriculture Faculty, Sultan Ageng Tirtayasa University, Pakupatan-Serang-Banten, 42121, Indonesia.
[2]Department of Physics Education, Faculty of Teacher Training and Education, Sultan Ageng Tirtayasa University, Serang Banten 42117, Indonesia.
[3]Department of Electrical Engineering, Engineering Faculty, Sultan Ageng Tirtayasa University, Cilegon, Banten 42435, Indonesia

*Corresponding author: upik_nurma@yahoo.co.id

Abstract. This study aims to determine the response of the addition of neem plant extract (*Azadirachta indica A. Juss*) and biosurfactant of diethanolamide olein palm to the mortality of cocoa moth pest (*Conopomorpha cramerella*). The research was carried out with experiments. The treatment consisted of 5 levels, namely (1) control (without treatment), (2) young leaf extract of neem plant consisted of 100 ml / 500 ml of water + 5% of biosurfactant dietanolamide palm olein, (3) old leaf extract of neem plant consisted of 100 ml / 500 ml of water + 5% of biosurfactant diethanolamide palm olein, (4) seed extract of neem plant consisted of 100 ml / 500 ml of water + 5% of biosurfactant diethanolamide olein palm, (5) young leaf extract of neem plant and seed extract of neem plant consisted of 100 ml / 500 ml of water + 5% of biosurfactant diethanolamide olein palm, (6) old leaf extract of neem plant and seed extract of neem plant consisted of 100 ml/500 ml of water + 5% of biosurfaktan dietanolamida olein sawit. The results showed that the treatment of neem plant extracts mixture of young leaves and seeds + biosurfactant dietanolamide olein palm gave the best influence in controlling cocoa moth pests. Furthermore, it was followed by the treatment of neem plant extracts mixed with old leaves and seeds + biosurfactant dietanolamide olein palm.

Keywords: Neem plant, Bio-surfactant diethanolamide olein palm, cocoa moth pest.

1. Introduction

The cocoa plant (*Theobroma cacao* L.) is one of the potential foreign exchange producing plant, besides that it is an important plantation commodity in Indonesia because it has been developed in the field of agro-industry, such as processing chocolate products, cosmetic raw materials, pharmaceuticals and other industries [1].

According to the [2], the development of cocoa bean production in 2012 to 2015 decreased by around 19.87%, from 740.5 thousand tons to 593.3 thousand tons. One of the contributing factors is the condition of plants that are old, damaged, and unproductive and exposed to pest attacks with severe attacks. According to the [3], as many as 235,000 ha of cocoa plants are less productive due to being attacked by pests and diseases with moderate attacks. One of the reasons for the decline in national cocoa production and productivity is caused by the attack of Conopomorpha cramerella (Snellen) cocoa pod borer [4]. Cacao fruit borer (PBK) C. cramerella is the main pest in cocoa which causes a lot of losses for farmers, especially in Indonesia [5]. This PBK pest is very destructive because it causes the cocoa beans to not develop; the seeds stick together and are black.

Efforts to control cocoa pod borer, farmers still depend on the use of synthetic pesticides (chemical), assuming that synthetic pesticides are more effective in controlling plant pest organisms (OPT). However, the use of synthetic pesticides has been shown to cause various negative impacts such as resistance, pest resurgence, the emergence of secondary pests, pesticide residues that pollute agricultural products and the environment which endanger the lives of non-target organisms (natural enemies / biological control agents) [6]. The presence of pesticide residues in agricultural products including cocoa beans will cause poisoning in humans, pets and the occurrence of other hazards in everyday life.

Vegetable pesticides are sourced from plants that have secondary metabolic compounds that are repellents, antifeedant/feeding, developmental inhibitors and oviposition repelent/deterrent and as chemicals that kill insects quickly [7] Vegetable pesticides are non-persistent in nature, so they can control plant pest organisms. One of the plants that are as a vegetable pesticide (biopesticide) is a neem plant (Azadirachta indica A. Juss). This plant contains the active ingredient such as azadirachtin (C35H44O16), meliantriol, salanin, neem, nimbidine and other ingredients.

According to [8] on the use of vegetable pesticides, surfactants that have the potential to be applied in neem pesticides are diethanolamide surfactant (DEA) derived from palm olein methyl ester. DEA surfactants are included in the group of nonionic surfactants that are biodegradable and environmentally friendly and also function as good dispersers and reducing surface tension and interface tension which is quite effective, so that DEA surfactants have the potential to increase the effectiveness of vegetable pesticides (biopesticides).

The purpose of this research was to determine the response of extracts of several parts of Neem Plant (*Azadirachta indica* A. Juss) plus Palm Olein Dietanolamide Biosurfactant on the Resistance Level of Cocoa Fruit Borer (*Conopomorpha cramerella*).

2. Methodology

The research was conducted from July 2018 to September 2018 at the Basic Science and Plant Protection Laboratory of the Faculty of Agriculture, Sultan Ageng Tirtayasa University, Serang, Banten, Indonesia.

The materials and tools used in this research consisted of cocoa fruit which had been attacked by cocoa stem borer (PBK) taken from Cikuya Village, Sindang Sari Village, Pabuaran District, Serang Regency, Banten Province, young leaves, old leaves and neem seeds, palm olein diethanolamide biosurfactant, jars, analytical balance sheets, microscopes, Erlenmeyer, Petridis

The research was conducted by experiment. The treatment consisted of 6 levels, namely (P0) control (without treatment), (P1) young leaf extract of neem plant 100 ml / 500 ml of water + 5% of palm olein diethanolamide biosurfactant, (P2) old leaves extract of neem plant 100 ml / 500 ml of water + 5% of palm olein diethanolamide biosurfactant, (P3) extract of neem plant seeds 100ml / 50 ml of water + 5% of palm olein diethanolamide biosurfactant, (4) young leaf extract and seeds of neem plant 100 ml / 500 ml of water + 5% palm olein diethanolamide biosurfactant, (5) old leaf extract and 100 ml / 500 ml neem plant seeds + 5% palm olein diethanolamide biosurfactant. The parameters observed consisted of (1) changes in morphology and physical pest after application. (2) Changes in morphology and pest behavior after application, mortality of cocoa fruit borer pests and (3) lethal time 50%

Application of Neem Extract plus Palm Olein Diethanolamide Biosurfactant in PBK pests

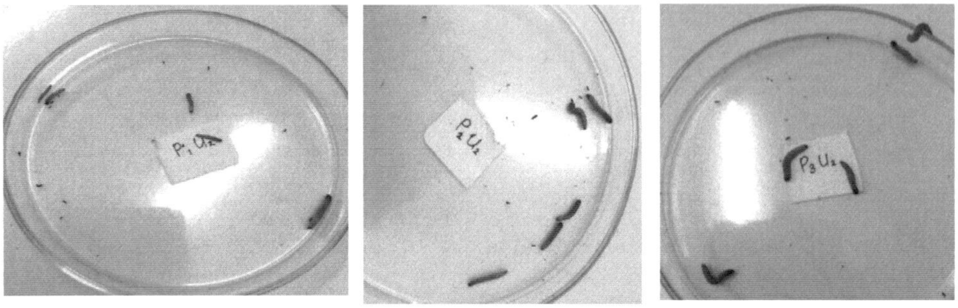

. **Figure 1.** Pest conditions before dripped of vegetable pesticides (active pests move)

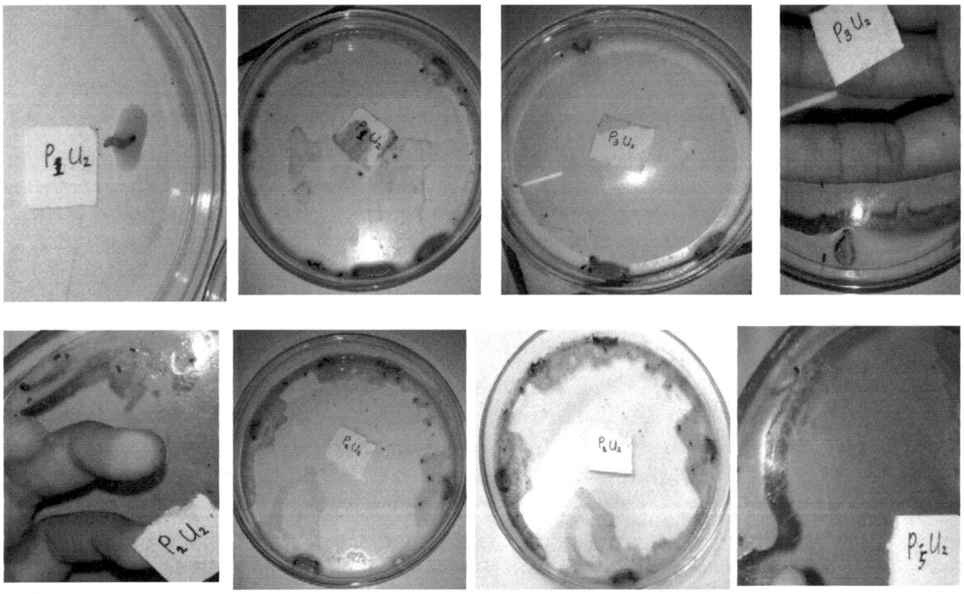

Figure 2. Application of vegetable pesticides to PBK pests (Application method is applied to the body of the cocoa fruit borer larvae (PBK), each of the larvae is drippeds 3 times)

3. Result and Discussion

3.1. Morphological changes and physical changes

PBK pests which have been dripped by various treatments of neem extract and observed in H + 1 to H + 5 changes in pest appearance, both physically, behavior, color and others. These PBK pests die on average within 38 seconds up to 01.06 seconds, this shows that the treatment of neem extracts of various treatments is very effective for controlling PBK pests. Before dead PBK pests came into contact with neem extracts of various treatments, resulting in convulsive pests or uncontrolled behavior, not long after PBK pests died.

In the treatment without the application of neem extract, the pest conditions were 5 days after the application was still alive and active, there were even larvae that had turned into pupae, but for treatment P1, P2, and P3 the condition was dead, this indicates that neem extract with various treatment is effective for PBK pest control.

A few days after the application of the neem extract, there was a change in the color of the PBK pest, starting from brown and then turning into black. In addition, the body of the PBK pest which was originally rather large turned soft, while the small-sized PBK pests were hardened and stiff.

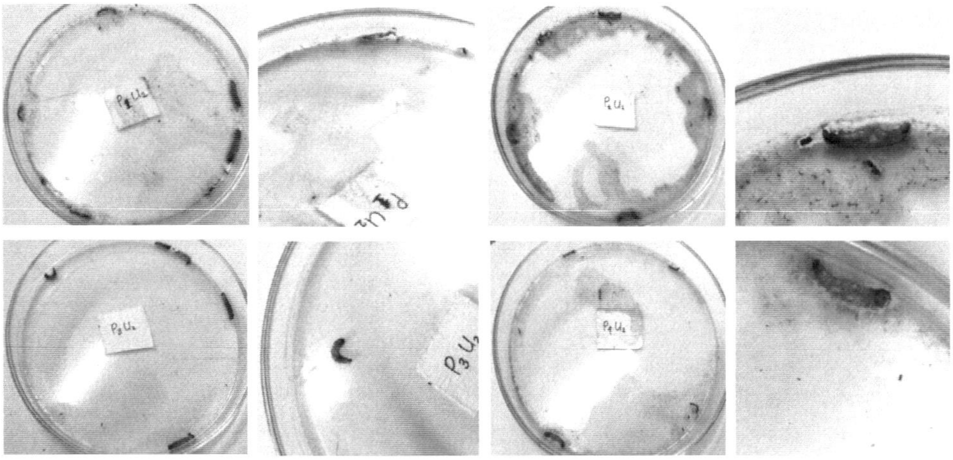

Figure 3. Observation of PBK Pest after H+5 Application

3.2. Morphological changes and pest behavior after application

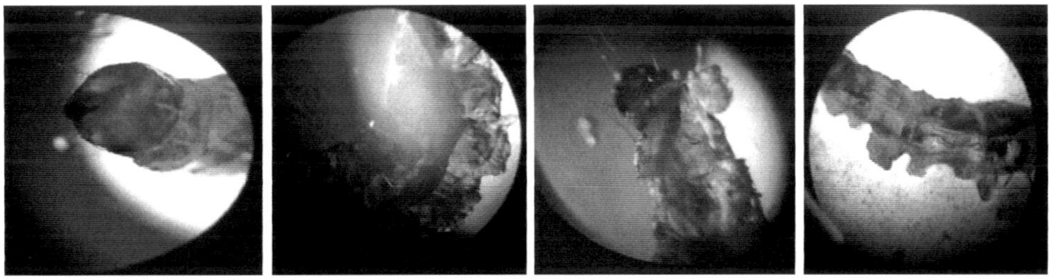

Figure 4. Observation of dead larvae using a microscope

The condition of the damaged larvae is seen to be flattened / thin, and it also softens in the middle and hardens on the sides of its body, so that its body becomes damaged and peels.

3.3. Lethal Time 50%

Table 1. 50% Lethal Time of PBK Larvae

Treatment					Average
P0 (Control / Without Treatment)					
1	2	3	4	5	
Life	Pupa	Pupa	Die	Pupa	-
P (Young Leaves Extract of Neem Plant)					
1	2	3	4	5	
0:00:31	0:00:39	0:00:40	0:01:17	0:02:24	0:01:06
P2 (Old Leaves Extract of Neem Plant)					
1	2	3	4	5	
0:00:36	0:00:29	0:00:44	0:00:37	0:00:17	0:00:33
P3 (Neem Seed Extract)					
1	2	3	4	5	
0:00:38	0:00:30	0:00:29	0:00:49	0:00:43	0:00:38

From the calculation results of the time of death of 5 treatments that have been applied, it shows that treatments P4 and P5 or leaf and seed mixture formulations produce the fastest death times compared to other pesticide formulations (P1 / P2 / P3). It can be concluded that the mixture of neem leaves and seeds has good pesticide content, that is, it is possible for the ingredients of each active ingredient of seeds and leaves to complement or react to each other well.

Cocoa fruit borer (CPB) generally attacks young cocoa fruits about 8 cm long. Stadium that causes damage is the larval stage. PBK larvae eat fruit flesh and food channels leading to seeds, but do not attack seeds. New symptoms appear from the outside when ripe fruit in the form of faded fruit skin and orange colored stripes and if shaken it does not ring. If split, the flesh will appear black, the seeds stick to each other in black, wrinkled, and light. As a result of this pest attack, the losses incurred can reach 80% of dry cocoa beans [9].

3.4. The content of neem leaves and seeds

The active ingredient in neem plants is azadirachtin, salanin, meliantriol, and neem, which are mainly found in seeds and leaves of plants. The azadirachtin substance is believed to have a killing power against insect pests. Neem leaves and seeds contain various chemical compounds, such as phenols, quinones, alkaloids and other nitrogen substances, acids and terpenes. Compounds that are believed to be bioactive ingredients of vegetable pesticides are neem (nimbinen), thionemon, meliantriol, azadirachtin, and salanin, which are chemical compounds from terpene groups [10]. [11] said that the content of the active ingredients, neem seeds and leaves containing azadirachtin as the main active compound, meliantriol, salanin, nimbidin, and neem, which are the results of secondary metabolites from neem plants.

Azadirachtin is a contact poison, stomach poison, and pest repellent. Neem extracts made from leaves, flowers, and neem seeds can be used to control various types of pests, such as Helopelthis sp., long caterpillar, Aphis sp., Nilarvata sp., And Sitophilus sp. [10]. Azadirachtin compounds can inhibit the growth of pests insect, reduce appetite, reduce egg production and hatching, increase mortality, activate infertility (function as antifertile), and reject pests around the neem tree.

Azadirachtin contained in neem seeds acts as a substance that can inhibit the work of ecdyson hormone, which is a hormone that functions in the process of metamorphosis of insects. Insects will be disrupted in the process of changing the skin, or the process of change from egg to larvae, or from larvae to cocoons or from cocoons to adulthood. Usually, failure in this process often results in death (Kuba et al., 1986 in [12]).

Salanin acts as a decrease in appetite which results in insect destructive power is greatly reduced, even though the insects themselves are not dead. Insect pests that have been exposed to the application of neem seed powder will be sprawled and the destructive power is greatly reduced because insects are sick [11].

Meliantriol acts as a barrier to pest insects which results in insect pests being reluctant to approach plants due to meliantriol substances. Neem and Nimbidine act as anti-microorganisms such as anti-virus, anti-bacterial, and anti-fungi. Neem and Nimbidine are very important and good for controlling plant diseases [11]. The application of neem seed extract can result in 60% mortality of red mite *Tetranychus urticae*, which is lower when compared to dicofol which reaches 81%; the same thing happened to predatory mites *Amblyseius longispinosus* [13].

Extracts from the leaves of the neem plant are reported to be able to control around 127 species of pests and are able to act as fungicides, bactericides, anti virus, nematicides and moluscicides [14]. The leaves of neem plants contain poison and are influential in the process of digestion of food, inhibiting intestinal contractions, so that the process of digestion of food cannot take place [15]. Besides being able to interfere with eating activities, poisons contained in the extracts of mimba leaves can also be absorbed through the body walls, in this case the leaves of the mimba leaves function as a contact poison [16].

3.5. Function of palm olein dea biosurfactant (dietanolamida)

One alternative that can be used for vegetable pesticides is to use diethanolamide surfactant (DEA) which is made from palm oil [17]. The advantage of using natural material based surfactants is that they are renewable and are more environmentally friendly in the production and application process than surfactants that use petroleum-based raw materials [18].

Based on research conducted by a research team of SBRC (Surfactant and Bioenergy Research Center) in 2012, DEA surfactants have the lowest surface tension (20.97 dyne / cm) compared to other surfactants such as APG (21-22 dyne / cm), ethoxylates (23-25 dyne / cm), and lauril betain (31.17 dyne / cm) which is widely used in the pesticide industry. Therefore, DEA surfactants have the potential to increase the effectiveness of insecticides [8]. Previous research conducted by [19] showed that DEA surfactant of palm olein was able to form good emulsions and increase the effectiveness of insecticides with active ingredients of buprofezin for brown planthopper pest control.

4. Conclusion

The best plant-based pesticide to control PBK pests are mixed vegetable pesticides from young seeds and leaves of Neem or P4 plants because the application results show that PBK pests are faster controlled or faster die. The reaction caused by PBK pests after dripped by vegetable pesticide liquid from a mixture of seeds and young leaves of the neem plant caused PBK pests to die within 24 seconds (the average time of 5 pests in one repetition). The reaction makes this pest move irregularly or convulsions and after that it dies.

5. References

[1] Wahyudi, T., T.R. Panggabean, and Pujianto. 2009. Panduan Lengkap Kakao, Manajemen Agribisnis dari Hulu Hingga Hilir. Penebar Swadaya. Jakarta.

[2] Directorate General of Plantations. 2017. Statistik Perkebunan Indonesia (Tree Crop Estate Statistics of Indonesia) 2015-2017 Kakao. 90p

[3] Ministry of Agriculture. 2012. Pedoman Umum Gerakan Nasional Peningkatan Produksi dan Mutu Kakao Tahun 2013. Jakarta: Direktorat Jendral Perkebunan.

[4] Valenzuela, I., Purung, H.B., Roush, R.T., and Hamilton, A.J. 2014. Practical Yield Loss Models for Infestation of Cocoa with Cocoa Pod Borer Moth, *Conopomorpha cramerella* (Snellen). Crop Protection. Vol. 66: 19-28.

[5] Adjinah, K.O., and Opoku, I.Y. 2010. The National Cocoa Diseases and Pest Control (CODAPEC): Achievement and Challenges Myjoyonline-Myyoyoonline.com. Feature Article

[6] Wiranti, E.W. 2005. Pemasyarakatan Penggunaan Pestisida Nabati dalam MEndukung Agribisnis. Jurnal Planta Tropika. Vol 1 (2): 84-88

[7] Prijono, D. 1999. PRospek dan Strategi PEmanfaatan Insektisida Alami *in* Dadang, B., W. Nugroho., amd D. Prijono (Eds.). Bahan Latihan Pengembangan dan Pemanfaatan Insektisida Alami. Pusat Kajian Pengendalian Hama Terpadu.

[8] Suryani, A., Dadang., Nisya F.N., and Nurkania A. 2012. Surfaktan Berbasis Minyak Nabati untuk Formulasi Herbisida Berbahan Aktif Glifosat. Laporan Penelitian. Bogor: SBRC LPPM IPB.

[9] Lukito, M., Tetty H., and Nofiandi. 2004. Budidaya Kakao. Jakarta: Pusat Penelitian Kopi dan Kakao Indonesia. 298p.

[10] Rukmana, R and Yuyun Y.O. 2002. Mimba Tanaman Penghasil Pestisida Nabati. Yogyakarta: Kanisius. 38p.

[11] Indiati and Marwoto. 2008. Potensi Ekstrak Biji Mimba Sebagai Insektisida Nabati. Buletin Palawija. 15: 9-14.

[12] Prakash, A and J. Rao. 1997. Botanical Pesticides In Agriculture. Lewis Publ. Baco Raton, New York. London. Tokyo. 460p.

[13] Singh, S.P and R.N. Singh. 2005. Efficacy of some Pesticides Against Spider Mite, Tetranychus Urticae Koch and its Predatory Mite, Amblyseius Longispinosus (Evans). Resistant Pest Management Newsletter. Vol 14(2): 7-10.

IOP Conf. Series: Earth and Environmental Science **383** (2019) 012034 doi:10.1088/1755-1315/383/1/012034

[14] Kardinan, A. 2002. Pestisida Nabati, Ramuan dan Aplikasi. Jakarta: Penerbit Penebar Swadaya. 80p.
[15]Nurtiati, H and T. Widya. 2001. Pemanfaatan Bioinsektisida Ekstrak Daun Azadirachta indica A. Juss. sebagai Pengendali Hayati Ulat Daun Kubis Plutella xyclostella. Jurnal MIPA. Vol. 6(1): 55-62.
[16] Ardiyansyah., Wiryanto., and Edwi, Mahajoeno. 2002. Toksisitas Ekstrak Daun Mimba (Azadirachta indica A. Juss) pada Anakan Siput Murbei (Pomacea canaliculata L.). Jurnal Bio SMART. Vol. 4(1): 29-34.
[17] Surya, N.A. 2015. Studi Efektivitas Surfaktan Dietanolamida (DEA) dari Metil Ester Olein Minyak Sawit Pada Proses Bioremediasi Tanah Tercemar Minyak Bumi [SKRIPSI]. Bogor: Fakultas Teknologi Pertanian IPB.
[18] Foster, N.C. 1996. Sulfonation and Sulfation Processes. In: Spitz L. (Ed). Soap and Detergents: A Theoritical and Practical Riview. Champaign, Illinois: AOCS Press.
[19] Ferdian, M.A. 2015. Kajian Stabilitas Emulsi Produk Insektisida Hama Wereng Coklat Menggunakan Surfaktan DEA Metil Ester Olein Sawit [TESIS]. Bogor: IPB.

IOP Conf. Series: Earth and Environmental Science **383** (2019) 012035 doi:10.1088/1755-1315/383/1/012035

Effect of extraction methods on the nutritional characteristics of milkfish (*Chanos chanos* Forsskal) bone powder

P Wulandari*) and S Kusumasari

Department of Food Science, Faculty of Agriculture, University of Sultan Ageng Tirtayasa Jl. Raya Jakarta km. 04 Pakupatan, Serang City, Banten Province, Indonesia
*)email puji.wulandari@untirta.ac.id

Abstract. Milkfish (*Chanos chanos* forskal) bone is one of underutilized part of milkfish and considered as a waste from ''satay bandeng'' home industry in Banten. The aim of this research was to determine the effect of different extraction methods on the nutritional characteristics of milkfish bone powder. Milkfish bone powder was prepared using two different extraction methods (alkaline and acid). Milkfish filleting waste was boiled with water for 30 min, removed the remaining flesh from the frames, the frames were boiled with water for 2 hours, autoclaved, boiled in water (MBPW), sodium hydroxide solution 1 N (MBPAl) and hydrocloric acid solution 1 N (MBPAc) for 30 min at 80-90 °C, rinsed with water, dried at 50°C in oven drying for 24 hours, and ground. The observed parameters were % yield, moisture-, protein-, fat-, carbohydrate-, and ash- content. Milkfish bone powder was contained 27.88% protein, 7.85% lipid 51.42% ash, and carbohydrate 7.36%. The milkfish bone power extracted by acidic and alkaline method treatment (MBAc and MBAl, respectively) showed statistical significant difference in yield, protein-, fat-, carbohydrate-, and ash- content. The alkaline and acid-extraction methods showed higher ash content than water extraction method so these method could be used to extract the mineral content from milkfish bone.

1. Introduction

Milkfish (*Chanos chanos* forskal) is one of popular raw material of Banten cullinary snack, such as milkfish satay, milk fish floss, milkfish bontot, milkfish crackers and chips. The most utilized part of milkfish are flesh and skin. Whereas, milkfish bone is one of underutilized part of milkfish and considered as a waste from these industry, especially from satay bandeng home industry. Based on data of Industrial and Trade Department Office in Serang City, there are 17 Small Medium Entreprises (SME) of milkfish satay in Serang City. Milkfish bone has been produced and discarded everyday as by product of the milkfish processing line which can cause environmental pollution. This by products as a result of processing represent more than 30% of the fish being processed. The amount of fish bone fraction is approximately 10 - 15% of the fish whole body weight [1].

Fish bone powder are known as one of sources of minerals such as calcium and phospor [2]. Fish bone powder usually extracted by boiling the fish frame with water, alkaline, acid or combination of these treatments. The alkaline extraction method of three kind fish bone powders from Grouper (*Epinephelus diacanthus*), Emperor (*Letrinus fraenatus*) and White snapper (*Pristipomoides filamentosus*) showed higher ash content than extraction in neutral conditin so this method could be used to extract the mineral content from fish filleting waste [3]. The acidic treatment is also used to extract calcium fish bone powder in Nila fish (*Oreochromis niloticus*) [4]. Previous study found that the alkaline and acidic treatments could affect the nutritional characteristic of the fishbone powder [3,4].

There are no studies addressed about the effect of extraction methods using alkaline, acidic, and neutral condition on the nutritional characteristic (macronutrients composition such as protein, fat, ash, carbohydrate) of milkfish bone powder. This study will help to effectively utilize the milkfish processing waste and reduce environmental pollution. The objectives of this research were to produce milkfish bone powder from milkfish filleting waste and to determine the effect of different extraction methods on the nutritional characteristics of milkfish bone powder.

2. Material and Methods

2.1 Sample & Chemicals

Milkfish (*Chanos chanos* forskal) filleting waste was obtained from Satay Bandeng home industry in Serang City, Banten Province. The sample was transported to the laboratory immediately and stored at -20°C until being processed. All chemicals (selenium mixture, HCl, H_2SO_4, NaOH, n-hexane, etc) were obtained from Merck.

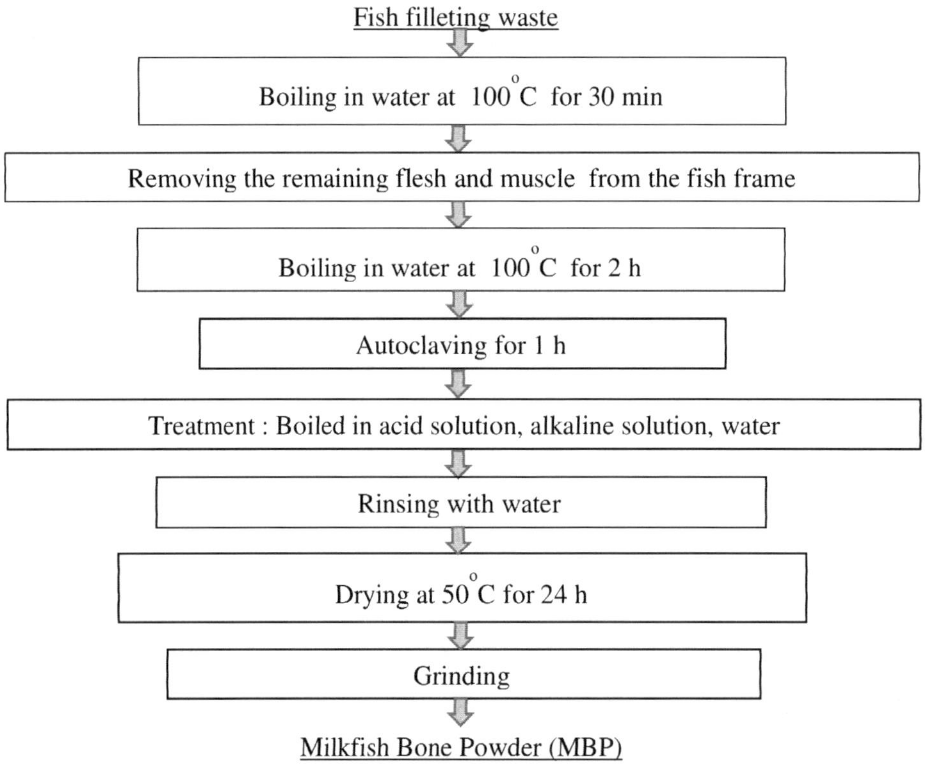

Figure 1. Flow chart of milkfish bone powder production.

2.2 Preparation of Milkfish Bone Powder

Milkfish filleting waste was further processed to produce milkfish bone powder. Milkfish filleting waste was boiled with water for 30 min, removed the remaining flesh from the frames, the frames were boiled with water for 2 hours, autoclaved, boiled in water, sodium hydroxide solution 1 N and hydrocloric acid solution 1 N for 30 min at 80-90 °C, rinsed with water, dried at 50°C in oven drying for 24 hours, and ground. Milkfish bone powder production steps was shown in figure 1. The product obtained, referred to as milkfish bone powder produced by alkaline condition, acidic condition, and neutral condition (further named as MBPAL, MBPAC, and MBPW, respectively). Water milkfish bone powder were packaged in polypropilene plastic bag, and stored at 4°C until analisys.

2.3 Nutritional Analysis
Milkfish bone powders were analized for their nutritional characteristics (macronutrients) such as protein-, fat-, carbohydrate- and ash- contents using the standard methods of AOAC [5]. Crude protein content was determined by the Kjeldahl method, moisture content was determined by thermogravimetry and total fat content by soxhlet extraction method, ash content was determined by combusting the samples in a muffle furnace at 550°C and carbohydrate content was determined by difference. The results were expressed in dry-basis.

2.4 Statistical Analysis
Statistical analysis was done using Excel spreadsheet and Statistical package for social scientists (SPSS) version 20. Analysis of variance (ANOVA) and Duncan test were used to test significant differences between samples ($P < 0.05$).

3. Result And Discussion

Milkfish filleting wastes from satay bandeng industry were taken as raw materials and processed to recover of milkfish frame then exposed to the alkaline and acidic treatment to extract of milkfish bone powder. The moisture content and yield of milkfish bone powder extracted by water, acidic, and alkaline condition are shown in table 1.

The moisture contents of all samples have no siginficantly differences. The moisture content range from 5.19 % to 5.49 %. The moisture content of fish bone powder are vary depends on the fish species. The moisture content of fish bone powder from Grouper, Emperor, White snapper, Nile Tilapia are 5.96%, 4.12%, 3.21% [3], 8.4% [6], respectively.

The yields are 67.57%, 39,67%, and 50,87%, for water, alkaline, and acidic extraction methods, respectively. The highest yield was obtained by water extraction method. Both of the alkaline and acidic treatments can cause loss in some compounds, such as protein and fat -contents as shown in Table 2. The alkaline extraction method cause more significant loss in yield than acidic extraction method, because greater loss of protein and fat in the alkaline- than the acidic- extraction method.

Table 1. Yield and Moisture Content of Milkfish Bone Powder.

Parameters	MBPW	MBPAl	MBPAc
Moisture content (%)	5.49^{a}	5.27^{a}	5.19^{a}
Yield (%)	67.57^{a}	39.67^{c}	50.87^{b}

Different lowercase superscripts within the same row indicatesignificant difference ($p < 0.05$)
MBPW = Milkfish Bone Powder extracted with water
MBPAl = Milkfish Bone Powder extracted with alkaline
MBPAc = Milkfish Bone Powder extracted with acid

Table 2 Macronutrients Composition of Milkfish Bone Powder.

Parameters (%)	MBPW	MBPAl	MBPAc
Protein content	27,88^{a}	1,74^{c}	4,68^{b}
Fat content	7,85^{a}	1,83^{c}	2,45^{b}
Ash content	51,42^{c}	80,84^{a}	73,35^{b}
Carbohydrate content (by difference)	7,36^{c}	10,62^{a}	14,33^{b}

Different lowercase superscripts within the same row indicatesignificant difference ($p < 0.05$)

The macronutrients composition of three milkfish bone powder are shown in table 2. The results of the macronutrients composition of MBPAl, MBPAc, and MBPW showed that the amount of the protein, fat, ash and carbohydrate is significantly different among all samples. The alkaline and acid treatment significantly decrease protein and fat of milkfish bone, and significantly increase ash content. Milkfish bone powder extracted with water extraction method was contained 27,88% protein, 7.85% lipid 51.42% ash, and carbohydrate 7.36%. The results of milkfish bone powder protein content obtained in the this study was comparable with previous studies [6, 7].The crude protein content of fishbone powder without alkaline and acidic treatments in tilapia fish bone powder and tuna frame powder were 31.5% [7] and 28.66% [6]. The alkaline and acidic extraction methods showed the the protein content in the Alkaline and Acidic were 1,74 % and 9,68%, respectively, while the fat content were 1.83 % and 2.45%, respectively. Both of the MBPAl and the MBPAc have lower protein content than the MBPW. The alkaline and acidic extraction methods are boiling bones with NaOH solution and HCl solution (respectively), so that it will remove the protein from fish bone matrices. The alkaline extraction methods with NaOH 1N solution as the boiling media not only can cause protein denaturation and hydrolisis, but also fat saponification, result in lower protein- and fat- content compared to water and acidic extraction methods. Previous study concluded that alkaline solution would be effective way to leach out protein from the bone [2]. The acidic or alkaline pretreatment can hydrolyze the crosslinking bonds between polypeptides in protein [8].

Similar results related to the content of protein and ash in fish bone powder was obtained by Nemati et al [9] in production of tuna bone powder. The alkaline extraction method using of NaOH solution will increase the ash content but decrease the protein content of the tuna bone powder Utilization of NaOH during MBP processing, dissolved and removed the organic materials from bone result in protein and fat contents in MBP decreased and ash content increased [9].

The significant difference in total fat content of fish bone powder was found in all treatments. Milkfish bone powder showed total fat content of 7,85% in MBPW, 1.83% in MBPAl, and 2.45% in MBPAc. These results are in agreement with several studies reported in tuna fish bone powder [6] and tia fish bone powder extracted by alkaline solution (NaOH) [9]. The fat content in tuna fishbone powder without pre treatments (alkaline) was 11 %, [6,9] and in tuna bone powder extracted by alkaline solution was 3.86% [9]. Having a low fat content in MBP is beneficial because the product can not easily oxidized so that it has a longer shelf life.

The major compound in the milkfish bone powder was ash content which was found to be 51.41% - 73.35%. The amount of the ash in MBPAl and MBPAc were higher than MBPW. The ash content in MBP were 51.42%, 80.84%, 73.35% for MBPW, MBPAl, MBPAc, respectively. The amount of the ash in MBP was considerable in comparison to the results reported from other fish bone in the literature such as tuna fish bone without treatment 53.43%, with alkaline treatment 77.97% [9], tilapia bone powder 75.3%, catfish bone powder 61.8% [10]. The alkaline treatment method provided rather high ash content because all of the protein and fat in bone was removed. After some amount of fat and protein in milkfish bone were removed, the minerals became concentrated [11]. The ash content usually corelated to the high mineral content in fish bone powder [2], especilally calcium content in tuna bone powder [9]. The alkaline solution (NaOH) usually used in production of biocalcium and calcined fishbone powder [11].

The ratio of ash/protein is the important criteria to indicate the bone mineralization, associated with hardness of the bone [2]. The ash/protein ratio in salmon, trout, and mackerel bone were lower than 1.00 [2]. In this study, the ratio ash/protein were 1.84, 46.46, 15.67 for MBPW, MBPAl, MBPAc respectively. The ash/protein ratio in MBPAl and MBPAc were higher than MPBW, its showed that both treatment with alkaline and acidic solution in extraction method of MBP could affect the ratio of ash/and protein. Similar result was found in production of tilapia bone powder and the value of ratio ash/protein could be used to indicate the purity of bone powder [12]. Both alkaline and acidic tretament in extraction methods could provide higher purity of milkfish bone powder.

4. Conclusion

The alkaline and acid- extraction methods showed higher ash content than water extraction method so these method could be used to extract the mineral content from milkfish bone so that both alkaline and acidic extraction methods could provide higher purity of milkfish bone powder. The results demonstrated that extracted milkfish bone powder are nutritious. Further, the recovered minerals of this milkfish bone powder may also have potential application in various pharmaceutical, and food industrial applications.

5. References

[1] Malde M K, Graff I E, Siljander-Rasi H, Venalainen E, Julshamn K, Pedersen J I and Valaja J 2010 Fish bones – a highly available calcium source for growing pigs *J. Animal Physiology and Animal Nutrition* **94** pp 66-76

[2] Toppe J, Lbrektsen S, Hope B and Aksnes A 2007 Chemical composition, mineral content and amino acid and lipid profiles in bones from various fish species *Comparataive Biochemistry and Physiology* **146** pp 395–401.

[3] Amitha, Raju C V, Lakshmisha I P, Kumar P A, Sarojini A, Gajendra and Jag Pal 2019 Nutritional composition of fish bone powder extracted from three different fish filleting waste boiling with water and an alkaline media International *Journal of Current Microbiology and Applied Sciences* **8(2)** pp 2942-2948

[4] Wijayanti I, Rianingsih L, Amalia U 2018 Karakteristik fisikokimia kalsium dari tulang nila (*Oreochromis niloticus*) dengan perendaman belimbing wuluh *JPHPI* **21(2)** pp 336-344

[5] AOAC 2000. *Official methods of analysis*. Association of Official Analytical Chemists, Washington, DC

[6] Abbey L, Glover-Amengor , Mary, Atikpo1 M O, Atter A and Toppe J 2017 Nutrient content of fish powder from low value fish and fish byproducts *Food Science and Nutrition* **5(3)** pp 374-379

[7] Vignesh R, and Srinivasan M 2012 Nutritional quality of processed head and bone flours of Tilapia (*Oreochromis mossambicus*, Peters 1852) from Parangipettai estuary, South East Coast of India. *Asian Pacific Journal of Tropical Biomedicine* pp S368-S372

[8] Yang H, Wang Y, Zhou P, Regenstein M, Joe 2008 Effects of alkaline and acid pretreatment on the physical properties and nanostructures of the gelatin from channel catfish skins. *Food Hydrocolloids* **22(8)** pp 1541-1550

[9] Nemati M, Huda N and Ariffin F 2017 Development of calcium supplement from fish bone wastes of yellowfin tuna (*Thunnus albacares*) and characterization of nutritional quality *International Food Research Journal* **24(6)** pp 2419-2426

[10] Luu P H and Nguyen M H 2009 Recovery and utilization of calcium from fish bones by-products as a rich calcium source *Tap Chí Khoa Hoc Và Công Nghe* **47** pp 91-103

[11] Benjakul S, Mad-Ali S, Senphan T, Sookchoo P 2017 Biocalcium powder from precooked skipjack tuna bone: Production and its characteristics *Journal of Food Biochemistry* e12 412

[12] Hemung B O 2013 Properties of tilapia bone powder and its calcium bioavailability based on transglutaminase assay *International Journal of Bioscience, Biochemistry and Bioinformatics* **3(4)** pp 306-309

6. Acknowledgments

The authors would like to acknowledge the support given by Sultan Ageng Tirtayasa University (Untirta) for financial support in this research.

The characteristics of "indigenous yeast mold" dried culture using tray dryer

R Rahmawati[1], D Hunaefi[2], I Basriman[1], D Saputra[3], W D Aozora[2] and B S L Jenie[2]

[1] Program Study of Food Technology, Sahid University, Jl Prof. Dr. Supomo SH. Nomor 84 Jakarta Selatan 12870, Indonesia

[2] Department of Food Science and Technology & Seafast Center, Bogor Agricultural University (IPB), Kampus IPB Darmaga PO box 220 Bogor 16002, Indonesia

[3] Department of Food Technology, Bina Nusantara University, Jl. Jalur Sutera Barat Kavling 21, Tangerang, 15326, Indonesia

E-mail rahmafarasara@usahid.ac.id

Abstract. This research studied the effect of temperature and drying time in making "Indigenous Yeast Mold" dried culture using tray drier. There are 2 dried culture, names AC and CC. AC consists of 3 mold and 1 yeast strains, while CC consists of 7 mold and 3 yeast strains. The temperature of drying used were 40°C and 50°C and the drying time were 0, 1.5, 3, 4.5, and 6 hours. The results showed that the quality of dried cultures AC and CC were influenced by the drying times and temperatures. The water content, viability of microorganisms, A_w, and pH value decreased with the increasing of drying time and temperature. The drying time at the temperature of 40°C affected the water content and water activity of AC and CC dried culture significantly. While the drying time at the temperature of 50°C affected the moisture content, water activity, viability of microorganisms, and pH value of AC and CC dried culture significantly. The desired water content in dried culture is 10% (wb) or 11.1 (db). To reach the water content of 10%, at the drying temperature of 40 °C, the AC dried culture required 4.46 hours, while CC required 4.44 hours. Meanwhile, at the drying temperature of 50 °C, the AC dried culture took 3.46 hours, while CC required 3.33 hours. Drying could be done at 50 °C without changing the quality of the dried culture.

1. Introduction

The results of fermented flour from local white corn flour [1] showed that the fermentation process changes the characteristics of native white corn flour. Likewise [2] have fermented corn grits using different types of starters produce different characteristics of white corn flour compared to the characteristics of their native flour. The use of indigenous yeast and mold in the form of starter cultures require special expertise and are difficult to apply in the community. To overcome this problem and facilitate the fermentation process, researcher [3] has made a dried culture consisting of indigenous yeast and mold. The use of this dried culture in the manufacture of fermented white corn flour produced corn flour which has the characteristics tend to fried products and products that are consumed hot such as cream soup and raw materials for instant porridge, sauce, custard, and bread.

The dried culture that contain indigenous yeast and mold have made using an oven as a dryer [3]. At present, the drying method of culture has increasingly developed such as the use of vacuum drying methods, spray drying, freeze drying, and the fluidized bed dryer method [4]. In addition, researcher [5] has dried a mixture of tempeh culture using tray dryer at a temperature of 50°C for 3 hours. Tray dryer can reduce the drying surface area and increase the efficiency of hot air contact to the material. This condition caused the air circulation in the drying chamber between the racks giving the same results [6]. Tray dryer was chosen to dry the indigenous yeast and mold culture because it is expected to be a more effective drying method. Based on the research of [3] that drying the indigenous yeast mold culture using oven at temperature of 40°C and solar drying produced the best dried culture with high water content (13.34% and 12.5%) and required a long time (48 hours and 7 days). Based on this, the purpose of this study was to determine the time length and temperature of the drying of indigenous yeast and mold culture using a tray dryer. The characteristics of the indigenous yeast mold dried culture that desired are the water content 10% or less, the total viability of yeast molds is at least 10^6 CFU/g, water activity values range from 0.51-0.89, and pH ranges from 2–8.5 [3].

2. Materials and Methods

2.1. Microorganisms

Microorganisms used as a CC starter culture were *Penicillium chrysogenum, Penicillium citrinum, Aspergillus niger, Rhizopus stolonifer, Rhizopus oryzae, Fusarium oxysporum, Acremoniumstrictum, Candida famata, Kodamaea ohmeri, Candida krusei/ incospicua*, while AC starter culture contain *Penicillium citrinum, Aspergillus niger, Acremonium strictum,* and *Candida famata*. The microorganisms used were previously isolated and identified from a spontaneous fermentation of corn grits [7].

2.2. Culture preparation and enumeration

One loop of each mold was streaked onto fresh Potato Dextrose Agar (PDA) slant and then incubated at 30°C for five days. After five days, molds were harvested by scrapping, suspended in 10 mL sterile water and appropriately diluted for enumeration using haemocytometer. Yeast culture was prepared as above, but incubation was carried out at 30°C for two days. Yeast enumeration was also carried out using haemocytometer [1].

2.3. The process of making AC-indigenous yeast mold culture (Modified [7] and [3])

Growing medium used in making yeast and mold culture was corn flour. According to [3] this medium produced the highest total yeast mold culture compared to other mediums (rice and tapioca flour). The technique of making yeast mold culture includes the stages: sterilizing corn flour, then put it into a sterile basin and adding sterile distilled water as much as 2/3 of the total weight of corn flour. Prepared culture suspensions (AC and CC) containing 10^6 CFU / mL per microorganism, then piped as much as 10% of the amount of water used. After that, all stirred until homogeneous and put ± 17 grams in each petri dish. Petri dishes were then incubated at 30 °C for 5 days. Furthermore, the dough is dried using a try dryer with a temperature of 40 °C dan 50 °C for 0, 1.5, 3, 4.5, and 6 hours. The dried AC-indigenous yeast mold culture is made powder using a blender that has been sprayed with 70% alkolhol. The AC-indigenous yeast mold culture powder was packed in plastic clips with silica gel, and then tested for water content, viability (total number of yeast molds), water activity, and pH. The process of making the AC and CC indigenous yeast mold culture powder can be seen in Figure 1.

2.4. AC-CC indigenous yeast mold culture analysis

The product produced is tested for the response. The responses observed were viability (CFU/g) [8], moisture content (thermo-gravimetry) [9], water activity (A_w meter Rotronic Hygrolab) [9], and degree of acidity (pH meter Orion Thermo-Scientific) [9]. The quality indicator was that (1) the water content was equal to or less than 10%; (2) the total viability of yeast molds was at least 10^6 CFU / g; (3) A_w values range from 0.51-0.89; and (4) pH ranges from 2–8.5 [3].

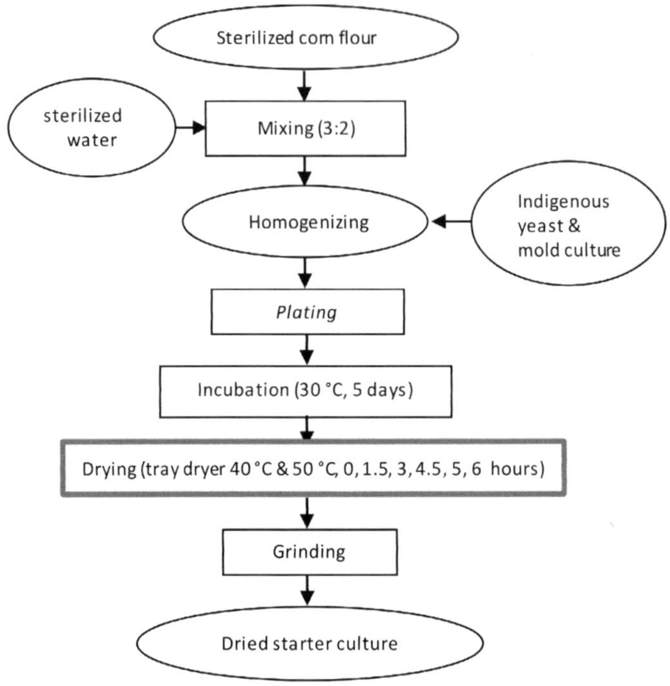

Figure 1. The process of making AC and CC-indigenous yeast mold dried culture (Modified [7] and [3].

2.5. Statistical analysis
Obtained data was then processed using T-test and ANOVA with SPSS (*Statistical Package for the Social Sciences*) software to analyze the influence of drying time and temperature of drying process.

3. Results and discussion
The characteristics of the indigenous yeast mold dried culture that desired are the water content 10% or less, the total viability of yeast molds is at least 10^6 CFU/g, water activity values range from 0.51-0.89, and pH ranges from 2–8.5 [3].

3.1. Influence of drying process on water content
The initial moisture content of the AC dried culture ranges between 55.97 - 56.81% and the CC is between 51.46 - 55.46%. After drying the moisture content of the AC ranged between 5.52-7.35% and CC between 4.90-8.40%. In general, the temperature and drying time reduce the water content. When the temperature increase, the water content will decrease. Likewise, when the drying time getting longer, the water content will decrease too (Figure 2 and 3).

Drying process of indigeous yeast mold culture was expected to reduce the water content to 10% or less. Water content determines the shelf life of the product because it determines the availability of water in food for microorganisms life [10]. The statistical test results showed the exponential equation for AC at temperature of 40°C is $y1 = 97.998e^{-0.512x}$ and the CC is $y2 = 78.556e^{-0.489x}$. The y variable represents the water content and x is the drying time. Based on these equations, it was found that the drying time needed to reach 10% of moisture content was x1 = 4.46 hours for AC and x2 = 4.22 hours for CC at the drying temperature of 40°C. Whereas the exponential equation for AC at a temperature of 50°C is $y3 = 95.162e^{-0.651x}$ and the CC is $y4 = 83.036e^{-0.636x}$. The drying time needed to reach 10% moisture content was x3 = 3.46 hours for AC and x4 = 3.33 hours to CC at a drying temperature of 50 °C.

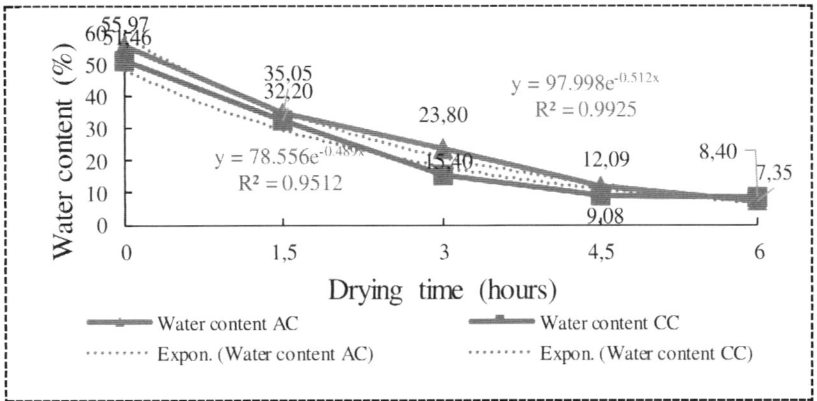

Figure 2. Changes in water content of AC and CC dried culture during drying time at temperature 40°C.

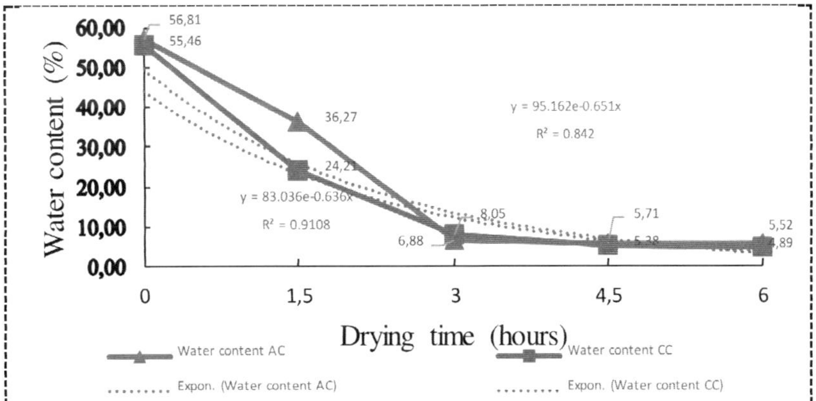

Figure 3. Changes in water content of AC and CC dried culture during drying time at temperature 50°C.

Drying process at 40°C results the drying rate of AC culture faster than CC. This can be seen at the value of the constant. Constants that have higher values will have a faster of the drying rate. The CC dried culture reached the 10% water content faster than AC, while the drying rate was slower (CC required 0.24 hours faster than AC). This was because the initial water content of the CC (51.46%) is lower than AC (55.97%). In addition, AC dried culture consisted of 4 isolates while CC 10 isolates. It caused the cell mass of CC dried culture larger and denser than AC, so it dried faster. Drying process at 50°C showed that the AC dried culture had a faster drying rate than CC, but the CC reaches a moisture content of 10% faster. CC drying time was 0.13 hours faster than AC. The initial water content of both dried culture relatively the same, so the difference in drying time was also relatively the same. Drying time of AC culture at a temperature of 50°C was faster than at a temperature of 40°C to reach 10% moisture content as well as CC dried culture. This was caused by the drying rate at a temperature of 50 °C was higher than at a temperature of 40 °C. A higher drying rate also caused the time needed is faster.

The results of the T-test showed that the drying temperatures of 40°C and 50°C produced different water levels which were not significant in the AC temperature, but were significantly different in the CC dried culture (p value = 0.031). This means that drying process can be done at temperatures of 40°C and 50°C because it produced a different water content that was not significantly different. Beside that, drying process of CC dried culture at temperatures of 40°C and 50°C produced water content that was significantly different. Therefore, choosing the drying temperature is needed to get the desired moisture content for CC dried culture.

Water content decreased during longer drying time. The ANOVA showed that the drying time affected the moisture content of AC dried culture at a temperature of 40°C, where water content at each drying time was significantly different (p value = 0.000) as well as the CC dried culture (p value = 0.000). Likewise drying at a temperature of 50°C, where the drying time affected the moisture content of AC and CC dried culture. This was indicated by the existence of significant differences at each drying time (p value = 0.000). This difference indicated the drying process is going well because the water content in the dried culture is reduced. Drying at temperature 50 °C reaches 10% moisture content faster than 40°C.

This was in line with research [11] where the temperature higher and the the drying time longer, the water molecules will evaporate more from the dried material so that the water content obtained gets lower. The ability of the material to release water from its surface will be greater with the higher the temperature of the dryer used and the longer the drying process, so the lower the water content produced.

3.2. Influence of drying process on viability of microorganisms

The initial viability of microorganims in AC dried culture ranged from 8.90 - 9.14 log CFU / g (db), where CC dried culture contain 8.82 - 9.20 log CFU / g (db). During the drying process the viability of microorganisms in the AC and CC dried culture were decreasing with higher temperatures and longer drying times (Figures 4 and 5). Table 1 showed the moisture content and viability of microorganisms in the AC and CC dried culture at temperatures of 40°C and 50°C during drying.

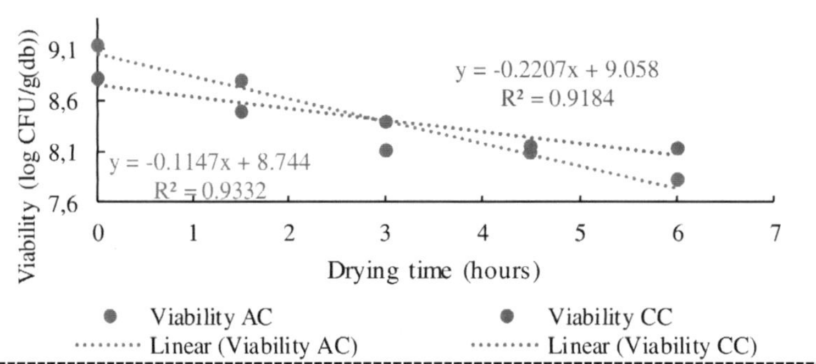

Figure 4. Curve of relationship between viabilities of treatment AC and CC with the drying time at temperature 40 °C.

Based on the exponential equations in Figures 2 and 3, the time needed to reach the water content of 10% at drying temperatures of 40 °C for AC dried culture is 4.46 hours and the CC is 4.22 hours, while at 50 °C for AC is 3.46 hours and CC 3.33 hours. Viability of microorganisms is around 8.10 log CFU / g (db) for AC and 8.16 log CFU / g (db) for CC dried culture to reach 10% moisture content with temperature of drying of 40 °C. Where at a temperature of 50 °C the viability of microorganisms in the AC dried culture between 7.60 - 8.14 log CFU / g (db) and CC contains microorganisms between 7.25 - 8.10 log CFU / g (db). The value of the viability is above the minimum requirement for the viability of microorganisms in dried culture (10^6 log CFU / g). Based on this, the drying of AC and CC can be carried out at 50 °C because the drying time is faster and the viability is still above the minimum requirement.

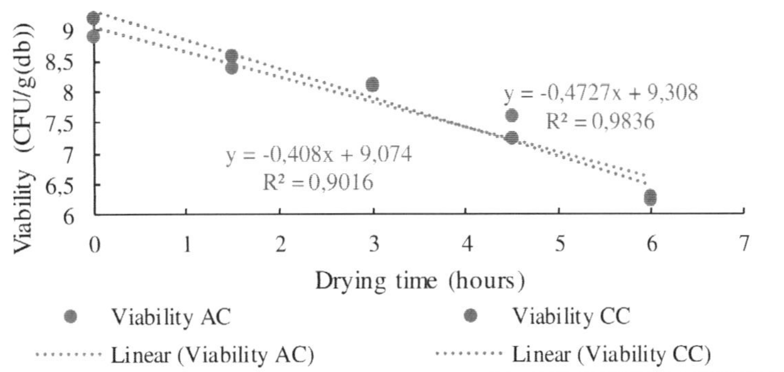

Figure 5. Curve of relationship between viabilities of AC and CC with the drying time at temperature 50 °C.

The T-test dependent showed that the viability of total yeast mold in the AC and CC dried culture over the two different drying temperatures is not significantly different. This showed that the drying of AC and CC dried culture at temperatures of 40°C and 50°C did not affect the viability of the AC and CC dried culture. The decreasing of microbial viability is influenced by the heat resistance of microbes, where each microbe has a different heat resistance. Microbial heat resistance, expressed as a D value. D value is the time in minutes at a certain temperature needed to reduce the number of spores or certain vegetative cells by 90% or one logarithmic [12]. The calculation results showed that the D value of AC and CC at temperatures of 40°C is 271.86 minutes (4.5 hours) and 523.10 minutes (8.7 hours) respectively, while at 50°C is 147.06 minutes (2.45 hours) and 127.93 minutes (2.13 hours) respectively. This D value is influenced by temperature. When the temperature higher, the D value will be smaller [13]. This was in line with the results of the calculation of the D value of AC and CC dried culture at a temperature of 40°C and a temperature of 50°C. Meanwhile, the rate of decline of microorganisms is determined by the value of k. The rate of decline of microorganisms in the AC and CC dried culture is faster at 50 °C with k values of 0.408 and 0.4727 while at 40 °C the rate of decline is slower with k values of 0.2207 and 0.1147 respectively. This showed that the AC and CC dried culture are more heat resistant at 40 °C.

Table 1. Viability (log CFU / g) and moisture content (% db) of AC and CC dried culture at different temperatures and drying times.

Drying time (h)	Temperature 40 °C				Temperature 50 °C			
	AC		CC		AC		CC	
	Water content (% db)	Viability (Log CFU/g)	Water content (% db)	Viability (Log CFU/g)	Water content (% db)	Viability (Log CFU/g)	Water content (% db)	Viability (Log CFU/g)
0	127.12^d	9.14	106.02^d	8.82	131.54^c	8.90^c	124.52^e	9.20^b
1.5	53.96^d	8.79	47.49^c	8.50	56.91^b	8.38^b	31.94^d	8.58^a
3	31.23c	8.12	18.20^b	8.39	7.39^a	8.14^a	8.75^c	8.10^a
4.5	13.75^b	8.10	9.98^a	8.16	6.05^a	7.60^a	5.70^b	7.25^a
6	7.93^a	7.83	9.18^a	8.13	5.84^a	6.23^a	5.15^a	6.32^a

The ANOVA results showed that at 40°C the viability of the AC dried culture was not significantly different at each drying time (p value 0.190), while the viability of the AC dried culture at 50°C was significantly different at each drying time (p value 0.000). This showed that the viability of the AC is stable during the drying process at a temperature of 40°C while it is not stable at 50°C. This was supported by a higher D value of AC dried culture at a temperature of 40°C and a small rate of decline. The viability of AC and CC dried culture is stable because molds and yeasts tend to be able to survive during the period of drying process. In this research *Penicillium*, *Aspergillus*, *Rhizopus*, *Fusarium*, and *Candida* were used. In which *Penicillium chrysogenum* and *P. citrinum* are mesophilic and grow at temperatures of 5 - 40 °C with optimal temperatures of 26-30 °C ([14]; [15]; [16]); Aspergillus, grows at a temperature of 8-45 ° C with an optimal temperature of 35-37 °C; Rhizopus grows at a temperature of 7-42 °C with an optimum temperature of 37 °C; while Fusarium grows at a temperature of 25-37 °C with an optimum temperature of 25-37 ° C; and Candida grows at a temperature of 8-47 °C [16]. Drying at a temperature of 40 °C and 50 °C is not expected to affect the viability of the AC and CC dried culture. This correlates with the ability of mold and yeast to survive in dry conditions. Sunatmo [17] stated that mold and yeast can survive in agricultural products with low water content with water activity 0.80 - 0.88. Beside that, the drying temperature and time are factors that influence the decrease in the amount of long viability [5].

3.3. Influence of drying process on water activity (A_w)

The value of water activity (A_w) is important in food ingredients because it is a safety characteristic of food ingredients. Product that has a low A_w value will have the shelf life longer, because microorganisms can only live in certain A_w conditions [10]. The value of the water activity (A_w) at the beginning of the drying process ranged from 0.925 - 0.940 for the AC dried culture and 0.946 - 0.958 for the CC dried culture. In general the value of water activity decreased with the longer drying time, likewise the A_w value of the AC and CC dried culture (Figures 6 and 7). As explained above, the characteristics of the indigenous yeast mold dried culture that desired are the water content 10% or less, where to reach this condition, at drying temperature 40 °C it takes 4.46 hours for AC and 4.22 hours for CC dried culture. On drying 50 °C, it takes 3.46 hours for AC and 3.33 hours for CC dried culture.

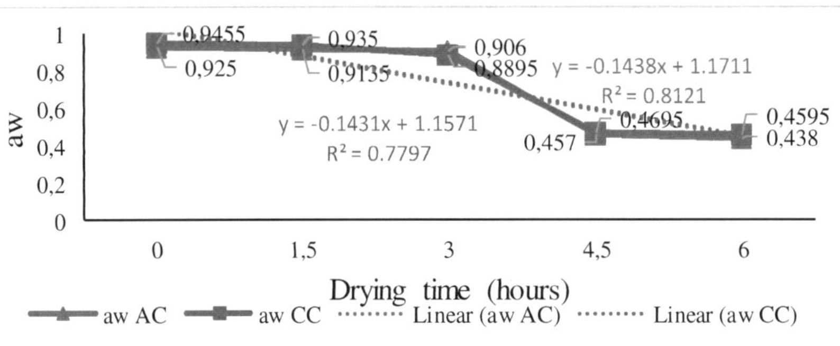

Figure 6. Changes in A_w value during drying time at temperature 40°C

Based on Figure 6 and 7, the linear regression equation for drying 40 °C for AC dried culture is y5 = -0.1431x + 1.1571 and then CC is y6 = -0.1438x + 1.1711. The linear regression equation for drying the temperature of 50 °C for AC line is y7 = -0.101x + 1.0763 and then CC is y8 = -0.0758x + 1.0622. From this equation, the A_w value of each dried culture under conditions of drying temperature when the water content reaches 10% can be calculated. The value of water activity when it reaches the water content of 10% at the drying temperature of 40 °C for the AC is y5 = 0.519 and for the CC is y6 = 0.469, while at the drying temperature 50 °C for the air AC is y7 = 0.727 and the CC is y8 = 0.809.

The T-test dependent results showed that the water activity value of the AC at the two drying temperatures was not significantly different (p value 0.225), as well as the value of water activity CC at both drying temperatures (p value 0.326). This showed that the drying time of AC and CC dried culture at temperature of 40 °C and 50 °C does not affect the characteristics of water activity dried culture. The ANOVA results showed that the drying time at 40 °C significantly affected the AC and CC water activity during drying process (p value 0.000). Likewise the drying time at 50 °C significantly affected the water activity value of AC and CC dried culture at each drying time (p value 0.000). Drying process of AC and CC dried culture at temperatures of 40 °C and 50 °C result the aw value significantly different (p value 0.000). This showed that the drying process of AC and CC dried culture at both drying temperatures resulted in a significant decrease in aw value at each drying time in accordance with [18] who stated that drying affects the decrease in moisture content of the material (A_w). In general, yeast molds can live at certain minimum A_w values. *Aspergillus* lives at a A_w value minimum of 0.98, *Rhizopus* 0.93, and *Penicillium* 0.99, where yeast can generally live at around 0.88-0.94 [19]. The results showed that yeast molds used were still able to live in aw conditions. This is supported by the high viability value that is still high.

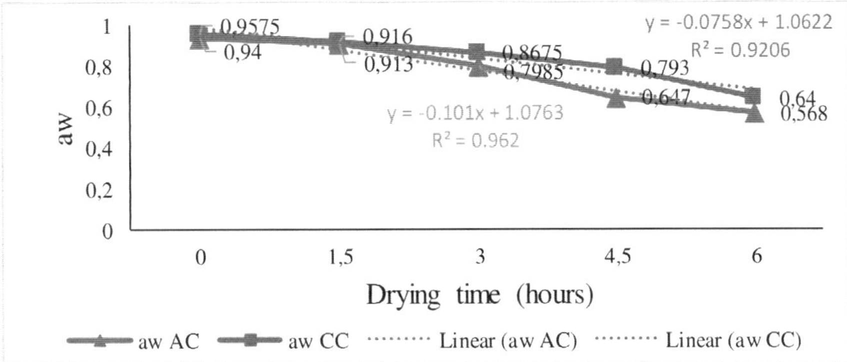

Figure 7. Changes in A_w value during drying time at temperature 50°C

3.4 Influence of drying process on the pH value

The pH value is the acidity degree used to express the acidity or alkalinity possessed by a substance, solution, or object. The initial pH value of AC dried culture ranges from 4.29 - 4.51 and the CC dried culture ranges from 4.00 - 4.22. During the drying process the pH of the AC dried culture decreased with the longer drying time, but the pH value increase in CC dried culture (Table 2).

The T-test dependent showed that the drying temperature of 40 °C and 50 °C had a significant effect on the pH value of the AC and CC dried culture (p value <0.05). Whereas the ANOVA test results in which the drying time at 40 °C significantly affected the pH value of the AC, but did not affect the pH of the CC dried culture. On drying 50 °C, the drying time affects the pH value of the AC and CC dried culture significantly.

Table 2. The pH value of AC and CC dried culture over each temperature and drying time.

Time (h)	Temperature 40 °C		Temperature 50 °C	
	AC	CC	AC	CC
0	4.51[ab]	4.00	4.29[a]	4.22[a]
1.5	4.37[c]	4.36	4.38[c]	4.32[a]
3	4.35[ab]	4.32	4.36[bc]	4.56[b]
4.5	4.32[bc]	4.24	4.32[ab]	4.73[c]
6	4.00[a]	4.20	4.35[b]	4.65[c]

Samples means with different superscripts in the same column are significantly different ($p < 0.05$) by Duncan's multiple range test.

The decreasing of pH of dried culture during drying time is related to the viability of microorganisms in dried culture. In general, yeast is more resistant to heat. *Candida krusei* can live at a temperature of 8-47 °C, while molds can grow optimally at a temperature range of 25-37 °C [16]. In addition, yeast is generally tolerant of acid and can grow at pH 4.0 - 4.5, with growth temperatures 0 °C - 50 °C, with optimum temperatures of 20 °C-30 °C [16]. The pH value of dried culture is in accordance with the pH of the dried culture when it reaches the water content of 10%. Low pH values can inhibit contamination of decomposing microorganisms and pathogenic microorganisms. Low pH values also affect the aroma of the final product [20]. The longer time of fermentation caused a decrease in pH value in fermented sweet potato flour. This is because in the fermentation process there is a metabolism of the activity of microorganisms that produce organic acids [21]. The pH value of the AC and CC was around the pH of 4.00 which is in the pH range for the growth of microorganisms. The pH value was produced because it was fermented at the same time, 5 days.

4. Conclusion

Drying of AC and CC dried culture could be done using a tray dryer. Drying temperature and time influenced the quality of AC and CC dried culture, where temperatures (40 °C and 50 °C) and drying times (0, 1.5, 3, 4.5, and 6 hours) influenced the water content, microorganism viability, water activity and pH value of the AC and CC dried culture. The higher the temperature and the longer the drying time caused the moisture content, the viability of microorganisms, the value of aw and pH to decrease. The length of drying time at a drying temperature of 40 °C caused the moisture content and water content of the AC and CC dried culture to decrease significantly, while viability and pH were not significant. The length of drying time at a drying temperature of 50 °C caused the moisture content, viability, aw and pH of the AC and CC to decrease significantly. Drying is expected to reach water content of 10% (bb) or 11.1% (bk). To reach 10% moisture content, drying at 40 °C takes 4.46 hours for AC and 4.22 hours for CC, while at drying temperature 50 °C the time needed for the AC is 3.46 hours and for CC is 3.33 hours. Drying could be done at a temperature of 50 °C without changing the specifications of the dried culture.

5. References

[1] Farasara R, Hariyadi P, Fardiaz D and Dewanti-Hariyadi D 2014 Pasting Properties of White Corn Flours of Anoman 1 and Pulut Harapan Varieties as Affected by Fementation Process *Food and Nutrition Sciences* **5** pp 2038-47

[2] Rahmawati R, Maulani R R and Saputra D 2018 Chemical properties, particle shape, and size of fermented local white corn flour of anoman fs variety *Jurnal Teknologi* **80 5** pp 155-61

[3] Rahmawati R, Maulani R R, and Saputra D 2017 Karakteristik Ragi Kapang Khamir Indigenus untuk pembuatan tepung jagung putih lokal fermentasi Prosiding Seminar Nasional PATPI: Peran ahli teknologi pangan dalam mewujudkan ketahanan pangan nasional *Bandar Lampung* Indonesia.

[4] Novelina 2005 Kajian Pengeringan Kemoreaksi dengan Kalsium Oksida Serta Dampaknya terhadap Stres dan Kerusakan Kultur Saccharomyces cerevisiae [disertasi]. *Bogor* (ID): Institut Pertanian Bogor.

[5] Mawasti T 2017 Ketahanan bakteri asam laktat terenkapsulasi selama pengeringan laru tempe campuran [Skripsi]. *Bogor* (ID): Institut Pertanian Bogor

[6] Sari D A , Hakim A, and Sukanta 2017 Pengeringan terasi lokal Karawang: sinar matahari- tray dryer. *Jurnal Sains dan Teknologi* **6 2** pp 311 – 20

[7] Rahmawati, Dewanti-Hariyadi R, Hariyadi P, Fardiaz D and Richana N 2013 Isolation and Identification of Microorganisms during Spontaneous Fermentation of Maize *Jurnal Teknologi dan Industri Pangan* **24** pp 38-44

[8] [BAM] Bacteriological Analytical Manual 2001 Gram Stain US FDA Center for Food Safety and Applied Nutrition

[9] [AOAC] Association of Official Analytical Chemist. 2006. Official Methods of Analysis. Washington DC: Association of Official Analytical Chemist.

[10] Barbosa-Canovas G V, Fontana A J, Schmidt S J, and Labuza T P 2007 Water Activity in Foods: Fundamental and Aplications. IFT Press, Blackwell Publishing.

[11] Alsmairat N, Al-Qudah T, El-Assi N, Mehyar G, Gammoh I, Othman Y A, Salah-Eddin A, Al-Antary T M 2019 Effect of Drying Process on Physical and Chemical Properties of 'MEDJOOL' Date Palm Fruits. *Fresenius Environmental Bulletin* **28 2A** pp 1552-59

[12] Hariyadi P 2014 Prinsip-prinsip Proses Panas untuk Industri Pangan Dian Rakyat.

[13] Kusnandar F 2010 Kimia Pangan Komponen Makro Jakarta (ID): Dian Rakyat.

[14] Meng L, Sun P, Tang H, Li L, Draeger S, Schulz B, Krohn K, Hussain H, Zhang W, Yi Y 2011 Endophytic fungus Penicillium chrysogenum, a new source of hypocrellins *Biochem Syst Ecol* **39** pp 163–5

[15] Van den Berg M A 2010 Functional characterisation of penicillin production strains *Fungal Biology reviews* **24** pp 73-8

[16] Pitt J I, Hocking A D 2009 Fungi and Food Spoilage 3rd Edition Springer

[17] Sunatmo T I 2009 Mikrobiologi Esensial Jakarta (ID): Ardy Agency

[18] Muchtadi T R and Ayustaningwarno F 2010 Teknologi Proses Pengolahan Pangan. Jakarta (ID): Alfabeta.

[19] Muchtadi T R, Sugiyono 2013 Prinsip Proses dan Teknologi Pangan. Bandung (ID): Alfabeta.

[20] Corsetti A and Settanni L 2007 Lactobacilli in sourdough fermentation *Review Food Research International* **40** pp 539-58

[21] Anggraeni Y P and Yuwono S S 2014 Pengaruh fermentasi alami pada chips ubi jalar (Ipomoea batatas) terhadap sifat fisik tepung ubi jalar terfermentasi *Jurnal Pangan dan Agroindustri* **2** **2** pp 59-69

6. Acknowledgements

The Authors would like to acknowledge to the Indonesian Ministry of Research and Higher Education – Directorate of Research and Community Empowerment for the grant research No. 107/SP2H/LT/DRPM/IV/2018.

Release pattern of ammonium, nitrate, and potassium from Slow-Release Fertilizer (SRF) in the Soil

F Rakhmad*, Suwardi, and T S Dyah

Department of Soil Science and Land Resources, Faculty of Agriculture, IPB University IPB Dramaga Campus, Bogor, Indonesia

*Corresponding author: fadillahrakhmad@gmail.com

Abstract

The low efficiency of nitrogen and potassium fertilization is one of the pressing problems in tropical agriculture. This is because the humid tropical climate, like what Indonesia has, which is characterized by high rainfall, causes a high rate of leaching of nitrogen and potassium in the soil. To improve fertilization efficiency, one practical option is to find a method of slowing down the release of available nutrients in the soil. Thus, the purpose of this study was to evaluate the pattern and rate of the release of nitrogen (in the form of ammonium and nitrate) and potassium from slow- release fertilizer (SRF). This study used three types of fertilizers, namely: SRF (13-8-10), SRF (19-12-15) and Mutiara granule fertilizer (16-16-16). Treatment dosages used corresponded to 600 kg/ha and 1200 kg/ha, with three replications, and observed over a 10-week incubation period. The results showed that ammonium release from SRF fertilizer in the first week that ranged 28.94% - 41.45%, then decreased progressively with incubation time, while that of nitrate ranged 5.77 - 7.21%, but increased over time. On the other hand, the potassium from SRF fertilizer was released by rates of 9.56% -24.58% during the first week, increasing with incubation time. The same pattern of nutrient release was observed in the case of Mutiara fertilizer, however at much greater rates compared to SRF, *viz.*: for ammonium (44.56 - 46.91%, at decreasing rate afterwards with incubation time), nitrate (26.33 - 30.10%, at increasing rates afterwards), as well as for potassium (15.39 - 15.75%, at increasing rates afterwards). In summary, the SRF (13-8-10), applied at a dosage corresponding to 1200 kg/ha yielded the slowest rate of nutrient release in the soil.

1. Introduction

Nitrogen is one of the most limiting factors in crop production, be it in the tropics or in the temperate regions. It is needed in relatively large quantities by plants, but its presence in the soil is always limited by its source and availability in nature. Nitrogen in the air, in the form of N_2 gas, occupies 78% of the volume of the atmosphere. It can be absorbed by plants in the form of ammonium or nitrate. However, nitrogen is not abundant in acidic soils [2], and, in raw form, it cannot be used by plants. Nitrogen gas must first be converted into a form of ammonium or nitrate, through the process of ammonification and nitrification, respectively, so that it can be used by plants. In fertilization, the nutrient N is usually given in the form of urea or ZA.

Similarly, potassium is a macro-nutrient that also plays an important role in plant growth and production. According to [5], potassium is the mineral nutrient that is most needed by plants after nitrogen. The total potassium present in the soil varies with the type of host rock and geographical location, but, in general, it is estimated to be between 0.5 and 2.5% [1]. However, around 90 - 98% of total potassium in the soil is in a form that is not available, meaning, it cannot be absorbed by plants [8]. Low absorption by plants of available potassium in the soil impacts on their growth, resulting in low-quality crop yields [4]. The amount of potassium absorbed by plants varies from 50 to 200 kg / ha, depending on the plant species and the desired level of production.

On the other hand, zeolites are microporous minerals with high cation exchange capacity (CEC), hence, they are very useful as adsorbers, binders, and cation exchangers. Its high CEC allows greater number of lattice exchanges, so that more ions can be adsorbed by the lattice. The adsorption of soil nutrient ions in the zeolite cavity / lattice is only temporary, and the nutrients are made available to plants only when needed [12]. Given their high exchange properties, zeolites can temporarily bind and store nutrients in the soil, and then release them back into the soil when plants need them. This is particularly true for N, because of the high zeolite adsorption selectivity to ammonium ions. Further, by adsorbing ammonium, zeolites inhibit the conversion of ammonium into nitrate, so that the loss of N in the form of nitrates, that is easily washed out by rainwater, is regulated. If the N level in the soil solution is decreased, the N adsorbed by zeolite would be released slowly for plant use [11].

In this context, the use of fertilizers that contain zeolite and humic acid is expected to increase the efficiency of nitrogen and potassium fertilization. The fertilizer can be prepared in pill form with a parabolic granulation device.

2. Materials and Methods

2.1 Preparation

This research was conducted from November 2017 to March 2018, at the Laboratory of the Department of Soil Science and Land Resources, Faculty of Agriculture, Bogor Agricultural University (IPB). Collected soil samples were incubated and stored at the Physical Resources Development Laboratory, while soil chemical analysis was carried out at the Soil Chemical and Fertility Laboratory. Dramaga Inceptisol soil, which is characterized by a rather thick solum (1-2 meters); black, or gray to dark brown; texture of dust, dusty clay or clay; crumb soil structure; loose consistency; pH 5.00 - 7.00; and high organic matter content of 10% -30% (Soepardi 1983), was used. Based on the analysis criteria for soil chemical properties (PPT 1983), the Dramaga Inceptisol soil has a moderate total N-content, but its K-available content is low, according to the initial analysis.

The tools used in the field included hoe, sack and spill to dry the soil, and a sieve to sift the soil; while the tools used for analysis in the laboratory were pH meter, EC meter, flamephotometer, soil oven, weigh bottles, weights, shake bottles, shaking machine, 250-ml Erlenmeyer flask, plastic tube, filter paper, test tube, 100-ml centrifuge tube, glass stirrer, pipette, boiling flask, analytical balance, 50-ml measuring flask, 100-ml measuring flask, distillation apparatus, distillation flask , and digital burette.

2.2 Incubation

Two types of slow-release fertilizers (SRF), with different NPK compositions (13-8-10 and 19-12-15), were tried in this study. In addition, Mutiara NPK (16-16-16) was used as a comparison fertilizer. Soil composites were collected at a depth of 0-20 cm, and then dried. For analysis of the chemical properties in the laboratory, the soil sample was ground and sieved on a 2-mm sieve. The measurement of the rate of fertilizer release was carried out by the incubation method, in an open space in the laboratory. Air-dried soil, weighing 129.77 g, or equivalent to 100 g absolute dry weight (ADW) with 11.96% ground water content, was inserted into a cylindrical tube-shaped plastic container (the incubation jar shown in Figure 1), with a diameter of 6.00 cm and a height of 6.70 cm. The SRF and Mutiara NPK fertilizer were weighed and then applied into the soil. Water was added to the fertilizers until field capacity (48.29%) moisture content was reached. Finally, the incubation jar was covered with a protective polyethylene plastic sheet. For this study, two fertilizer dosage levels, corresponding to 600kg/ha (A) and 1200 kg/ha (B), were tried, and each treatment was replicated three times. The material composition of the two SRF fertilizers are summarized in Tables 1 and 2 below.

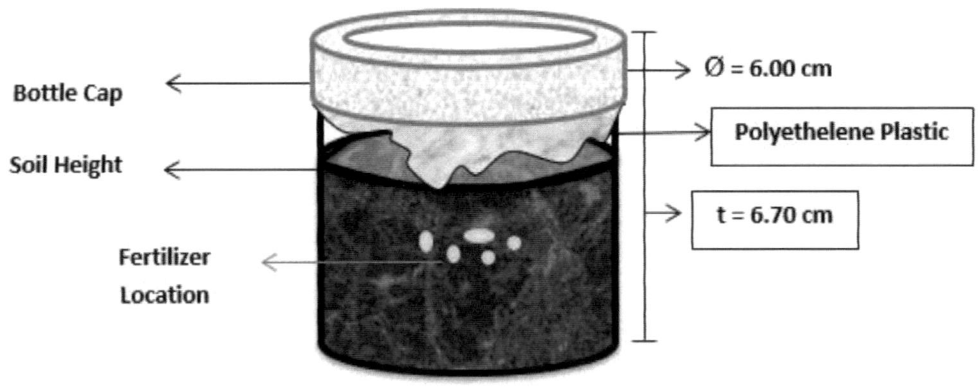

Figure 1. Illustration of incubation jar containing soil and fertilizer for incubation over 10 weeks

Incubation was carried out at room temperature in the incubator for 10 weeks. Observations and measurements on ammonium, nitrate, nitrogen, potassium, pH, EC and soil water content were undertaken weekly. The study data were used in the subsequent statistical and regression analyses.

Table 1. SRF 13-8-10 nutrient content

Materials	Amount (mg)	% Nutrient
Urea	119.45	13.00
DAP	173.91	8.00
KCl	166.67	10.00
ZA	208.33	5.00
$MgSO_4$	25.00	2.00
ZnO	4.44	0.40
CuO	3.33	0.30
$MnSO_4$	3.75	0.30
$CaCO_3$	72.73	4.00
Zeolite 11	222.38	
Total	1000.00	

Table 2. SRF 19-12-15 nutrient content

Materials	Amount (mg)	% Nutrient
Urea	234.88	19.00
DAP	260.87	12.00
KCl	250.00	15.00
ZA	166.67	4.00
ZnO	4.44	0.40
CuO	3.33	0.30
$MnSO_4$	3.75	0.30
Zeolite 12	76.06	
Total	1000.00	

2.3 Soil Analysis

At the laboratory, a soil sample was taken for analysis from each incubation jar for each treatment (including control). The remaining fertilizer was removed from the incubation jar, and the soil in the middle of the incubation jar, that was not directly in touch with the fertilizer, was stirred and taken for analysis. The amount of soil taken was 10 grams for N-available analysis, 10 grams for pH and EC measurement, and 1.5 grams for K-available analysis. Using the same soil sample, pH and EC was determined using a calibrated pH meter and EC meter. Available N (in the form of ammonium and nitrate) was determined by the Distillation method. In this method, 0.1N HCl, 1N KCl, 40% NaOH, 4% boric acid, Conway indicator, devarda alloy, and distilled water were used.

On the other hand, K-available was determined by the Bray-1 method with a flamephotometer measuring instrument, at a wavelength of 660 micrometers. The analysis of N-available K-available was calibrated by reducing the nutrient content in the soil without fertilizer every week (control).

2.4 Data Analysis and Interpretation
The rate of nutrient release from fertilizer can be determined by using a number of kinetic models, including zero-order kinetics, first-order kinetics, and Elovich models. The zero-order kinetics model is expressed using the following equations:

$$Co - Ct = ko \cdot t \quad (1)$$

the integration form of Equation (1) is:

$$Ct = Co - ko \cdot t \quad (2)$$

where Ct is the nutrient concentration in the soil at time t, Co is the initial concentration of fertilizer applied, ko is the zero-order rate with a unit of concentration per unit time. By plotting Ct against t, zero-order kinetics is obtained.

The first-order kinetics model is expressed by equations

$$\frac{dC}{dt} = -k_1 \cdot C \quad (3)$$

The integration form of Equation (3) is

$$\ln Ct = \ln Co - k_1 \cdot t \quad (4)$$

where Ct is the concentration of nutrients in the soil at time t, kt is a first-order rate constant in units per unit time. By plotting ln (Ct) against t, a first-order kinetics is obtained.

The Elovich model is expressed by equations (Chien and Clayton, 1980)

$$\frac{dCt}{dt} = a \cdot \exp(-bq1) \quad (5)$$

The integration form of Equation (5) is

$$Ct = \frac{1}{b}(\ln(ab) + \frac{1}{b}\ln t \quad (6)$$

where Ct is the amount of nutrients in the soil at time t, where a and b are constants. By plotting Ct with ln t, we get constants a and b.

For each treatment, average values for all observations were obtained, corrected with the respective control, and range values (lowest to highest percentages for each week) were generated. The sets of average values for each Treatment over the 10-week incubation period were used to construct the prediction graphs, using the Microsoft Excel 2013 application. The resulting graphs (Figures 2, 3, and 4) display the pattern in the rates (percentage) of cumulative values of released ammonium, nitrate, and potassium, respectively. The percentage values for each nutrient were obtained from the ppm values (mg / kg), which were first converted to mg /100g, and then multiplied by the maximum level of fertilizer in mg.

3. Results and Discussions

3.1 Ammonium release pattern

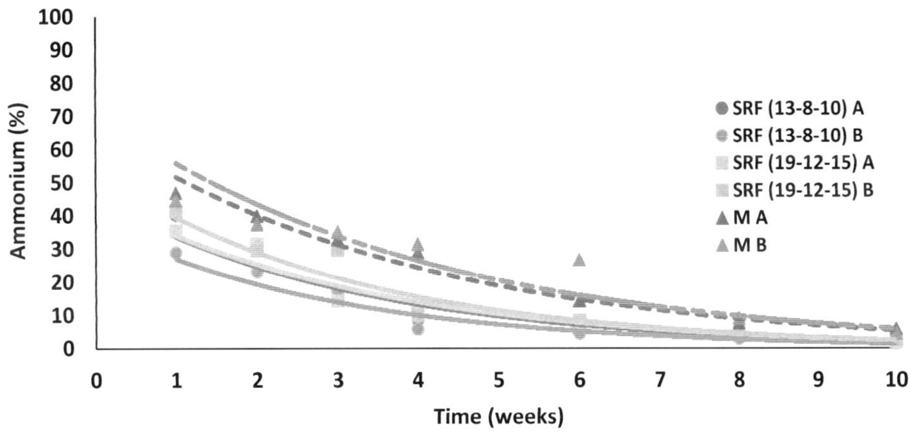

Figure 2. Ammonium release pattern for SRF and Non-SRF fertilizers

Figure 2 above shows the pattern of the release of ammonium from the soil samples applied with fertilizers, namely: SRF (13-8-10), SRF (19-12-15), and Mutiara fertilizer (16-16-16), each of which was applied in two treatments dosages - corresponding to 600 kg/ha (A) and 1200 kg/ha (B). The lowest ammonium release percentage from the first week to the 10th week was exhibited by Treatment SRF (13-8-10) B, with release rate 28.94% during the first week, and 1.54% in the 10th week. Following closely was Treatment SRF (13-8-10) A, with release rate 40.38% during the first week, and 2.24% in the 10th week. A similar trend was shown by Treatment SRF (19-12-15), both Treatment B and Treatment A, in that order. Treatment SRF (19-12-15) B, had a release rate 41.45% during the first week, and 2.85% in the 10th week, followed by Treatment SRF (19-12-15) A, with release rate 35.53% during the first week, and 1.97% in the 10th week. These results are consistent with the findings of [3], who found that the release rate of ammonium decreased progressively every incubation week to near zero with the use of SRF fertilizer mixed with urea, zeolites, and humic acid. The same trend of ammonium release was observed in the case of Mutiara fertilizer, however at much greater rates compared to SRF, *viz.*: for ammonium (44.56 - 46.91% during the 1st week, and 4.47% - 5.81% in the 10th week). The above data implies that SRF fertilizer can slow down the release of ammonium, as compared to Mutiara NPK (16-16-16) fertilizer. The ability of the SRF fertilizer to regulate the release of ammonium is likely brought about by their zeolite content (as shown in Tables 1 and 2). Zeolite has the natural ability to absorb $NH4^{+}$ ions; the exchange grating in the zeolite binds the hydrolyzed $NH4^{+}$, so the ion can readily be adsorbed by the zeolite. It is detached from the zeolite only when the concentration equilibrium in the soil substance is lower than the concentration equilibrium in the fertilizer [10].

3.2 Nitrate release pattern

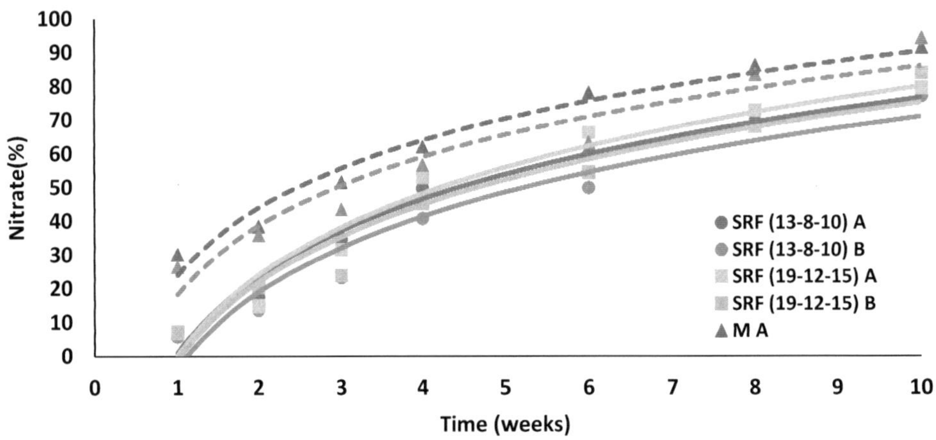

Figure 3. Nitrate release pattern for SRF and Non-SRF fertilizers

Figure 3 shows an increase in the rate of nitrate release from the first to the 10th week, which was the opposite pattern in the ammonium release. Nitrate release in the first week ranged from 5-30% and increased to as much as 100% in the 10th week. The lowest nitrate release was exhibited by the SRF (13-8-10) B treatment, at 5.92% in the first week and reaching 77.05% in the 10th week. A similar trend was shown by Treatment SRF (19-12-15), both Treatment B and Treatment A, in that order. Treatment SRF (19-12-15) B, had a nitrate release rate 6.80% during the first week, and 83.84% in the 10th week, followed by Treatment SRF (19-12-15) A, with nitrate release rate 7.21% during the first week, and 79.49% in the 10th week. In comparison, Mutiara treatment had the highest nitrate release percentage range of 26.33% - 30.10% in the first week and 91.46 - 94.17% in the 10th week.

3.3 Potassium release pattern

In Figure 4 below, the release rate of potassium in the SRF fertilizer shows an increasing trend every week, reaching a range of 42.08% - 57.75% in the 10th week. This is likely because the zeolite that was added to SRF bound the very small K^+, and since zeolite minerals possess a high selectivity in absorbing K^+ cations [6], then K^+ is adsorbed by the zeolite. The pattern of potassium release in the Mutiara fertilizer likewise showed a weekly increase, but Mutiara fertilizer released potassium only at a rate of 30.08% to 39.19% in the 10th week. This is due to the fact that Mutiara fertilizers do not contain zeolites, unlike the SRF fertilizers. As stated earlier, SRF fertilizers have high-CEC zeolites, and high CEC can affect the availability of potassium in the soil.

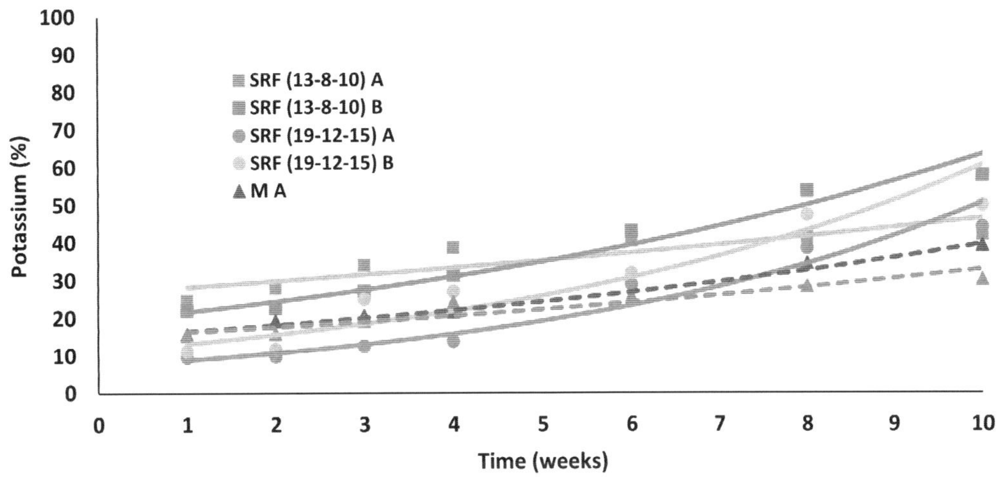

Figure 4. Potassium release pattern for SRF and Non-SRF fertilizers

Table 3 below shows the linearized (1st-order) regression equations for each fertilizer Treatment unit, all of which produced highly reliable prediction estimates, based on the high values of the linear regression coefficients (r^2). It can be observed that the coefficients of the variable x, denoting the slope of each graph, also represent the coefficients (k) that express the respective rates of release of ammonium, nitrate, and potassium from the soil for the different treatment units in this study. As demonstrated by the prediction equations, the coefficient value (0.015) for SRF (13-8-10) B is the lowest among the ammonium regression equations, implying that this Treatment (SRF (13-8-10) B) produces the slowest rate of ammonium release. This can be explained by the fact that this SRF contained a much higher amount of zeolite. A similar result can be seen in nitrate release rate, with the lowest value of coefficient (0.0693), again for the same reason. However, a different result can be observed in the case of potassium release, in which the lowest coefficient value (-0.0327) occurs also in Treatment SRF (13-8-10) B but with a negative sign, meaning the release of potassium is increasing but at a decreasing rate over the 10-week observation period.

Table 3. Predicted patterns in the rate of release of ammonium, nitrate, and potassium

Treatment	Ammonium Equations	Ammonium r^2	Nitrate Equations	Nitrate r^2	Potassium Equations	Potassium r^2
SRF (13-8-10) A	y = 0.0209x + 1.711	0.7121	y = -0.0697x + 1.9224	0.9886	y = -0.0124x + 1.6454	0.804
SRF (13-8-10) B	y = 0.015x + 2.0615	0.7704	y = -0.0693x + 2.263	0.9867	y = -0.0327x + 2.0279	0.9774
SRF (19-12-15) A	y = 0.0186x + 1.8881	0.7481	y = -0.0762x + 2.0956	0.973	y = -0.0257x + 1.9619	0.9573

SRF (19-12-15) B	y = 0.023x + 2.1533	0.7787	y = -0.0797x + 2.4399	0.9618	y = -0.0293x + 2.2389	0.9582
M 1	y = 0.0281x + 1.7109	0.9267	y = -0.1048x + 1.9651	0.997	y = -0.0154x + 1.9289	0.9531
M 2	y = 0.0259x + 2.0136	0.9712	y = -0.1151x + 2.3589	0.9215	y = -0.0098x + 2.2179	0.933

4. Conclusion

Slow-release fertilizer (SRF) can significantly restrict the release of nitrogen (in the form of ammonium and nitrate) and potassium in the soil, that would otherwise be washed away by rainwater, hence, made unavailable to plants and simply wasted away. This is mainly due to the presence of high-CEC zeolite that can bind, adsorb, and keep the mineral nutrients over a period of time, releasing the adsorbed nutrients only when needed by plant roots. This natural nutrient release mechanism in the soil is governed by the balance in the nutrient concentration equilibrium in the zeolite with the nutrient concentration equilibrium in the soil. The rate of ammonium release in the SRF-fertilized soil goes down progressively over time, while that of nitrate increases over time. Potassium release from SRF-fertilized soil likewise goes up, but at a decreasing trend over time. In summary, the SRF (13-8-10) fertilizer, applied at a dosage corresponding to 1200 kg/ha, yields the best results in terms of regulating nutrient release rates in the soil.

5. References

[1] Havlin JL, Tisdale SL, Beaton JD, Nelson WL. 2005. Soil Fertility and Fertilizer: An Introduction to Nutrient Management 7th ed. New Jersey (US): Pearson Education.
[2] Hardjowigeno S. 2003. Ilmu Tanah. Jakarta: Akademika Pressindo.
[3] Hidayatussitah N. 2018. Pola Pelepasan Amonium, Nitrat, dan Kalium dari *Slow Release Fertilizer* (SRF) pada Epiaquept [skripsi]. Bogor (ID):Institut Pertanian Bogor.
[4] Jones JB, Wolf B, Mills HA. 1991. Plant Analysis Handbook: A Practical Sampling, Preparation, Analysis, and Interpretation Guide. Georgia (US): Micro-Macro Publishing.
[5] Leiwakabessy FM. 1988. Diktat Kuliah Kesuburan Tanah. Bogor (ID) : Institut Pertanian Bogor.
[6] Nurwiyanto. 1991. Pengaruh perlakuan dosis pupuk N (urea) dan pupuk K (KCl) serta waktu inkubasi terhadap jerapan kation amonium (NH_4^+) dan kation (K^+) oleh mineral zeolit [skripsi]. Bogor (ID):Institut Pertanian Bogor.
[7] Pusat Penelitian Tanah (PPT). 1983. Survey kapabilitas tanah dalam klasifikasi kesesuaian lahan. *Term of Reference. Type A. No. 59/1983*. Bogor (ID): P3MT. PPT.
[8] Rehm G, Schmitt M. 1997. Potassium for Crop Production. Minnesota Extension Service, Minnesota (US): University of Minnesota.

[9] Soepardi G. 1983. Sifat dan Ciri Tanah. Bogor (ID): Fakultas Pertanian Institut Pertanian Bogor.

[10] Suwardi, Darmawan. 2009. Peningkatan efisiensi pupuk nitrogen melalui rekayasa kelat urea-zeolit-asam humat. Prosiding seminar hasil-hasil penelitian. Bogor (ID): Institut Pertanian Bogor. hlm 516-524.

[11] Suwardi. 2002. Pemanfaatan Zeolit untuk Meningkatkan Produksi Tanaman Pangan, Peternakan, dan Perikanan. Makalah Seminar Teknologi Aplikasi Pertanian. Bogor (ID): IPB University.

[12] Suwardi. 1991. The Mineralogical and Chemical Properties of Natural Zeolite and Their Application Effect for Soil Amandement. A Thesis for the Degree of Master. Laboratory of Soil Science. Departement of Agriculture Chemistry, Tokyo University of Agriculture.

Factors affecting participation of women farmers in supporting family food security: case study in Pandeglang regency, Indonesia

R Megasari [1]*, Y Budiawati[1], A Mulyaningsih

Department of Agribisnis, Faculty of Agriculture, Universitas Sultan Ageng Tirtayasa, Jl. Raya Jakarta Km 4 Pakupatan Serang Banten Indonesia

*Corresponding author: ratnatanjung@ymail.com

Abstract. The potential of women farmers in achieving household food security must be supported by adequate female farmer capacity, optimal participation of women farmers at every stage, and eliminating and finding solutions to a number of obstacles faced by women in order to strengthen their capacity and increase their participation in achieving home food security stairs. The purpose of this study is to analyze the factors that influence the participation of female farmers in supporting family food security. The population in this study were female farmers in two sub-districts in Pandeglang District, Banten Province, namely Munjul and Cibaliung Districts. The number of samples is 125 female farmers. The research activities were carried out for three months (April to June 2019). Data analysis with. The inferencing statistics used are SEM using LISREL 8.73 (Linear Structural Relationships). The results showed that the factors that influence the participation of female farmers in supporting family food security are access to resources, physical and socio-economic support and extension support. For this reason, government support is needed, by giving attention to women farmers as the main actors in achieving food security in rural households.

Keywords: participation, women farmers, family food security

1. Introduction

Socio-culturally, peasant women in Pandeglang Regency are responsible for achieving household food security. Therefore, the success of achieving household food security in Pandeglang Regency cannot neglect the role of women farmers. The potential of peasant women in achieving household food security must be supported by adequate capacity of peasant women, optimal participation of peasant women at each stage, and eliminating and finding solutions to a number of obstacles faced by women in order to sqtrengthen their abilities and increase their participation in achieving home food security stairs.

Farmer women play an important role in household food security, both in the components of food availability, food access and food utilization. Farmer women contribute as workers in the agricultural sector. In 2010, in Latin America and the Caribbean there were approximately 20 percent of women in the agriculture sector, in Near East and North Africa and Sub-Saharan Africa nearly 50 percent, in South Asia around 35 percent, and in East and Southeast Asia nearly 50 percent. percent. Farmer women are also responsible for planning, processing, preparing, and serving food for the family [1], [9].

Farmer women are exclusively responsible for family nutrition, as producers and providers of food for families [6]. Farmer women also buy food using the income they get from working [7], [14], [4], [5], [9]. The International Center for Research on Women / ICRW states that farm women are also an integral part of efforts to reduce hunger and malnutrition because it is women who are responsible for ensuring the availability of balanced, nutritious food for their families.

Farmer women also carry out household strategies to meet food shortages (coping ability indicator) [3], [1], [7], [17], [2], [5]. The above description reinforces that the participation of women farmers in achieving household food security is not only related to increasing food production and food providers but also with increasing food and nutrition of household members. Alignment must occur between increased food production and income with improved food and nutritional status of households [2], [9]. This supports the opinion that household food security is the ability of households to meet the adequacy of food for their members to be able to live healthy lives and be able to carry out daily activities that are reflected in the consumption of nutrients (energy and protein) that meet the adequacy norms [12], [8] According to [10] efforts to achieve household food security can be pursued through increasing food and nutrition knowledge to maintain harmony between food availability and the quality of community food consumption.

To increase one's participation requires certain capability requirements in their implementation and analyzing social, economic and cultural factors (personal characteristics of farm women, household socioeconomic households, socio-cultural environmental support, access to resources, and support for counseling) that affect ability and participation and find a way out. Supporting the implementation of food security counseling for farm women is needed to succeed household food security, because extension workers have a very strategic role [13].

The instructor should not only understand the production aspects but also need to have adequate knowledge, attitudes and skills in terms of consumption patterns, food distribution in order to provide correct information to farmer women in relation to food consumption and distribution patterns. The achievement of household food security gives meaning to the fulfillment of the rights of all individuals in the household to quality food (quality, balanced nutrition) at all times and can live a healthy and active life, as mandated by [15] Law of the Republic of Indonesia Number 18 2012 concerning Food. Household food security will determine the achievement of national and even global food security. Food security and nutrition are the common thread connecting various elements of sustainable future development. Food insecurity can have a long-term negative impact on the growth prospects of the whole community. Therefore, the government must continue to pay attention to farm women as the main actors in achieving household food security in rural areas. The purpose of this study is to analyze the factors that influence the participation of women farmers in supporting family food security.

2. Materials and Methods

This research is quantitative research supported by qualitative data. The research design is descriptive which takes a sample in one population using a questionnaire as a primary data collection tool [16] and aims to make a description of the situation or event from sample to population so that conclusions can be drawn. The research activities will be carried out for 3 months from April to June in 2019 in two sub-districts in Pandeglang District namely Munjul District and Cibaliung District. The sample in this study was 125 peasant women from two sub-districts.

The research uses survey methods and data analysis is done by descriptive statistics and inferential statistics. Inference statistics are used to estimate or estimate the population (generalization) in order to see the extent to which independent variables influence the dependent variable and to see the compatibility of the research model designed with the actual model. Statistical inference used is SEM by using LISREL 8.73 (Linear Structural Relationships). Determination of the sample in the study was carried out through two stages, namely the first stage was to determine the number of research samples. The size of the research sample follows the rules in the analysis of Structural Equation Modeling (SEM) with a maximum likelihood estimation method requiring a minimum sample of 100-150 respondents or as much as five times the indicators in the model [11]. In this study there were 25 indicators and the number of samples taken was 5 x 25 = 125 people. The number of samples in this study met the requirements for SEM testing. The second stage is to determine respondents in each selected village. The instrument used in this study was in the form of a questionnaire containing a list of questions made on the basis of predetermined variables.

Questionnaire in the form of a list of questions and guideline questions in the form of a number of key questions to find out qualitative phenomena from respondents and informants related to the problem and research objectives. The collected data was analyzed using inferential statistics. To give an idea of the factors that influence the participation of female farmers, the variables studied are: farmer characteristics (X1), access to resources (X2), economic and physical support (X3), extension support (X4), participation rates, (Y1) and food security (Y2).

The hypothesis in this study is that the level of participation of female farmers in supporting family food security is influenced by the characteristics of farmers, access to resources, physical and socio-economic environmental support, and extension support. Structural Equation Model and measurement model with the following formula:

(1) Structural Equation Model, is:

$$\eta = B\eta + \tilde{I}\xi + \zeta \quad (1)$$

Description :

η = eta, vektor from *endogenous* variable (Y)
B = beta, a coefficient matrix describing the effect of other endogenous variables
Ĩ = gamma, a coefficient matrix describing the effect from the exogenous variable to the endogenous variable
ξ = xi, a vector of exogelous variables (latent variables X)

ζ = zeta, a vector of residuals or errors in the equation

(2) *Measurement Model*, is:

$$X = \Lambda x\,\zeta + \delta\, Y = \Lambda y \zeta + \varepsilon \quad (2)$$

Description :

X = a vector of measurement of free variables
Λx = lambda X, a matrix of loading X on the latent variable exogenous which is not observed
δ = delta, a vector of related measurement errors with X variables
Λy = lambda Y, a matrix of loading X on the endogenous variable not observed
ε = epsilon, a vector of related measurement errors with Y variables

3. Results and Discussion

The results of hypothesis testing indicate that the factors that influence the level of participation of women farmers in supporting family food security are access to resources (X2), physical and socio-economic environmental support (X3), and extension support (X4). However, the level of participation of female farmers was more dominantly influenced by access to resources by 0.53, when compared to the influence of physical and socio-economic environmental support, and extension support by 0.33 and 0.39, respectively. The positive influence shows that the higher access to resources, physical and socio-economic environmental support, and extension support will further increase the participation of women farmers in supporting family food security.

The estimation of structural model parameters between the tested variables, namely: Y1 = 0.53 * X2 + 0.13 * X4, with R2 = 0.57 meaning that the simultaneous influence of the two variables on the level of participation of women farmers in supporting family food security is 0.57. This means that the diversity of data that can be explained by the model is 57 percent, while the rest (43%) is explained by other variables (which are not yet in the model) and errors.

Figure 1 shows the factors that influence the level of participation of women farmers in supporting family food security. Factors affecting the level of participation of female farmers were analyzed using structural equation modeling (SEM) with the help of the LISREL 8.72 program, which was initially estimated or tested on the parameters of the model (framework of thought). Testing the model using a two-step procedure (two step approach). First, testing the Goodness of fit model. Second, the results of processing for testing the goodness of fit by using alternative testing for the use of SEM indicators, namely the results of processing for testing the goodness of fit by using an alternative testing for the use of SEM indicators, namely the results of several measurement indicators test results in the conclusion that the model meets the criteria for goodness of fit.

According to the criterion size RMSEA (Root Mean Square Error of Approximation) produces a value of 0.070 ≤ 0.080 which means that the resulting model is fit. The use of other good fit criteria namely GFI (Goodness of Fit Index), AGFI (Adjusted Goodness of Fit), CFI (Comparative Fit Index), IFI (Incremental Fit Index), and NFI (Normed Fit Index) produce values ≥ 0.90 which means the resulting model is already good fit (Table 1). The test results of several measurement indicators produce the conclusion that the model meets the criteria for goodness of fit, so that hypothesis testing can be done.

Table 1. The results of testing the goodness of fit model of factors that influence the level of participation of women farmers in supporting family food security.

Goodness-of-Fit	Cutt-off-Value	Result	conclusion
RMSEA	≤ 0.08	0.070	Fit
GFI	0,80≥GFI> 0.90	0.87	Fit
AGFI	0,80≥GFI> 0.90	0.83	Fit
CFI	≥ 0.90	0.97	very Fit
IFI	≥ 0.90	0.97	very Fit
NFI	≥ 0.90	0.94	very Fit

The proposed construct measurement model is fit with the data. This means that the model is able to estimate the covariance matrix of sample data. In other words, the model can be used as a basis for making generalizations about the phenomena under study. In summary based on the estimation of structural model parameters between the tested variables,

it can be seen that access to resources directly influences the level of participation of women farmers in supporting family food security, this is because the input factors and farming technology support the participation of farmers in managing their farming. Physical social economic support affects the participation of women farmers, especially in terms of government policy support, support of farmer leaders, institutional support, infrastructure support, and expert support. In addition, extension support also affects the participation of farmers, especially in supporting family food security. Dal is supported by material that is suitable to the needs of farmers, high intensity of counseling and good extension agent. The participation of women farmers is reflected by the participation variable with indicators of implementation (λ = 0.93) and implementation (λ = 0.80). Farmer participation, especially in the implementation phase, is very much needed for improvement. The more farming is carried out properly and in accordance with the rules, the more weaknesses will be known and can later be corrected in the next planting season.

3.1. Level of participation

The level of participation of women farmers in supporting family food security in both districts is classified as moderate. This can be seen in Figure 1 where it can be seen that farmers from Cibaliung and Munjul sub-districts have a moderate level of participation, however, Munjul sub-district is relatively more participatory than farmers from Cibaliung sub-district.

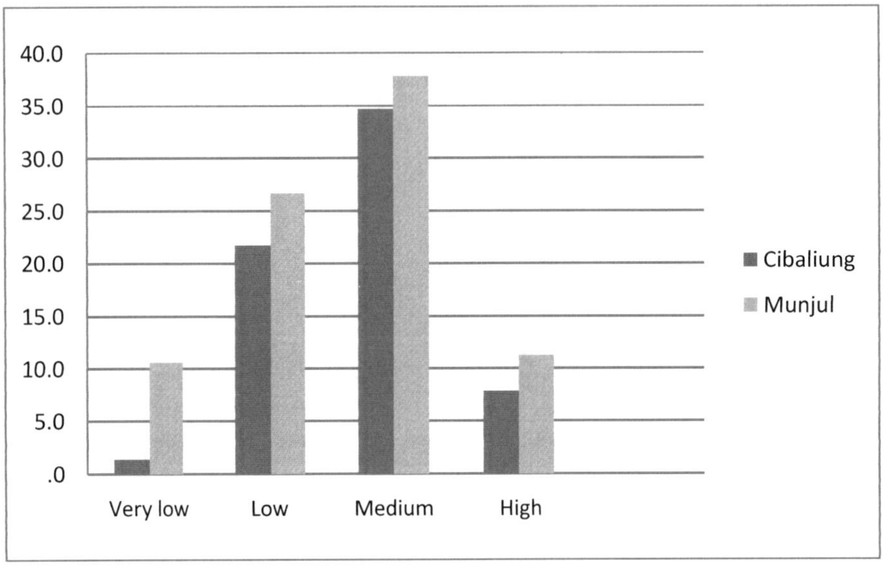

Figure 1. Participation level of women farmers

3.2. Farmers Women's Participation Rate

Based on Figure 1, it can be seen that the element of participation that supports family food security from the aspect of availability and distribution aspects is implementation and benefits. This shows that in general women farmers from two districts conduct farming activities to support the food security of their families so that the availability of food in their homes can be guaranteed. In addition, some of the results of his farming are distributed to be sold and bought other food needs besides rice. The participation of women farmers from the aspect of benefits is to use the results of their farming for family food sufficiency. Thus it can be concluded that the involvement of farm women in farming can support family food security, especially in the aspects of availability and distribution aspects.

4. Conclusion

Factors that influence the participation of women farmers in supporting family food security are access to resources, physical and socio-economic support and extension support. To support family food security, support from the government is needed, especially access to resources, support for physical and socio-economic facilities and extension support.

5. References

[1] Arumsari V, Rini WDE. 2008. Peran perempuan dalam mewujudkan ketahanan pangan pada tingkat rumah tangga di Kabupaten Sleman Daerah Istimewa Yogyakarta. Jurnal Ekonomi Pembangunan 13(1): 71-82.

[2] Baliwati YF, Khomsan A, Dwiriani CM. 2010. Pengantar: Pangan dan Gizi. Depok (ID): Penebar Swadaya. Berg A.1986. Peranan Gizi dalam Pembangunan. Jakarta (ID): Rajawali.

[3] Brown LR, Feldstein H, Haddad L, Pena C, Quisimbing A. 2001. Generating Food Security in the Year 2020: Women as Producers, Gatekeepers, and Sock Absorbes. Di dalam: The Unfinished Agenda: Perspective on Overcoming Hunger, Poverty, dan Environmental. Washington DC (US): International Food Policy Research Institute.

[4] [FAO] The Food and Agricultural Organization. 2011. The Role of Women in Agriculture. ESA Working Paper. No. 11-02.

[5] Hubeis AVS 2002. Tantangan dan Prospek Teknologi Informasi dan Komunikasi dalam Otonomi Daerah. Di dalam: Pambudy R, Adhi AK, editor. *Pemberdayaan Sumber Daya Manusia Menuju Terwujudnya Masyarakat Madani*. Bogor (ID): Pustaka Wirausaha Muda.
[6] Hubeis AVS 2010. Pemberdayaan Perempuan dari Masa ke Masa. Bogor (ID): IPB Press.
[7] Ibnouf FO. 2009. The role of women in providing and improving household food security in Sudan: Implication for reducing hunger and malnutrition. Journal of International Women's Studies 10(4) [Internet].
[8] Jayaputra. 2001. Ketahanan Pangan Rumahtangga Petani di Daerah Kawasan Pertambangan PT Newmont Nusa Tenggara. [Tesis]. Bogor (ID): Program Pascasarjana IPB.
[9] Karl M. 2013. Inseparable: The Crucial role of women in food security revisited. ProQuest Agriculture Journals.
[10] Kusharto CM, Hardinsyah. 2012. Ketahanan dan Kemandirian Pangan. Di dalam: Merevolusi Revolusi Hijau. Pemikiran Guru Besar IPB. Bogor (ID): IPB Press.
[11] Kusnendi. 2008. "*Model-model Persamaan Struktural, Satu dan Multigrup Sampel dengan LISREL*". Bandung: Alfabeta.
[12] Maxwell S, TR Frankenberger. 1992. Household Food Security: Concepts, Indicators, Measurements, A Technical Review. Rome: International Fund for Agricultural Development/United Nation Children's Fund.
[13] Ndraha T. 1990. Pembangunan Masyarakat. Jakarta (ID): Rineka Cipta.
[14] Ogunlela, Mukhtar. 2009. Gender issues in agricultural and rural development in Nigeria; The Role of Women. Humanity and Social Sciences Journal. 4 (1): 19—30. Tersedia pada: http://www.idosi.org/.
[15] [UU] Republik Indonesia Nomor 18 Tahun 2012 tentang Ketahanan Pangan.
[16] Singarimbun M, Effendi S. 2006. *Metode Penelitian Survai*. Jakarta: LP3ES.
[17] World Bank. 2009. Gender in Agriculture Sourcebook. The World Bank, FAO, IFAD. Washington DC (US).

The Characteristics of Shrimp Caught by Trammel net in The Waters of Lontar, Banten

R S Mardiah* and T D Pramesthy

Department of Capture Fisheries, Marine and Fisheries Politechnic of Dumai
St. Wan Amir No. 01, Pangkalan Sesai, Dumai City, Riau Province, Indonesia

*Corresponding author: ratu.sarimardiah2@gmail.com

Abstract. Shrimp is one of the high commodities caught in Lontar waters. The Fishing gear used is trammel net. The type of trammel net based on slackness partition has been discussed in previous research. The aim of this study is to identify shrimp caught by trammel net, analyze the legal size of shrimp and analyze the relationship between length and weight of shrimp. The study was conducted in November and December 2018 in waters of Lontar, Banten. The trammel net used is divided into 3 types. The first trammel net is trammel net control (TK) which is not insulated and the distribution of slackness is only under the webbing. The second trammel net is divided into 2 parts (TN A) and the third is divided into 3 parts (TN B). The number of settings is 32 times. The result showed that there were two species of shrimp caught, namely white shrimp (*Penaeus merguiensis*) and mantis shrimp (*Squilla mantis*). The total catch of white shrimp is 478 shrimp (14 kg) and 235 shrimp of mantis (7 kg). The length of white shrimp caught in TK is 11-18 cm, TN A is 11-20 cm and TN B is 10-20 cm. The shrimp caught in TK have a size between 10-16 cm. In TN A and B, the size of shrimp caught is 10-23 cm. The total of white shrimp which has legal size is 420 shrimp (88%). The number of mantis shrimp has a smaller than white shrimp, which are 49 shrimp (21%) spread on three types of trammel net. The length and weight relationship between two species of shrimp are the same. The relationship is negative allometric.
Key Words: Length, Mantis shrimp, Trammel Net and White Shrimp

1. Introduction

Trammel net is a fishing device designed no capture demersal organism. Example are shrimp (*Penaeus* sp.), common pony fish (*Leiognathus equulus*), croaker fish (*Nibea albiflora*) and groper fish (*Epinephelus* sp.) [9 and 10]. The catch caught in the webbing will be twisted at the bottom and entangles at the top. The organism that is often entangled are fish that have vertical swimming, such as pepetek and gulamah. Many organism are twisted in the the lower of the webbing, especially shrimp. The reason is the passive behaviour of the shrimp, the way of swimming shrimp that jerking the shrimp body back and the shape of the body shrimp is curved. Trammel net fishermen consider the main catch is shrimp, because a lot of caught and have high economic value.

Shrimp caught by trammel net fishermen consist of several species. The species are white and mantis shrimp. The morphometrics of both are very different. The striking difference is carapase. Mantis shrimp have a harder carapase and have claws. The two body lengths are also different. Adult mantis shrimp can grow up to 20 cm. This difference is the basis of research of the Characteristics of the Shrimp Caught by Trammel Net in The Waters of Trammel Net, Banten. The aim of the study was to identify shrimp caught by trammel net, analyze the size of shrimp suitable for capture and analyse the relationship between the length and weight of shrimp.

2. Material and Method

The study was conducted in November and Desember 2018 in waters of Lontar, Banten. The first type is trammel net which is not parttioned and the distribution of slck is only under the webbing called trammel net control (TK). The second trammel net has 2 partition (TN A) and the third is a 3 parted net (TN B). Total of settings is 32 times.

This research was conducted using experimental fishing. The way with operating the three types of trammel net using one boat. The operation of fishing gear is carried out between 04.00-14.00 WIB. The data collected is primary and secondary. The primary data of the study included the size and weight of shrimp. Secondary data obtained from a literature. Analysis of trammel net capture using analysis of comparative descriptive. It was carried out on the identification of caught shrimp and analized the legal size of main catch. Then an analysis of the relationship of shrimp lengths is done using the following equation [7]:

$$W = a\, L^b$$

Where:
W = Weight of shrimp (g);
L = Length of shrimp (cm);
a dan b = Constant.

If the value of b < 3, the character is allometric negative, which means that the carapace length increase is more dominant than its weight gain. If the value of b > 3, the character is positive allomatric, which means that shrimp weight gain is more dominant than the increase in carapace length. If the value of b = 3, the character is isometric, which means the weight gain of the shrimp is balanced by the increase in length [7].

3. Result and Discussion

3.1. Identify shrimp caught by trammel net

Shrimp caught are white shrimp (*Penaeus merguiensis*) and mantis (*Squilla mantis*). The total catch of 704 shrimp consist of 478 white shrimp and 216 mantis. Both species of shrimp have the same class, that is malacostraca. Ordo of two species is different. White shrimp included in ordo of decapoda and mantis has stomatopoda [13]. Shrimp classification presented in Table 1.

Table 1. Classification of shrimp caught.

No	Klasifikasi	White Shrimp [8]	Mantis Shrimp [15]
1.	Kingdom	Animalia	Animalia
2.	Phylum	Arthropoda	Arthropoda
3.	Class	Malacostraca	Malacostraca
4.	Sub Class	Eumalacostraca	Hoplocarida
4.	Ordo	Decapoda	Stematopoda
5.	Family	Penaeidae	Squilidae
6.	Genus	Penaeus	Squilla
7.	Spesies	*Penaeus merguiensis*	*Squilla mantis*

3.2. Length of shrimp

Length of White shrimp. The ranges of caught white shrimp are from 10-20 cm. The smallest size is between 10-11 cm are 8 shrimp. The size most caught is between 14-15 cm were 336 shrimp. Shrimp are spread on TK was 56 shrimp, 163 shrimp on TN A and 117 shrimp on TN B. TN A can capture more and a longer size. The reason is that TN A of construction has been modified in webbing. The center of webbing is given a partition. So that the webbing has more space to entangle the shrimp [5]. The length specification and total of shrimp are presented in Figure 1.

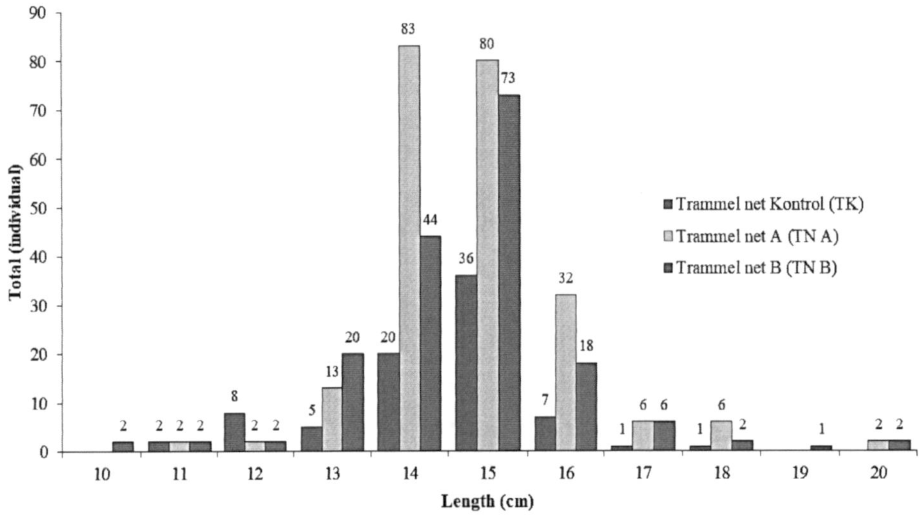

Figure 1. The length specification and total of white shrimp (*Penaeus merguensis*).

Length of Mantis Shrimp. The length of mantis ranges from 10-23 cm. The smallest size is between 10-11 cm are 21 shrimp. The size most caught is 12-13 cm, which is 105 shrimp. Mantis is caught a lot in webbing, because mantis lives in sand and muddy habitats [2]. This is in accordance with Lontar waters. The length of mantis has a shorter size than white shrimp. Mantis has longer growth than white shrimp. Mantis can live up to 20 years and spawn up to 20-30 times [17]. At a certain length, mantis will use energy it has for growing claws and shells. The length of mantis is attachted in Figure 2.

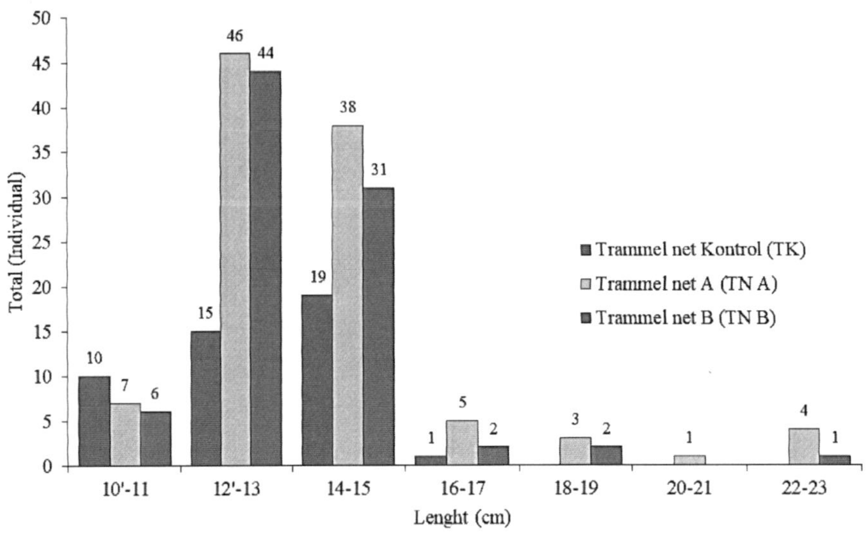

Figure 2. The length specification and total of mantis (*Squilla mantis*).

3.3. Weight of shrimp

Weight of White shrimp. The total weight of white shrimp during research was 14 kg. Range of weight from 15-86 g. The smallest weight is between 15-22 g which is 70 shrimp. The most weight that caught is 23-30 g of 274 shrimp and more caught on TN A. Adult shrimp have the power to jump, so they can be caught in the middle of webbing [5]. White shrimp was caught had enough weight. The cause is November and Desember is the time for spawning white shrimp. The size of spawn is more than 13 cm or 16 g [3]. The weight specification and total of shrimp are presented in Figure 3.

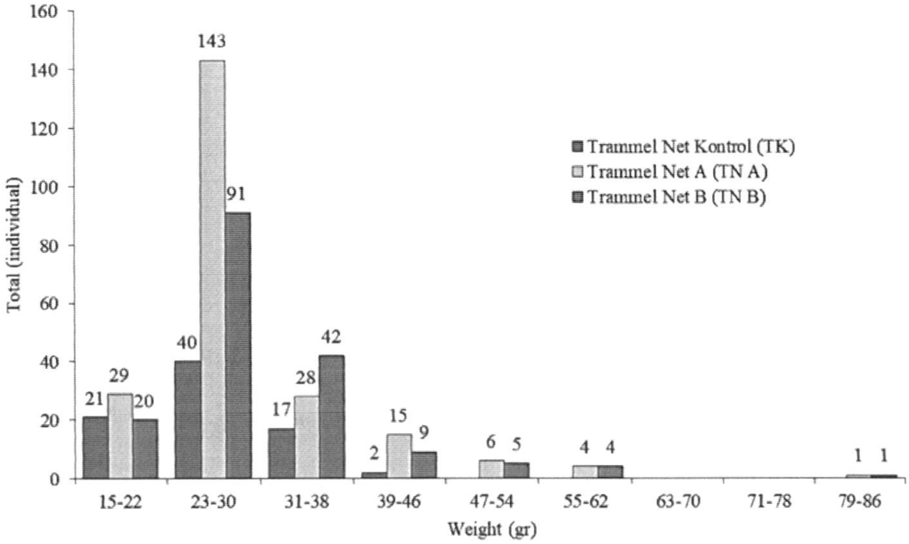

Figure 3. The weight specification and total of white shrimp (*Penaeus merguensis*).

Weight of Mantis shrimp. Range of mantis weight from 14-101 g. The smallest weight is between 14-24 g, which is 135 shrimp. Many mantis caught with that range. The largest weight of shrimp range from 91-101 g. That weight caught by TN A. Trammel net A can catch more and larger size of shrimp. The weight of mantis is greatly influenced by the morphology. Mantis body consists of 60% carapase and others are both claws. Clawed mantis like grasshopper claws. The shape will elongate and enlarge [17]. The weight specification and total of mantis are presented in Figure 4.

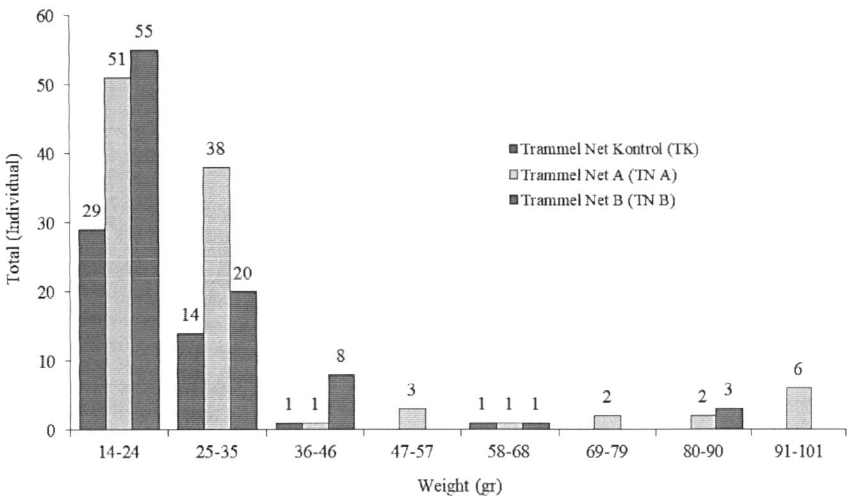

Figure 4. The weight specification and total of mantis shrimp (*Squilla mantis*).

3.4. The legal size of shrimp

The legal size of white shrimp has a length >14 cm [11]. Total of shrimp caught with legal size as much 420 shrimp (88%) and the number of illegal size is 58 shrimp (22%). Legal size of total white shrimp is higest than illegal size. Factors that influence the legal size catch is fishing season, life cycle and environmental conditions [1]. Shrimp spawning time in June and catching time in October-Desember [7]. Some research in carpentaria show shrimp migration to enlargement area (sea) in rainy season (October-Desember) [18]. This is in accordance with the time of research is November and Desember. Persentage of white shrimp legal size catching presented in Figure 5.

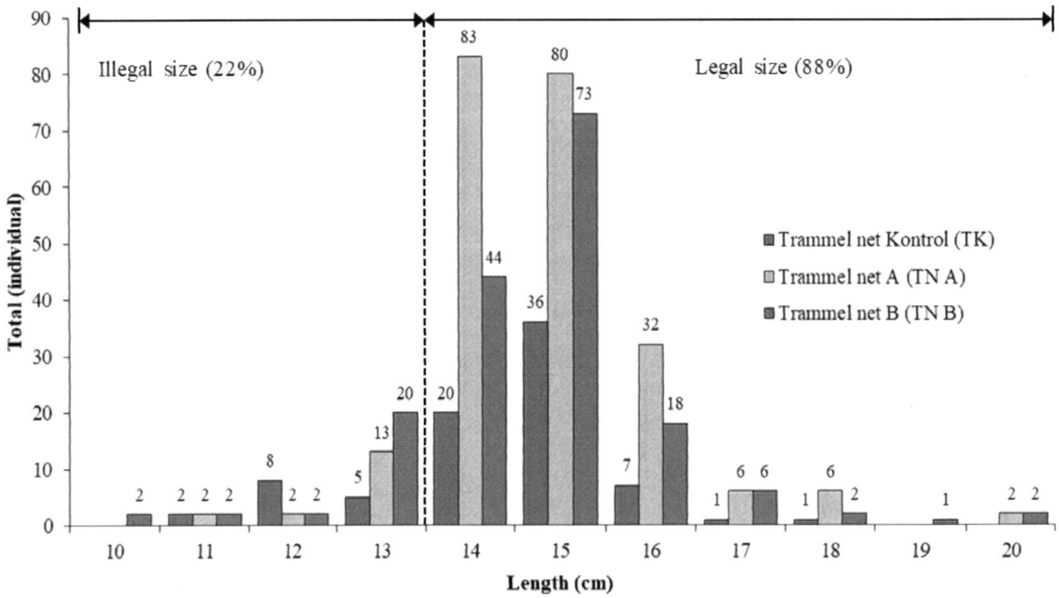

Figure 5. Percentage of white shrimp legal size.

The legal size of mantis shrimp has a length >15 cm [9]. Total of mantis caught with legal size as much 19 shrimp (92%) and the number of illegal size is 216 shrimp (8%). Legal size shrimp is less than illegal size. The existence of mantis depends on the season. The peak season for spawning mantis was declared in October [12]. When November and Desember are thought to be in a period of growth anad are in transition. Persentage of mantis shrimp legal size catching presented in Figure 6.

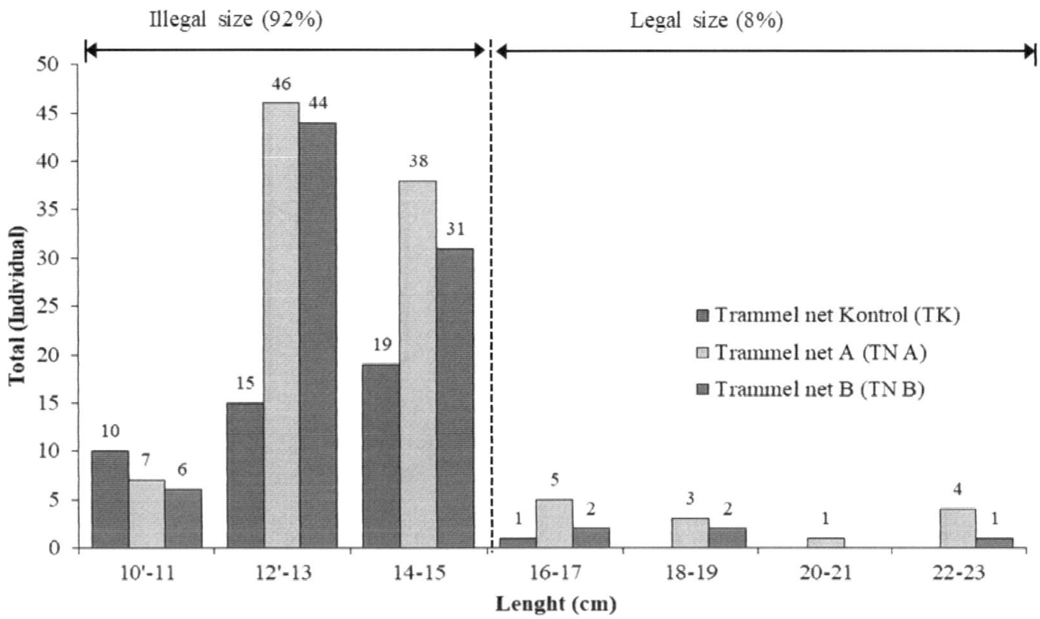

Figure 6. Percentage of mantis shrimp legal size.

Mantis is the dominant catch on trammel net capture in Lontar waters. Lontar waters including Banten Bay. It is a distribution of mantis shrimp habitat. Other studies report that mantis are found with the same species in Banten Bay [15]. Mantis distribution areas in Indonesia include the waters of Malacca strait, the east and west coast of Sumatra, the java sea and south java [19].

3.5. The relationship of length and weight shrimp

The relationship between the length and weight of white and mantis shrimp has different R^2 values. Determination (R^2) of white shrimp is 0,8847. The coefficient of determination of white shrimp is greater, so the analysis of the relationship between carapase length and weight of white shrimp has closer correlation than the correlation of mantis shrimp. The coefficient of b for the relationship between carapase length and weight is 2,6049. That is, the coefficient of b is significantly different from 3. The coefficient of b indicates that the relationship of length and weight has negative allometric, then the addition of length is more dominant than the weight of white shrimp.

Characteristics of white shrimp when researching in lontar water is the same as the characteristic of white shrimp in Segara Anakan is negative allometric [20], Iran Bay [14], mangrove forest of Percut Sei Tuan, North Sumatra [6]. Shrimp with the same characteristic of the relationship between length and weight can have different values of *b*, because it is relative and can change with time. Variation in value of b is due to various environmental factors such as food, sex, life cycle phase (gonad maturation stage) and habitat conditions [4 and 16].

The relationship of length and weight mantis has different values. Determination of mantis is 0,8722. The coefficient of b for mantis is the same as white shrimp. coefficient of b is significantly different from 3. The coefficient of b indicates that the relationship between length and weight has negative allometric, then the addition of length is more dominant than the weight of mantis. The result of length and weight relationship of white and mantis shrimp are attached to Figure 7 and 8.

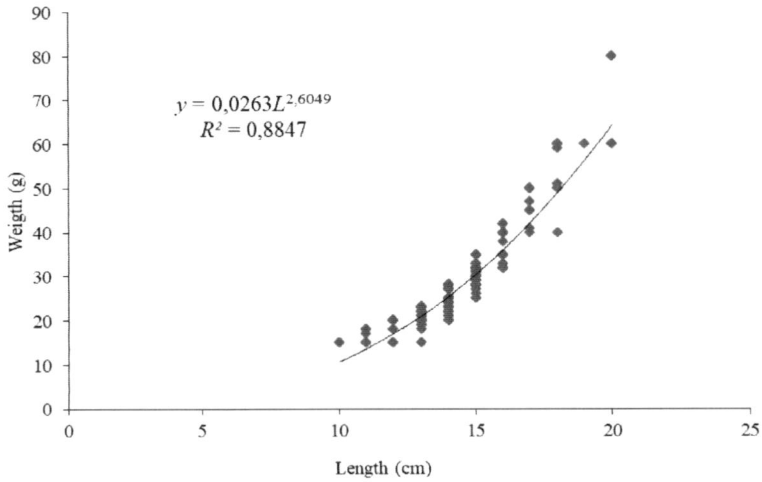

Figure 8. The relationship between length and weight of white shrimp.

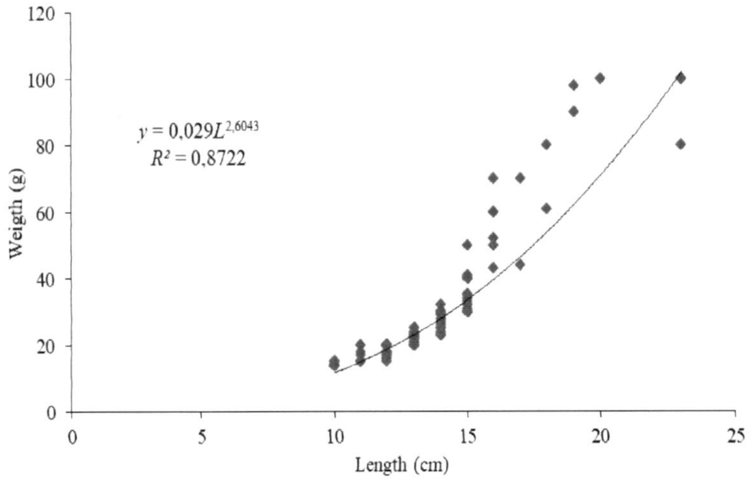

Figure 9. The relationship between length and weight of mantis.

4. Conclusions

The identification showed that there were two species of shrimp caught, namely white shrimp (*Penaeus merguiensis*) and mantis shrimp (*Squilla mantis*). The total catch of white shrimp is 478 shrimp (14 kg) and 235 shrimp of mantis (7 kg). The length of white shrimp caught in TK is 11-18 cm, TN A is 11-20 cm and TN B is 10-20 cm. The shrimp caught in TK have a size between 10-16 cm. In TN A and B, the size of shrimp caught is 10-23 cm. The total of white shrimp which has legal size is 420 shrimp (88%). The number of mantis shrimp has a smaller than white shrimp, which are 49 shrimp (21%) spread on three types of trammel net. The length and weight relationship between two species of shrimp are the same. The relationship is negative allometric.

5. References

[1] Adnan NA, Loneragan NR, Connolly RM 2002 Marine Biology (141) 241-251
[2] Astuti and Ariestyani 2013 *Media Akuakultur* 8(1) 39-44
[3] Chodrijah U and Faizah R 2018 *Bawal* 10(1) 49-55

[4] Gulland JA 1983 *Fish of Stock assessment: A Manual of Basic Metods* (New york: John Wiley and Sons Chicester) p 233
[5] Mardiah RS, Puspito G, Mustaruddin 2016 *Jurnal Teknologi Perikanan dan Kelautan* 7(1) 1-10
[6] Mulya MB, Bangen DG, Kaswadji RF, Riani E 2011 Omni Akuatika 10(13) 49-56
[7] Mulya MB 2012 Kajian Bioekologi Udang Putih di Ekosistem Mangrove Percut Sei Tuan Sumatera Utara (Bogor: Sekolah Pascasrjana, Institut Pertanian Bogor) p 127
[8] Myers P, Espinosa R, Parr CS, Jones T, Hammond GS, Dewey TA 2008 *The Animal Diversity* (USA: University of Michigan Museum of Zoology)
[9] Najamuddin, Hajar MAI, Sarira M 2017 *Jurnal IPTEKS PSP* 4(7) 79-94
[10] Nanlohy AC 2018 *Jurnal PAPALELE* 2(2) 58-65
[11] Palomares MLD nad Pauly D 2019 *Journal of Fish Biology* (http://www.sealifebase.ca)
[12] Purnamaningyas SE, Tjahyp DWH 2018 *Omni-Akuatika* 14(1) 87-95
[13] Putra DF, Muhammadar AA, Muhammad N, Damora A, Waliul A, Abidin MZ, Othman N 2018 *IOP Conference Series: Earth Environmental Science* 216 012022
[14] Safei M 2015 *Tropical Zoologi* 1-14
[15] Situmeang NS, Purnama D, Hartono D 2017 *Jurnal Enggano* 2(2) 239-248
[16] Sparre P and Venema SC 1992 *Introduksi Pengkajian Stok Ikan Tropis* (Jakarta: Badan Penelitian dan Pengembangan Pertanian) p 438
[17] Sukarni, Rina, Samsudin A, Purna Y 2018 *Jurnal Penyuluhan dan Kelautan* 12(3) 174-188
[18] Staples DJ and Vance DJ 1989 *Ecol. Prog. Ser* (27) 239-252
[19] Sumiono B and Priyono BE 1998 *Potensi dan Penyebaran Sumberdaya Ikan Laut di Perairan Indonesia* (Jakarta: P2O-LIPI)
[20] Wagiyo K, Damora A, Pane ARP 2018 *Jurnal Penelitian Perikanan Indonesia* 24(2) 127-136

6. Acknowledgement

We are grateful to Education and Culture Agency of Banten Province which has provided research assistance fund in "Beasiswa Putra Daerah Banten". We thank to who have helped a lot from beginning until its completion.

IOP Conf. Series: Earth and Environmental Science **383** (2019) 012040 doi:10.1088/1755-1315/383/1/012040

Microbial consortia effects on the yields of water spinach in milkfish aquaponics system

R O Khastini[1,2*], F R Idaryanto[3], Alimuddin[4], I Wahyuni[1], I J Sari[1], N Puspita[1], and A F Wahyuni[1]

[1]Department of Biology Education, Faculty of Teacher Training and Education University of Sultan Ageng Tirtayasa, Jl. Raya Jakarta km. 04 Pakupatan, Serang City, Banten Province, Indonesia
[2]Indonesia Center of Excellence for Food Security, University of Sultan Ageng Tirtayasa, Jl Raya Jakarta Km 4, Pakupatan, Serang, Banten, 42124. Indonesia.
[3]Department of Fisheries, Faculty of Agriculture, University of Sultan Ageng Tirtayasa, Jl. Raya Jakarta km. 04 Pakupatan, Serang City, Banten Province, Indonesia
[4]Department of Electrical Engineer, Faculty of Engineer, University of Sultan Ageng Tirtayasa, Indonesia Jl. Jenderal Sudirman Km 3, Kotabumi, Kec. Purwakarta, Kota Cilegon, Banten 42435

*Corresponding author: rida.khastini@untirta.ac.id

Abstract. Aquaponics is a technology combining fish, plants, and microorganisms into one system. In order to better understanding concerning the role of microorganism upon plant growth, this research was focused on the effect of microbial consortia inoculation on the yield of hydroponically water spinach (*Ipomea aquatica* Forssk) integrated with milk fish (*Chanos chanos*). The plant growth quality were determined to assess the efficacy of the beneficial microorganism. The experimental design was randomized blocks with five treatments which corresponded to the microbial consortia inoculation containing species of *Lactobacillus*, species of *Bacillus*, species of *Nitrosomonas* and *Nitrobacter* and combination of *Lactobacillus casei* and *Saccharomyces cerevisae* with three replicates. The result indicated that the water spinach with microbial consortia inoculation higher growth and yields than that of the uninoculated control significantly. However, it was not significantly different on plant yields from the species composition of microbial consortia. The results support the hypothesis that the use of microbial consortia can increase the yields efficiency of aquaponically water spinach. In interesting implication of result findings was that in the aquaponics system, the milk fish can be coupled with the water spinach growing system since its benefits were not dependent upon continuous recirculation between the plant and fish system but also its living biotic community.

Keywords: Aquaponics, microbial consortia, water spinach, milk fish

1. Introduction

Milkfish (*Chanos chanos*) is one of the most important species in Asian countries, including Indonesia. The fish is the most extensively cultivated fish in outdoor-based systems in coastal ponds and therefore have to cope with daily and/or seasonally changing environmental conditions. Milkfish is a source of food with high quality of protein, vitamins and minerals such as Omega-3 and also easily digested and absorbed by the body. In Banten province, milkfish is usually processed into traditional culinary, Sate Bandeng. Production of milkfish aquaculture in Banten in 2015 reached 10,638 tons [1].

Good water quality must be maintained at all times for milkfish to maintain optimal growth conditions and health of the fish. Accumulation of fish waste as an organic waste and toxic compounds is a significant problem inherent in aquaculture. Solid wastes consisted of faeces, and uneaten food can increase BOD (biochemical oxygen demand) and ammonia toxicity when it decomposes. Nitrite is a naturally occurring intermediate product of the nitrification process. The nitrate ion (NO_3^-) is the most oxidized form of nitrogen in nature and is relatively non-toxic to fishes [2]. Microorganisms, in general, are a crucial component in aquaponics systems because they enable the transformation of rough organic matter into molecules absorbable by the plants [3]. Microorganisms also clean the water from the accumulation of compounds such as ammonia which, when reaching high concentration levels, can become toxic to the fish. If the bacteria communities are not healthy and balanced, the whole aquaponics system can crash.

Usually milkfish are traditionally cultured in shallow brackish water pond. Nowadays it is a supporting technology that integrated in aquaculture, therefore milk fish can be cultured together with vegetable in aquaponics system. Networks connection in aquaponics systems act as a symbiotic relationship, in which plants grow by utilizing nutrients released by fish [4] a]nd plants acts as bio filter, remove the metabolites present in the water, which are harmful to the fish, allowing their development [5] before it is returned, cleaned back into the fish tank. The systems can be sustainable to be applied since the system with less labour, less land, and fewer chemicals, and takes up a fraction of the water usage and has the potential for higher yields of produce and protein.

Many types of plants can grow successfully in aquaponics systems. It is not only leafy green vegetable and herb crops could be grown, but a wide variety of fruiting crops, beans, and flowers can be grown effectively. Among them, water spinach, a semi-aquatic, tropical plant grown as a vegetable for its tender shoots and leaves can be cultivated in aquaponics system. A research proved that water spinach could be used as new sources of natural antioxidants as well as other bioactive compounds of human health benefits [6]. Water spinach grown on the wild environment can be harmful to the health since this plant is susceptible to aquatic pollution, water carrying pathogens, and various chemicals such as pesticides. Therefore the aquaponics system will provide a safe environment for growing water spinach. Aquaponics is suitable for limited land and water because it produces about three to six times the vegetables [7] and uses about 1% of the freshwater used by traditional aquaculture [8]. A small scale aquaponics can be used as a solution to the need for an inexpensive, nutritious and reliable food source for a family that has the capacity to provide a full meal both protein and vegetables with low inputs [9].

2. Materials and Method

2.1 Time and site

This study was conducted at the Department of Biology Education, Faculty of Teacher Training and Education, University of Sultan Ageng Tirtayasa from May-September 2018.

2.2 Aquaponics system preparation

The aquaponics system was designed by placing a container as a fish tank connected to the plant tray with inlet and outlet channels. The inlet channel directly connected to the pump AR 650 filter (Armada), for watering the plant (Figure 1). Seeds of the water spinach var. Sutera was surface sterilized by immersed in a 70% solution of ethanol for 30 seconds, followed by a solution of sodium hypochlorite (2% available chlorine) for 1 minute. Then seeds were rinsed three times with sterilized distillate water, dried overnight. The seed was sown and sprouted in the sowing trays before being transferred to the aquaponics system. Rockwool used as the growing medium for the seedlings. Parameter for water qualities i.e. water temperature (°C), dissolved oxygen (mg L^{-1}) pH and ammonia were measured during the study. Observation of temperature; pH; and DO was performed daily at 4 pm, while for ammonia levels were performed every seven days. Ammonia measurement used the reference SNI 06-6989. 30-2005.

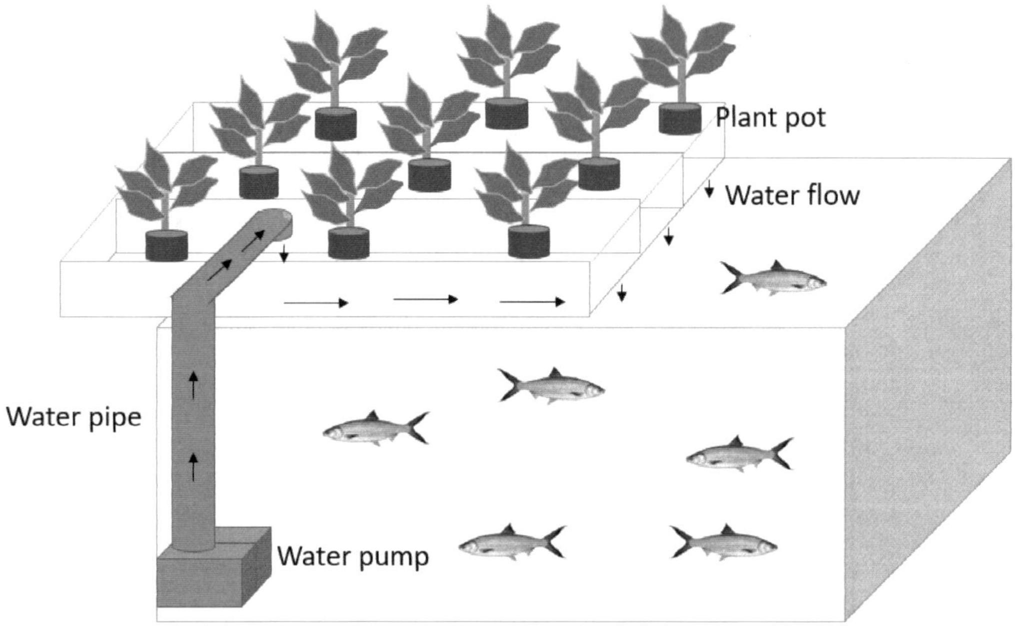

Figure 1. Schematic diagram of milk fish aquaponic system integrated with water spinach plant.

2.3 Experimental design
The complete random design and non-factorial were used on a small scale of the aquaponics system. The tested factor was the microbial consortia inoculation containing species of *Lactobacillus* (L), species of *Bacillus* (B), species of *Nitrosomonas* and *Nitrobacter* (N) and combination of *Lactobacillus casei* and *Saccharomyces cerevisiae* (LS), while for the control was aquaponics system without inoculation. Each treatment was conducted by three replicates. Data on the water spinach morphological parameters were collected at 30 days intervals from planting to harvesting date. At harvest, the plant was washed carefully in tap water to remove from the growing media. Plants were then separated into roots, and the fractions weighed and dried at 80°C until constant weight.

2.4. Data analysis
The data obtained in this study were analysed using oneway ANOVA. The significance of the difference among the means will be analysed by Duncan's multiple ranges. The statistics software was using SPSS version 22.

3. Result and Discussion
3.1. Aquaponics water quality parameters
Microbial inoculation in a milkfish aquaponics system can improve water quality. It can be seen through the water quality parameters (Table 1). The temperature both in controls and by microbial consortia inoculation as the treatment in the range of 27°C-30°C. The pH with a range of 6.0-8.0. DO range was 4.58-7.83 mg/l and 4.18-8.97 mg/l for the treatment and control, respectively.

Tabel 1. Milkfish aquaponics water quality parameters

Treatments	Parameters			
	Temperature (°C)	Dissolve Oxygen (mg.L^{-1})	pH	Ammonia (mg. L^{-1})
C	27-30	4.18-8.97	7.8-8.0	10.00-5.815
L	27-30	4.58-7.83	6.8-7.0	10.00-0.553
B	27-30	4.58-7.83	6.5-7.0	10.00-0.010
N	27-30	4.58-7.83	6.2-7.0	10.00-0.532
LS	27-30	4.58-7.83	6.0-7.0	10.00-0.545

The quality of water in aquaponics systems become the biological limits for milk fish sustainable production. A co-culture organisms such as plants and microorganism affected the welfare conditions through complex interactions between parameters of water qualities. The range of temperature in aquaponics system support the growth of fish and microorganism. The good temperature range for the fish in the tropics is 25-32°C [10]. The milkfish in their natural habitat or pond environment encounter daily or seasonal temperature fluctuations which greatly influence their growth. As [11] reported that milkfish in ponds behave normally at temperatures between 20 to 33 °C.

Dissolved oxygen is one of important aquaponics parameter which is required by the milkish for respiration, water spinach for respiration, health, and strength of their roots and nutrient uptake and microorganism for respiration and nitrification process through the decomposition of organic matter. The first product resulted in ammonia. In addition, the urine and faeces from the aquatic animals can cause high content of ammonia nitrogen and an increase of BOD. Ammonia is the main nitrogenous waste produced by fish via metabolism and is excreted across the gills [12] Ammonia is steadily released into the water through the excretion of the fishes as a product of their metabolism, must be filtered out of the water since higher concentrations of ammonia (commonly between 0.5 and 1 ppm) can kill fish [13]. Although plants can absorb ammonia from the water to some degree, but nitrates are assimilated more easily [14], thereby efficiently reducing the toxicity of the water for fish

Previously research reported that bacteria isolated from marine environments could remove certain levels of nitrogen [15, 16]. Strains of *Lactobacillus* spp. has been used as probiotics and can the removal of up to 400 μM NH_4^+,NO_2^-, and NO_3^- [17]. According to [18] *Bacillus* inoculation in aquaponics system can affected nutrient dynamics showed by faster decreases in ammonia concentration and faster increase in nitrite and nitrate concentrations than the control. The two most common species of nitrifying bacteria are *Nitrosomonas* sp. and Nitrobacter sp. *Nitrosomonas* sp. Converts ammonia (NH_3) to nitrite (NO_2) while *Nitrobacter* sp. uses nitrites for their energy source during its conversion to nitrate (NO_3). Nitrogen in nitrate form is absorbed and used as a nutrient by plants

3.2 Performance of water spinach influenced by microbial consortia inoculation

The effects of microbial consortia inoculation with on the growth of 30-day-old water spinach are shown in Table2. Inoculated water spinach produced more leaves, taller plants with more developed and bigger leaves than those without inoculation. There was an increment of number of leaves, shoot height shoot and fresh root weight and dry weight to inoculation over control. The enhancement of plant growth resulted in a significant increase of fresh and dry weight of shoots by up to 79 and 75 %, respectively, particularly for those inoculated with microbial consorcia combination of species *Lactobacillus casei* and *Saccharomyces cerevisae*. Fresh weight of roots was also increased by 45-53% in all microbial consortia inoculation

Tabel 2. Plant growth characteristics of non-inoculated and inoculated water spinach with microbial consortia inoculation

Treatments	Number of Leaves	Shoot height (cm)	Shoot Fresh weight (g. L^{-1})	Root Fresh Weight (g. L^{-1})	Shoot Dry weight (mg. L^{-1})
Control	5^a	7.4^a	0.370^a	0.013^a	0.02^a
L	7^b	10.08^b	1.410^b	0.026^b	$0{,}056^b$
B	7^b	10.31^b	1.490^b	0.028^b	$0{,}056^b$
N	7^b	10.36^b	1.703^b	0.026^b	$0{,}066^b$
LS	8^b	12.25^b	1.816^b	0.024^b	$0{,}083^b$

Values in a column with similar superscript letter are not significantly different, LSD, $P<0.05$

Bacillus, Nitrosomonas and *Nitrobacter* are categorized as PGPR microorganism, act through N_2 fixation, extracting nutrients from the soil, production of various kinds of plant hormones and biological controls, and promotion of plant growth [19, 20, 21]. *Lactobacillus* sp. and *Saccharomyces cerevisiae* has been used as probiotics in aqua culture that can help the process decomposition of ammonia as a bioremediation agent for water quality control. Then the ammonia is converted to the nitrate and use for the plant growth. All species used in this research significantly affected water spinach growth compared to the non-inoculated control. The species composition of microbial consortia inoculated to the aquaponic system was not significantly affected by plant yields.

4. Conclusion

The research on milkfish aquaponics system integrating with water spinach has been successfully conducted. The result indicated that the water spinach with microbial consortia inoculation showed higher growth and yields than that of the uninoculated control significantly. However, it was not significantly different on plant yields from the species composition of microbial consortia.

5. References

[1] Kementrian Kelautan dan Perikanan. 2016. Laporan Tahunan Direktorat Produksi Tahun 2016. Direktorat Jenderal Perikanan Budidaya, KKP.

[2] Zhang J Z, and Chen R P 2004. The danger of too much sediment in fish pond. *Guangdong feed* 12, 39–40

[3] Somerville C, Cohen M, Pantanella E, Stankus A, and Lovatelli A 2014 Small-scale Aquaponics Food Production: Integrated Fish and Plant Farming. FAO. Rome

[4] Roosta H R and Afsharipoor S 2012 Effects of Different Cultivation Media on Vegetative Growth, Ecophysiological Traits And Nutrients Concentration In Strawberry Under Hydroponic And Aquaponic Cultivation Systems. *Adv Env Bio*, 6(2): 543-555

[5] Hundley G C, Navarro R D, Figueiredo C M G, Navarro F K S P, Pereira M M, and Filho S J. T. 2013. Aproveitamento do efluente da produção de tilápia do Nilo para o crescimento de manjerona (*Origanum majorana*) e manjericão (*Origanum basilicum*) em sistemas de Aquaponia. *Revista Brasileira de Agropecuária Sustentável*, *3*(1), 51-55

[6] Lakshmi B, Vimala V 2000. Nutritive value of dehy-drated green leafy vegetable powders. *J Food SciTech*, 37: 465-471

[7] Resh H M *2004*. Hydroponic Food Production. Newconcept Press Inc. New. Jarsey.

[8] Rakocy J E. 1989. Hydroponic lettuce production in a recirculating fish culture system. University of the Virgin Islands, Agricultural Experiment Station, *Island Perspectives* 3:4-10

[9] Connolly K and Trebic T 2010 Optimization of a backyard aquaponic food production system, Faculty of Agricultural and Environmental Sciences, Macdonald Campus, McGill University, BREE 495 Design 3, Bioresource Engineering.

[10] Boyd C E *1990*. Water Quality in Ponds for Aquaculture. Birmingham. Publishing Co. Birmingham

[11] Lin Y M, Chen C N, and Lee T H 2003 The expression of gill Na, K-ATPase in milkfish, (Chanos chanos) acclimated to seawater, brackish water and fresh water. *Comparative Biochemistry and Physiology Part A*, 135: 489-497
[12] Cao L, Wang W, Yang C, Yang Y, Diana J, Yakupitiyage A 2007 Application of microbial phytase in fish feed. *Enzyme and Microbial Technology* 40: 497–507.
[13] Durborow R M, Crosby D M and Brunson M W, 1997. Ammonia in Fish Ponds. SRAC Publication No. 463 (link:www.hccfl.edu/media/520120/srac%20ammonia.pdf)
[14] Rakocy J E, Shultz R C, Bailey D S, Thoman E S. 2003 Aquaponic Production of Tilapia and Basil: Comparing a Batch and Staggered Cropping System. In Proceedings of the South Pacific Soilless Culture Conference, Palmerston North, New Zealand, 10–13 February 2003; Nichols, M.A., Ed.; International Society for Horticultural Science: Leuven, Belgium, 2004; pp. 63–69.
[15] Kim S K, Kong I S, Lee B H, Kang L S, Lee M G, Suh K H 2000 Removal of ammonium-N from a recirculation aquacultural system using an immobilizednitrifier. *Aquac. Eng.* 21, 139–150
[16] Grommen R, Van Hauteghem I, Van Wambeke M, Verstraete W 2002 An improvednitrifying enrichment to remove ammonium and nitrite from freshwater aquariasystems. *Aquaculture* 211, 115–124
[17] Maa C, Chob YS, Oh K H 2009Removal of pathogenic bacteria and nitrogens by Lactobacillus spp. JK-8 and JK-2009 *Aquaculture* 287 266–27
[18] Cerozi B D, Fitzsimmons K 2016The effect of pH on phosphorus availability and speciation in an aquaponics nutrient solution. *Bioresource Technology* 219 778–781
[19] Kidoglu F, Gül A, Tüzel Y, Özaktan H 2009 Yield enhancement of hydroponically grown tomatoes by rhizobacteria. *Acta Hortic*. 807, 475–480.
[20] Lee S, Ge C, Bohrerova Z, Grewal P S, Lee J 2015 Enhancing plant productivity while suppressing biofilm growth in a windowfarm system using beneficial bacteria and ultraviolet irradiation. *Can. J. Microbiol*. 61 (7), 457–466.
[21] Woitke M, Schitzler W H, 2005 Biotic stress relief on plants in hydroponic systems. *Acta Hortic*. 697, 557–565

The role of organic fertilizer and plant material in the growth of pepper seeds for sustainable agriculture

R Rusiva[1)]* and Suherman[2)]

[1)]Graduate school of Sultan Ageng Tirtayasa University, Serang, Banten, Indonesia
[2)]Department of Agribusiness, Sultan Ageng Tirtayasa University, Serang, Banten, Indonesia

*Corresponding authors: rifaipb@gmail.com

Abstract. Pepper (*Piper ningrum* L.) is one of Indonesia's most important spice commodities with high economic value. Pepper plants are propagated by vegetative methods using cuttings. Factors The origin of planting material and the application organic fertilizer can support the development of technology to provide good pepper propagation material. This study used Completely Randomized Group Design with two factors: the origin of planting material and organic fertilizers. The origin of planting materials was 2nd (S2), 3rd (S3), 4th (S4) and 5th (S5) node cuttings, while organic fertilizer used were fulvic acid (P1) and biological fertilizer. The results showed that organic fertilizers did not affect all observed parameters, while the origin of plant material affected the percentage of life cuttings, seed height, number of leaves, number of internodes and nodes, and number of branching plants. The origin of plant material that had the best results was the 3rd node cuttings (S3), while the best interaction of fertilizers and cuttings showed by combination of fulvic acid and 3rd node cuttings (P1S3). Further research was needed on the dose and frequency application of organic fertilizer.

Keywords: Growth Optimization; Piper nigrum L; Plant material; Organic fertilizer.

Introduction
Pepper (*Piper ningrum* L.) is one of Indonesia's most important spice commodities with high economic value. Indonesia is one of the main producers of pepper. There was a decrease in pepper crop plantation area and production in 2012-2015. The decline in pepper plantation area in Indonesia in 2012, 2013, 2014 and 2015 was 177,787 thousand ha, 171,920 thousand ha, 16,751 thousand ha and 167,590 thousand ha, respectively. Similarly, the decline in production was 87,841 thousand tons, 91,039 thousand tons, 87,448 thousand tons and 81,501 thousand tons, respectively. The decline was probably because this commodity was cultivated by smallholder farmers in the form of small plantations which was cultivated for generations and highly labour intensive. So that pepper cultivation activities became less intensive. Several regions that were the production centres of pepper commodities were Bangka Belitung, Lampung, South Sumatra, East Kalimantan and South Sulawesi [1].

Pepper as one of mainstay exported commodities requires technological support that can improve its cultivation efficiency and increase national productivity. Pepper cultivation needs high quality planting material as the determining aspect of growth and production. Pepper planting material can be propagated vegetatively and generatively.

Propagation with stem cuttings is generally easier and cost effective because the stem provides a sufficient source of energy to support growth in the early stage. [2] explained that the selection of stem cuttings should consider the age of the stem. When stem used as planting material is too old, it will be difficult to form roots, when it is too young, the rate of transpiration is high so that the cuttings will be weak and die. In addition, there are several factors that influence the success of stem cuttings, namely the origin of cuttings, cuttings length, and environment (rooting media, humidity, temperature and light).

The best source of pepper planting material is climbing tendrils, originating from plants less than 3 years old (have not producing), free of pest and pathogen. The planting material of climbing pepper cuttings comes from stem cuttings that have sticky roots. Single leafed, one node cutting has the advantage of efficient planting material utilization. Seeding and transplant technique is used when using one node cutting as planting material, so the use of short cuttings can reduce mortality rate by up to 10% [3].

Nursery activities are important preparing good planting material. One of the supporting factors for the success of planting materials to grow well is to use the appropriate composition of the planting media. [4] explained that a good planting medium must have the ability to hold water, have loose structure, aeration and good drainage, a pH that is suitable for the type of plant and contains nutrients to support plant growth. [5] added that good planting medium for the growth of pepper seedling is the combination of soil, sand, and manure with ratio of 2: 1: 1, that composition of planting medium produces 96% cuttings viability.

One technology in accelerating the growth of pepper seeds is by adding fertilizer as nursery activity. The use of fertilizer which is considered safe for the growth of pepper cuttings is organic fertilizer. Organic fertilizers are fertilizers which mostly or entirely consist of organic material derived from plants and or animals that have been through an engineering process, can be solid or liquid which is used to supply organic materials to improve the physical, chemical and biological properties of the soil [6]. The application of humus and liquid biological fertilizer is expected to accelerate the growth of plants.

The origin of planting material and the application of organic fertilization are expected to be a consideration in conducting pepper nurseries. Therefore, information regarding research on the origin of cuttings and the application of appropriate organic fertilization is needed. It is expected that the growth and success of vegetative pepper propagation in nurseries with a single leafed node can be accelerated, thus supporting the development of pepper cultivation business.

2. Materials and Methods

2.1. Experimental Site

This research was conducted from January to June 2018, located at Gunung Gede Experimental Field, Bogor Agricultural University. Planting medium analysis was conducted at ICBB Environmental Biotechnology Laboratory, while leaf nutrient analysis, chlorophyll destructive observation, and dry weight were conducted at Agronomy and Horticulture Department Laboratory, Faculty of Agriculture, Bogor Agricultural University.

This research used cuttings of Natar pepper variety and the cuttings were originated from different position on mother plant. Planting medium used was a combination of soil, sand and manure with ratio of 2:1:1. The soil type used was red-yellow podzolic soil from Jasinga Bogor area. The fertilizers used as treatment were organic fertilizer 1 contained fulvic acid and organic fertilizer 2 that was biological fertilizer containing microbes. The tools used were common agricultural tools, measuring tapes, scales, polybags sized 15 x 20 cm, cutting scissor, camera, and laboratory equipment.

2.2. Experimental Design

Experimental design used was Completely Randomized Block Design with two factors. First factor was organic fertilizer, with three treatments that were without application of organic fertilizer (P0), application of organic fertilizer 1 (P1), and application of organic fertilizer 2 (P2). The second factor was origin of planting material (four treatments) which was 2nd (S2), 3rd (S3), 4th (S4), and 5th (S5) node.

There were 12 combination of treatments, with three replications, which made 35 experimental unit. Each experimental unit consisted of 15 sample plants, which made 540 observed sample plants.

The research was conducted in nursery building with area of 10 x 5 m. There were three confinement units with area of 4 x 1.2 m each for pepper seedling initiation stage. Planting medium was mixed in closed condition inside the nursery building with composition of soil, sand and manure (2:1:1), and then put inside polybags sized 15 x 20 cm.

Pepper seedling cuttings were planted in prepared medium and arranged in accordance with the experimental design. Seedling initiation was conducted under hoods for 4 weeks. Maintenance included checking medium humidity by watering the medium twice a day and opening the hood for one hour in the morning to keep temperature from going too hot. At 3 weeks after planting (WAP), one side of the hoods were opened gradually, until opened completely at 4 WAP. Fertilizer application was conducted every week from 4 WAP to 19 WAP, with dosage of 10 ml per seedling.

Observations were conducted from 5 WAP to 20 WAP or after one week from fertilizer application. Observations were conducted on agronomic aspects such as original cutting length, percentage of live plants, seedling height, number of leaves, number of branching plants, number of internodes for 20 WAP. Environmental observations included humidity (RH), temperature (°C), sunlight intensity (%), luminous emittance (lux), and pest and disease found in pepper plant. Obtained data was analyzed using SAS 9.4 software for analysis of variance test-F with alpha of 5 % and 1 %. If the data was significantly different, further test of Duncan Multiple Range Test (DMRT) was conducted to compared average of every variable [7].

3. Results And Discussion

3.1. Pepper Seedling Growth Response

Pepper propagation usually use single node cutting that was planted under hood for 4 weeks. Environmental requirements for climbing pepper seedling nursery based on BSN Indonesia are sunlight intensity of 50 – 70%, temperature of 22-30 °C, and humidity (RH) > 80% (BSN 2006). Observation of micro-climate showed that research location had met those requirements (Table 1). The diseases found were stem rot caused by fungi *Phytophthora capsici* and leaf curls cause by green or white aphids (*Aphis* sp.) as vector. Pests found were stem borer (*Lophobaris piperis*) and grasshopper that caused rotten and broken tip. The severity of pest and disease attack in nursery was low, but pest management was conducted by spraying fungicide and insecticide every two weeks with dosage of 2 g l^{-1} to prevent outbreak.

Table 1. Average value of environmental measurement of pepper nursery

Location	RH (%)	Temperature (°C)	Sunlight intensity (%)	Luminous emittance (lux)
Nursery hood (0-4 MST)	93.2	23.0	32.0	1704
Nursery building (5-20 MST)	83.2	25.3	59.6	3494

Result of anova of growth and physiology of climbing pepper seedling was not significantly different on original cutting length, number and density of stomata. These parameters were not affected as was shown by the low coefficient of determination (R^2) that was $\leq$ 50%. Significant difference from treatments was shown on percentage of live plants, seedling height, number of leaves, number of branching plants, and number of internodes and nodes. Those parameters were significantly affected as was shown by high low coefficient of determination (R^2) that was $\geq$ 50% (Table 2). Value of R^2 determined the effect of other variables (error) outside of treatments free variables (x) against result (y). Parameters with $R^2 \leq$ 50% were mostly affected by other variables (error) thus had insignificant value on every treatment, while parameters with $R^2 \geq$ 50% were dominantly affected by treatment (x) thus had significant value on one of the treatments.

3.2. Percentage of live plants

Surviving pepper seedling at 20 WAP had different percentage of live plants. The effect of treatments showed that the average of live plants was higher than 77% (Table 3). Treatment of organic fertilizer 2 (P2) had only improved percentage of live plants by 7.38% compared to without application of organic fertilizer (P0). The improvement was because of higher availability of C-organic, nutrients, and microorganism in the P2. This was in accordance with [8] that showed application of rhizobacteria *Bacillus cereus* in pepper cutting could improved percentage of live plants to 96% compared to control of no application of *B. cereus*.

[9] explained that biological fertilizer was utilization of active biological product containing soil enhancer bacteria to increase fertilizer efficiency, soil health and fertility. Fertilizer efficiency could be improved by using N2 fixation, P and K solvent microbes. Application of soil enhancer bacteria could improve the availability of soil nutrients and plant metabolites for plants, and also protect plant roots from pest and disease.

Treatment of the origin of cutting materials had a significant effect on the percentage of live plants. Table 3 shows that the lowest percentage of live plants came from the 5th node (S5) cutting, which was 77.38%, while the highest percentage of live plants came from the 3rd node (S3) cutting which was 95.56%. The low percentage value of live plants on the S5 was thought to came from the length of the cutting. This could be seen in the length of the original cutting, S5 was the shortest compared to the others. The ability of cuttings to survive was affected by the availability of nutrients in the cutting material in the early stage of growth, where the shorter cuttings would lose nutrients faster. [10] added that the pepper cuttings originated from the middle, that were 2nd and 3rd node were the optimum cuttings, and were able to initiate roots better and survive better.

Table 2. Recapitulation of analysis of variance result on growth and physiology of pepper seedling

Observation parameter	Fertilizer (P)	Cutting (S)	Interaction (PxS)	CV	R^2
Original cutting length (cm)	ns	ns	ns	15.65	0.38
Percentage of live plants (%)	ns	*	ns	12.14	0.52
Seedling height (cm)					
10 WAP	ns	**	*	23.21	0.80
12 WAP	ns	**	ns	17.22	0.73
14 WAP	ns	**	*	12.37	0.81
16 WAP	ns	**	ns	13.08	0.68
18 WAP	ns	**	ns	11.27	0.70
20 WAP	ns	**	ns	10.52	0.69
Number of leaves					
14 WAP	ns	**	ns	18.88	0.73
16 WAP	ns	**	ns	15.33	0.71
18 WAP	ns	**	ns	10.96	0.68
20 WAP	ns	**	ns	11.52	0.61
Number of internodes					
14 WAP	ns	**	ns	14.77	0.77
16 WAP	ns	*	ns	12.34	0.71
18 WAP	ns	*	ns	10.02	0.65
20 WAP	ns	*	ns	9.01	0.65
Number of branching plants					
14 WAP	ns	**	ns	25.24 (t)	0.58 (t)
16 WAP	ns	**	ns	17.51 (t)	0.68 (t)
18 WAP	ns	**	ns	16.41 (t)	0.61 (t)
20 WAP	ns	*	ns	16.40 (t)	0.56 (t)

Notes: (ns): not significant, (*): significant at alpha 5%, (**): significant at alpha 1 %, CV: coefficient of variation, (t): transformed value of $\sqrt{(x+0.5)}$.

IOP Conf. Series: Earth and Environmental Science **383** (2019) 012041 doi:10.1088/1755-1315/383/1/012041

Table 3. Average of percentage of live plants and original cutting length

Treatment	Percentage of ive plants (%)	Original cutting length (cm)
Organic fertilizer		
P0	84.70	3.93
P1	88.61	4.44
P2	92.08	4.14
Notation	ns	ns
Origin of plant material		
S2	91.67 a	4.38
S3	95.56 a	4.21
S4	89.25 a	4.13
S5	77.38 b	3.96
Notation	*	ns

Notes: (ns): not significant, (*): significant at alpha 5%, numbers followed by the same letter was not significantly different on DMRT 5%.

3.3. Seedling Height

Organic fertilizer treatment did not affect seedling plant, while origin of planting material significantly affected seedling height from 10 WAP to 20 WAP. Interaction between fertilizer and origin of planting material affected seedling height from 10 WAP to 14 WAP (Table 2). Organic fertilizer treatment did not significantly affect seedling height. This was shown by the small difference on variables on fertilizer treatment. Significant difference was observed on origin of planting material treatment from 10 WAP to 20 WAP. Cutting that originated from 3rd node had the best seedling growth from 12 WAP. Significant difference also observed on cutting originated from 5th node (S5) that had the lowest seedling height (Table 4). The use of S3 cutting that was relatively young can increase seedling height by 5.07 cm.

Table 4. Average of climbing pepper seedling height

Treatment	Seedling height (cm)					
	10 WAP	12 WAP	14 WAP	16 WAP	18 WAP	20 WAP
Organic fertilizer						
P0	6.00	10.34	13.77	17.97	24.34	31.05
P1	5.65	9.90	14.04	18.34	24.07	31.05
P2	5.98	10.19	14.10	18.11	23.63	29.49
Notation	ns	ns	ns	ns	ns	ns
Origin of planting material						
S2	7.05 a	11.50 a	15.85 a	19.40 ab	25.53 ab	32.24 ab
S3	6.84 a	11.57 a	15.60 a	20.13 a	26.29 a	32.89 a
S4	5.65 a	9.53 b	13.24 b	17.40 bc	22.88 bc	29.17 bc
S5	3.99 b	7.96 b	11.19 c	15.63 c	21.35 c	27.82 c
Notation	**	**	**	**	**	**

Notes: (ns): not significant, (**): significant at alpha 1%, numbers followed by the same letter was not significantly different on DMRT 5%.

3.4. Number of Leaves

Fertilizer treatment did not affect the number of leaves, shown in Table 5. The difference of number of pepper leaves was observed on origin of planting material treatment. Very significant difference was observed on 14 WAP to 20 WAP. Cutting from 3rd node (S3) had the highest number of leaves while S5 had the least. High number of leaves on S2 and S3 was because younger planting material origin, making faster cell and organ development.

3.5. Number of Nodes and Internodes

The average number of nodes and internodes did not appear to be significantly different in the treatment of organic fertilizer, but has a very significant difference in the treatment of origin of planting material. Cutting from the 3rd node had higher number of nodes and internodes than the others, amounting to 8 at 20 WAP, while the number of internodes in fertilizer and other cutting treatments amounted to 7 (Table 6). The number of nodes and internodes became a reference in the selection of seedling that would be ready for planting. The quality of the climbing pepper seedlings ready for planting according to BSN Indonesia is to have 5-7 segments [11].

Table 5. Average of number of pepper leaves

Treatment	Number of leaves			
	14 WAP	16 WAP	18 WAP	20 WAP
Organic Fertilizer				
P0	3.14	4.89	6.94	9.04
P1	3.16	5.26	7.18	9.27
P2	3.28	5.19	7.09	9.02
Notation	ns	ns	ns	ns
Origin of planting material				
S2	3.68 a	5.61 a	7.59 a	9.71 a
S3	3.64 a	5.82 a	7.74 a	9.88 ab
S4	2.93 b	4.81 b	6.77 b	8.75 bc
S5	2.52 b	4.20 b	6.18 b	8.10 c
Notation	**	**	**	**

Notes: (ns): not significant, (**): significant at alpha 1%, numbers followed by the same letter was not significantly different on DMRT 5%.

Table 6. Average of number of nodes and internodes of pepper seedling

Treatment	Number of nodes and internodes			
	14 WAP	16 WAP	18 WAP	20 WAP
Organic Fertilizer				
P0	3.12	4.55	6.01	7.64
P1	3.12	4.71	6.17	7.81
P2	3.09	4.56	6.00	7.34
Notation	Ns	ns	ns	ns
Origin of planting material				
S2	3.47 a	4.75 a	6.30 a	7.89 a
S3	3.55 a	5.05 a	6.48 a	8.00 a
S4	2.94 b	4.56 ab	5.99 ab	7.51 ab
S5	2.48 c	4.07 b	5.47 b	6.98 b
Notatiom	**	*	*	*

Notes: (ns): not significant, (*): significant at alpha 5%, (**): significant at alpha 1%, numbers followed by the same letter was not significantly different on DMRT 5%.

3.6. Number of Branching Plants

Pepper seedling started to branch at 14 WAP. Number of branches was mostly one, but there was small amount of seedling that had 2 to 3 branches at 16 to 20 WAP. Average of number of branching plant was not significantly different in organic fertilizer treatment and was significantly different in origin of planting material treatment. Highest number of branching plants was found on S3 cuttings. Showed in Table 7, in the end of observation at 20 WAP, number of branching plants of S3 was 11 seedlings. Branches in pepper plant was affected by the age of cutting origin, where the younger the faster cell differentiation and development.

Table 7. Average of number of branching plants in pepper

Treatment	Number of branching plant			
	14 WAP	16 WAP	18 WAP	20 WAP
Organic fertilizer				
P0	3.92	6.83	7.92	8.58
P1	5.00	7.17	7.58	8.00
P2	5.25	7.67	9.00	9.17
Notation	ns	ns	Ns	Ns
Origin of planting material				
S2	6.22 a	8.22 a	8.56 a	8.67 ab
S3	6.44 a	9.44 a	10.44 a	11.00 a
S4	4.00 ab	7.56 a	8.33 a	8.56 ab
S5	2.22 b	3.67 b	5.33 b	6.11 b
Notation	**	**	**	*

Notes: (ns): not significant, (*): significant at alpha 5%, (**): significant at alpha 1%, numbers followed by the same letter was not significantly different on DMRT 5%.

Table 8. Recapitulation of highest average value from treatments interactions of climbing pepper seedling

Interaction	Parameters				
	1	2	3	4	5
Interaction code	P2S3	P1S2	P1S3	P1S3	P1S3
Value	98.33	32.51	10.50	8.34	11.00

Notes:
1. Percentage of live plants (%);
2. Seedling height (cm);
3. Number of leaves;
4. Number of internodes and nodes;
5. Number of branching plants;

Recommendation from this research was found in treatment of origin of planting materials. Cutting from 3rd node (S3) was the best recommendation for origin of planting materials. The highest value of number of leaves, number of nodes, and number of branching plants was found in the interaction between fulvic acid and cutting from 3rd node (P1S3) (Table 8). The effect of cutting from 3rd node (S3) was supported by [12] who stated that there was plenty undifferentiated meristematic tissue in younger plant part, making this tissue easily differentiated into root primordial and shoot formation. [13] added that the growth of root and shoot was determined by the condition of cutting used. Cutting from older stem could ended up with yellowish and then dry seedling. In contrast, cutting from younger stem was light green and often ended up rotten.

4. Conclusion

There was no effect of organic fertilizer treatment to observed variables. There was a significant effect of origin of planting materials to percentage of live plants, seedling height, number of leaves, number of nodes and internodes, and number of branching plants. Cutting from 3rd node was the best recommendation for origin of planting materials, while the best interaction was the combination between fulvic acid and cutting from 3rd node (P1S3). The interaction effect was observed in number of leaves, number of internodes and nodes, and number of branching plants.

5. References

[1] Direktorat Jenderal Perkebunan. 2016. *Luas Areal, Produksi dan Produktivitas Perkebunan di Indonesia Tahun 2011-2015*. Jakarta (ID): Direktorat Jenderal Perkebunan.

[2] Susilo E, Pujiwati H. 2015. Pertumbuhan *Tithonia diversifolia* pada beberapa asal bahan tanam dan konsentrasi ZPT yang berbeda. *Prosiding seminar perbenihan tanaman rempah dan obat.* Jakarta (ID): Badan Penelitian dan Pengembangan Pertanian.

[3] Manohara D, Wahyuno D, Rivai A. 2013. *Teknologi Unggulan Lada.* Bogor (ID): Pusat Penelitian dan Pengembangan Perkebunan.

[4] Acquaah G. 2009. *Horticulture: Principles and Practices.* Ed ke-4. New Jersey (US): Prentice Hall.

[5] Saefudin, Listyati D. 2012. Pengaruh media tumbuh dan interval penyemprotan fungisida terhadap viabilitas, pertumbuhan dan harga pokok benih lada. *Buletin RISTRI* 3(2): 135-142.

[6] Suriadikarta DA, Simanungkalit RDM. 2006. *Pupuk Organik dan Pupuk Hayati.* Bogor (ID): Balai Besar Litbang Sumberdaya Lahan Pertanian, Badan Penelitian dan Pengembangan Pertanian.

[7] Gomez KA, Gomez AA. 2007. *Prosedur Statistika untuk Percobaan Pertanian.* Sjamsudin E, Baharsjah JS penerjemah. Jakarta (ID): UI Pr. Terjemahan dari *Statistical Procedures for Agricultural Research.*

[8] Aziz ZFA, Halimi MS, Kundat FR, Jiwan M, Wong SK. 2015. Rhizobacterium *Bacillus cereus* induces root formation of pepper (*Piper nigrum* L.) stem cuttings. *Research in biotechnology* **6(2)**: 23-30.

[9] Saraswati R. 2012. Teknologi pupuk hayati untuk efisiensi pemupukan dan keberlanjutan sistem produksi pertanian. *Prosiding Seminar Nasional Teknologi Pemupukan dan Pemulihan Lahan Terdegradasi*; hlm. 727-738. Bogor (ID): Badan Penelitian dan Pengembangan Pertanian. Kementerian Pertanian.

[10] Sharangi AB, Kumar R, Sahu PK. 2010. Survivability of black pepper (*Piper nigrum* L.) cuttings from different portion of vine and growing media. *Journal of Crop and Weed* **6(1)**: 52-54.

[11] Badan Standarisasi Nasional. 2006. *Benih lada* (*Piper nigrum* L.)- SNI 01-7155-2006. Jakarta (ID): Badan Standarisasi Nasional.

[12] Weaver JW. 1972. *Plant Growth Substances in Agriculture.* San Fransisco (CF): W.H. Freeman and Co.

[13] Rochiman K, Harjadi SS. 1973. *Pembiakan Vegetatif.* Bogor (ID): Departemen Agronomi. Fakultas Pertanian. Institut Pertanian Bogor.

The effect of seaweed fertilizer and compost combination on the growth of lettuce head

R Kusumawati*, I Munifah, and J Basmal

Research Center for Marine and Fisheries Product Processing and Biotechnology, KS. Tubun Petamburan VI, 10260, Jakarta, Indonesia

*Corresponding author: tanjung.1979@gmail.com

Abstract. Research on the effect of organic fertilizer combination, seaweed fertilizer and compost, on the growth of head lettuce plants has been conducted. The treatment in the study was head lettuce fertilizing with compost and seaweed fertilizer in ratio 2:0, 1:1, and 0:2. Land used for head lettuce without fertilizing treatment as control. The study was intended to get the best combination of organic fertilizer for head lettuce growth. The results showed that the best fertilizer formula was obtained from a combination of compost and seaweed fertilizer in ratio 2:0. Growth indicator as a result are average of leaf length of 26.64 cm, leaf width of 25.48 cm, 40 number of leaves, weight 311.75 g, and moisture content of 93.51%.Compost can stabilized acidity of planting land but the present of seaweed fertilizer can stabilized humidity of planting land and does not indicate the presence of fungi and parasites.

1. Introduction

Head lettuce or iceberg lettuce cultivation is one of the agricultural activities that is increasingly in demand along with the growth of the culinary and tourism industry. Lettuce is one of the most widely consumed vegetables worldwide. This vegetable is low in calories, fat and sodium, a good source of fiber, iron, folate, and vitamin C. Lettuce reported has anti-inflammatory, cholesterol-lowering, and anti-diabetic activities attributed to the bioactive compounds in it. More nutritious lettuces are leaf type lettuce and romaine with folate content comparable to other rich leafy vegetable sources.

The challenge in cultivating head lettuce before harvest is leaf spot disease caused by the fungus Cercospora and Alternaria. Besides that, the caterpillars that attack during the growth period of 1-2 weeks. Farmers generally use pesticides and their residues can endanger consumers. Research on the growth of bacteria supporting plant growth has been carried out and will be applied to the head lettuce cultivation land. the presence of fungi and parasites will be observed qualitatively, while the growth of head lettuce will be observed quantitatively. The study expected to get the best combination of fertilizers with the highest growth indicator of head lettuce and not indicated the presence of fungi and parasites.

2. Material and method

Materials used in the study were seaweed liquid biological fertilizer, seaweed solid biological fertilizer, compost, and head lettuce seedlings.

2.1. *Seaweed liquid fertilizer preparation*
Biological liquid fertilizer is prepared by homogenizing 93% of solid waste extraction so that in sap liquid cottonii and 7% bacterial consortium. The bacteria used are the same as in making solid fertilizer.

2.2. *Biological seaweed solid fertilizer preparation*
Biological solid fertilizer is prepared by homogenising 2% fish meal, 1% molasses, 73% Sargassum flour, 2% cottonii processed into pasta, 5% livestock manure, 10% dolomite and 7% bacterial consortium. Bacteria used include Azotobacter sp., Azospirillium sp., Pseudomonas sp., Lactobacillius sp., Bacillus sp., Rhizobium sp., and Streptomyces sp.

2.3. *Compost fertilizer preparation*
Compost fertilizer is prepared by homogenising the results of fermentation of the leaves waste with animal manure. Aerobic fermentation was carried out for 3 months by arranging layers of 1 part of leaves and 2 part of animal waste sequentially.

2.4. *Head lettuceseedlings head*
Seed used in this research produced by Bina Sarana Bakti Agatho Organic Farm as a result of the adaptation of the original seed from Switzerland. Poly bag filled with 1 part of sand and 1 part of compost, then added with 4 seeds. Seeding is done in 7 to 10 days or until 4 leaves grow from each seed. After that the seedlings from the seeding are sorted, about 164 seedling need for each plot.

2.5. *Preparation of planting land*
Planting land measuring in Bina Sarana Bakti Agatho Organic Farm area is 10 m^2 per plot and each treatment uses 1 plot. Liquid seaweed biofertilizer is applied to all plots after the planting land has been processed, then allowed to stand for 10 days. After that, planting seeds is carried out with a spacing of 25 cm.

2.6. *Observation and data analysis*
Measurements include head lettuce growth parameters (leaf length, leaf width, number of leaves, plant weight, and plant moisture content), as well as environmental parameters (pH and soil moisture). The experimental design used in this study was completely randomized design with 5 replications. Data obtained were analyzed using MS-excel program.

3. Results and discussion
Characteristics of fertilizer used in the study showed in Tabel 1. The fertilizer contain macro and micro nutrients, also growth hormones. Micro nutrients in solid biological seaweed fertilizer contain ferum (Fe), cuprum (Cu), mangan (Mn), and cobalt (Co) higher than others but it has zink (Zn), plumbum (Pb), cadmium (Cd), and sulphur (S) lower than others. All macronutrients content in compost is higher than others, it indicate that the growth of head lettuce might be better if treat by fertilizer containing compost. For growth hormones parameter, both solid and liquid biological seaweed fertilizer has similar content, except Gibberellin. The liquid fertilizer has higher content of Gibberellin.

Planting land prepared for the study was in Bina Sarana Bakti Agatho Organic Farm area, in rainy season. There was special treatment to control water volume entry in planting land and humidity also such as using plastic mask over the planting land and minimize water. The growth of head lettuce can follow by measuring length and width of the leaf, weight and increasing number of leaf. It might support by using seaweed fertilizer which contain growth hormones such as auxin, cytokinin, and gibberellin.

Table 1. Characteristics of fertilizer used in study.

Parameter		Fertilizer		
		Solid biological seaweed fertilizer	Liquid biological seaweed fertilizer	Compost
Micro nutrients	Fe (ppm)	8105.85	6534.25	-
	Cu (ppm)	43.41	42.1	-
	Mn (ppm)	279.92	275.24	-
	Zn (ppm)	68.56	75.85	-
	Pb (ppm)	31.79	37.28	-
	Cd (ppm)	2.85	3.45	-
	Co (ppm)	11.15	10.37	-
	S (%)	6.63	11.87	-
Macronutrients	N (%)	0.13	0.1	2.0 – 2.5
	P (%)	0.35	0.41	0.01 – 0.14
	K (%)	2.19	2.55	0.39 – 1.35
	Na (%)	0.78	0.76	-
	Ca (%)	13.66	13.3	0.13 – 1.32
	Mg (%)	4.53	3.57	0.04 – 0.21
	C-org (%)	15.99	16.88	19.0 – 40.0
	pH	7.25	7.52	-
Growth hormones	Sitokinin (%)	0.01	0.01	-
	Giberelin (%)	0.01	0.10	-
	Auksin (%)	0.02	0.02	-
	Kinetin (%)	0.02	0.02	-

Apparently, lettuce head with A3 fertilizer has increased the length of leaves (picture 1) and roots that are longer (picture 1and 2) after a maintenance period of 14 days, but lower in width (picture 3 and 4). Among all the combination fertilizer, A3 fertilizer has shown a maximum shoot and root length compared to C and others. Growth hormone in seaweed fertilizer both liquid and solid plays a role, including auxin, cytokines (zeatin and kinetin), and gibberellin.

Figure 1. The growth at 14 days

Figure 2. The growth of head lettuce at 28 days

Auxin as one of growth hormones plays a role in cell division, stem and root growth and other developmental change [11]. Auxin can be obtained from fresh and extracted seaweed thallus [3, 4].

The presence of microbes in seaweed fertilizer does not interfere with the presence of auxins [3]. Cytokines (zeatin & kinetin) are also involved in plant cell division, especially during embryogenesis, shoots, young leaves, cambium and leaf aging, and root formation, but also play a role in terms of environmental stress tolerance and nutrient signaling [10, 12, 13, 15]. Higher P content in seaweed fertilizer (Table 1) also played a role in this regard. Phosphorus has functions to stimulate root growth of seeds or young plants, accelerate and strengthen the growth of young plants.

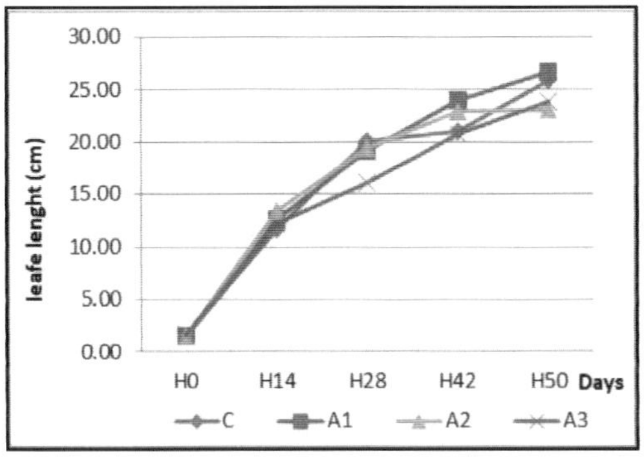

Figure 3. Length of head lettuce leaf

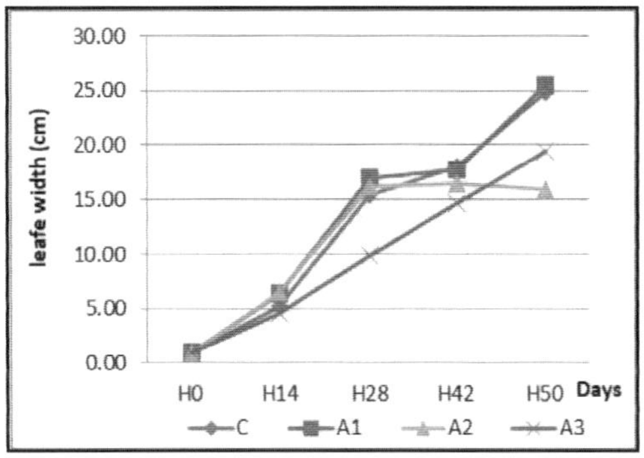

Figure 4. Width of head lettuce leaf

After a 28-day maintenance period, the roots no longer elongate and grow more towards the widening of the leaves. Besides that, the leaves appear to move in a circle in one direction. Apparently, the role of the gibberellin hormone can be seen in this condition. Gibberellin plays a role in leaf expansion and trichome development [6, 1, 8, 10, 14]. Leaf widths in plants with A1 fertilizer appeared to increase larger and plants with C fertilizer were slightly lower. Leaves on plants with A3 fertilizer appear to be increasing linearly but still lower than C and A12, while plants with A2 fertilizer appear to no longer widen.

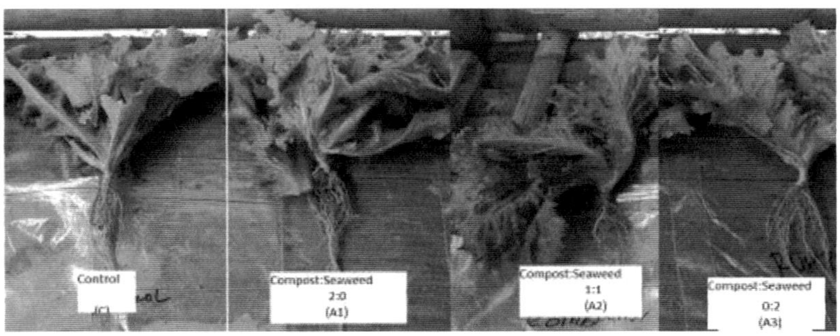

Figure 5. the growth of head lettuce at 42 days

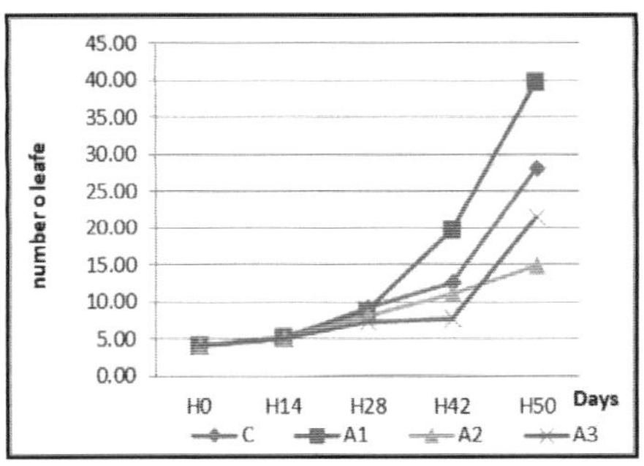

Figure 6. Number of head lettuce leaf

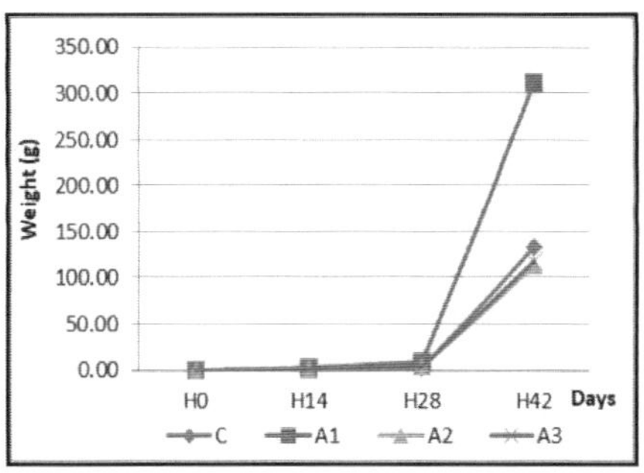

Figure 7. Weight of head lettuce leaf

Plant weight and number of leaves appear maximum after maintenance 42 days. Picture 5, 6, and 7). It has increased significantly. This is in accordance with the increase of leaf strands and leaf widening. The treatment with the lowest number of leaves was obtained in plants with A1 fertilizer, but it was heavier than the other treatments. N content in compost is higher than seaweed fertilizer (Table 1), this macro nutrient element is needed in the formation of vegetative parts of plants, such as leaves, stems and roots, the formation of green leaves as a place of photosynthesis for the formation of proteins, fats and various organic compounds, and improving the quality of leaf-producing plants. The green color of the lettuce head qualitatively seems unchanged, especially in plants that receive sea rumpt fertilizer (A2 and A3). The intake of Mn from seaweed fertilizer seems to play a role this element serves to maintain the green condition of the leaves on old leaves, as an enzyme activator of the assimilation process.

After 42 days of observation, lettuce head takes 8-10 days to be harvested. This period is a critical period where leaf spot disease often appears. On day 42 appeared brown spots on the lower leaves of the C treatment plants, on the 50th day these patches increased in the lower leaf layers and the same symptoms appeared in the A1 treatment plants. Apparently, the use of a bacterial consortium on seaweed fertilizer works well. Lettuce heads do not experience leaf spot disease on maintenance until the 50th day. The role of mineral K is also predicted to play a role. This element can function to strengthen the body of the plant so it is not easy to fall and higher resistance to disease. As in table 1, K content in seaweed fertilizer is higher than compost, as well as the Ca. K works together with the elements Ca and B in multiplying nodule branches so that it can provide a lot of growth-supporting bacteria so as to prevent parasitic bacteria.

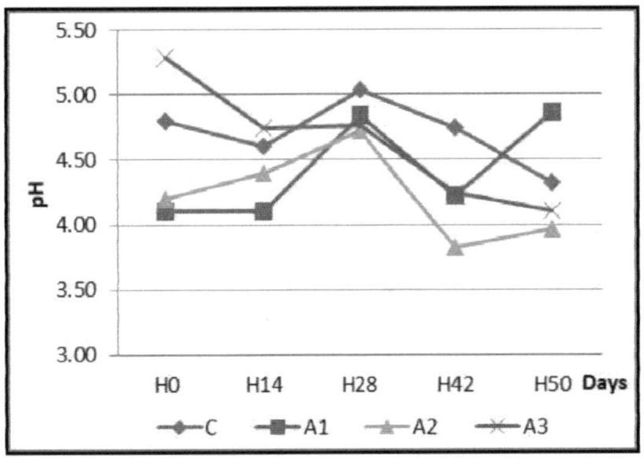

Figure 8. acidity of planting land during fertilizing head lettuce

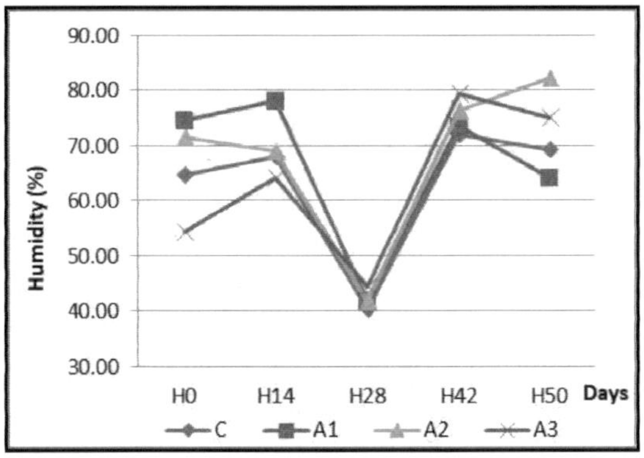

Figure 9. humidity of planting land during fertilizing head lettuce

Soil acidity indicates conditions that are not yet in accordance with the needs of lettuce head growth, i.e. pH 5.5 - 7.5. shown in Figure 8, the acidity of the soil in the planting area is quite low (below 5.5). One of the ways to handle soil with high acidity is by using compost [7]. However, the consistency of watering land also affects. on the 28th day, the acidity of the soil decreased because watering was reduced from 2 times a day to 1 time a day because it was associated with high rainfall. if it is not a reduced watering schedule, the watering volume may be able to maintain the acidity of the soil around pH 5. This is possible because in general between H-42 and H-50 where watering is back to normal, the acidity of the soil can increase again.

Besides seasons or rainfall, watering plants also affects soil moisture. Cultivation of this type of plant requires 75% soil moisture. As with soil acidity, soil moisture seems to decrease significantly due to reduced watering, then return to the expected conditions after watering again as usual. the soil moisture seemed stable maintained at around 75% in the A2 and A3 treatments using seaweed fertilizer.

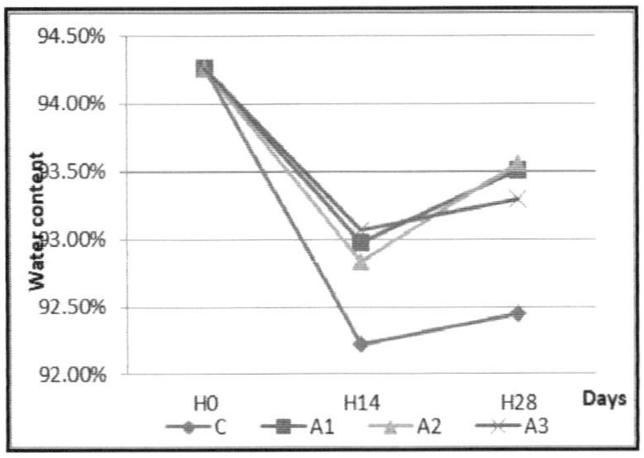

Figure 5. Water content of leaf head lettuce

The results of the study in the form of indicators of plant height, leaf width, number of leaves, and plant weight showed the highest value on head lettuce with A1 treatment (combination of compost and solid biological seaweed fertilizer are 2:0). But qualitatively, the use of seaweed fertilizer can provide a better green color until harvest time, soil moisture is well maintained, and leaf spot disease is not visible. Further research is needed to obtain the right stages and portions in the application of compost and seaweed fertilizer in the cultivation of head lettuce as well as to observe the stability of leaf green and other bioactive compounds i.e. chlorophyll, carotenoids, and tacochromanol [5], phenolic compounds and vitamin [9].

4. Conclusion

The best growth of head lettuces obtained from maintenance using a combination of compost and solid seaweed biofertilizers in the ratio of 2:0. Growth indicator shows that yields of head lettucefertilized with compost have characteristics of average leaf length of 26.64 cm, leaf width of 25.48 cm, 40 number of leaves, weight 311.75 g, and moisture content of 93.51%. Compost can stabilize acidity of planting land but the combination fertilizer can stabilize humidity of planting land and does not indicate the presence of fungi and parasites.

5. References

[1] Achard, P and Genschik, P 2009 J. of Exp. Botany, 1085-1092.
[2] Argueso, C T, Ferreira, F J, and Kieber, J J 2009 Plant, Cell, and Environment, 1147-1160.
[3] Basmal, J, Kusumawati, R, and Nurhayati 2016 Penelitian Pembuatan Pupuk Bio Padat Dan Cair dari Rumput Laut (Jakarta: Balai Besar Riset Pengolahan Produk dan Bioteknologi Kelautan dan Perikanan).
[4] Basmal, J, Munifah, I, Rimmer, M, and Paul, N 2019 Identification and Characterization of Soil Waste from Gracilaria Extraction (Jakarta: BBRP2BKP-ACIAR Report pp 1-16.
[5] Cruz, R, Gomes, T, Ferreira, A, Mendes, E, Baptista, P, and Cunha, S 2014 Food Chem. 95-101
[6] Daviere, J M and Achard, P 2013 Gibberelin signaling in plants (Development).
[7] Djamaan, A 2006 Pemberian Nitrogen (Urea) terhadap Pertumbuhan dan Hasil Selada (*Lactuca sativa* L) (Padang: Balai Pengkajian Teknologi Pertanian).
[8] Fos, M, and Garcia-Martinez, J L 2000 Plant Phy. 471-479.
[9] Kim, J G, Takami, Y, Mizugami, T, Beppu, K, Fukuda, T, and Kataoka, I 2006 Sci. Holticulture 110, 219-222.
[10] Matsuo, S, Kikuchi, K, Fukuda, M, Honda, I, and Imanishi, S 2012 J. of Exp. Botany 5569-5579.
[11] Pramanik, K and Mohapatra, P P 2017Int. J. Curr.Microbiol. App. Sci. 1624-1636.
[12] Sakakibara, H 2006 Cytokinin: Activity, Biosynthesis, and Translocation (Annual review of plant biology) pp 431-449.
[13] Schmulling, T 2004 Cytokinin In Lennarz, W. Lennarz, & M. D. Lane 2004 Encyclopedia of Biological Chemistry (Academic Press/Elsevier Science) pp 1-16.
[14] Serrani, J C, Foss, M, Atare's, A, and Garcia-Martinez, J L 2007 J. Plant Growth Reg. 211-221.

[15] Werner, T, Motyka, V, Laucou, V, Van Onckelen, H, and Schmulling, T 2003The Plant Cell 2532-2550.

6. Acknowledgments

This research was support from Bina Sarana Bakti Agatho Organic Farm for seedlings, planting land, compost, and agriculture maintenance.

Growth and yield of soybean plants on intercropping systems with different phosphorus dosage

Rusmana1* and E P Ningsih

Department of Agroecotechnology, Faculty of Agriculture, University of Sultan Ageng Tirtayasa. Jl. Raya Jakarta KM 04, Pakupatan, Serang, Banten

*Corresponding Author: rusmana@untirta.ac.id

Abstract. The intercropping system with the right dose of phosphorus will show higher yields than the monoculture planting system. The study was carried out with the aim of: (i) comparing crop yields between intercropping and monoculture planting systems, (ii) comparing the efficiency of land use of each planting system, and (iii) phosphorus doses which provide the highest yield for each cropping system. The study was conducted at the Experimental Garden of the Faculty of Agriculture, University of Sultan Ageng Tirtayasa Cikuya Karang Kitri District Serang, starting from April to July 2019. The experiment used a factorial randomized block design. The first factor is the planting system consisting of three levels, namely the system of soybean monoculture planting, maize monoculture planting system, and intercropping system of soybean+corn. The second factor is the dose of phosphorus fertilizer consisting of two levels, namely 25 $kg.ha^{-1}$ and 100 $kg.ha^{-1}$. Each treatment combination was repeated three times. To find out the purpose of the study used the Land Equity Ratio (LER) indicator and Variant Analysis of observational data which included: plant height, number of leaves, weight per plant, crop weight per plot and seed dry weight. The plant growth with corn monoculture is better than intercropping system. The yield of plants with intercropping systems is higher than that of monoculture systems. Plant yields in both cropping systems (intercropping and monoculture) were higher by giving phosphorus doses of 100 $kg.ha^{-1}$ compared to 25 $kg.ha^{-1}$.

Keywords: intercropping, monoculture, phosphorus doses

1. Introduction

Indonesia's population continues to grow at a rate of population growth for seven years (2010-2016) at an average of 1.36% [1]. The increase in population causes the demand for agricultural products both in terms of type, number and quality also increases. National corn demand continues to increase reaching 8.6 million tons per year or around 665 thousand tons / month [2]. According [3] that in 2005-2015, the growth rate of corn harvest area fluctuated, while for the average corn production in Indonesia in 2005-2015 amounted to 43.42 ku/ha. [4] stated that in 2016 soybean supply experienced a deficit of 1.61 million tons. Therefore, in an effort to increase crop productivity, it is necessary to use land efficiently. Efforts to improve the efficiency of land use include cultivation of intercropping systems such as corn intercropping with beans. Intercropping is planting more than one plant at the same time or during the planting period in the same place [5].

To increase the production of soy and corn required fertilizer. One of the fertilizers needed is Posfor. [6] states that P plays a role in plant growth (stem, roots, twigs and leaves). Phosphate is needed by plants for the formation of cells in the root tissue and the stem that grows as well as strengthens the stem, so it is not easily collapsed [7]. In soybean, phosphates are required for maximum root star activity

IOP Conf. Series: Earth and Environmental Science **383** (2019) 012043 doi:10.1088/1755-1315/383/1/012043

than is necessary for the formation of root nits. This fact shows that the maximum seed yield requires phosphate fertilizer sufficient to ensure the fixation process of N2 maximum [8]. A serious phosphate deficiency can slow down and postpone the primordial, so that the seed is produced wrinkles, mild, sprouts small and ripe early. It should be considered in order to direct the production of both quality and quantity [9].

The study was carried out with the aim of: (i) comparing crop yields between intercropping and monoculture planting systems, (ii) comparing the efficiency of land use of each planting system, and (iii) phosphorus doses which provide the highest yield for each cropping system.

2. Materials and Method

The materials used are: SP-36, urea, KCl, compost, corn and soybean. Tools used include: analytic scales, ruler, and camera.

The study was conducted at the Experimental Garden of the Faculty of Agriculture, University of Sultan Ageng Tirtayasa Cikuya Karang Kitri District Serang, starting from April to July 2019. The experiment used a factorial randomized block design. The first factor is the planting system consisting of three levels, namely the system of soybean monoculture planting, maize monoculture planting system, and intercropping system of soybean+corn. The second factor is the dose of phosphorus fertilizer consisting of two levels, namely 25 $kg.ha^{-1}$ and 100 $kg.ha^{-1}$. Each treatment combination was repeated three times. If the variance analysis shows effected, will continue with test DMRT 5%.

3. Results and Discussion

Table 1. High crop with intercropping system treatment and fertilizer posfor at age 14 DAP, 21 DAP, 28 DAP, 35 DAP, 42 DAP and 49 DAP (on centimetre).

Days After Planting	Phosphorus Doses	Planting System			Average
		T1 (monoculture corn)	T2 (monoculture soybean)	T3 (intercropping system of soybean+corn)	
14	P1 (25 $kg.ha^{-1}$)	23.57	7.10	27.45	19.37
	P2 (100 $kg.ha^{-1}$)	30.66	8.15	23.73	20.84
	Average	27.12 a	7.62 b	25.59 a	
21	P1 (25 $kg.ha^{-1}$)	35.80	12.45	39.47	29.24
	P2 (100 $kg.ha^{-1}$)	49.60	9.30	32.55	30.48
	Average	42.70 a	10.87 b	36.00 a	
28	P1 (25 $kg.ha^{-1}$)	61.00	16.00	55.90	52.38
	P2 (100 $kg.ha^{-1}$)	66.79	11.00	43.78	48.96
	Average	63.89 a	13.50	49.84 a	
35	P1 (25 $kg.ha^{-1}$)	74.95	26.50	71.65	66.61
	P2 (100 $kg.ha^{-1}$)	104.40	16.50	63.90	74.48
	Average	89.67 a	21.50 c	67.77 b	
42	P1 (25 $kg.ha^{-1}$)	101.96	42.52	100.65	92.90
	P2 (100 $kg.ha^{-1}$)	128.37	38.60	105.47	105.72
	Average	115.16 a	40.56 b	103.05 a	
49	P1 (25 $kg.ha^{-1}$)	121.33	54.68	130.18	115.60
	P2 (100 $kg.ha^{-1}$)	142.50	41.12	119.17	118.02
	Average	131.92 a	47.90 b	124.67 a	

Description : numbers followed by the same letters in the same line showed no significant according to Duncan Multiple Range Test 5%

Table 2. Number of leaves with the intercropping system treatment and fertilizer posfor at age 14 DAP, 21 DAP, 28 DAP, 35 DAP, 42 DAP and 49 DAP (on centimetre).

Days after planting	Phosphorus Doses	Planting System			Average
		T1 (monoculture Corn)	T2 (monoculture soybean)	T3 (intercropping system of soybean+corn)	
14	P1 (25 kg.ha^{-1})	2.20	1.15	2.18	1.84
	P2 (100 kg.ha^{-1})	2.21	2.57	1.93	2.23
	Average	2.20 a	1.86 b	2.05 a	
21	P1 (25 kg.ha^{-1})	6.26	2.45	5.33	4.68
	P2 (100 kg.ha^{-1})	6.06	5.20	5.33	5.53
	Average	6.16 a	3.82 b	5.33 a	
28	P1 (25 kg.ha^{-1})	8.53	3.60	7.06	7.20
	P2 (100 kg.ha^{-1})	8.33	8.00	7.46	7.91
	Average	8.43 a	5.80 b	7.26 ab	
35	P1 (25 kg.ha^{-1})	10.60	2.40	5.93	7.42
	P2 (100 kg.ha^{-1})	8.13	5.00	6.40	6.94
	Average	9.36 a	3.70 b	6.16 ab	
42	P1 (25 kg.ha^{-1})	9.66	6.80	7.93	8.51
	P2 (100 kg.ha^{-1})	9.06	8.40	9.06	8.97
	Average	9.36	7.60	8.50	
49	P1 (25 kg.ha^{-1})	10.13	9.00	8.06	9.08
	P2 (100 kg.ha^{-1})	9.40	8.80	9.00	9.14
	Average	9.76	8.90	8.53	

Description : numbers followed by the same letters in the same line showed no significant according to Duncan Multiple Range Test 5 %

Table 3. Effect of intercropping and fertilizer system treatment on weight per plant (on gram).

Phosphorus Doses	Planting System			Average
	T1 (monoculture Corn)	T2 (monoculture soybean)	T3 (intercropping system of soybean+corn)	
P1 (25 kg.ha^{-1})	71.67	10.00	35.66	35.04
P2 (100 kg.ha^{-1})	51.11	12.29	44.03	37.99
Average	59.33 a	10.92 b	39.01 a	

Description : numbers followed by the same letters in the same line showed no significant according to Duncan Multiple Range Test 5 %

Table 4. Effect of intercropping and phosphorus fertilizer treatment on crop weight per plot(on gram).

Phosphorus Doses	Planting System			Average
	T1 (monoculture Corn)	T2 (monoculture soybean)	T3 (intercropping system of soybean+corn)	
P1 (25 kg.ha^{-1})	860.0	700.1	1069.9	878.8
P2 (100 kg.ha^{-1})	613.3	860.0	1320.7	885.9
Average	712.0	764.1	1170.2	

Description : numbers followed by the same letters in the same line showed no significant according to Duncan Multiple Range Test 5 %

Tabel 5. Effect of intercropping system and fosfor fertilizer on weight 100 seeds (on gram).

Phosphorus Doses	Planting System			Average
	T1 (monoculture Corn)	T2 (monoculture soybean)	T3 (intercropping system of soybean+corn)	
P1 (25 kg.ha^{-1})	20.83	18.61	20.18	19.87
P2 (100 kg.ha^{-1})	27.50	20.94	29.27	25.90
Average	24.16	19.77	24.72	

Description : numbers followed by the same letters in the same line showed no significant according to Duncan Multiple Range Test 5 %

The planting system treatment affects the crop height in 7-49 DAP. Foshor treatment did not effect on the crop height parameters in the DAP 7-49. There is no interaction between the treatment of the planting system and nutrient fertilizer on the high parameter of the plant (table 1). Table 1 shows that the height of the plant inside the corn monoculture is greater than the intercropping system (soybean and corn) and the soybean monoculture system. In the high parameters of the plant in both the planting system (intercropping and monoculture) higher by giving a dose of phosphorus 100 kg. Ha^{-1} compared to 25 kg. Ha^{-1}.

Treatment system of planting and phosphorus fertilizer did not effect on the number of leaf parameters. There is no interaction between the processing of the planting system and phosphorus fertilizer on the number of leaves (Table 2). Table 2 shows that the number of leaves in corn monoculture is much more than the intercropping system (soybean and corn) and soybean monoculture system. On the number of leaves in both the planting system (intercropping and monoculture) higher by giving a dose of phosphorus 100 kg. Ha^{-1} compared to 25 kg. Ha^{-1}.

This may be because in the intercropping system of corn and soybeans there is a competition in taking nutrients, water and different light between the corn plant and the soybean plant in the vegetative phase. This is in line with the statement [10] stating that increased cereal growth will make the growth of nuts depressed. The research of [11] states that there are inter-specific rivalries such as water, light, air and nutrients other than corn (C4) are more aggressive than soy (C3). [12], shows that corn plants are more dominant and more competitive compared to Cowpea with competition ratio value of 1.89 vs 0.52.

Phosphorous fertilizer at a dose 100 kg. Ha^{-1} increases plant growth compared to 25 kg. Ha^{-1} because of the provision of phosphate fertilizer supports the growth of plants in both monoculture systems and intercropping. The high growth of plants is a plant growth in the vegetative phase where cell division and enlargement occur in a special tissue called meristems. [13] state that eventually meristem produce new cells at the end of the roots or stems, causing the plant to grow taller or longer. P fertilizer works to stimulate root growth and development. The roots will absorb water and nutrients into the leaves into carbohydrates that will be transplanted into the parts of plants that require food and energy reserves.

The planting system treatment affects weight per plant. Phosphorus fertilizer treatment did not affect the weight per plant. There is no interaction between the processing of the planting system and the phosphorus fertilizer in the weight parameter per plant (Table 3). In the weight parameters of plants per plot treatment of planting system and phosphorus fertilizer does not affect the weight parameter of the plant per plot. There is no interaction between the processing of the planting system and phosphorus fertilizer on the weight parameter of the plant per plot (Table 4). The weight per plant in the processing of corn monoculture is greater than that of soybean monoculture and the intercropping system (soy + corn) which is suspected because corn plants do not experience much competition or competition in sunlight or Nutrients in the soil, so photosynthesis of the monoculture system would be good to increase the growth of corn. According to [14] that the dry weight of plants shows the efficiency of the photosynthesis process. The larger the photosynthate is produced, the greater the dry weight is generated. The greater the dry weight of the plant, the better the growth of plants and nutrients and the water absorbed by the plant is also better. It is powered by [15] The larger the heavy dry plant, the better the growth of plants and nutrients and the water absorbed by the plant is also better. [16] adds that the more sunlight received by plants can add photosynthetic products utilized by plants for growth and metabolism.

Dose of phosphorus at a dose 100 kg. Ha^{-1} and 25 kg. Ha^{-1} did not effect on weight parameters per plant and plant weight per plot. It is suspected that the dose of phosphorus fertilizer is not high enough and did not been absorbed completely by the roots to increase the dry weight of the plant so it did not help significantly improve photosynthesis. According to [17] dry plant material is strongly influenced by the optimal process of photosynthesis. The dry weight formed reflects the amount of photosynthate as a result of photosynthesis, because the dry material depends heavily on the rate of photosynthesis. [18] argue that plant growth can be shown to one or more organs, expressed in dry weight.

In table 5 indicates that the weight of 100 seeds with a higher intercropping system than soybean monoculture system and corn monoculture system. The weight of 100 seeds in both the planting system (intercropping and monoculture) is higher by providing a dose of phosphorus 100 kg. Ha-1 compared to 25 kg. Ha-1. It is suspected that soybeans can increase corn production as N fixation is provided by soy for the availability of corn production. According to [19] One of the important elements in crop production is N. Most of the nitrogen is transferred in the generative phase which is able to stimulate the formation of cob on corn (Zea Mays). The translocation of nitrogen nutrients that take place well on the plant affects the growth, cob size and weight of corn seeds. According to [20], the planting of intercropping will give higher production between nuts and food crops.

Phosphorus fertilizer at a dose of 100 kg. Ha^{-1} produces a weight of 100 seeds compared to 25 kg. Ha^{-1}. It is suspected that phosphate fertilizer affects crop production in both monoculture systems and intercropping systems. According to [21], element P can increase the protein content and weight of the seeds which in turn affects the crop yield. Large seed size delivers high total dried seed yield. [22] states the accumulation of proteins in the seed prescribed by the adequacy of P-plants, therefore the deficiency of P in plants will produce low growth rates and inhibit the formation of nodules. According to [23] the replenishment of the seeds comes from photosynthates produced after flowering and relocation of the stored photosintat.

Conclusion

The plant growth with corn monoculture is better than intercropping system. The yield of plants with intercropping systems is higher than that of monoculture systems. Plant yields in both cropping systems (intercropping and monoculture) were higher by giving phosphorus doses of 100 $kg.ha^{-1}$ compared to 25 $kg.ha^{-1}$.

References

[1] BPS. 2019. Population growth rate by province. https://www.bps.go.id/statictable/2009/02/20/1268/laju-pertumbuhan-penduduk-menurut-provinsi.html. Accessed on 5 Agustus 2019.

[2] KEMENPERIN. 2016. RI Imports Corn at 2.4 Million Tons. Tersedia online di http://kemenperin.go.id/artikel/13892/2016,-RI-Impor- Jagung-2,4-Juta-Ton. Accessed on 5 Agustus 2019.

[3] Kementrian Pertanian. 2015. Corn Farm Food Commodity Outllook. Pusat Data dan Sistem Informasi Pertanian. Jakarta.

[4] Pusat Data dan Sistem Informasi Pertanian. 2015. Outlook on Agricultural Commodities for Soybean Food Crops. Pusat Data dan Sistem Informasi Pertanian. Jakarta.

[5] Indriati, T. R. 2009. *Effect of Organic Fertilizer Doses and Plant Populations on Growth and Yields of Soybean Glycine max L and Corn* (*Zea mays* L.). Tesis. Sekolah Pascasarjana, Universitas Sebelas Maret, Surakarta.

[6] Embleton, T.W., W.W. Jones, C.K. Lebanauskas, and W. Reuther. 1973. *Leaf Analysis as a Diagnostic Tool and Guide to Fertilization.* In W. Reather (Ed.). The Citrus Industry. Rev. Ed. Univ. Calif .Agr. Sci. *Barkely*. **3**:183-210.

IOP Conf. Series: Earth and Environmental Science **383** (2019) 012043 doi:10.1088/1755-1315/383/1/012043

[7] Aleel, K.G. 2008. *Phosphate Accumulation in Plant: Signaling. Plant Physiol.* **148**:3-5.
[8] Somaatmadja, S. 1985. Increasing Soybean Production Through Assembling Varieties. Balai Penelitian Tanaman Pangan. Bogor.
[9] Lingga, P. 1989. *Instructions for use of fertilizers.* Penebar Swadaya. Jakarta.
[10] Ariel, C. E., Eduardo, O. A., Benito, G. E., & Lidia, G. 2013. Effects of two plant arrangements in corn (Zea mays L .) and soybean (Glycine max L. Merrill) intercropping on soil nitrogen and phosphorus status and growth of component crops at an Argentinean Argiudoll. *American J. of Agriculture and Forestry.* **1**(2): 22-31.
[11] Muoneke CO, Ogwuche MAO, Kalu BA. 2007. Effect of maize planting density on the performance of maize/soybean intercropping system in a Guinea savannah agro-ecosystem. African. *J. of Agricultural Research.* **2**(12): 667-677.
[12] Alla, W.A.H., E.M. Shalaby, R.A. Dawood, and A.A. Zohry. 2014. *Effect of cowpea (Vigna sinensis L.) with maize (Zea mays L.) intercropping on yield and its components. International Scholarly and Scientific Research & Innovation.* **8**(11): 1258–1264.
[13] Gardner, F. P., R. B. Pearce daan R. L. Mitchell. 1991. *Physiology Cultivation Plant.* Translation H. Susilo. UI Press. Jakarta.
[14] Musfal. 2010. Potential of arbuscular mycorrhizal fungi to increase crop yields of corn. *Jurnal Litbang Pertanian.* **29** (4):154-158.
[15] Zuchri, A. 2007. Optimization of Peanut and Corn Crop Results in Intercropping through Planting Row Arrangement and Corn Leaf Composting. Jurusan Budidaya Pertanian Fakultas Pertanian Unijoyo. Embryo 4(2):157-163.
[16] Warsana, E. 2009. Potential Kerandang (*Canavalia virosa*) as Alternative Animal Feed and Food Sources. Balai Pengkajian Teknologi Pertanian, Yogyakarta. Seminar Nasional Teknologi Peternakan dan Veteriner: 765-769.
[17] Dwijosepoetro, D. 1981. *Introduction Plant physiology.* PT. Gramedia Pustaka Utama. Jakarta.
[18] Tisdale, S.L. dan W.L. Nelson. 1975. Soil Fertility and Fertilitizers. Third Edition. Mac. Millian Publishing Company, Inc. New York.
[19] Sirajuddin, M., dan S. A. Lasmin. 2010. *Response to growth and yield of sweet corn (Zea mays saccharata) at various times of nitrogen fertilizer and straw mulch thickness.* Jurusan Budidaya Pertanian, Fakultas Pertanian, Universitas Tadulako, Palu, Sulawesi Tengah. *Jurnal Agroland.* **17**(3):184-191.
[20] Mwangi, D. M., G. Cadisch., W. Thorpe And K. E. Giller. 2004. *Harvesting Management Options for Legumes Intercropped in Napier Grass in The Central Highlands of Kenya. Tropical Grasslands.* **38**:234-244.
[21] Saerong. 2008. Effect of Phosphorus and Potassium on the Growth and Yield of Peanuts in Dry Land. Balai Pengkajian Teknologi Pertanian. Temu Teknis Fungsional Non Peneliti. Malang.
[22] Mitsuhashi, N., M. Ohnishi, Y. Sekiguchi, K. Y.U. Kwon, Y.T. Chang, S.K. Chung, Y. Inoue, R.J. Reid, H. Yagisawa,and T. Mimura. 2005. *Phytic acid synthesis and vacuolar accumulation in suspensioncultured cells of Catharanthus roseus induced by high concentration of inorganic phosphate and cations. Plant Physiol.* **138**:1607-1614.
[23] Goldworthy dan Fisher. 1992. The *Physiology of Tropical Fields Crops* (Translation dari The Physiology of Tropical Fields Crops by Tohari). Gadjah Mada University Press. Yogyakarta.

Sensory evaluation of synbiotic yoghurt with Banten taro flour as prebiotic

S Kusumasari and V Y Pamela

Department of Food Technology, Faculty of Agriculture, University of Sultan Ageng Tirtayasa, Indonesia
Indonesia Center of Excellence for Food Security, University of Sultan Ageng Tirtayasa, Indonesia
E-mail septariawulan@gmail.com

Abstract. Banten taro is indigenous food in Banten Province, Indonesia. The functional value of yogurt can be increased by the addition of Banten "beneng" taro flour to stimulate the growth of probiotic bacteria in the digestive tract. Banten taro had potential as prebiotic. Sensory quality was an importance aspect of food product. Flavor change caused by adding Banten taro flour in yoghurt was undetermined. This study was aimed to determined the effect of variation lactic acid bacteria starter for the sensory quality of Banten taro yoghurt synbiotic. A seven-point facial hedonic scale was used to evaluate six yogurt samples. Variation of lactic acid bacteria starter and fortification of modified taro flour was not given significant effect for organoleptic quality (taste, texture, and smell) of synbiotic yogurt. The results showed that the best combination starter to produce yogurt was *Lactobacillus bulgaricus* + *Lactobacillus achidophillus* + *Streptococcus thermophillus*.

1. Introduction

Fermentation products such as yoghurt had been recognized for its beneficial effects on human health, longevity, and wellness [1]. The addition of lactic acid bacteria into the food can improving digestion, optimizing absorption of nutrition, and enhancing food safety [2,3]. Fermentation was a low cost cooking technique and easy to apply in household level. Probiotics are live microorganisms when administered in adequate amounts confer health benefits on the host. The reported beneficial effects of probiotic consumption include improvement of intestinal health, amelioration of symptoms of lactose intolerance, enhancement of the immune response, reduction of serum cholesterol, control of urogenital and respiratory tract infections, and reduction of the risk of colon cancers [2,3].

Synbiotics consist of both probiotic microorganisms and prebiotic compounds. Synbiotic products contain probiotics (useful bacteria) and prebiotics (indigestible carbohydrates) that stimulating beneficial bacteria growth. Synbiotics have antimicrobial, anticancer, antiallergic and immunestimulating properties. Synbiotics can improving absorption of minerals, preventing incidence of diarrea, and optimizing assimilation of nutrients [4].

Prebiotics can stimulate selective probiotics growing in health human colon [5]. One of the most familiar prebiotic was inulin. Inulin is polysaccharides (mostly fructan) that consist of fructose with glycosidic bond β-(2-1) and glucose terminal at the ends [6]. Inulin is expensive so synbiotic products with inulin have higher price. Cheaper prebiotics was needed such as dahlia flour [6], taro starch [7], gembili [8], banana [9] and others. Taro was one of potential alternative prebiotic that contain of inulin, rafinosa and resistant starch [7]. Banten taro (*Xanthosoma undipes* K. Koch) had bigger tuber than other taro and easily cultivated. Banten taro (figure 1) is indigenous food from Juhut village

Pandeglang districts Banten province. Banten taro called Beneng (besar koneng) because of its huge tubers and yellowish colour. Banten taro had produce 20 kg tubers for 2 years. Banten taro had more starch (75.62%) than hijau taro, mentega taro, and semir taro [10]. Banten taro flour can be cheaper alternative but flavor changes of synbiotic yoghurt was undetermined.

Figure 1. Banten taro.

The objective of this study was to determined the effect of variation lactic acid bacteria starter for the sensory quality of Banten taro yoghurt synbiotic and to determined consumer acceptability of synbiotic yoghurt with Banten taro flour as prebiotic.

2. Materials and Methods

2.1. Preparation of Yoghurt Starter

Yoghurt starter was freezed dried. Five grams of starter was used for one litre of milk. S1 was *Lactobacillus bulgaricus* + *Streptococcus thermophillus*, S2 was *Lactobacillus bulgaricus* + *Lactobacillus achidophillus* + *Streptococcus thermophillus*, and S3 was *Lactobacillus bulgaricus* + *Lactobacillus achidophillus* + *Streptococcus thermophillus* + *Lactobacillus casei* + *Bifidus longum.*

2.2. Yoghurt Manufacturing

Standardized milk with 2.4% fat was heated 140°C for 4 seconds then mixed with 5% sugar and 1% Banten taro flour, cooled to 40°C, then add yoghurt starter. Fermented in yoghurt maker for 8 hours. Yoghurt was refrigrated before served. Y1 (milk without Banten taro flour, S1), Y2 (milk with 1% Banten taro flour, S1), Y3 (milk without Banten taro flour, S2), Y4 (milk with 1% Banten taro flour, S2), Y5 (milk without Banten taro flour, S3), Y6 (milk with 1% Banten taro flour, S3).

2.3. Panel selection

To be eligible for participation, subjects had to be over the age of 18 years, capable of understanding the intent of the study and able to provide informed consent. Fifty untrained panelists were selected to evaluate variations of yogurt. A total of 12 untrained male panelists and 38 untrained female panelists between the ages of 18 - 61 years old completed the sensory evaluation.

2.4. Sensory Evaluation

Sensory evaluations were privately conducted while participants were seated in a quite area with privacy. Tastings occurred between the hours of 16:00 - 20:00 to correspond with regular meal times. Prior to the tastings, all ingredients were disclosed to the participants to ensure cultural or religious restrictions would not be violated by participation. Each panelist was given a pen for recording on her/his evaluation sheet and a glass of water to cleanse the mouth between tastings.

A seven-point facial hedonic scale in which 7 = "liked extremely", 4 = "neither liked nor disliked" and 1 = "disliked extremely" was used by each participant for sample evaluation. Approximately 20 ml of each homogen sample were presented in a plastic cup with a plastic spoon and a plastic straw. Six samples were served one by one in a random order. Hedonic organoleptic parameters were taste, texture, and smell [11].

3. Results and discussion

3.1. Banten taro flour

Banten taro flour was made by cleaning (soak in NaCl 10% for 1 hour), chopping, drying (50°C-60°C for 6-12 hours), milling, and shieving. Banten taro flour yield is 10,24 % [12]. Banten taro contain high oxalic acid. Oxalic acid gives itchy feel. Banten taro in Juhut had oxalic acid lesser than others because cultivated in less acidic soil [13]. Step soak in NaCl 10% was to reduced oxalic acid in Banten taro. This step can reduce oxalic acid 50% [14]. Banten taro flour characteristics was shown in Table 1 [15].

Table 1. Banten taro flour characteristics.

Characteristics	% w/w
Water content	6.10
Ash content	6.11
Fat	0.39
Protein	6.70
Carbohydrate	80.70

3.2. Sensory properties

Sensory properties (taste, smell, and texture) obtained by hedonic aspect. In this research, consumer acceptability change was determined by compared yoghurt with and without Banten taro flour and compared variety of starter for synbiotic yoghurt (yoghurt with Banten taro flour as prebiotic). Overall, participants preferred yogurt products which were sweet and little bit sour with fruity sour aroma and those which maintained a smooth and homogenous texture.

Hedonic value (taste, smell, and texture) of all yoghurt sampel (Y1, Y2, Y3, Y4, Y5, and Y6) had 5-6 poin from 7 scale (slightly like to like). All samples had given good acceptability (above 4 point). Y1 and Y2 showed no difference in texture and smell (Figure 2). Y2 (with Banten taro flour) had lower acceptability score than Y1 for taste. Yoghurt with Banten taro flour had more sourness than the original, but panelists able to handle the sourness. Comparation Y3 and Y4 or Y5 and Y6 were given the same result. Banten taro flour in yoghurt changed taste but texture and smell was similar with the original. Fermentation time had effect of sourness yoghurt. Yoghut with Banten taro flour was fermented for 5, 6, and 7 hours. Yoghurt that fermented for 6 hours had similar sourness with original yoghurt that fermented for 8 hours. Then, added Banten taro flour to yoghurt did not change the taste just made fermentation process faster. Adding Banten taro flour changed yoghurt colour. Yoghurt synbiotic had more yellowish darker colour than the original. This colour change had no effect in organoleptic evaluation because panelis evaluated sample one by one and cant compared between samples.

Comparation hedonic score Y2, Y4, and Y6 was to determine effect of different starter in synbiotic yoghurt (figure 3). Y2 with standart yoghurt culture had the lowest score for taste, smell, and texture. Adding more lactic acid bacteria (Y4 and Y6) improve acceptability hedonic score for taste, smell, and texture. For taste and smell Y4 had the higest hedonic score. Y4 had taste sweet and sour that given sour fresh fruity like aroma. For texture Y6 had the higest hedonic score. Y6 had liquid texture that the easiest to swallowed. Y4 and Y6 had similar organoleptic features, but Y6 was more fluid than Y4. In commentary column, most panelist preferred Y5 from all sample because it had slightly sour taste with fruity aroma and easy to swallowed. For acceptability and economic reason, starter 2 was the best starter to produce synbiotic yoghurt with good acceptability. Starter 2 produced Y3 and Y4 that had organoleptic score for taste 5.8; 5.5 (slightly like to like), smell 5.1;5.3 (slightly like to like), and texture 5.9;5.8 (slightly like to like). Viability test of lactic acid bacteria need to be done to decide the best starter for synbiotic yoghurt in health aspect and optimation for amount of Banten taro flour that added.

IOP Conf. Series: Earth and Environmental Science **383** (2019) 012044 doi:10.1088/1755-1315/383/1/012044

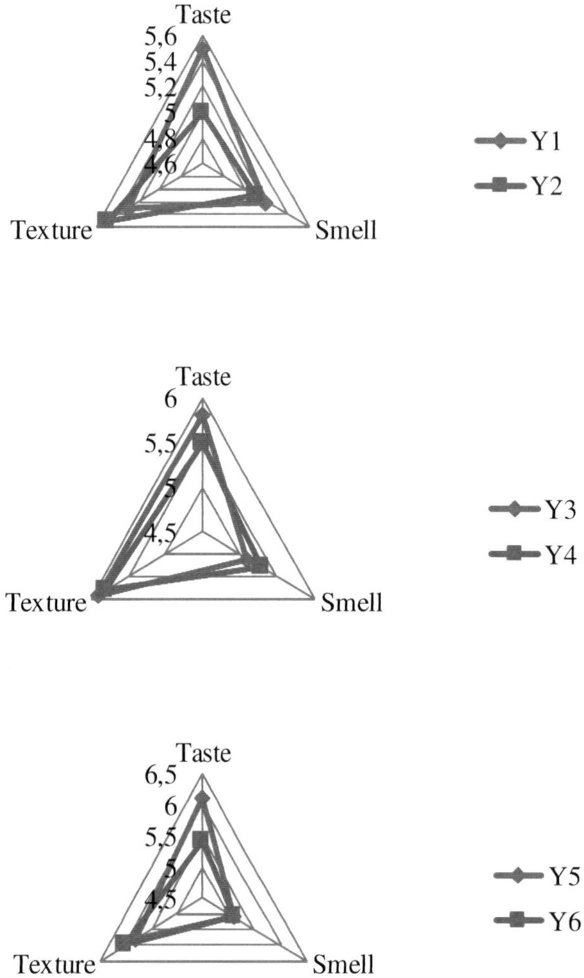

Figure 2. Organoleptic hedonic score (compared yoghurt with and without Banten taro flour).

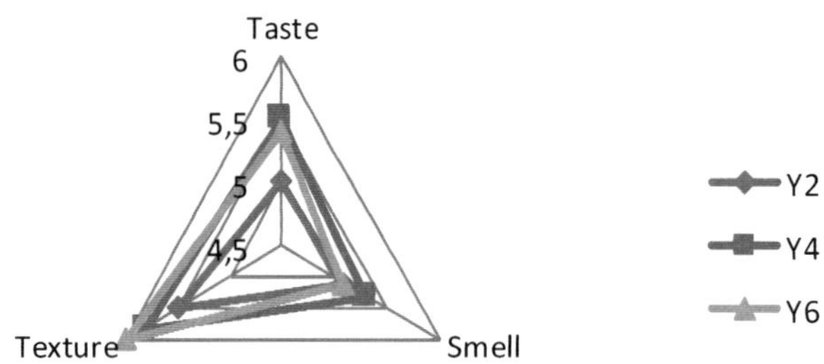

Figure 3. Organoleptic hedonic score (synbiotic yoghurt with starter variation)

4. Conclusion

Yoghurt with Banten taro flour 1% had more sour taste than the original yoghurt. The sour taste of synbiotic yoghurt was still acceptable (hedonic 5-5.5 poin from 7 scale) in this community. Texture and smell of the original yoghurt and synbiotic yoghurt was indistinguishable. Variation of starter affect the texture and smell of yoghurt. S2 and S3 made yoghurt with texture like fluid. S1 made creamy like yoghurt. S3 made yoghurt with smell like sour fresh fruity that the most acceptable aroma of all yoghurt (Y5). For acceptability and economic reason, the results showed that the best combination starter to produce yogurt was *Lactobacillus bulgaricus* + *Lactobacillus achidophillus* + *Streptococcus thermophillus* (S2).

5. References

[1] Conly J M, Johnston L B. 2004. Coming Full Circle: From Antibiotics to Probiotics and Prebiotics. *The Canadian J of Infectious Disease and Med Microbiology*.**15** pp 161-163
[2] Vasudha S, Mishra H N. 2013. Non-dairy probiotic beverages. *International Food Research Journal*. **20** pp 7-15
[3] Prasanna P H P, Grandison A S, Charalampopoulos D. 2014. Bifidobacteria in milk products: An overview of physiological and biochemical properties, exopolysaccharide production, selection criteria of milk products and health benefits. *Food Research International*. 55 pp 247-262
[4] Antarini A A N. 2011. Sinbiotik antara prebiotik dan probiotik. *J Ilmu Gizi*. **2** pp 148-155
[5] Venica C I, Bergamini C V, Rebechi S R, Perotti M C. 2015. Galacto-oligosaccharides formation during manufacture of different varieties of yogurt. Stability through storage. *LWT - Food Science and Technology*. **63** pp 198-205
[6] Faridhi K K, Lunggani A T, Kusdiyantini E. 2013. Penambahan Filtrat Tepung Umbi Dahlia sebagai Prebiotik dalam Pembuatan Yoghurt Sinbiotik. *BIOMA*. **15** pp 64-72
[7] Setiarto R H B, Nunuk W, Iffat F. 2017. Pengaruh starter bakteri asam laktat dan penambahan tepung talas termodifikasi terhadap kualitas yogurt sinbiotik. *Jurnal Riset Teknologi Industri*. **11** pp 18-30
[8] Febriansyah R, Pramono A. 2015. Yoghurt Sinbiotik Tanpa Lemak dengan Penambahan Tepung Gembili terhadap Kadar Trigliserida Tikus Hiperkolesterolemia. *J of Nutrition College*. **4** pp 57-61
[9] Rahmawati F C, Djamiatun K, Suci N. 2017. Pengaruh Yoghurt Sinbiotik Pisang terhadap Kadar Glukosa dan Insulin Tikus Sindrom Metabolik. *J Gizi Klinik Indonesia*. **14** pp 10-18
[10] Apriani R R N, Setyadjit, M. Arpah. 2011. Karakterisasi Empat Jenis Umbi Talas Varian Mentega, Hijau, Semir, dan Beneng serta Tepung yang Dihasilkan dari Keempat Varian Umbi Talas. *Jurnal Ilmiah dan Penelitian Ilmu Pangan* **1** pp 2-6
[11] Gonzalea N J, Adhikari K, Sancho-Madriz M F. 2011. Sensory characteristics of peach flavored yogurt drinks containing prebiotics and synbiotics. *LWT – Food Science and Technology*. **44** pp 158-163
[12] Lestari S, Susilawati P N. 2015. Uji Organoleptik Mi Basah Berbahan Dasar Tepung Talas Beneng (Xantoshoma undipes) untuk Meningkatkan Nilai Tambah Bahan Pangan Lokal Banten. *Prosiding Seminar Nasional Masyarakat Biodiversitas Indonesia*. **1** pp 941-946
[13] Pancasasti R. 2016. Pengaruh Elevasi Terhadap Kadar Asam Oksalat Talas Beneng (*Xanthosoma undipes* K.Koch) di Sekitar Kawasan Gunung Karang Provinsi Banten. *Jurnal Ilmiah SETRUM*. **5** pp 21-25
[14] Muttakin S, Muharfiza, Lestari S. 2015. Reduksi Kadar Oksalat pada Talas Lokal Banten melalui Perendaman dalam Air Garam. *Prosiding Seminar Nasional Masyarakat Biodiversitas Indonesia*. **1** pp 1707-1710
[15] Budiarto M S, Rahayuningsih Y. 2017. Potensi Nilai Ekonomi Talas Beneng (*Xanthosoma undipes* K.Koch) Berdasarkan Kandungan Kandungan Gizinya. *Jurnal Kebijakan Pembangunan Daerah*. **1** pp 1-12

Evaluation of land suitability for cultivation of java patchouli (*Pogostemon heyneanus* Benth) in the dry land of northern Serang, Banten province

S Ritawati*, P Utama, E Pancaningsih, Ismawati

Agroecotechnology Department, Faculty of Agriculture, University of Sultan Ageng Tirtayasa, Jl. Raya Jakarta Km.04, Pakupatan, Serang City, Indonesia

*Corresponding author: sri.ritawati@untirta.ac.id

Abstract. This research aimed to determine (1) land suitability limiting factors for Javanese patchouli cultivation in North Serang Regency, Banten Province, (2) the level of land suitability for Javanese patchouli cultivation in North Serang District, Banten Province, (3) land improvement efforts to be suitable for Java Patchouli Plant cultivation in North Serang Regency, Banten Province. This type of research was descriptive research. The study was conducted in the Dry Land of North Serang Regency, Banten Province including Tirtayasa District, Kramatwatu District, and Waringinkurung District. The research was conducted from October 2018 to March 2019. The research consisted of 5 stages: preparation and secondary data collection, preparation of land unit maps, data collection stage in the field (primary data), analysis of land suitability data and stages making land suitability maps. The method of data analysis was done by matching between the characteristics and quality of land in the study area with the criteria for growing patchouli plants. The results showed that : (1) Land suitability limiting factors for Java Patchouli cultivation included the availability of water, rooting media, nutrient availability, terrain conditions. (2) Land use in Tirtayasa sub-district includes land suitability classes S2 (moderately suitable) and S3 (marginally suitable). Land use in Kramatwatu District includes land suitability classes S2 and S3. Land use in Waringinkurung District includes land suitability classes S3 and S2. (3) Efforts to improve land can be made by making terraces, providing efficient additional irrigation, making drainage channels, liming, fertilizing, adding organic matter.

Keywords : land suitability evaluation, patchouli plant, dry land.

1. Introduction

Patchouli (*Pogostemon* sp.) is one of the essential oil-producing plants known as patchouly oil. Patchouli oil is widely used in cosmetics, perfume, soap, insect killers and medicines industries. In Indonesia there are 3 types of patchouli oil, namely Java (*Pogostemon heyneanus* Benth), Soap (*Pogostemon hortensis* Back), and Aceh (*Pogostemon cablin* Benth.) which are distinguished by morphological character, oil content and quality as well as resistance to biotic and abiotic environmental stress [1]. Patchouli oil from both domestic and foreign countries is increasing, but the increase in the quantity of production of patchouli raw materials is still not optimal. Increasing patchouli crop production can be achieved by extending the patchouli cultivation on marginal lands/ dry land.

Indonesia is an agricultural country with a large potential of land resources. Various crops can be used to develop the potential of farmers so that it is beneficial for improving the standard of living of the community, one of which is in the District of North Serang, Banten Province [2]. The potential of farm land (Huma) which includes dry land in Banten is quite extensive, so that it can be used as an alternative land expansion/extensification for agricultural cultivation, including patchouli. Javanese patchouli is cultivated in East Java and Central Java, which have the advantage of being more tolerant of bacterial nematodes and wilt than Aceh patchouli.

Dry land is a stretch of land that has never been flooded or flooded for most of the year. The term dry land is often used for matching upland, dryland or unirrigated land. The last two terms indicate land use for rainfed agriculture. Upland shows land in a higher-lying area cultivated without water flooding such as wetland rice fields [3]. Dry land is an expanse of land that has never been flooded or flooded most of the time of the year [4].

Land evaluation is part of the land use planning process. The essence of land evaluation is to compare the requirements requested by the type of land use to be used [5]. Patchouli plants can grow in the area of land between the lowest plains to a fairly high plateau, which is up to 2000 masl. The yield of oil produced in the highlands is relatively lower compared to plants grown in lowland areas. Patchouli plants need an ideal temperature between 22 - 28°C or between 22 - 28 capacity of water vapor (g / m^3) with humidity above 75%. Patchouli plants need water availability at the beginning of planting until the growth process takes place to achieve optimal growth. Adequate sunlight at the age of more than 3 months until nearing the harvest is also needed [6].

Land evaluation is considered important in the use of agricultural land, because the reality of each type of plant requires different requirements. Besides land varies greatly in various factors such as topography, climate, geology, soil and vegetation. Optimal and sustainable productivity and sustainability of land resources can be expected from the interpretation of plant species according to the quality of the land. The success of patchouli plant cultivation is closely related to the requirements for patchouli plant growth and limiting factors for land suitability. Therefore it is necessary to conduct an evaluation study of land suitability for patchouli cultivation, one of which is Java Patchouli in the dry land of North Serang, Banten province.

2. Methods

This type of research was descriptive research. This research was carried out in Drylands in Tirtayasa Subdistrict, Kramat watu, and Waringin Kurung Sub district, North Serang Regency, Banten Province and in the Land and Agro-climate Laboratory of the Agriculture Faculty of Sultan Ageng Tirtayasa University in August to October 2018.

The tools used in this research were hoe, plastic, sample ring, oven, desiccator, sitting balance, ruler, stationery, tray, clamping pliers, petri dish, dry soil test device, analytic balance, digestion tube and block digestion, boiling 250 ml pumpkin, 100 ml erlenmeyer, 10 ml or 5 ml burette, 25 ml measuring cup, magnetic stirrer, test tube, distillation device or spectrophotometer, GPS, and 4 in 1 soil survey instrument.

The material used was primary data such as soil physical properties data (texture, structure, volume weight, specific gravity) and soil chemical properties (pH, N, P, K, C-Organic), secondary data, namely land use maps, topographic maps, climate data (rainfall, temperature, humidity, irradiation intensity), concentrated sulfuric acid (95-97%), selenium, 1% boric acid, 40% sodium hydroxide (NaOH), Conway pointer and 0.05 N. H2SO4.

Data collection methods were carried out by taking soil samples, laboratory tests, observations, and documentation. Data analysis method is done through map overlay and matching between land characteristics and quality in the study area with criteria for growing patchouli.

3. Results and Discussion

3.1. Field conditions

3.1.1. Slope

Based on Table 1, it was found that land use that has S2 (suitable) land suitability class which was in the slope of 2-8% consists of moor in Tirtayasa district. Land use that had S3 land suitability class (Marginally suitable) on a slope of 8-15% was the use of moor and mixed garden in Kramatwatu sub district as well as the use of mixed garden and vacant land in Waringinkurung sub district. Land use that had land suitability class N (Not suitable) that was on a slope> 15% which consisted of mixed garden land use and vacant land in Tirtayasa sub district, vacant land use in Kramatwatu and moor land use in Waringinkurung sub district.

Table 1. Slope suitability

Sub district	Land use	Value (%)	Suitability class	Limiting criteria
Tirtayasa	Moor	3	S2	The slope becomes a non permanent boundary.
	Mixed garden	16	N	
	Vacant land	16	N	
Kramatwatu	Moor	13	S3	
	Mixed garden	13	S3	
	Vacant land	16	N	
Waringinkurung	Moor	18	N	
	Mixed garden	12	S3	
	Vacant land	13	S3	

Slopes affect erosion in relation to steepness and slope length. Land with a steep slope has a greater gravity effect than land with a slope rather steep and gentle slope. This is due to the greater gravity in line with the more sloping ground surface from the horizontal plane, so that the eroded topsoil will be more and more. This gravity is an absolute requirement for the erosion, transportation and settling process. So, if the slope is getting bigger then the number of soil grains sprinkled down by the collision of raindrops will be more and more [5].

3.1.2 Altitudes

Table 2. Altitudes Suitability

Sub district	Land use	Value (masl)	Suitability class	Limiting criteria
Tirtayasa	Moor	40	S2	
	Mixed garden	50	S2	The altitudes becomes permanent boundary.
	Vacant land	47	S2	
Kramatwatu	Moor	30	S2	
	Mixed garden	28	S2	
	Vacant land	28	S2	
Waringinkurung	Moor	63	S2	
	Mixed garden	78	S2	
	Vacant land	60	S2	

The ideal altitudes for growing patchouli plants was 100-400 meters above sea level. In general, the district of Serang is located at an altitude of less than 500 meters above sea level. The altitudes of the place was measured in the field using GPS Map 78S, the results obtained that the land use in the District Tirtayasa, District Kramatwatu and District Waringinkurung were in the range of altitudes of 0-100 meters above sea level so that it had a suitability class S2 (moderately suitable) for cultivation of patchouli plants. According to [7], altitude is very influential on plant growth. The altitudes of the place affects changes in air temperature. The higher place, for example mountains, the lower the temperature of the air or the cooler the air. Otherwise, the lower area is the higher temperature of the air.

3.2. Temperature
Temperature in Serang Regency based on data from the Climatology and Geophysics Meteorological Agency from 2009-2018 had an average temperature of 27°C. Temperature in Serang Regency had S2 suitability level because it was in the range of 26-28°C, the temperature conditions in the area were moderately suitable for patchouli plant growth requirements. Temperature affects the yield content of patchouli plants.

Table 3. Temperature suitability

Sub district	Land use	Value (o C)	Suitability class	Limiting criteria
Tirtayasa	Moor			
	Mixed garden			
	Vacant land			The temperatur
Kramatwatu	Moor	27	S2	becomes
	Mixed garden			permanent
	Vacant land			boundary.
Waringinkurung	Moor			
	Mixed garden			
	Vacant land			

Temperature affects the growth and development of plants. Because temperature affects the rate of metabolism, photosynthesis, respiration and plant transpiration. However, high temperatures can damage the enzymes so that metabolism does not work well. Even low temperatures cause enzymes to be inactive and metabolism to stop. Therefore, plants have an optimal temperature between 10-38°C. The plants will not survive on temperatures below 0°C and above 40°C [8].

3.3. Water avaibility

Table 4. Rainfall suitability

Sub district	**Land use**	**Value (mm/yr)**	**Suitability class**	**Limiting criteria**
Tirtayasa	Moor			
	Mixed garden			The rainfall
	Vacant land			becomes a
Kramatwatu	Moor	1549,3	S3	non
	Mixed garden			permanent
	Vacant land			boundary.
Waringinkurung	Moor			
	Mixed garden			
	Vacant land			

Water availability is influenced by rainfall. Rainfall in Serang Regency from 2009-2018 which is average 1549.3 mm / yr [9]. Rainfall in Serang Regency has S3 suitability class that is marginal according to the requirements for patchouli plant growth. Rainfall is a non-permanent barrier for Java Patchouli Cultivation so that improvements can be made to make it more suitable for Java Patchouli Cultivation.

Patchouli plants will grow and produce well in areas with relatively high rainfall, evenly distributed throughout the year and require good drainage because if the stagnant roots become rotten and the plants will die. The water content of plants is influenced by environmental factors, especially the weather. If transpiration evaporates excessively for a long time, for example due to a dry season of 3-5 months causing plants to experience stress or stress resulting in low leaf and plant water potential, leaf surface area is reduced due to depletion of food reserves due to decreased photosynthetic activity due to decreased the surface of the leaf stomata [10].

3.4.Rooting media

3.4.1. Drainage

Table 5. Drainage suitability

Sub district	Land use	Value	Suitability class	Limiting criteria
Tirtayasa	Moor	Little good	S2	The drainage becomes a non permanent boundary.
	Mixed garden	Little good	S2	
	Vacant land	Bad	N	
Kramatwatu	Moor	Bad	N	
	Mixed garden	Little good	S2	
	Vacant land	Bad	N	
Waringinkurung	Moor	Little good	S2	
	Mixed garden	Little good	S2	
	Vacant land	Little good	S2	

The drainage condition in the research area for Tirtayasa vacant land, Kramatwatu moor, Kramatwatu vacant land was bad so it had a suitability class N which is bad for Java Patchouli Cultivation. Of the 3 districts, land use for drainage in Waringinkurung District was more in accordance with the suitability class compared to Tirtayasa District and Kramatwatu District because the land in Waringinkurung District did not have yellow, brown or gray patches in the upper and lower layers.

The function of drainage is very different from the function of irrigation. If irrigation delivers water to the planting media, drainage removes excess water from a medium or agricultural land. With a good drainage system land use can be optimized and also minimizes damage to soil structures for roads and other buildings, besides that with a planned drainage system it can be optimized water management that functions to control the presence of abundant water in the rainy season and drought in the dry season [11].

3.4.2. Soil Texture

Table 6. Suitability of soil texture

Sub district	**Land use**	**Value**	**Suitability class**	**Limiting criteria**
Tirtayasa	Moor	Dusty clay loam	S3	Soil texture becomes permane nt boundary .
	Mixed garden	Dusty loam	S2	
	Vacant land	Dusty clay	S3	
Kramatwatu	Moor	Clay	S3	
	Mixed garden	Clay	S3	
	Vacant land	Dusty clay	S3	
Waringinkurung	Moor	Dusty loam	S2	
	Mixed garden	Dusty clay	S3	
	Vacant land	Dusty loam	S2	

Soil texture in the study area for Tirtayasa moor was dusty clay loam, while Tirtayasa vacant land was dusty clay, Kramatwatu moor and mixed garden were clay, Kramatwatu vacant land was dusty clay, and Waringinkurung mixed garden was dusty clay. From the results of the analysis that had been done, the area had a level of conformity S3 that was marginal according to the requirements for patchouli plant growth. Moreover, Tirtayasa soil texture mixed gardens, Waringinkurung moor and vacant land that were dusty loam which had a level of suitability S2 or moderately suitable for patchouli plant growth requirements.

According to [12], good soil is reflected by dust and clay textured soil which has optimum availability for plants, but in terms of nutrient clay soil is better than dust textured soil. According to [13], clay-textured soils make plants unable to grow properly. Clay is a soil that has many micro or porous pores. Micro pores in clay are caused by solid soil structure. Between the aggregates of land there are very few gaps or spaces. This causes very limited air and water trapped easily, so that the clay is difficult to pass water, or in other words the permeability is low. Clay is less supportive to the development of plant roots because of its low porosity so that it affects interferes with respiration carried out by the root.

3.5. Nutrient retention
3.5.1. Soil acidity (pH)

Table 7. Suitability of Soil acidity (pH)

Sub district	Land use	Value	Suitability class	Limiting criteria
Tirtayasa	Moor	6,45	S1	
	Mixed garden	5,31	S2	Soil acidity becomes a non permanent boundary.
	Vacant land	4,87	S3	
Kramatwatu	Moor	6,48	S1	
	Mixed garden	6,54	S1	
	Vacant land	7,63	S1	
Waringinkurung	Moor	6,72	S1	-
	Mixed garden	6,14	S1	
	Vacant land	6,65	S1	

Based on Table 7 it is found that the acidity (pH) of the soil in the study area is more dominated by land suitability class S1 (Very suitable) which is in the range of 5.5-7, but besides S1 suitability class there are land uses that are classified as S2 land suitability class (moderately suitable) which is in the range of 5.5-5, namely the use of Tirtayasa mixed garden and for the use of vacant land Tirtayasa sub district has a land suitability class S3 (Marginally suitable) which is in the range of 4.5-5.

Based on matching results for soil acidity, Tirtayasa's vacant land has a pH < 7 which means it is rather acidic. According to [14], the pH level of the soil itself is influenced by several things, one of which is the decomposition of organic material in which microorganisms in the soil will continuously decompose soil organic matter into other forms, namely organic acids, carbon dioxide (CO_2) and water, carbonic acid forming compounds. Then Ca and Mg carbonate in the soil will react to form more soluble bicarbonate, which can be leached out and finally can make the soil become more acidic.

3.5.2. C-Organic

Table 8. Suitability of C-Organic

Sub district	**Land use**	**Value (%)**	**Suitability class**	**Limiting criteria**
Tirtayasa	Moor	3,92	S2	
	Mixed garden	1,56	S3	C-Organic becomes a non permanent boundary.
	Vacant land	3,13	S2	
Kramatwatu	Moor	5,49	S2	
	Mixed garden	3,91	S2	
	Vacant land	3,91	S2	
Waringinkurung	Moor	3,12	S2	
	Mixed garden	0,78	S3	
	Vacant land	1,56	S3	

Based on Table 8, it was found that the use of land containing C-Organic was suitable for Java Patchouli Cultivation because it belongs to the suitability class S2 (moderately suitable) in the range < 3-5%, namely the use of moor, mixed gardens, and vacant land in the District Kramatwatu; the use of moor and vacant land in Tirtayasa District; and the land use of moor in Waringin Kurung District.

According to [15], C-Organic states the number of organic compounds as a source of carbon elements contained in the soil, including litter, fraction of lightweight organic matter, biomass microorganisms, organic matter seen in water and stable organic matter or humus. According to [16], C-Organic content tends to decrease with increasing soil depth because organic matter is only applied or falls on the ground. So that the organic material accumulates in the top soil layer and part of it is washed into a deeper layer (sub soil).

3.5.3. Cation Exchange Capacity (CEC)

Table 9. CEC Suitability

Sub district	Land use	Value (cmol(+)kg-1)	Suitability class	Limiting criteria
Tirtayasa	Moor	16,53	S2	CEC becomes a non permanent boundary.
	Mixed garden	13,14	S2	
	Vacant land	7,92	S2	
Kramatwatu	Moor	14,25	S2	
	Mixed garden	0,13	S3	
	Vacant land	7,43	S2	
Waringinkurung	Moor	7,09	S2	
	Mixed garden	8,07	S2	
	Vacant land	6,79	S2	

Based on Table 9, it was found that the CEC content of the three districts which had a high CEC content, namely moor use in Tirtayasa Subdistrict with CEC value of 16.53 (cmol (+) kg-1) included in the S2 (moderately suitable) land suitability class. Whereas the land use in Kramatwatu and Waringinkurung sub-districts was dominated by land suitability class S2 (moderately suitable) which was in the range of 5-15 (cmol (+) kg-1), but in the mixed garden land use Kramatwatu sub-district had land suitability class S3 (Appropriate marginal) which was in the range <5 (cmol (+) kg-1).

According to [13], cation exchange capacity is a chemical characteristic that is closely related to soil fertility. Land with high CEC is able to trap and provide nutrients better than land with low CEC. Soils with high CEC when dominated by base cations, Ca, Mg, K, Na (high base saturation) can increase soil fertility, but if dominated by acid cations, Al, H (low base saturation) can reduce soil fertility.

3.6. Nutrient availability

3.6.1. K_2O suitability

Table 10. K_2O

Sub district	Land use	Value	Suitability class	Limiting criteria
Tirtayasa	Moor	Low	S3	K_2O content becomes a non permanent boundary.
	Mixed garden	Low	S3	
	Vacant land	Medium	S2	
Kramatwatu	Moor	Low	S3	
	Mixed garden	Low	S3	
	Vacant land	Low	S3	
Waringinkurung	Moor	Low	S3	
	Mixed garden	Low	S3	
	Vacant land	Medium	S2	

The content of K_2O in the study area for Tirtayasa vacant land and Waringinkurung vacant land had a level of suitability S2 that is moderately suitable for patchouli plant growth requirements. The content of K_2O in Tirtayasa moor, Tirtayasa mix garden, Kramatwatu moor, Kramatwatu mixed garden, Kramatwatu vacant land, Waringinkurung moor, Waringinkurung mixed garden have a level of S3 suitability which is marginal according to patchouli growing requirements. The source of potassium contained in the soil comes from weathering minerals that contain K. These minerals when weathered release K soil solubility or soil absorption in the form of exchange. The location of potassium clay is generally in the surface surface (internal surface) which is often occupied by ions Mg2 +, Fe3 +, Al4 + and H_2O molecules.

Potassium loss in the soil can occur in a number of ways such as being transported by plants together with harvesting, washing, erosion and fixation. Potassium loss that is removed by plants is caused by the nature of potassium which can be absorbed by plants excessively beyond the actual needs. This excessive uptake no longer increases crop production, resulting in a waste of the use of soil potassium.

3.6.2. P_2O_5 suitability

Table 11. P_2O_5 suitability

Sub district	Land use	Value (ppm)	Suitability class	Limiting criteria
Tirtayasa	Moor	41,83	S3	P_2O_5 content becomes a non permanent boundary.
	Mixed garden	30,24	S3	
	Vacant land	26,28*	S3	
Kramatwatu	Moor	13,66	S2	
	Mixed garden	3,92	N	
	Vacant land	10,09	S2	
Waringinkurung	Moor	8,51	N	
	Mixed garden	7,45	N	
	Vacant land	5,35	N	

Based on Table 11 it was found that the use of dry land and the use of vacant land in Kramatwatu Subdistrict had S2 suitability class because it was in the range of 10-15 ppm. The 3 land uses in Tirtayasa Subdistrict had S3 conformity class because they were in the range > 25 ppm. The mixed land use of Kramatwatu Subdistrict and the 3 land uses of Waringinkurung Subdistrict had N suitability class because they were in the range of < 10 ppm.

According to [17], the availability of soil P for plants is mainly influenced by the nature and characteristics of the soil itself. P becomes unavailable and insoluble due to fixation by clay minerals and the highly soluble Al, Fe, Mg or Ca ions, forming complex and insoluble compounds. There are several factors that also influence the availability of soil P, clay type, soil pH, reaction time, temperature and soil organic matter. In addition flooding can also affect. There are several ways to lose P from the soil, which are transported by plants, washed and eroded. This substance is absorbed by plants in the form of H_2PO_4- and HPO_4-. The main function of this element is to accelerate root growth especially lateral roots and hair roots, accelerate and strengthen the growth of young plants into adulthood, accelerate flowering and ripening of seeds and increase seed production.

From the description above it is obtained that the suitability of land for Java Patchouli Cultivation in Tirtayasa, Kramatwatu, and Waringinkurung District is as follows (Table 12):

Table 12. Land suitability class for Java Patchouli cultivation at land use in the research area

Sub district	Land use suitability class		
	Moor	Mixed garden	Vacant land
Tirtayasa	S2	S3	S3
Kramatwatu	S2	S2	S3
Waringinkurung	S3	S2	S2

Based on Table 12, it is found that those suitable for Java Patchouli Cultivation in North Serang with Class S2 (moderately suitable) were the land use of moor in Tirtayasa District, moor and mixed garden land use in Kramatwatu District, vacant land and mixed garden land use in WaringinKurung District. The results of the land use suitability class in the study area were obtained from consideration of efforts to improve based on the matching results of the limiting factors. Based on the results of the analysis it was found that the limiting factors of land use in the study area were water availability, which included Tirtayasa District, Kramatwatu District and Waringinkurung District. The Media of Incense which included vacant land Tirtayasa Subdistrict, and vacant land Kramatwatu Subdistrict. The availability of nutrient covered mixed garden in Kramatwatu Subdistrict, moor, mixed garden and vacant land in Waringinkurung Subdistrict. The condition of field which included mixed gardens and vacant land in Tirtayasa Subdistrict, vacant land Kramatwatu Subdistrict and Waringinkurung District. Therefore, efforts to improve land with moderate and high management are needed for the land of Tirtayasa, Kramatwatu and Waringinkurung sub District to make it more suitable for patchouli cultivation and can increase productivity. Efforts to Improve the Limiting Factors of Land Suitability for Java Patchouli Cultivation are presented in Table 13 below :

Table 13. Efforts to Improve Limiting Factors of Land Suitability for Java Patchouli Cultivation

Sub district	**Parameter**	**Improvement efforts**		
		Moor	**Mixed garden**	**Vacant land**
Tirtayasa	Slope (%)	√	Terrace making	Terrace making
	Altitude (masl)	-	-	-
	Average Temperature (°C)	-	-	-
	Rainfall (mm)	Irrigation	Irrigation	Irrigation
	Texture	-	-	-
	Drainage	√	√	Drainage channel
	pH	√	√	Liming
	C-organic (%)	√	Fertilization	√
	CEC (me/100g)	√	√	√
	K_2O (me/100g)	Fertilization	Fertilization	√
	P_2O_5 (ppm)	Fertilization	Fertilization	Fertilization
Kramatwatu	Slope (%)	Terrace making	Terrace making	Terrace making
	Altitude (masl)	-	-	-
	Average Temperature (°C)	-	-	-
	Rainfall (mm)	Irrigation	Irrigation	Irrigation
	Texture	-	-	-
	Drainage	Drainage channel	√	Drainage channel
	pH	√	√	√
	C-organic (%)	√	√	√
	CEC (me/100g)	√	Organic matter	√

	K_2O (me/100g)	Fertilization	Fertilization	Fertilization
	P_2O_5 (ppm)	√	Fertilization	√
Waringinkurung	Slope (%)	Terrace making	Terrace making	Terrace making
	Altitude (masl)	-	-	-
	Average Temperature (°C)	-	-	-
	Rainfall (mm)	Irrigation	Irrigation	Irrigation
	Texture	-	-	-
	Drainage	√	√	√
	pH	√	√	√
	C-organic (%)	√	Fertilization	Fertilization
	CEC (me/100g)	√	√	√
	K_2O (me/100g)	Fertilization	Fertilization	√
	P_2O_5 (ppm)	Fertilization	Fertilization	Fertilization

Information :
√ : No repair needed
- : Cannot be repaired

Table 13 shows land improvement efforts that can be made to improve land quality limiting factors for Java Patchouli Plants in Tirtayasa District, Kramatwatu and Waringinkurung District. Improvement efforts to improve the nature of land suitability constraints for the cultivation of patchouli plant on the slopes by making terraces to facilitate land management and prevent erosion. Improvement efforts to improve the nature of land suitability constraints for the cultivation of patchouli plant in the availability of water by providing irrigation water. According to [18], the purpose of irrigation in an area is an effort to supply and regulate water to support agricultural activities, from water sources to areas that need and distribute technically and systematically. While the benefits obtained from irrigation activities are helping soil wetting in areas where rainfall is lacking or erratic, increasing soil fertility by flowing water and sludge containing dissolved nutrients needed by plants in agricultural areas so that the area gets additional soil nutrients . Improvement efforts to improve the nature of land suitability constraints for the cultivation of patchouli plant on the root media (drainage) by making drainage channels. According to [19], there are several benefits of making drainage channels including lowering the surface of ground water at an ideal level so that it does not make the plants experience excess water and control soil erosion, good soil will greatly affect the plants that grow on it. Improvement efforts to improve the limiting nature of land suitability for the cultivation of patchouli plant on nutrient retention (pH) of acidic soil by liming. According to [14], Liming is an effort made to increase soil pH by adding lime to the soil. The purpose of liming is to increase the acidic pH to neutral pH and reduce or eliminate Al poisoning.

Improvement efforts to improve the nature of land suitability constraints for the cultivation of patchouli plant in nutrient retention (C-Organic) by fertilizing using manure, compost or green manure. According to [20], very poor soils should be fertilized with organic fertilizer. Sand soil or soil that is much eroded is better fertilized with organic fertilizer than with artificial fertilizer, because the application of artificial fertilizer on the soil will be easily washed away by rain water. With manure given, the power to hold water and soil cations increases, so that if it is also given artificial fertilizer, washing by rain water and erosion can be inhibited. Improvement efforts to improve the nature of land suitability constraints for the cultivation of patchouli plant on nutrient retention (CEC) with the addition of organic matter. According to [21], the effect of organic matter is to increase the absorption capacity and cation exchange capacity (CEC). About half of the cation exchange capacity of the soil comes from organic matter. Organic matter can increase cation exchange capacity two to thirty times greater than colloidal minerals which account for 30 to 90% of the absorbent power of a mineral soil. The increase in CEC due to the addition of organic material due to weathering of organic material will produce humus (organic colloids) which have a surface that can hold nutrients and water so that it can be said that the

IOP Conf. Series: Earth and Environmental Science **383** (2019) 012045 doi:10.1088/1755-1315/383/1/012045

provision of organic material can store fertilizer and water given in the soil. The increase in CEC increases the ability of the soil to retain nutrients.

The K_2O content can be improved by adding K-containing fertilizers and compound fertilizers containing P and K elements. According to [22], fertilizing with potassium can be done by spreading or sprinkling directly on plants. This type of fertilizer can be combined with other types of fertilizers such as urea and SP-36. Considering that this potassium is soluble, potassium fertilizer is administered the same as N fertilizer. Cereals are given three times, where 1/3 part of K_2O is given consecutively at planting and at primordial time. According to [23], the addition of P into the soil can be derived from various sources, namely from phosphate fertilizer, weathering minerals containing P and from the remains of plants and animals. The addition of P to the soil from the remains of plants and animals is very small, because the consumption of phosphate by plants and animals is also small.

4. Conclusion

1. The limiting factor of land suitability for Java Patchouli Cultivation in North Serang Regency, Banten province in Tirtayasa District, Kramatwatu District and Waringinkurung District is Water Availability. The limiting factor in vacant land of Tirtayasa subdistrict, moor and vacant land of Kramatwatu subdistrict namely rooting media. The limiting factor in mixed garden of Kramatwatu Subdistrict, moor, mixed garden, and vacant land of Waringinkurung District is availability of nutrient. The limiting factor in mixed garden and vacant land of Tirtayasa Subdistrict, vacant land of Kramatwatu Subdistrict and moor of Waringinkurung Subdistrict is field conditions.
2. The level of land suitability for Java Patchouli Cultivation in North Serang District, Banten Province, namely the land suitability class S2 (moderately suitable) consists of moor land use in Tirtayasa District, moor and mixed gardens land use in Kramatwatu Subdistrict, vacant land and mixed garden land use in Waringinkurung Sub-district. While for S3 land suitability class (marginally suitable) consists of vacant land and mixed garden land use in Tirtayasa District, vacant land use in Kramatwatu District and moor land use in Waringinkurung District.
3. Efforts to improve land for the cultivation of patchouli plants in the northern Serang Regency, Banten Province are the construction of terraces, provision of additional irrigation water, drainage, liming and fertilizing.

5. References

[1] Pitono J, Marizka I, Syakir M., Ragapadmi H, Nurhayati, Setiawan, Kuswadi, Zaenuddin, and Santoso T 2007 *Selection of Resistance to Drought Stress on Some Patchouli Somaklon Numbers*. Technical Research Report for Fiscal Year 2007

[2] Nuryani Y, Emmyzar, and Wiratno 2005 *Patchouli Cultivation* (Bogor : Research Institute for Medicinal and Aromatic Plants)

[3] Notohadinegoro T 2000 *Chemical and Biological Physical Diagnostics of Land Damage* Paper on Seminar on Investigation of Land / Land Damage Criteria" Asmendep I Environment / Bapedal 1-3 Juli 1999 Yogyakarta

[4] Adimihardja A, and Agus F 2000 *Post NWMCP Conservation Technology Development* Proceedings of the National Workshop on the Discussion of the Results of Watershed Management Research. Alternative Soil Conservation Technology. Bogor page 25–38

[5] Hardjowigeno, Sarwono and Widiatmaka 2007 *Land Suitability Evaluation and Land Use Planning* (Yogyakarta: Gadjah Mada University Press.)

[6] Mangun HMS, 2009 *Patchouli* (Jakarta : Penebar Swadaya)

[7] Hidayat A, Hikmatullah and P. Santoso 2000 *Potential and Management of Lowland Dry Land in Indonesian Land Resources and Its Management* (Bogor : Center for Soil and Agro-climate Research)

[8] Kartasapoetra AG, 2008 *Climatology : Effect of Climate on Land and Plants* Revised Edition. (Jakarta: Earth Literacy)

[9] Class I Meteorological Station Serang 2018 *Serang Climate Data and Surrounding*

[10] Nuryani Y, Mustika I, and Syukur C, 2001 Phenol and Lignin Content of Patchouli Hybrid Plants (Pogostemon sp.) Protoplast Fusion Results *Journal of Industrial Plant Research* Vol **7** (4) : 104-108.
[11] Gunawan T and Hartono 2000 *Watershed Management* (Jakarta: Bakosurtanal Natural Resource Survey Center)
[12] Hanafiah KA 2005 *Fundamentals of Soil Science* (Jakarta : PT. Rajagrafindo Persada)
[13] Poerwidodo 2000 *Study of Soil Fertility* (Bandung : Space)
[14] Prijono S, 2013 *Work Instructions on Measurement of pH, Organic Materials, KT and KB* (Malang: Brawijaya University)
[15] Novizan 2002 *Effective Fertilization Instructions* (Tangerang : Agromedia Reader)
[16] Rosmarkam A, and N.W. Yuwono 2002 *The Science of Soil Fertility* (Yogyakarta: Kanisius)
[17] Sukristyonubowo, Mulyadi, P. Wigena, and A. Kasno 2001 *The Effect of Addition of Organic Material, Lime and NPK Fertilizer on the Chemical Properties of Soil and Peanut Results* (Jakarta : Erlangga)
[18] Center for Soil and Agro-climate Research 2003 *Farming on Dry Land* (Bogor : Agricultural Research and Development Agency Ministry of Agriculture)
[19] Rachman A, and A. Dariah 2008 *Soil Conservation in the Conservation of Dry Land* (Jakarta : Agency for Agricultural Research and Development)
[20] Setyorini D, R. Saraswati and E. Kosman 2006 *Compost in Organic Fertilizers and Biofertilizers* (Jakarta: Agricultural Research and Development Agency)
[21] Suriadikarta DA, T. Prihatini, D. Setyorini, and W. Hartatik 2005 *Management of Soil Organic Materials* (Jakarta: agricultural research and development agency)
[22] Syarif S, 2003 *Fertility and Fertilization of Agricultural Land* (Bandung: Pustaka Buana)
[23] Sutedjo MM, 2002 *Fertilizers and Fertilizing Methods* (Jakarta: PT. Rineka Cipta)

Agrobacterium mediated transformation of Banten local rice (cv. Pare Gajah) for Folate gene

Susiyanti[1,2,*], T A Larasati[1], Nurmayulis[1,2], S Isminingsih[1,2], Sjaifuddin[1,3], and S Amin[4]

[1]Department of Agroecotechnology, Faculty of Agriculture, University of Sultan Ageng Tirtayasa, Indonesia
[2]Center of Excellent for Food Security University of Sultan Ageng Tirtayasa
[3]Department of Science Education, University of Sultan Agung Tirtayasa, Serang City, Banten Province, Indonesia
[4]Department of Agroecotechnology, Agriculture Faculty, University of Padjajaran, West Jawa Province, Indonesia

*Corresponding author: susiyanti@untirta.ac.id

Abstract. Folate (vitamin B9) deficiency is a common problem that often occurs in developing countries including Indonesia. Rice which is a staple containing folate is very low. So it is necessary to find an effort so that the rice consumed can contain folate which is good for health.The aim of this study was to transform the folate gene to plant Banten local rice (*cv.* Pare Gajah) using *Agrobacterium tumefaciens*. This research was conducted from January to Mei 2019 in the Laboratory of Physiological and Plant Biotechnology, Agriculture faculty, Sultan Ageng Tirtayasa University. Transformation research used a Completely Randomized Design with two factors, the first factor is the transformation method (I) consist of 2 levels; namely in vitro and in planta. The second factor is hygromycin concentration (H) consist of 4 levels; namely 25, 50, 75, and 100 ppm. Each treatment was replicated six times. The results showed, in vitro transformation method had a lower lethal dose, causing average death of explant reaches 93,75%. In planta transformation method giving the best average efficiency transformation reach 13,33%. Hygromycin concentration 100 ppm causing the higher average death of explant reaches 98,34%. Hygromycin concentration 25 ppm causing the higher average efficiency transformation reach 17,50%.

Keywords: 2.4-D, Transformation, in vitro, in planta, hygromycin, Rice, folate, CGH1 gene.

1. Introduction

Folic acid (vitamin B9) is needed by children and adults to produce red blood cells and prevent anemia. Folic acid plays a major role in cell growth and development, and tissue formation. Folic acid deficiency, the body will be susceptible to diseases such as depression, anxiety, fatigue, insomnia, difficulty remembering, red tongue and sores to digestive disorders. Folic acid deficiency in pregnant women increases the risk of premature delivery, babies with low birth weight or with neural tube defects [1], [2] said that someone who lacks folate is more at risk of getting cancer because folate deficiency can increase chromosomal damage and incorporation of uracil into DNA.

The human body cannot synthesize the structure of folate so it requires intake from food. About 24-60% of women, both in developing countries and in developed countries experience folic acid deficiency because the content of folic acid in their daily diet is not sufficient.

To fulfill the need for folic acid, the assembly of rice varieties that can contain folic genes is needed. This is due to the high level of rice consumption in Indonesia, so that by assembling rice varieties containing folic genes can increase folic acid intake and prevent folic acid deficiency in the body. According to [3] this development effort can be carried out with plant breeding, one of which is biotechnology. Assembling rice varieties using genetic transformation techniques is expected to produce superior rice varieties with desired characteristics.

The use of local varieties of rice in breeding programs has often been advocated with the aim of broadening the genetic background of the superior varieties to be produced [4]. According to [5] Pare Gajah variety is one of the local varieties of Banten which is cultivated by the Kasepuhan community of Cisungsang Village because it is known for its long and rather large grains. By doing the breeding process to improve the weaknesses of local varieties of rice without changing other traits that have been favored such as the taste of rice and good adaptation in areas where rice is widely cultivated, local varieties can be obtained which are more beneficial while adding to its economic value.

Agrobacterium tumefaciens is one of the transformation techniques that many researchers have done and has been successful. *Agrobacterium* is a bacterium that plays a role in helping to insert genes into plant genomes. Rice is one of the agricultural crops that has been carried out a lot of genetic transformation to get superior varieties of rice. Some genes that have been successfully inserted in rice plants and are being evaluated for expression and function are genes for resistance to yellow stem borer pests, fungal disease, and drought-tolerant genes [6].

Transformation with *Agrobacterium* is divided into 2 methods, namely in vitro and *in planta*. *In vitro* transformation is a transformation method that uses tissue culture techniques, namely by callus culture or callus induction. While the *in planta* transformation method is a new method that is simpler than in vitro because it is done in ex vitro (outside the lab).

Based on the description above, it is necessary to do research on the Transformation of Folate Gene in the Banten Local Rice Varieties of Pare Gajah (*Oryza sativa* L.). The purpose of this study was to carry out the transformation of the folate gene into the genome of rice plants of the Banten local variety Pare Gajah (*Oryza sativa* L.).

2. Materials and Methods

This type of research is experimental research. This research was conducted at the Biotechnology Laboratory, Faculty of Agriculture, Sultan Ageng Tirtayasa University from November 2018 to January 2019.

2.1. Treatments

2.1.1. Callus Induction

The design used in the manufacture of callus as a preparation for transformant material consisted of one factor, namely the concentration of 2,4 Dichlorophenoxyacetate (D). The 2,4-D concentration factor consists of 3 levels, namely:

D_1 : 1 ppm
D_2 : 2 ppm
D_3 : 3 ppm

Each treatment was repeated 9 times to obtain 27 experimental units. Each experimental unit contained 6 seed explants. Then the total explants needed are 162 seed explants.

2.1.2. Folate Gene Transformation

The design used in this transformation research consisted of two factors. The first factor is the transformation method (I) and the second factor is the concentration of hygromycin (H). The first factor transformation method (I) consists of 2 levels, namely:

I_1 : *In vitro* transformation method
I_2 : *In planta* transformation method

The second factor is the concentration of hygromycin (H) consisting of 4 levels, namely:
H_1 : 25 ppm
H_2 : 50 ppm
H_3 : 75 ppm
H_4 : 100 ppm

Each treatment was repeated 6 times to obtain 48 experimental units. Each unit of experiment contained 10 callus / sprouts. Then the total explants needed are 240 callus and 240 sprouts.

2.2. Observation parameters:

2.2.1. Callus Induction

1) Time when Callus Formed (DAP / *days after planting*)
   Observation of the emergence of callus is done every day by counting how many days the callus has begun to appear or grow.
2) Number of Explants Forming Callus (%)
   Observations were made in the 3rd week after planting explants, with the following formula :
   $$\% \text{ Callus} = \frac{\text{The number of explants forms a callus}}{\text{Total explants}}$$
3) Callus Texture
   Callus texture is visually observed in the appearance of callus which includes compact, intermediate or crumb callus textures. Callus texture was observed at 35 DAP (5 weeks after implantation).
4) Callus Colour
   The color of the callus is seen from the color formed. Callus color was observed at 35 DAP (5 weeks after implantation).
5) Callus Diameter (mm)
   Callus diameter is calculated by measuring the longest diameter of the side of the callus. Observation of callus diameter was carried out at the end of observation of callus induction (5 weeks after explant planting).
6) Callus Area (mm^2)
   The callus area was calculated using millimeter block. Callus area observation was carried out at the end of callus induction observation (5 weeks after explant planting).

2.2.2. Folate Gene Transformation

1) Lethal hygromycin dose
   Observation of lethal hygromycin dose test on transformation in vitro and *in planta* was done at the end of explant selection using hygromycin antibiotics, with the following formula [9]:
   $$LD_{50} = a - (b/c)$$
   Notes:
   a = The smallest dose that causes the highest death of explants in one dose group.
   b = Total number of multiplications between different doses and the average mortality at the same interval
   c = The number of explants in one group
2) Efficiency of Putative Transgenic Transformation (%)
   Calculation of transformation efficiency *in vitro* and *in planta* is done using the following formula [6]:
   $$\text{Transformation efficiency (\%)} = \frac{\text{The amount of hygromycin - resistant explants}}{\text{The number of initial explants of transformation}} \times 100\ \%$$

2.3. Research Implementation

2.3.1. In Vitro Transformation

1) Making Embryogenic Callus

Rice seeds are washed clean and soaked for 30 minutes in distilled water. After that the seeds are peeled and sterilized with 2 g.L^{-1} detergent solution, 2 g^{-1} agrymicin, and 2 g^{-1} dithane for 15 minutes, then the seeds are rinsed with distilled water and sterilized again with 30% chlorine solution for 15 minutes in LAF. After that the seeds are rinsed with sterile distilled water 6 times, and then soaked in betadine solution for 5 minutes [8]. Then the seeds are planted on 2,4-D callus induction media. Media that has been planted with explants is then placed on a culture rack at 25-28°C. Every 2 days the culture bottle is sprayed with 70% alcohol to avoid contamination. Callus began to form 2-3 weeks after planting explants in the media. The resulting callus was transferred to the same callus induction medium by cutting buds and endosperm. The first 2-week embryogenic callus can be used as an explant for transformation.

2) *Agrobacterium tumefaciens* Culture

Agrobacterium strain LBA4404 which carries the construction of the folate gene is grown on solid LB media containing 100 ppm kanamycin antibiotic and 50 ppm hygromycin antibiotic. Bacteria culture that has been distreaked on solid LB media is stored in an incubator of 28°C for about 3 days or until the bacteria grows [6, 7].

3) Transformation of *Agrobacterium tumefaciens* to Embryogenic Callus

The infection solution to be used was inserted into corning as much as 15 mL, *Agrobacterium tumefaciens* culture was added to OD (optical density) 0.03-0.05 at 600 nm wave, acetosiringone was also added and allowed to stand for 15 minutes before use. The embryogenic callus that has been prepared is soaked with an infusion solution that has been added with acetosiringon first for 15 minutes for adaptation, then soaked in a bacterial suspension infectious solution for 15 minutes, after which the suspension is discarded and the callus is drained on filter paper for 10 minutes [11].

4) Co-cultivation

Co-cultivation media were made with 2 sheets of filter paper that were given liquid infection and acetosiringone that were previously made as much as 2 mL. Callus that has been infected by *Agrobacterium tumefaciens* was carried out co-cultivation in the dark for 3 days on the previously created co-cultivation media [11].

5) Selection of Transgenic Plants

Co-cultivation callus was subsequently grown on selection media with modified hygromycin concentrations of 25 ppm, 50 ppm, 75 ppm, and 100 ppm, then grown in dark conditions for 2 weeks (until the callus grows with callus enlarged, callus colorless black or necrosis) [11].

2.3.2. In planta Transformation

1) Rice Sprouts Preparation

Rice seeds used as explants are washed and soaked for 30 minutes in distilled water. After that the seeds are peeled and sterilized with 2 g/L detergent solution, 2 g.L^{-1} agrymicin, and 2 g.L^{-1} dithane for 15 minutes, then the seeds are rinsed with distilled water and sterilized again with 30% chlorine solution for 15 minutes in LAF. After that the seeds are rinsed with sterile distilled water 6 times, and then soaked in betadine solution for 5 minutes [8]. Furthermore, the seeds are soaked in sterile water for 2 days at 20°C. Water is replaced once during the soaking process. After 2 days of immersion, the embryo will turn white [10].

2) *Agrobacterium tumefaciens* Culture

Agrobacterium strain LBA4404 which carries the construction of the folate gene is grown on solid LB media containing 100 ppm kanamycin antibiotic and 50 ppm hygromycin antibiotic. Bacteria culture that has been distreaked on solid LB media is stored in an incubator of 28°C for about 3 days or until the bacteria grows. [6, 7].

3) *Agrobacterium tumefaciens* inoculation to rice shoot

The infection solution to be used was inserted into corning as much as 15 mL, added *Agrobacterium tumefaciens* culture to OD 0,3-0,6 at 600 nm, also added acetosiringone and allowed to stand for 15

IOP Conf. Series: Earth and Environmental Science **383** (2019) 012046 doi:10.1088/1755-1315/383/1/012046

minutes before use. For bacterial inoculation of rice shoots is done by pricking the embryo (1-1,5 mm depth) with a sterile needle tip and then dipping in bacterial suspension for 15 minutes, after that the suspension is removed and rice sprouts are drained on filter paper for 10 minutes [10].

4) Co-cultivation

Co-cultivation media were made with 2 sheets of filter paper that were given liquid infection and acetosiringone that were previously made as much as 2 mL. Rice sprouts that have been infected by *Agrobacterium tumefaciens* were co-cultivated in the dark for 5 days on previously cultivated co-cultivation media [11].

5) Transgenic Plants Selection

Rice seeds were grown under sterile conditions in water (2-3 mm depth) containing hygromycin antibiotics with concentrations of 25 ppm, 50 ppm, 75 ppm and 100 ppm at 28°C. Resistance to hygromycin antibiotics was assessed after 7 days later [10].

6) DNA Isolation.

Leaves of 23 F1 from cross-bred Kewal Bulu Hideung X Mira-1 and the two parents (21 day after planting) were used to isolate the DNA with CTAB method. DNA from each samples was amplified with specific CGH1 SSR primer. The PCR product was separated using 2% agarose and visualized with Chemidoc gel system.

3. Results and Discussion

3.1. Callus Induction

The results of the variance recapitulation in Table 1 show the 2.4-D concentration significantly affected the time callus parameter (DAP) and had a very significant effect on the number of explants forming callus (%), but gave no significant effect on the callus diameter parameter (mm) and callus area (mm2).

Table 1. Recapitulation of Variance Effects of Growth Regulatory Substances 2.4-D Concentration on Banten Local Callus Pare Gajah (*Oryza sativa* L.)

No.	Observation Parameters	Concentration 2,4-D	KK (%)
1.	Time when Callus Formed (DAP)	*	6,52
2.	Number of Explants Forming Callus (%)	**	13,83
3.	Callus Diameter (mm)	tn	5,20[a]
4.	Callus Area (mm^2)	tn	5,82[a]

Notes :
- tn : Have no influential effect
- * : Significant influential effect at the 5% level
- ** : Very significant influential effect at 1% level
- KK : Coefficient of Diversity
- DAP : Days After Planting (*Hari Setelah Tanam)*
- a : Transformed data $\sqrt{x + 0,5}$ once

3.1.1. Time of Callus Formation

Based on the data presented in Table 2 it was known that the 2.4-D concentration significantly affects the time parameters of callus formation. The 2.4-D concentration gives an average callus formation time at 13.37 DAP (days after planting), and the best 2.4-D concentration is produced at the D2 level (2 ppm) which is 12.67 DAP. The process of callus formation in Pare Gajah rice embryos induced using 2.4-D can be seen in Figure 1.

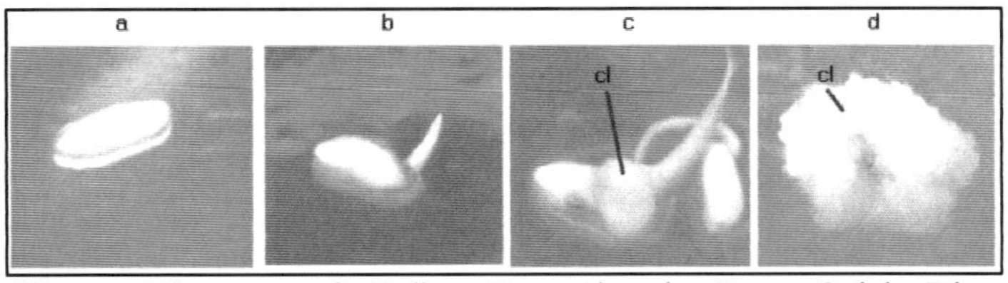

Figure 1. Process of Callus Formation in Pare Gajah Rice Embryo Induced using 2,4-D. Notes: a. Rice seeds were planted on N6 media with the addition of 2,4-D on day 0. b. Shoots on the 7th day after planting. c. Callus formed at the base of the bud of Pare Gajah rice embryo on the 12th day after planting. d. Callus shape on the 35th day. Information: (bj) seeds, (tn) shoots, and (kl) callus.

Table 2. The Effects of Growth Regulatory Substances 2,4-D Concentration towards Time when Callus Formed (DAP)

2,4-D Concentration	Time when Callus Formed (DAP)
D1 (1 ppm)	14,00c
D2 (2 ppm)	12,67a
D3 (3 ppm)	13,44b
Average	13,37

Note : the numbers that followed by the same letters in the same row or column show no significant difference according to the DMRT test of 5% level.

3.1.2. Number of Explants Forming Callus (%)

Based on the data presented in Table 3, the 2.4-D concentration at D2 level (2 ppm) is the best concentration with the percentage of explants forming callus that is 94.44%.

Table 3. The Effects of Growth Regulatory Substances 2.4-D Concentration towards Number of Explants Forming Callus (%)

2,4-D Concentration	Number of Explants Forming Callus (%)
D1 (1 **ppm**)	72,22a
D2 (2 **ppm**)	94,44b
D3 (3 **ppm**)	88,89b
Average	85,18

Note: the numbers that followed by the same letters in the same row or column show no significant difference according to the DMRT test of 5% level.

3.1.3. Callus Texture

Based on the data presented in Table 4. it is known that the three concentrations of growth regulators 2,4-D have no significant effect on the callus texture, because all three concentrations of 2,4-D produce a callus that is crumbly (friable)

Table 4. The Effects of Growth Regulatory Substances 2,4-D Concentration towards Callus Texture

2.4 D concentration	(1)	(2)	(3)
		%	
D1 (1 ppm)	0	0	100
D2 (2 ppm)	0	0	100
D3 (3 ppm)	0	0	100

Note: Scoring of callus texture : (1) solid, (2) intermediate, dan (3) friable.

3.1.4. Callus Colour

From the data in Table 5. it is known that administration of different 2,4-D growth regulators produces different callus colors. At the beginning of the formation of the callus, the resulting callus is yellowish white. But over time there are even calluses that change color to white, yellow, and even brownish white at the end of the observation (Figure 2). In the treatment D1 (1 ppm), the color of the callus is dominated by yellowish white and brownish white. Whereas the treatment of D2 (2 ppm) and D3 (3 ppm) was dominated by a yellowish white color.

Table 5. The Effects of Growth Regulatory Substances 2,4-D Concentration towards Callus Colour

2.4 D concentration	(1)	(2)	(3)	(4)	(5)	(6)	(7)
				%			
D1 (1 ppm)	22,22	11,11	0	0	33,33	0	33,33
D2 (2 ppm)	33,33	0	0	0	66,67	0	0
D3 (3 ppm)	22,22	0	0	0	44,44	0	33,33

Notes: scoring of callus colour : (1) white, (2) yellow, (3) green, (4) brown, (5) yellowish white, (6) greenish white, dan (7) brownish yellow.

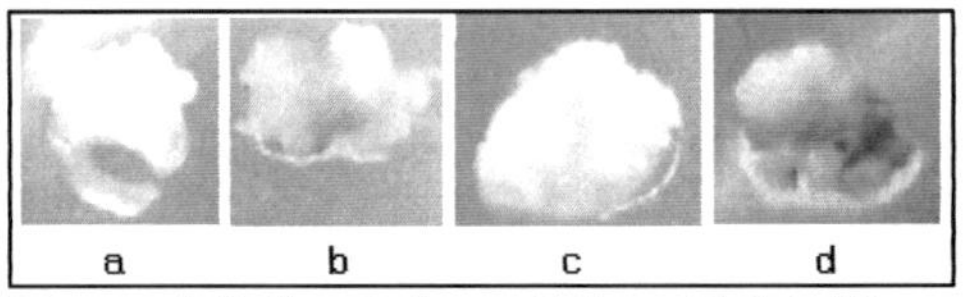

Figure 2. Callus Colour of Pare Gajah Rice on 35 DAP (days after planting). Notes: (a) White, (b) Yellow, (c) Yellowish White, and (d) Brownish Yellow

3.1.5. Callus Diameter (mm)

Table 6. Shows the average callus diameter decreases with increasing 2,4-D concentrations given. The average callus diameter tends to be longer to be found at the D1 level (1 ppm) which is 0.49 mm and which tends to be smaller at the D3 level (3 ppm) which is 0.37 mm.

Table 6. The Effects of Growth Regulatory Substances 2,4-D Concentration towards Callus Diameter (mm)

2.4 D concentration	Callus diameter (mm)
D1 (1 ppm)	0,49
D2 (2 ppm)	0,44
D3 (3 ppm)	0,37
Average	0,44

3.1.6. Callus Area (mm^2)

Table 7. The Effects of Growth Regulatory Substances 2,4-D Concentration towards Callus Area (mm^2)

2.4 D concentration	Callus area (mm^2)
D1 (1 ppm)	43,32
D2 (2 ppm)	38,14
D3 (3 ppm)	30,25
Average	37,24

The diameter of the callus is directly proportional to the area of the callus, the greater the diameter of the callus produced, the greater the callus area. Based on Table 7 it is known that the 2.4-D concentration gives an average callus area of 37.24 mm2. Callus tends to be more widely produced at the D1 level (1 ppm) which is 43.32 mm2 and callus tends to be smaller at the D3 level (3 ppm) which is 30.25 mm2. Calculation of callus diameter and callus area using millimeter block can be seen in Figure 3.

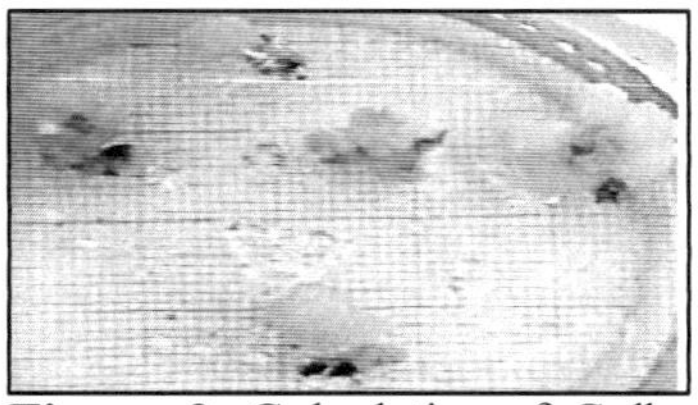

Figure 3. Calculation of Callus Diameter and Area Using Millimeter Block.

3.2. Folate Gene Transformation

The results of the variance recapitulation in Table 8. show that the transformation method treatment significantly affected the hygromycin lethal dose parameters (LD50) and the transgenic putative transformation efficiency (%), while the hygromycin concentration treatment had a very significant effect on the parameters of lethal hygromycin dose (LD50) observations and the efficiency of putative transformation transgenic (%). transgenic putative transformation efficiency (%), but there is no interaction between the transformation method treatment and hygromycin concentration.

Table 8. Recapitulation of the Variance effects of the transformation method and Hygromycin concentration towards the results of the transformation of folate genes in local Banten rice explants Pare Gajah

No.	Observation parameters	Treatment			cv (%)
		Transformation Method (I)	Hygromycin concentration (H)	Interaction (I*H)	
1.	Lethal Hygromycin Dose (LD_{50})	*	**	ns	10,97
2.	Efficiency of Transformation (%)	*	**	ns	10,08[a]

Notes: ns : Not significant
* : Significant at 5%
** : Very significant at 1%
cv : Coefisien of variation
a : Transformed data $\sqrt{x+0,5}$ twice

3.2.1. Lethal Hygromycin Dose

Table 9. Percentage of dead Pare Gajah Rice Explants in Hygromycin Lethal Antibiotic Test Trials on Selection Media

Hygromycin Concentration	Transformation Method		Average
	In vitro (I1)	*In planta* (I2)	
H1 (25 ppm)	83,33	81,67	82,50[a]
H2 (50 ppm)	95,00	83,33	89,16[b]
H3 (75 ppm)	96,67	85,00	90,84[b]
H4 (100 ppm)	100,00	96,67	98,34[c]
Average	93,75[b]	86,67[a]	

Notes : the numbers that followed by the same letters in the same row or column show no significant difference according to the DMRT test of 5% level.

The table shows the percentage of dead callus and dead or non-growing sprouts increasing with increasing concentrations of hygromycin antibiotics. In the in vitro transformation the lowest percentage of callus was 25 ppm which was 83,33%, while the largest percentage of callus died was 100% at 100 ppm hygromycin concentration. *In planta* transformation data obtained dead sprouts tend to be lower at 25 ppm hygromycin concentration that is 81,67% and tend to be higher 96,67% at 100 ppm.

From the table shows LD50 hygromycin using the Karber Method formula for in vitro transformation which is 29,17 ppm. This figure is smaller than the LD50 transformation *in planta* which is 35,63 ppm.

The results of transformation carried out *in vitro* and *in planta* using Agrobacterium tumefaciens were tested based on the effectiveness of hygromycin in inhibiting the growth of explants that failed to be transformed. Explants that have successfully grown and which have died in the selection media can be seen in Figure 4. and Figure 5.

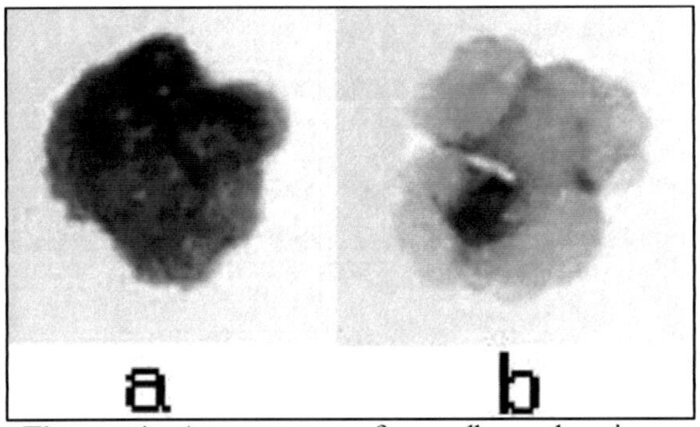

Figure 4. Appearance of a callus that is not resistant to hygromycin (a); kalus resisten terhadap higromisin (b).

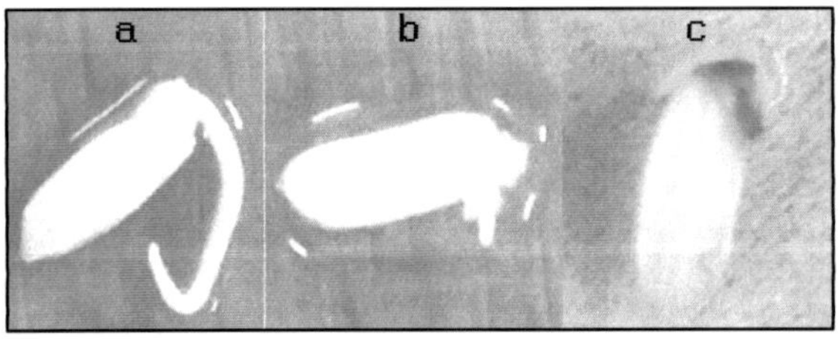

Figure 5. (a) Transgenic putative rice seeds are still able to germinate in hygromycin-containing media, (b) Non-transgenic rice seeds are germinated in hygromycin-containing media, (c) Non-transgenic rice seeds blackened after selection on hygromycin media.

3.2.2. Efficiency of Putative Transgenic Transformation (%)

Table 10. The Effect of Transformation Method and Concentration of Hygromycin towards the Efficiency of Putative Transgenic Transformation of Pare Gajah Rice Explants (%)

Hygromycin Concentration	Average Number of Callus Resistant	Average Number of Sprouts Resistant	Efficiency of *In Vitro* Transformation (%)	Efficiency of *In Planta* Transformation (%)	Average Transformation Efficiency (%)
25 ppm	1,67	1,83	16,67	18,33	17,50[c]
50 ppm	0,50	1,67	5,00	16,67	10,84[b]
75 ppm	0,33	1,50	3,33	15,00	9,16[b]
100 ppm	0,00	0,33	0,00	3,33	1,66[a]
Average			6,25[a]	13,33[b]	

Notes : the numbers that followed by the same letters in the same row or column show no significant difference according to the DMRT test of 5% level.

Based on Table 10. it is known that the transformation efficiency of putative transgenic *in planta* method is greater than *in vitro* method, which is 18.33% and the percentage of transformation efficiency of putative transgenic *in vitro* and *in planta* decreases with increasing hygromycin concentration in the selection media. Putative transgenic that pass in hygromycin media are then maintained and PCR tested with a phen-specific CG CG1 primeer. The PCR results can be seen in Figure

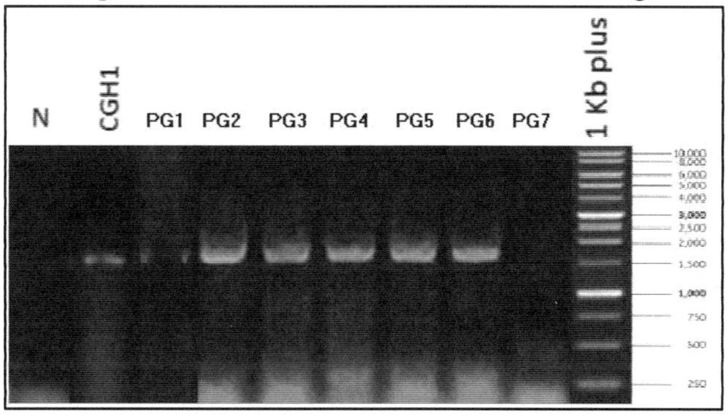

Figure 6. PCR result of putative transgenic rice folate gene

Notes:

N = Pare Gajah (Non-transgenegenic)
PG1,2,3,4,5,6,7 = Pare Gajah putative transgenic folate gene
CGH1 = *Agrobacterium tumefaciens* which contains the folate gene construct
1 kb plus: Ladder

Figure 7. Appearance of Pare Gajah transgenic rice in vegetative phase that has been inserted with the folate gene (a); starts flowering (b); starts the grain filling phase (c)

Positive rice has a 1200 bp folate gene band, treated and implanted in soil media. Based on information from PCR results it is known that putative GMOs PG1, PG2, PG3, PG4, PG5, PG 6 have a DNA band size of 1200 bp (the same as the control which is *Agrobacterium tumefaciens* which carries 'construct folate gene'). Putative transgenic samples of PG7 did not have a 1200 bp DNA band (the same as the negative control which is Pare Gajah rice that is not transformed). The appearance of the transgenic plant can be seen in Figure 7.

4. Conclusions

4.1. Callus induction

Callus induction namely: 2,4-D concentration gives a significant effect on the time parameters of callus formed (DAP) and significantly affects the number of explants forming callus (%), but not significantly on callus texture parameters, callus diameter (mm) and callus area (mm^2). Of the three concentration levels used, a 2,4-D concentration of 1 ppm produced the best callus diameter (0.49 mm) and callus area (43,32 mm^2). 2,4-D concentrations of 2 ppm for the time the callus formed (12,67 days), the number of explants formed a callus (94,44%) and the best callus color (yellowish white).

4.2. Transformation of folate genes

1. The transformation method significantly affects the lethal parameters of hygromycin dose and the efficiency of transgenic putative transformation (%). Of the two transformation methods used, the *in vitro* transformation method has a lower lethal dose (more toxic), causing an average explant mortality of 93.75% with an average transgenic putative transformation efficiency of 6,25%. In the in planta lethal method the dose is higher with an average explant mortality of 86,67%, and the best efficiency of transgenic putative transformation is 13,33%.
2. Hygromycin concentration has a very significant effect on lethal hygromycin dose and the efficiency of transgenic putative transformation (%). Of the four levels of hygromycin concentration used, the 100 ppm hygromycin concentration had the most toxic lethal dose which caused the highest mortality rate (98,34%) with the lowest average level of transgenic putative transformation efficiency (1,66%). Hygromycin concentrations of 25 ppm had lethal doses causing the lowest explant death rates (82,50%), with the highest average transgenic putative transformation efficiency (17,50%).
3. There is no interaction between the treatments of transformation method with the treatment of hygromycin concentration on the results of the transformation of folate genes in the Banten Pare Gajah local rice.
4. PCR results show DNA samples of Pare Gajah rice PG1, PG2, PG3, PG4, PG5, PG 6 have a folate gene encoding DNA band with a size of 1200 bp.

5. References

[1] Bender, D.A. 2002. *Introduction to Nutrition and Metabolism, 3rd edition*. New York: Taylor Francis Inc,.

[2] Juba, N.C. 2011. Metabolic Engineering of The Pterin Branch of Folate Synthesis by Over-expression of a GTP cyclohydrolase I in Peanut. [Disertasi] Blacksburg: Virginia Polytechnic Buku Kedokteran EGC.

[3] Santosa, B., Sobir, Sriani, S., dan Kurniawan, R.T. 2011. Penyisipan Gen *OsDREB1A* pada Tanaman Padi untuk Regenerasi Sifat Toleran Kekeringan. Penelitian Pertanian Tanaman Pangan. Vol. 30. No. 2. Hal. 95-100.

[4] Cooper, H.D., C. Spillene, dan T. Hodgken. 2001. *Broadening The Genetic Base of Crops: an Overview.* pp. 1-23. H.D. Cooper, C. Spillene, and Hodgken (eds.). Broadening the genetic base of crops. IGRI, FAO. UK: CABI Publishing.

[5] Wardah. 2005. Pemanfaatan Tumbuhan Pada Masyarakat Kasepuhan Desa Cisungsang Dikawasan Taman Nasional Gunung HalimunKabupaten Lebak Banten. Berita Biologi. Vol.7. No. 6.Hal. 323-332.

[6] Rahmawati, S., dan Inez , S. 2006. Introduksi Gen *cryIB-cryIAa* ke dalam Genom Padi (*Oryza*

sativa) cv. Rojolele Menggunakan Transformasi Agrobacterium. Hayati. Vol. 13, No. 1. Hal. 19-25.

[7] Rahmawati, S. 2006. Status Perkembangan Perbaikan Sifat Genetik Padi Menggunakan Transformasi *Agrobacterium*. Jurnal AgroBiogen 2(1): 36-44.

[8] Sumadji, A.R., Ahmad, Y., dan Sunarto. 2014. Induksi Kalus Padi (*Oryza sativa L.*) Varietas IR64, Menik Wangi, dan Rojolele melalui Kultur *In vitro*. EL-VIVO. Vol.2. No.1.

[9] Manggung, R.E.R. 2008. Pengujian Toksisitas Akut Lethal Dose 50 (LD_{50}) Ekstrak Etanol Buah Belimbing Wuluh (*Averrhoa bilimbi* L.) pada Mecit (*Mus musculus albinus*). [Skripsi]. Bogor: Institut Pertanian Bogor.

[10] Suparthana, I.P., Masahiro, N., dan Mineo, K. 2014. Identifikasi Transgene pada Tanaman Padi (*Oryza sativa* var. Koshihikari) yang Ditransformasi dengan Bantuan *Agrobacterium tumefaciens*, Menggunakan Metode Tanpa Teknik Kultur Jaringan. Media Ilmiah Teknologi Pangan. Vol. 1. No. 1. Hal. 24-30.

[11] Wang, Kan. 2015. Agrobacterium Protocols: Volume 1, Methods in Molecular Biology. Volume 1223.

6. Acknowledgements

Thank you for the Islamic Development Bank which has funded this research as a part of the Center of Excellent for Food Security of University of Sultan Ageng Tirtayasa.

Utilization and Improvement of Marginal Soils for Agricultural Development in Indonesia

Suwardi

Department of Soil Science and Land Resources, Faculty of Agriculture, IPB University
Jl. Meranti, Dramaga IPB Campus, Bogor, Indonesia

corresponding author: suwardi-soil@apps.ipb.ac.id

Abstract. Indonesia is a humid tropical country that has very intensive weathering of soils resulting acidic soils such as Oxisols and Ultisols. In swampy ecosystems, organic matters accumulated become peat soils or Histosols. Reclamation of shallow peat for agricultural land will induce the forming of acid sulfate soils (Sulfaquept and Sulfaquent). Due to low fertility soils, the three of above soils are classified as marginal soils and now they are used for agricultural activities and some of them are still in forests. The use of marginal soils should refer to land suitability whether it is retained as a forest or used as agricultural land. The development of agricultural commodities on marginal soils must consider soil characteristics and agro-climate conditions. Oxisols and Ultisols have characteristics of low pH, low organic C, low macro and micro nutrients. In areas where rainfall is higher than 2,000 mm/year and evenly distributed throughout the year, oil palm plantations can be well developed. Meanwhile in the rainfall between 1,750-2,000 mm/year, rubber, cacao, coffee plants can be cultivated. Agricultural development in acidic soils need to be followed by the addition of soil amendments such as organic materials, humic substances, lime, fly ash, steel slag as well as fertilizers. Acid sulfate soils have a pH value of around 3 due to pyrite oxidation and usually low macro and micro nutrient content. *Acacia mangium* plants as raw material for pulp and oil palm plantation can grow quite well if followed by application of fertilizers and appropriate soil amendments. Peat soils are very fragile ecosystem so that in the upstream area and peat dome, they must be retained as forests. Peat can be developed for *Acacia crassicarpa*, oil palm, and rubber plants with good water and fertilizer management. Thin peat underneath mineral soil from river sediments can be developed rice field.

Keywords: humic substances, marginal soils, soil ameliorant, water management.

1. Introduction

In general, Indonesia has a tropical monsoon type climate, featuring slight changes in seasons and temperatures, low winds, a high degree of humidity and periodic heavy rainfall. There are two Monsoons: East Monsoon or dry season running from May to September which is influenced by the Australian Continental air masses, and the West Monsoon or rainy season from December to March, which is influenced by the Asian Continental and Pacific Ocean air masses passing over the ocean. Temperatures are largely uniform throughout the year, i.e. between 22 until 27°C, with a uniform relatively humidity between 70 to 90 percent. The rainfall distributions vary between the regions, depending on topography, air mass movement, etc. Generaly the country annual rainfall is relatively abundant and varies between 700 to 4,000 mm. On most area of Sumatra, Kalimantan and West Java, the dry period is however not completely rainless, and there is still moisture over almost the whole year. The climax vegetation on this part of Indonesia is therefore typical rainforest vegetation, and the area enjoys optimum climatic condition for growing a wide range of crops. East Java, Nusa Tenggara and part of Sulawesi are much drier, and are covered by vast grasslands. However, in those less favored areas, rain fed agriculture for at least 4 to 5 months each year, while tree growth is almost continuous [1]. Oil palm plantations occupy the largest area, reaching 14.5 million ha spread across Sumatra, Kalimantan and Papua Islands. Then followed by rubber, sugar cane, cacao, coffee plantations, etc. Food crops that are widely available in Indonesia are 7.4 million ha of rice, followed by corn, soybeans, cassava, etc. Horticultural crop development is mostly done on fertile land, especially Andisols in the highlands.

Various types of soils indicate that Indonesia support many type of ecosystem, because soil characteristics that determine the type of soil are the results of interaction of some land ecosystem parameters. Generally, Indonesia has both land and marine ecosystems, in which the land ecosystems include wetland and dry land ecosystems. Those ecosystems can still be differentiated into some sub-ecosystems. Various types of ecosystem in the tropical and humid climate of Indonesia are responsible for luxuriant vegetation of tropical rain forests containing timber and other forest products. Within the forests grow a large variety of epiphytes, and many kinds of flowers, including the great number of orchids, various kinds of bamboo, rattan, palm trees, etc.

With a humid tropical climate, macro and micro nutrients in the soils are easily leached out and organic matter is easily decomposed so that the soil becomes acidic and poor of nutrients and organic matter. This condition causes most of the soils have a low soil fertility. These soils require good management and proper fertilization in order to provide high production. Soils of this nature are classified as marginal soils which include acidic soils, peat soils and acid sulfate soils. In order marginal soils can provide great benefits to the community for agricultural business, the selection of agricultural commodities must be appropriate by choosing commodities that are adaptive to the characteristics of the soil and agro-climate.

If marginal soils are used for forests, then forest growth is very good because the nutrient cycle the forest soil takes place in a balance between what is absorbed by the forest trees and the entry of nutrients from decomposed litter fall. Within this region make the flora of Indonesia completely different from the flora of the neigh boring continents, Asia and Australia, as well as from the tropical areas within these. The richness of the Melanesian region of which Indonesia represents the major portion, is reflected in the accommodation of close to 40,000 species of plants or about 10 to 12 percent of the estimated number of global plant species. Those tropical rain forests are the host of various fauna species. According to [2], 12 percent of mammals, 17 percent of birds, and 17 percent of reptiles and amphibian species of the world are found in Indonesia. Marine ecosystems are even more diverse, whereas Indonesia lies in the middle of a global bull' eyes. Dealing with such richness of flora and fauna, Indonesia is known as the world's mega-biodiversity country.

In this paper, we will describe how to manage marginal soils in the form of acidic soils, peat soils and acid sulfate soils based on soil and climate characteristics and the author's considerations during fieldwork. The description that will be explained includes the characteristics of the soil, soil improvement, selection of agricultural commodities, and management recommendations so that marginal soils can provide great benefits for humans while considering sustainability.

2. Materials and Method

This paper is written and summarized from various sources from several papers and other sources such as discussions in various seminar forums and Focus Group Discussion (FDG). Discussions and conclusions are based on the author's experience during field surveys in various locations on acidic soils, peat soils and acid sulfate soils. The research studies have been carried out in various locations in Sumatra and Kalimantan Islands. For peat soils, the research study was carried out in Sumatra, such as in Jambi, Riau and South Sumatra. Meanwhile, for acid sulfate soils were carried out in South Kalimantan, Central Kalimantan and Jambi.

The agricultural commodities that the authors examine in acidic soils are rice, oil palm, rubber, and *Acacia mangium*. Likewise, for peat and acid sulfate soils, many research studies were carried out for the commodity *Acacia crassicarpa*, oil palm, and rubber, and rice. By linking the characteristics of soil, agro-climate, agricultural commodities and agricultural production, recommendations can be made to improve acid soils, peat soils and acid sulfate soils. Then by connecting the soil and agro-climate will be recommended types of plants that can be recommended for each soil type in a particular agro-climate

3. Results and Discussion

3.1. Characteristic, problems, and management of acidic soils

Acidic soils are soils that have a pH of less than 5.5 in both dry and wet soils. According to Soil Taxonomy [3], acidic soils are mostly classified into Oxisols, Ultisols, and Inceptisols. The soils of Indonesia are dominated by Inceptisols, Ultisols, and Oxisols so that it can be said that soils in Indonesia are mostly acidic soils. The three orders scattered mainly at Sumatra, Kalimantan, Sulawesi, and Papua Islands occupying 73% of the land area (Table 1). The soils generally have low pH less than 5.5, low CEC, and low macro and micro nutrients. The soils are grown tropical-rain forest, plantations, and small area for food crops.

Table 1. The distribution of soil order in Indonesia

Soil Order	Area (million ha)	Percentage	Areal Distribution
Inceptisols	59.69	31.99	Sumatra, Kalimantan, Sulawesi, Papua, Java
Ultisols	54.20	29.05	Kalimantan, Sumatra, Sulawesi, Papua
Oxisols	23.08	12.37	Sumatra, Kalimantan, Papua
Entisols	14.54	7.79	Sumatra, Kalimantan, Java, Nusa Tenggara
Histosols	11.89	6.37	Sumatra, Kalimantan, Papua
Mollisols	9.75	5.23	Java, Kalimantan, Sulawesi, Papua
Andisols	7.77	4.16	Sumatra, Java
Vertisols	3.40	1.82	Central and East Java
Spodosols	2.07	1.11	Kalimantan
Alfisols	0.20	0.05	Java, Nusa Tenggara, Sulawesi
Total	186.59	100.00	

Source: Soil Research Institute, 2006 [4].

The main problem with acid soils is low pH and is followed by low levels of organic matter. To overcome these problems, the technique by addition of liming materials such as calcite, dolomite, steel slag, and fly ash. The choice of materials and how much is the dosage will be applied, depend on the soil properties such as exchangeable Al, initial pH, organic matter and soil texture as well as the crops that will be cultivated.

Most of acidid soils have very low soil organic matter content. As have been stated by many researchers, soil organic matter plays very important role in the soil ecosystem. Soil organic matter influenced soil physical, chemical and biological properties of the soil. Therefore lacking of soil organic matter will disturb many soil processes and inhibit plant growth. Addition of organic matter into the soil is one of best practice in agriculture. Organic matter addition can be accomplished in the forms of plant residues, manure, compost, and humic substances. Humic substance is the essence of organic matter. Example of the characteristic of humic substances is shown in Table 2. From the solubility characteristic in alkali or acid, humic substance can be divided into four category [5], namely: Fulvic (soluble in alkali and acid), Humic (insoluble in acid but soluble in alkali); Himatomelanic (soluble in alcohol), and Humin (insoluble in acid and alkali).

Table 2. Chemical characteristics of humic substances

Parameter	Unit	Value
C-ash	%	28.54
Ash content DTA	%	41.94
Carbon	%	29.76
Nitrogen	%	1.85
C/N	%	16.10
Solid	%	9.58

To improve the nutrients availability, fertilizers must be applied to the soils. However, because of soil properties and natural condition (low organic matter, low active clay therefor low CEC, and high rainfall), ordinary fertilizers are not enough. Special fertilizers technology is needed to overcome the problem of disadvantages in soil properties that cause low soil fertility. Soil analysis is one key success factor in determining the nutrients need in this particular soil. Another factor is special form of fertilizers both for macro and micro nutrients that will be added to the soil base on the soil analysis. The way of the fertilizers application is also very important in increasing the efficacy and the efficiency of the fertilizers.

The main plantations planted at those soils are oil palm, rubber, coffee, cacao, sugar cane, etc. The oil palm is planted in area where rain is fall all over the year with the rainfall at least 2,000 mm/year. Sumatra, Kalimantan, and Papua are the main Islands where oil palm is planted. Meanwhile in the rainfall between 1,750-2,000 mm/year, rubber, cacao, coffee plants can be cultivated. Beside rice, the other Indonesian main food crops are corn and cassava. The central corn plantation is Gorontalo Province of Sulawesi Island whereas the central for cassava is Lampung Province, Central Java, and East Java.

3.2. Characteristic, Problems, and Management of Peat Soils (Histosols)

Histosols are peat soils that developed from organic matters accumulated in swampy ecosystems. These soils have organic soil materials that extend from the surface to any depth if the organic soil material rests on fragmental material and the interstices are filled with organic materials or rest on a lithic or paralithic contact. The peat soils in Indonesia is very typical, because these soils developed from organic accumulation of the tropical rain forest, therefore their bulk densities are very low (about 0.05 - 0.40 g/cm^3) and the content of the macro-nutrients for plant such as N, P, K, Na, Ca, Mg, and micro-nutrients such as Fe, Cu, Zn, Mn and others are very low. Related to these characteristics, plant cultivation on the peat soils need a lot of fertilizers.

Histosols occupy 6.37% of the total Indonesian land. These soils distributed at east seashore of Sumatera Island, Kalimantan, and Papua. Histosols are mostly grown lowland tropical forest. Some Histosols near the rivers have been reclaimed by local peoples for paddy field. Since 1980's government opened some Histosols at Sumatra and Kalimantan Islands for transmigration of Java people for paddy field. In the first 5 years, the production was very high but in the next years the production decreased gradually and abandoned by farmers. In 1990'0 some private companies used Histosols for forest plantation especially acasia.

Since the year of 70's systematic reclamation of wetland ecosystem is intensified and extended for transmigration program and for plantation sponsored by the Indonesian government. The peak of the wetland reclamation activities was the implementation of Mega Rice Project 1 million hectares in Kalimantan. Recently among 6 million hectares that have been open up, about 3-4 million hectares wetland are degraded and abandoned due to among others acid sulfate. The degraded wetland are mainly in Kalimantan (ex Mega Rice Project) and Sumatra. Increasing demand of energy and depletion of fossil energy source increase the challenge to produce green energy materials. Improvement of degraded wetlands provides opportunity to produce green energy material in industrial scale. There are some possibility of technology that can be used for degraded wetland improvement, among others is using of steel slag for ameliorant.

Indonesian peat soils were developed from woody materials, have many disadvantages characteristic as compared to peat soils from temperate areas. Peat soils in Indonesia are very acid (pH < 4.00), have very low ash content that mean also very low in nutrient, have very low bulk density and contain a lot of phenolic compounds that are toxic to plants especially food crops. Since 1970's government of Indonesia has reclaimed millions of hectares of peat lands as transmigration area, however most of the reclaimed area were failed due to many factors such as low fertility of the soils, unsustainable agricultural practice by transmigrant (burning the peat to increase soil fertility), water mismanagement etc. But nowadays, several private companies have been successfully open peat land for Acacia plantation. The key success of acacia plantations in peat lands are: good water management, fertility improvement of the soil through boiler ash addition and good fertilizer management.

Around 6 million hectares of Histosols have been exploited for rice field, plantation, and forest plantation. The use of wetland in Indonesia has been carried out since a long time before world war II, especially for agriculture activities in traditional manners. Some researches dealing with traditional agriculture on wetland ecosystem in Kalimantan and Sumatra have been conducted by some researchers such as [6, 7, 8, 9].

Peat can be developed for *Acacia crassicarpa*, oil palm, and rubber plants with good water and fertilizer management. Shallow peat underneath mineral soil from river sediments can be developed rice field. Farmers cultivate rice, rubber, coconut on the reclaimed swampy areas.

3.3. Characteristic, problems, and management of acid sulfate soils

Sulfaquents are soils without development that have specific characteristics i.e. have sulphidic materials within 50 cm of the mineral soil surface, or are permanently saturated with water and have in all horizons below 25 cm dominant hue that is neutral or blue than 10Y and color that change on exposure to the air. These characteristics indicate that the sulphidic materials are only stable below water table under reduced environment. If water table is lowered, the soil environment becomes aerated and the sulphidic materials are under oxidation condition and produced strong acid. Within such condition the Sulfaquents are converted to Acid sulfate Soils, whereas the acid should be neutralized before cultivation.

Considering steel slag has high pH and base cations such as CaO, MgO and contain of micro nutrients needed by plants such as Cu, Zn, Fe dan Mn, CS has good prospective for soil amendment in acid soils having low in Si, macro and micro nutrients. The soils can be meet in peat soil and degraded lowland such as acid sulfate soils. The soils have pH of about 3 so that plants can not grow normally. Efforts to utilize peat land for production of crops should be accompanied by soil amendments such as lime, fertilizers of NPK, sulfate salts such as Fe, Cu, Zn, etc. Considering the price of fertilizers and soil amendment are very high, efforts to utilize S for improvement of peat and acid sulfate soils are very attractive.

Utilization of steel slag for improvement of degraded wetland such ad acid sulfate soils is very important because these soils are widely spreading in Sumatra and Kalimantan Islands and having very low productivity. For example farmer in Delta Berbak, Jambi Province only obtained 1 ton/ha/ year. In acid sulfate soils which have very low pH (<3), solubility of Al ion in water is very high and become toxic for plants. The content of Al in the canal water can reach >100 ppm. Sumawinata *et a*l 2008 showed that acid sulfate soil in Rantau Rasau, Jambi besides low pH also contained of very low macro and micro nutrients. Acid parent material and intensive leaching in very acid environment (pH <3) causing most Fe ion and other ions in the soils are leached. Improvement of acid sulfate soil by application of lime, macro and micro nutrients increase the rice production become 4-5 ton/ha [10].

Efforts to utilize slag for soil amendment in degraded acid sulfate soils have good prospect. However the real application in commercial scale are still very limited because of slag sizing and transportation problems. Therefore, researches for utilization of steel slag for improvement of degraded wetland to support biomass production are needed. Through this research collaboration we hope steel slag produced from Sumitomo Corporation can be used for improvement of degraded wetland in Indonesia in order to increase the income and poverty of farmers as well as for production of green energy materials. Steel slag approved as a silicate fertilizer by Ministry of Agriculture, Forestry, and Fisheries of Japan in 1955

[11]. The type of steel slag commonly used as silicate fertilizer is blast furnace slag and this slag also called as calcium silicate slag.

Research on utilization of slag in Indonesia for rice cultivation mostly conducted in rice field at Java Island. Soils in Java Island develop from volcanic ash materials or soil develop from old sedimentary rocks contain of enough Si content. Soils in Java Island periodically still enriched with volcanic ash material from the explosion of volcano rich in Si. With the above reasons application of steel slag in Si rich soils did not significantly increase the production of rice. On the other hand, application of steel slag in the rice field at peat soils showed the prospective results. [12] showed that application of Indonesian electric furnace slag on peat soil significantly improved the availability of Si as well as increased soil pH, exchangeable Ca and Mg. On the other hand, it significantly decreased soil organic matter content, total N, and the availability of Fe, Mn, and Zn. Wetland rice grown on peat soil highly responded to the slag application. The number of productive tiller, number of panicle, and weights of filled and total spikelets were significantly raised with the slag application. Increasing the growth and yield of wetland rice after application of the slag was associated with increasing the availability of Si, soil pH, and exchangeable Ca and Mg. However, utilization of steel slag did not widely applied in the field because of particle size that difficult in application and transportation problems.

Acid sulfate soils actually have more complexes problem. Mostly, increasing soil pH does not enough to solve the problem. To overcome the problems in acid sulfate soils, several approaches have to be accomplished altogether. Improvement technology for acid sulfate soil include:

(1). Fertilizer technology. Acid sulfate soils usually contain very low nutrient, because most of the nutrients are dissolved and then washed away due to very acid condition of the soils. Therefor special form of fertilizer is needed. The way the fertilizer is applied and when the fertilizer is applied also must be considered to overcome the extreme condition that lead to nutrients leaching.

(2). Red soil application. Although several researches though that in acid sulfate soil, there is iron toxicity but actually that is not the case. This thinking maybe due to the fact that pH of acid sulfate soil is very low therefor the dissolution of iron will be very high and this will cause iron toxicity. Of course at the beginning of acid sulfate soil formation that will be the case, however in the long run, the dissolve iron will be washed away and this process continue for several years, and at last most of the iron will be depleted from the system leaving the soil deficient of iron. To add the iron into the acid sulfate soil, red soil that is rich of iron has to be added.

(3). Lime application. As has been mention above, most of the nutrients in the acid sulfate soil have been washed away, that include Calcium and Magnesium. To add these nutrients into the soil, liming materials such calcite and or dolomite have to be applied in acid sulfate soil. Addition of the liming materials can also increase the pH of the soil so decreasing the acidity of the soil.

(4). Water management. The low pH of acid sulfate soil was caused by the oxidation of pyritic sediment in the soil. The oxidation can occur if the pyritic sediment is exposed to the air. This can happen because mismanagement of surface water level in that area. This lead to over drainage in that area so the soil surface become too dry that lead pyrite oxidation. To prevent this oxidation, the pyritic sediment has to be inundated by water. Therefor water management is very important in preventing pyrite oxidation that cause soil acidification.

(5). Planting system. As mention above, pyrite oxidation must be restricted, that mean pyritic sediment has to kept inundated by water. This lead to less choice of plant that can be cultivated in that area. Only crop that can live in always saturated or flooded condition that can be choosen such as paddy. In some places like in South Kalimantan, Banjarese People using multiple tranplanting technique to plant their paddy.

Many areas of acid sulfate soils are developed for oil palm plantation, rubber, and *Acacia mangium*. Those plants still adapted with very acidic condition. Good water and fertilizer management are the key for successful of agricultural practices in acid sulfate soil.

4. Conclusions

The development of agricultural commodities on marginal soils especially acidic soils, peat soils, and acid sulfate soils must consider characteristics and agro-climate conditions of the soils. Oil palm and rubber are the most suitable for acidic soils. Peat soils can be developed for *Acacia crassicarpa*, oil palm, and rubber plants, rice, and horticulture. The most important of peat development for agriculture is water management and fertilization. Development of acid sulfate soils are prioritized for plantation such as oil palm and rubber. Development of rice and other food crops should be followed by high doses of fertilizers and liming.

5. References

[1] FAO-UNEP, 1994. Advisory Services to Jamaica and Indonesia on the Formulation of. National Soil Policies. Rome, FAO the United Nation: 71 p.

[2] Tobey, A. and Dutton, I. 2002. Tens Years Effort on Conservation in Indonesia. The Nature Conservancy. Jakarta.

[3] Soil Survey Staff. 1999. Soil taxonomy: A Basic System of Soil Classification for Making and Interpreting Soil Surveys. 2nd edition.

[4] Soil Research Institute. 2006. Map of Distribution of Soil in Indonesia, Bogor.

[5] Tan, KH., J.C. Lobartini, D.S. Himmelsbach & L.E. Asmussen. 1991. Composition of humic acids extracted under air and nitrogen atmosphere. Commun. Soil Soi. Plant Anal. 22:861-877.

[6] Mulyanto B., B. Sumawinata, G. Djajakirana, Suwardi. 1999. Micromorphological Characteristics of (Potential) Acid Sulphate Soil Under The Banjarese Traditional Land Management (BTLM) System. Proc. Seminar toward Sustainable in Humid Tropic Facing 21 St Century: 227-292. Bandar Lampung, Indonesia September 27-28, 1999.

[7] Mulyanto,B, dan B. Sumawinata. 2002. Soil Characteristic Chang-es of Wetland Under Banjarese Traditional Land Management System. The International Symposium on Sustainable Land Management, Kuala Lumpur, August 8-10, 2002. ISBN 967-9945-22-7

[8] Sumawinata, B. and B. Mulyanto. 2002. Evaluation of Peatland Development after 30 Years Reclamation Program in Indonesia. International Symposium on Land Management an Biodiversi ty in Southeast Asia. Bali.

[9] Suwardi, B. Sumawinata, H. Wijaya, G. Djajakirana, H. Furukawa. Pilot Project of Rice Culture in Acid Sulfate Soil Collaborated with Provincial Government of Jambi Province. Proceeding of International Symposium on Nature and Land Management of Tropical Peat Land in South East Asia. Bogor, 20-21 September 2006.

[10] Suwardi, Darmawan, Gunawan Djajakirana and Basuki Sumawinata. 2007. Some Efforts for Improving Degraded Peatland in Berbak Delta-Jambi, Indonesia. International Seminar Tropical Peat Seminar. Yogyakarta, 27-29 August 2007..

[11] Ma, J. F. and E. Takahashi. 2002. Soil, Fertilizer, and *Plant Silicon Research in Japan*. Elsevier, Amsterdam.

[12] Suwarno. 2002. Utilization of steel slag in wetland rice cultivation on peat soil, pp 211 - 215. In Proceedings of the International Symposium on Land Management and Biodiversity in Southeast Asia, Bali, September, 17-20, 2002.

A strategy of assistance (guidance) for farmers to manage Bali cattle (Bos javanicus) improvement technology in South Sulawesi, Indonesia

S Bahar[1]*, R Rachman[2], J Corfield[3], and B Pengelly[4]

[1)]Assessment Institute for Agricultural Technology, Jakarta, Indonesia. Jalan Raya Ragunan No. 30, Pasar Minggu, Jakarta 12520, Indonesia
[2)]Assessment Institute for Agricultural Technology, S. Sulawesi, Indonesia.
[3)]CSIRO Sustainable Ecosystems, Townsville, Queensland, Australia.
[4)]CSIRO Sustainable Ecosystems, Brisbane, Queensland, Australia.

*Corresponding author: syamsubahar@litbang.pertanian.go.id

Abstract. The Australian Centre for International Agricultural Research (ACIAR) to finance research projects in Eastern Indonesia, especially in South Sulawesi through the project number. SMAR/2006/061 that have been conducted for the renewal of Bali cattle management technology. Bali cattle (Bos javanicus) are a vital component of Indonesia's crop-livestock farming systems. This paper describes the method of participatory approaches to management of Bali cattle through the renewal component approach to production technology, namely: 1) Better use of existing forage/ feed sources, 2) Introduction and use of new forage, 3) Seasonal mating at the best time, 4) Early weaning and preferential feeding, 5) The calculation and the feed planning. Points 1 and 2 are important because often the main obstacles experienced by farmers. All of these components must interact with each other. Field tests conducted in farmers' land by 60 people scattered guided farmers in 12 villages in 3 districts in South Sulawesi is the regency of Bone, Barru and Gowa. Specific case studies carried out case by case basis experienced farmers in these three districts. The impact is being tested now and for the propagation in the system of plants and livestock farms in South Sulawesi.

Keywords: improving, management, Bali cattle, smallholders

1. Introduction

1.1. Background

Program conducted by the Australian Centre for International Agricultural Research (ACIAR) in Indonesia are concerned in terms of research and application of agricultural technology development aimed at increasing farmers' income, especially the Eastern area of Indonesia. This concern has been reinforced through participation in the fund ACIAR AusAID Smallholder Agribusiness Development Initiative (SADI) for 10 years from the year 2006 which aims to achieve higher income households in rural farmers through increased productivity, access better markets and the value of activities outside farming. This activity is conducted in 4 provinces in the eastern region of Indonesia South Sulawesi, Southeast Sulawesi, West Nusa Tenggara and East Nusa Tenggara. ACIAR SADI role in the implementation of a sub component of the research program of Support for Market-Driven Adaptive Research (SMAR) which aims to improve access for farmers and Micro-Small-Medium Enterprises

(SMEs) in terms of productivity and knowledge base of marketing agricultural products at the production level and higher quality. Research priorities during consultation for this is the SMAR in 4 provinces in late 2006 was given a high priority is to build a sustainable system of crop and livestock farming as the core of this program.

1.2. Basic Considerations

Increased demand in the Indonesian cattle for meat and live cattle to the present transmigration areas. Role of Bali cattle is dominant, especially in South Sulawesi as a producer of meat. The results of research / survey by [1] shows several important points are:

a) Bali cattle represent 90% of the total population of cattle in South Sulawesi.
b) Approximately 70-80% cows sold for slaughter, the rest to be maintained as seeds or cattle for fattening. There is also a cow that was sent to the inter-island as Kalimantan.
c) The total population of Bali cattle in South Sulawesi in 2005 is 628,000 head, and high population centers in Bone regency, Barru, Enrekang, Sidrap, Gowa, Sinjai and Bulukumba.
d) The rate of slaughter of cows increased lately, so the reliability of supplies have stabilized.
e) The possession of an average 2-3 cattle per family farmers, but there are some farmers who have a commercial business that can have up to 30 cattle.
f) The primary obstacle to cattle production is the supply of feed, especially in the dry season, the calving interval, health of livestock, availability of venture capital, marketing and maintence workers.

From the above findings clearly the role of Bali cattle in crop and livestock farming in South Sulawesi. Types of Bali cattle is very familiar among farmers and proved resistant to extreme environmental conditions such as dry climate, the drought conditions were still able to bear children in good health. Animal Husbandry Department Program of South Sulawesi Province which target the achievement of cattle population of one million head in 2013. In the year 2007 population of 668,000 head in 2008 and the population to 704,303 head [2]. Application of improved technology package cattle feed will show the optimal results and good if the farmers not only as objects but as subjects of technology actors. Thus, researchers and farmers together to do research on farmers' land and direct benefits felt by the farmers. This has been done previously by [3].

1.3. Purpose and Output

The goal is to develop a strategy of assistance (guidance) for farmers to manage Bali cattle improvement technology in South Sulawesi. The output is a package of assistance strategies Bali cattle management innovation.

2. Methodology

2.1. Structure and Roles

Implementation of the program will be overseen by a Project Steering Committee (PSC) consisting of officials and the Office of Technical Agency and the personal from the project leader and project coordinator Indonesia and Australia. Project Management Team (PMT) consists of project leader and project coordinator of Indonesia and Australia which will directly manage project activities. Project is also supported by researchers and extension workers to form a multi-disciplinary technical teams from both sides Indonesia and Australia, called Project Specialist Team (PST). Then there On the Ground Team (OGT), each of which is formed separately for the work and responsibility in each district 3. The OGT is a personal effort Strata 1 young graduates who were recruited through newspaper notices and then media were selected oral and written strictly. Varied backgrounds and disciplines are trained in a training before being placed in the village with the farmers.

2.2. Preparation

a) The recruitment of younger staff OGT Strata 1 graduates from different disciplines but mainly agriculture and husbandry. In each district OGT placed 1 team consists of 4 people. The number of districts as much as 3, so that the OGT was formed 3 teams for each district. The total number of personnel are recruited 12 people.

b) Training for the OGT is a technical training as a preparation for the farm team for the implementation of activities in the field, especially in interacting with villagers (farmers).
c) Build relationships between OGT becomes an inseparable part to establish relationships between team members because of the varied disciplines of science, agriculture, livestock, agribusiness, and there are socio-economic all of which need to interact in solving problems arising in the farmers.
d) Fix/clarify the methodology to be applied when necessary to revise the method, but does not alter the purpose and objectives to be achieved.
e) Identify 3 districts of each district 4 villages per district bringing the total village was built 12 villages.

2.3. Identification and Testing

a) Conducting basic surveys and research areas interviews with each farmer built.
b) Identify the target farmers randomly selected with certain criteria that is expected to represent the farmers in his village.
c) Testing to the farming land of several components that offer the technology as a technology choice.
d) Monitoring and Evaluation of each activity in the field by regularly visiting each location.

2.4. Dissemination

a) Through the media outreach conducted by the counselors at each of their meetings with farmers.
b) Through the media write good media such as newspapers, brochures, journals and electronic media.
c) Conduct training involving researchers, extension workers and farmers who simply packaged and interact with each other so that the impression is more relaxed but still the target.

3. Results and Discussion

The results of the preparation phase is recruiting OGT (On The Ground Team) has been carried out based on the written test data and results of interviews and tests which qualify as many as 13 people selected and placed in the respective regions coachee. Performed after determining the location or selection of survey villages in 3 district in South Sulawesi, aim to facilitate the gathering of information desired and the location of the election results: District in Barru: Mattirowalie village, Lompo Riaja village, Tompo village and Anabanua village. In Bone district: Bune village, Mattirowalie village, Tappale village and Laburaseng village. In Gowa district: Bontomanai village, Mangempang village, Pabbentengang village and Maccini Baji village.

OGT has conducted training among other things: 1) Basic training is a training facilitator who_aims to provide insight to the OGT of the facilitator's self-image, duties and functions of ethics facilitators and facilitators. The training was also trained how to do OGT and also facilitating the development of training OGT in drafting work. The training is conducted as stock preparation for the OGT join community. 2) Training on livestock feed, nutrition and breeding is in order to increase knowledge about cattle management by OGT.

Training is conducted in the form of material space and practiced in the field. These activities make the participants better understand the feed livestock both theory and practice in the field. This training is more effective because of what is presented in the form of material and then practiced in the field so that they not only learn in theory, but also learn to recognize directly the conditions on the ground.

Activities farmers identification of prospective farmer will be built in 12 villages has been completed. Basic data (Benchmarking data) were collected from 12 villages in 3 districts, then in the analysis. Identification of farmers as many farmers built about 30 people, then the farmers selected 10 candidates who identified the next in an interview by the Specialist Team and the OGT, and the interview was given a score to determine the 5 people who selected two target farmers per village.

The number of farmers selected guided through the selection of Barru is 20 people, spread in 4 villages. The number of farmers selected built in Bone is a people scattered in 20 villages 4. The number of farmers selected built in Gowa were 20 people scattered in 4 villages.

IOP Conf. Series: Earth and Environmental Science **383** (2019) 012048 doi:10.1088/1755-1315/383/1/012048

Application of the technology package of integrated livestock management of cattle through the 5 approaches: 1) Better use of existing forage/ feed sources, 2) Introduction and use of new forage, 3) Seasonal mating at the best time, 4) Early weaning and preferential feeding, 5) The calculation and the feed planning. Components 1 and 2 more on the priorities to be applied as the main constraints experienced by farmers or cattle ranchers.

Component 1, 2 and 5 has been adopted by farmers all built, but the components 3 and 4 was partially due to the limitations of adopting a bull that is difficult to regulate mating and early weaning.

Table 1. Adoption of the technology package per component technologies cattle management

Technology package	Componet 1) Better use of existing forage / feed sources		Component 2) Introduction and use of new forage		Component 3) Seasonal mating at the best time		Componet 4) Early weaning and preferential feeding		Component 5) The calculation and the feed planning	
District	Number of farmers built	Farmers who adopt	Number of farmers built	Farmers who adopt	Number of farmers built	Farmers who adopt	Number of farmers built	Farmers who adopt	Number of farmers built	Farmers who adopt
BONE	20	20	20	20	20	17	20	8	20	20
BARRU	20	20	20	20	20	6	20	3	20	20
GOWA	20	20	20	20	20	10	20	4	20	20

Of the five components of the applied approach are: 1) Better feed utilization with an existing, 2) Use of new forage plants, 3) Control mating time, 4) Early weaning followed by improvement of feed, and 5) Planning the feed so it appears that the components 1, 2 and 5 has been adopted by all farmers had guided and disseminate to other farmers nearby. However, components 3 and 4 not all farmers adopt coachee. This is because cattle are very limited male cattle or bull as most farmer to sell cattle when old enough. Bulls are left only young bulls that have not been good to serve as a male cattle parent.

Component 1 of the farmers had to make the most of existing feed include a field of grass, elephant grass and plant waste like straw and peanut hay shortage during the dry season. As for component 2 is the use of new forage plants also have been conducted with farmers in fields to plant their grass species and legumes include *Paspalum atratum, Brachiaria hybrid* cv Mulato, *Setaria sphacelata, Panicum maximum, Clitoria ternatea, Centrosema pubescens* and *Stylosanthes guianensis*. These types are able to adapt and was able to produce forage feed for cattle [4]. The fifth component of the feed meant planning was already counting the needs of farmers needed to feed cattle maintained, so that the amount of feed can be prepared for some cows. It is important that feed availability can be guaranteed throughout the year.

Dissemination to other farmers by farmers each coachee. There are two target farmers who quickly spread to other farmers, some are slow. Depending on how farmers built to promote its activities. Observations on technology dissemination component Bone faster due to the role of farmers built in Bone more intensively and actively disseminate its activities. In Figure 1 presented dissemination by farmers for farmers built around it. Dissemination of technology components appear very rapidly in the Bone, this is due to farmers' average built very actively provide information to other farmers, both in formal forums or in-formal, so that other farmers who know about it more and more. Although the Barru and Gowa still a little spread, but it can be seen that there is an increase from time to time, this means that farmers built remains an example and motivate other farmers nearby. Another thing is also due in Gowa and Barru late onset of the start of October and November 2008, while Bone began in July 2008.

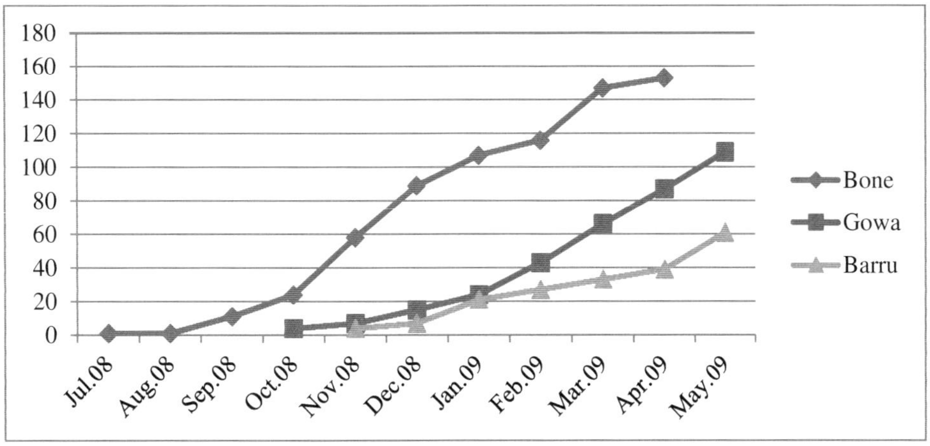

Figure 1. Dissemination of technology components from peasant farmers built into other

From the ACIAR assistance strategy for the management of Bali cattle in South Sulawesi by the approach taken for the implementation of the technology package comes several important questions are :

1) What factors that encourage dissemination of technology?
2) What obstacles are found so far in terms of dissemination?
3) How this can be accelerated deployment?

Answers to these questions can be described and analyzed as described below:

1. What factors that encourage dissemination?
   a) Success farmers built
   b) Strategic location and good for access
   c) Farmers have the talent to interact and communicate
   d) The relationship between family
   e) Good initial impression
   f) Curiosity about the new
   g) The experience of farmers built
   h) Trying new things
   i) Location of farmers built
   j) The social status of farmers
   k) The success of farmers built
   l) Cooperation with OGT
   m) Socialization by OGT and PPL
   n) Farmers actively guided to inform other farmers
2. What obstacles are found so far in terms of dissemination?
   a) Some farmers built the farmers did not inform the other
   b) There is a competition between farmers
   c) Farmers are very busy
   d) Location of farmers who are too far
   e) Farmers coachee inactive and less confident
   f) Self-centered attitude of farmers
   g) Remote field conditions
   h) Lack of land resources
   i) Information submitted farmers too convoluted
3. How this can be accelerated deployment?
   a) Support from local government
   b) There are special incentives for farmers built
   c) There is a public activity (safety, parties, etc.)

d) Visits from farmers to farmers
e) Formal meeting (workshop)
f) There is cooperation with the formal institutions
g) Training for the PPL
h) Create a pilot plot in the village
i) Cooperation with local authorities for dissemination and information
j) To make the location a successful farmer built as a place to visit
k) Meeting of farmer groups
l) The meeting between farmers 'groups with other farmers' groups

4. Conclusion

By ACIAR assistance strategy for the management of innovation, Bali cattle in South Sulawesi seems applicable is the mentoring begins an intensive to target farmers and motivate farmers to spread the coachee to other farmers. Thus occurred the transfer of technology from farmers to farmers.

5. Acknowledgements

Project funding support from the Australian Centre for International Agricultural Research (ACIAR) is gratefully acknowledged. The authors wish to acknowledge all Austalian and Indonesian members of the ACIAR_SMAR/2006/061 project team involved in this study and also their Indonesian partner organizations including BPTP Sulsel, Dinas Peternakan Sulsel and Hasanuddin University Makassar. Finally we wish to express our appreciation to all 60 case study farmers and their village communities who invested their time, resources and trust in helping to make this study a success through working and learning together. We hope the outcomes achieved help to improve both their farming systems and livelihoods into the future.

6. References

[1] Padjung, R and A. Natsir. 2005. Preliminary Survey on Beef Cattle Production in South Sulawesi. Survey Results Report Hasanuddin University, Makassar.
[2] David, A. 2009. Movement achievement cattle population of one million head in South Sulawesi. Sinar Tani. 1-7 April 2009 edition. No. 3297. Pp. 7.
[3] Corfield, J., S. Bahar, S. Lisson and R. Rachman. 2008. Improving forage and feeding management options for smallholders: Recent lessons from Eastern Indonesia. Proceeding of National Seminar on Beef Cattle. Cooperation between the University Tadulako with Animal Husbandry Department Palu in Central Sulawesi. Palu, 24 November 2008.
[4] Bahar, S. 2008. Productivity of forage plant for beef cattle production in South Sulawesi. Proceeding Naional Beef Cattle Seminar. Cooperation between the University Tadulako with Animal Husbandry Department Palu in Central Sulawesi. Palu, 24 November 2008.

Selecting Traits for Drought Tolerance Screening in Rice

U Susanto*, W R Rohaeni, and P Sasmita

Plant Breeding Division, Indonesian Center for Rice Research (ICRR) of Indonesian Agency for Agricultural Research and Development (IAARD).
Street 9, Desa Sukamandijaya, Kec. Ciasem, Kab, Subang, West Java, Indonesia

Corresponding author: untungsus2011@gmail.com

Abstract. Drought is one of major constrains for rice production in Indonesia. Development of drought tolerant rice variety is believed as an efficien and effective strategy to face the challange. Nevertheless, drought is a complex trait thus many approach had been implemented in the selection of drought tolerant rice genotypes. This study aimed to search traits correlated with drought tolerance in rice plant. Two screening methods, i.e. seedling box and mini pot methods was conducted into 60 rice genotypes. The experiments were conducted during dry season 2016. The results showed that based on % recovery, TIL24 had the highest (82.5 %) while number of green leaves had strong positive correlation with root length, root number in zone 1 (horizontal), 2 (medium), and 3 (vertical), total root number, leaf rolling, and plant recovery. Nevertheless, proportion of root in zone 3 negatively correlated with shoot length, green leaf number, root number in zone 1 and 2, total root number, and plant recovery. It seem that proportion of vertical root compensate with other traits in developing drought tolerance, while number of green leaf under stress and root number in zone 2 correlated more with traits supporting drought tolerant in rice.

1. Introduction

Drought is one of major constrains for rice production globally and include Indonesia. Drought stress mostly happen in unbunded upland, bunded upland, shalow rainfed, mid-lowland rainfed and in water-short irrigate. Drought affects around 23 million ha of rainfed rice accross the world [1]. The condition is predicted to be worse in teh future, due to global climate change. It would also interact with other abiotic and biotic stresses [2]. In Indonesia, it is reported that around 50.000 to 300.000 ha areas are affected by drough every year [3]. It is predicted that due to climate change, there will be also reduction on rainfall during June – August [4].

Development of drought tolerant rice variety is believed as an eficient and effective strategy to face the challenge. Nevertheless, drought is a complex trait thus many approach had been implemented in the selection of drought tolerant rice genotypes [5]. This study aimed to search traits correlated with drought tolerance in rice plant at the early vegetative growth.

2. Methodology

The materials of this research is 60 rice genotypes which is simultaneously selected using two screening methods, i.e. (1) minipot for root angle pattern and (2) Seedling box for seedling dry. Both methods are used for early vegetative growth phase drought tolerance selection. The experiment was conducted in ICRR green house in Sukamandi during October to December 2016.

The seedling box experiment was foolowed randomized complete block design with two replications. The seeds of the 60 rice genotypes we sown in petridished with weth paper inside for around two days to let seed for initiating germination. The well inititally germinated seeds were transfered into the seedling box made of wooden materials, each row contained 20 seeds with the distance of around 1 cm within row and 5 cm between row. IR20 as susceptibel check and Salumpikit as tolerant check were planted in every 20 tested entries. IR20 was also planted surrounding the tested genotypes. Watering was conducted for seven days then wait until IR20 died or severely dried. Scoring was done into leaf rolling (Table 1) and drying (Table 2) [6], and plant recovery (percentage of survive plants). After both scoring, watering was given again to the plants for three days. Plant receovery was counted one week after the watering. Soil moistutre content/soil osmotic was monitored using Irrometer moisture meter in 15 cm depth. Observation was executed into leaf rollong score, leaf drying score, and revcovery (percentage of recover plants after stopping the drought treatment.

Table 1. Leaf rolling scoring for drought screening at seedling stage using Seedling box method

Score	Symptom
0	Leaves healthy
1	Leaves start to fold (shallow)
3	Leaves folding (deep V-shape)
5	Leaves fully cupped (U-shape)
7	Leaf margins touching (0-shape)
9	Leaves tightly rolled

Table 2. Leaf drying scoring for drought screening at seedling stage using Seedling box method

Score	Symptom
0	No symptoms
1	Slight tip drying
3	Tip drying extended up to ¼ length in most leaves
5	One-fourth to 1/2 of all leaves dried
7	More than 2/3 of all leaves fully dried
9	All plants apparently dead

The minipot experiment was managed according to randomized complete block design of three replications. A single plant in a pot is representing one unit data. Media was prepared as mixed of soil and sand in an equal ratio. Minpots were filled up with the prepared media with sieve at the top of the media in each pot. The pots were made of paralon pipe with the diameter of 2 inch with 20 cm depth. Sieve was made in sunken dimention with the dept of 5 cm. Sieve diameter was manged to be fitted with the pot diameter. The sieve was marked by red colored string weaving into the sieve in two lines horisontally separating the sieve proportionally into three zone, i.e. representing vertical, medium, and horizontal root growth direction. The seeds were placed on the media on the sieve on the pot. The pots were placed in box with around 3 cm standing water to keep the humidiy of the media. The plant was establish for 42 days, so that the final observasion was done at the end of vegetative plant growth phase. Observation was executed into shoot length, tiller number, root length, and root number at zone 1 (horizontally growth), zone 2 (medium), and zone 3 (vertically growth). Observations were conducted to Shoot length (cm), Tiller Number, Root length (cm) Zone 1, Root Number Zone 2, Root Number Zone 3, Total Root Number, and proportion of Zone 3 Root Number to Total Root Number.

Correlation coefficient and T value of the coefficient were analyzed among the observes traits in both experiments. The Minitab Release 14 Software Analysis is used for the analysis.

3. Results and Discussion

The results showed that leaf rolling of the genotypes ranged from 5 to 9 (average of two replications). It showed that the stress level is very high (Figure 1). Based in leaf rolling, Kelimutu (5), Bahbutong (5), Huanghuazhan (6), and HHZ5-SKI-9-3-0Kr-JK-IND (6) had the best performance, while IR 20 had score of 8 and 19 genotypes had score of 9 (Table 3). Leaf rolling is one of plant response to drought that might be corelated with reducing leaf expansion, lower stomatal conductance, and transpiration rate [7]. It showed that leaf rolling may correlate with drought tolerance of the plant.

Based on leaf drying, Inpari21, TIL24, IR64IRRI, and BP16188M-2D-SKI-19-2-1 had the best performance (score 1). IR 20 along with other 37 genotypes had relatively good score (2), including Kelimutu, Huanghuazhan, Gajang Mungkur, and Bahbutong (Table 3). IR 20 in some cases showing relatively good performance comparable with the tolerant check. Nevertheless, there are some lines having lower performance compare to IR 20. It indicate that the selection was very thight. Leaf rolling and drying are significantly associated with rapid ability to recover after drought stress [8]. This trait is still widely used to screen drought tolerant mateials of rice.

Figure 1. Performance of 60 rice genotypes under drought stress treatment using seedling box method, Sukamandi, 2016

Table 3. Leaf rolling, leaf drying, and plant recovery of 60 rice grnotypes under drought screening at seedling stage using seedling box method, Sukamandi, 2016

No	Genotype	Leaf Rolling	Leaf Drying	Plant Recover (%)
1	Dro1	9	3	30.00
2	GajahMungkur	9	2	15.00
3	Lipigo4	7	2	40.00
4	Inpari10	8	2	60.00
5	Inpari18	9	3	55.00
6	Towuti	7	3	67.50
7	HHZ12-SKI-1-2-0Kr-JK-IND	8	2	30.00
8	TIL3(IR84636-13-12-2-6-3-3-2-2-B)	7	3	37.50
9	Huanghuazhan	6	2	7.50
10	Kelimutu	5	2	2.50
11	BatangLembang(6661)	8	3	22.50
12	Dular	8	4	15.00
13	HHZ5-SKI-9-3-0Kr-JK-IND	6	3	5.00
14	BP14342f-7	7	3	52.50
15	Limboto	7	2	10.00
16	IR20	7	2	17.50
17	IR64IRRI	9	1	2.50
18	BP17552-1c-SBY-0-CRB-0	7	3	57.50
19	Inpari21	7	1	12.50

No	Genotype	Leaf Rolling	Leaf Drying	Plant Recover (%)
20	Bahbutong(1182)	5	2	20.00
21	HHZ5-SKI-7-1-0Kr-JK-IND	9	2	7.50
22	IR83142-B-20	8	2	65.00
23	TIL2	9	2	25.00
24	TIL24	7	1	82.50
25	BP16734e-3	8	4	45.00
26	BP14262e-1-1	8	2	55.00
27	IR83383-B-B-11-4	8	2	65.00
28	PR40781b-3-4-SBY-0-CRB-2-SKI-0-5	8	3	65.00
29	BP17300M-54D-SKI-8-2-2	9	3	12.50
30	BP17298M-53D-SKI-1-4-2	7	2	65.00
31	BP15994M-4D-SKI-18-1-3	9	3	12.50
32	BP16146M-1D-SKI-2-13-1-3	8	3	50.00
33	BP16178M-4D-SKI-16-6-2-2	7	2	42.50
34	BP16178M-1D-SKI-13-8-1	8	2	35.00
35	BP16188M-2D-SKI-19-2-1	9	1	50.00
36	GSRIR1-11-D3-S3	9	3	45.00
37	GSRIR1-11-D1-Y1	9	2	45.00
38	GSRIR1-11-D3-S2	9	4	65.00
39	BP17282M-41D-1-SKI	9	4	52.50
40	BP17280M-46D-IND	9	4	75.00
41	IR83142-B-20	8	3	57.50
42	IR61336-4B-14-3-2(PSBRC94)	9	2	30.00
43	BP14262e-2-8	8	2	42.50
44	BP14262e-1-1	8	2	32.50
45	BP14352e-1-1-2Op-JK-0	7	2	32.50
46	BP14352e-2-2-0Kr	7	2	75.00
47	HHZ5-SKI-7-3-0Kr-JK-0	8	3	42.50
48	BP14352e-2-3-3Op-JK-0	9	2	57.50
49	BP17280M-1	9	2	52.50
50	BP17280M-66-1	8	3	32.50
51	BP17292M-22	8	2	47.50
52	BP14342f-7	8	2	47.50
53	BP14352e-2-3-1Op-JK-0	9	4	55.00
54	HHZ9-SKI-19-8-0Kr-JK-0	8	5	60.00
55	BP14476e-3-4-2Kr-JK-0	7	4	10.00
56	BP14476e-3-9-1Kr-JK-0	8	3	65.00
57	BP14352e-1-2-2Kr-JK-0	8	3	22.50
58	HHZ18-SKI-2-1-0Kr-JK-0	8	3	20.00
59	CT18510-23-4-4-1-MP-JK-0	9	4	7.50
60	WAB99-2-1-JK-0	8	2	5.00
	5%LSD	2.58	3.04	60.74
	CV (%)	16.30	62.00	78.70

Based on % recovery, TIL24 had the highest value (82.5 %). It is an IR64 isolines improved on root characteristics. Analysis variance of recovery of the genotypes showing no significance among the genotypes (Table 2). It frequently happen for drought research. Nevertheless, the value seem to be reasonable showing the variation among the genotypes. We could disect the difference among the genotypes at the confidential level of 44%. It migh due to the complexity of the complexity of drought tolerance traits.

The frequency distribution of the genotypes are scatterred in the upper midle of the population (Figure 2). It indicate that the genotypes had relatively good tolerance to drought through percentage of plant recovery mechanism. Other genotypes following TIL 24, such as BP14352e-2-2-0Kr (75.00%), BP17280M-46D-IND (75.00 %), Towuti (67.50 %), BP17298M-53D-SKI-1-4-2 (65.00 %), IR83142-B-20 (65.00 %), IR83383-B-B-11-4 (65.00 %), PR40781b-3-4-SBY-0-CRB-2-SKI-0-5 (65.00 %), BP14476e-3-9-1Kr-JK-0 (65.00 %), GSRIR1-11-D3-S2 (65.00 %), and Inpari10 (60.00 %). Inpari 10 had been reported to have drough tolerance and water use efficiency. IR83142-B-20 and GSRIR1-11-D3-S2 (65.00 %) are GSR lines, which were designed to be tolerance to multiple stress [9; 10]. IR83383-B-B-11-4 is also dedicated for drough tolerance in rainfed lowland area. This results indicating that the drought tolerant systems which developed by many groups are working.

Table 4. Analysis of variance of plant recovery of 60 rice genotypes under drought Screening at Seedling Stage using Seedling Box Method, Sukamandi, 2016

No	Source of Variance	Degree of Freedom	Sum of Square	Mean Square	F Value	Probability
1	Replication	1	213.334	213.334	5.79	0.018
2	Genotype	59	2264.37	38.3791	1.04	0.439
	Residual	59	2174.67	36.8588		
	Total	119	4652.37	39.0955		

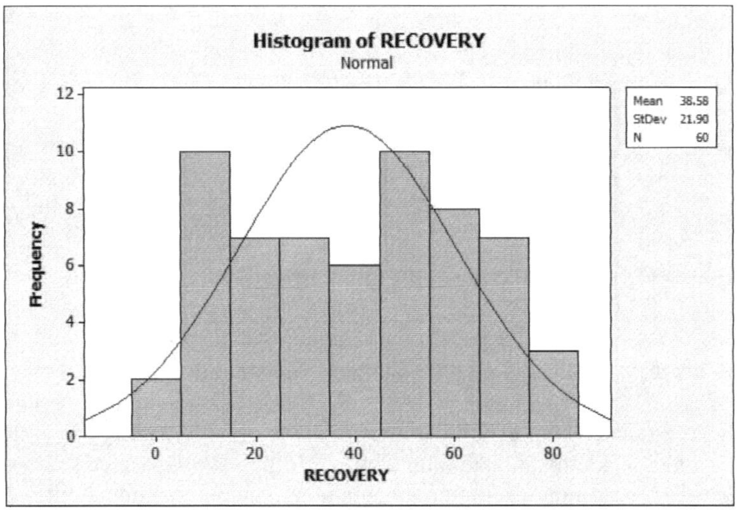

Figure 2. Frequency distribution of recovery of 60 rice genotypes under drought Screening at Seedling Stage using Seedling Box Method, Sukamandi, 2016

The second experimen Inpari21, BP16178M-1D-SKI-13-8-1, HHZ5-SKI-7-1-0Kr-JK-IND, Inpari18, Towuti, and Bahbutong (1182) had either high number of vertical root growth (more than 10 roots) as well as high proportion of the vertical root growth (more than 40%). The genotypes were suspected to have good dorught tolerant due to the root architecture [11]. The data had been discussed in the publication.

IOP Conf. Series: Earth and Environmental Science **383** (2019) 012049 doi:10.1088/1755-1315/383/1/012049

Correlation analysis among the traits of experiment 1 and 2 showed that the number of green leaves had strong positive correlation with root length, root number in zone 1 (horizontal), 2 (medium), and 3 (vertical), total root number, leaf rolling, and plant recovery (Table 3). It indicated that number of green leaves under drought condition might be used as indireect selection criteria for drought tolerance screening material. This finding is relevant with the finding [12] which definig Drought Tolerant Degree (DTD) as ratio of green leaf length to total leaf length of the top three leaves in every seedling after drought treatment. DTD is significantly correlated with drougth tolerant considerable traits such as water potential, survival rate, panicles per plant, spikelets per panicle, seed setting rate, yield per plant, and contents of proline, chlorophyll, and malondialdehyde (MDA). This study concluding that DTD is a simple method to directly select drought tolerant material. The number of green leaves after drought stress treatment found in this study correlated with some drought tolerant related traits also indicating that the trait could be used as new criteria in selecting drought tolerant materials. This trait is relatively simple and it is possible to do high throughput selection using this trait. On the other hand, the ability of plant to keep the leaves green might be correlated with the vigoring ability of the plant at early plant growth. This early vigor is also positively correlated with yieling ability either under optimum or drought stress condition [13].

Number of zone 3 root (grow vertically to the basal) in this study is positively correlated with shoot length, tiller number, and green leaf number, total root number, proportion of zone 3 root number from the total root number. It is also relevant twith [14] study which conclude that deep root architechture is reported to increase plant capability to uptake water from the deeper soil zone. Under drought stress, it would benefit to plant in collecting the deep. *Dro 1* gene has been proven for this mechanism.
Among the root zones, number of root in zone 2 had higher correlation coefficient with the number of green leaf (0.625) followed by Zone 3 (0.415) and zone 1 (0.400). It indicated that the existence of root in zone 2 is also important for the plant to survive. Nevertheless, proportion of root in zone 3 negatively correlated with shoot length, green leaf number, root number in zone 1 and 2, total root number, and plant recovery. It seem that proportion of vertical root compensate with other traits in developing drought tolerance. It indicated that developing cultivars having deep root would helpful for drought, but should manage the proportion from the total roots. Finally, it could be seen that root trait strongly correlated wit drought toleance [15]. Thus the study is continuosly being conducted by many scientists. Selection for drought tolerant during vegetative growth phase is still widely being used with new modificatios, and it shows promising results [16; 17]. This method is relatively simple, quickly done, and could suitable for hugh throughput running .

Single trait selection might not be enough to find robust durable drught tolerance variety. The rice plants response to drought stress in many traits, including morphological, physiological, biochemical, and molecular [18]. It makes the trais very complicated and a single approach may not be enough to overcome the stress. More over, drought stress may come with other stresses all together on field.
Final selection on yield under drought condition might work better in selecting drought tolerant lines [19]. Nevertheless, on field selection under natural condition would be very dificult to control. The selection might be succesfull during dry season only (once a year), still the risk of unpredictable rainfall occurence. On the other hand, a quick request to cope with drough incidence is almost every year to come.

Screening of breeding materials and finding a new donors while studying physiological mechanism of drough tolerant should be conducted continuously. Molecular tools, in such stage would also be very helpful [20; 21]. Variety combining drought tolerance with drought avoidence like very early maturing traits migh be feasible for this immediate requirement recently. Agronomical effort such as developing ponds and modifieng irrigation system would help more to overcome the drought problem.

4. Conclusion
Number of green leaf under stress and root number in zone 2 correlated more with traits supporting drought tolerant in rice. Both traits are supposed to be useful for drought stress tolerance selection of rice plants. Between both of the traits, number of grean leafes seem to be simples and more practive to be applied. Further confirmation is needed and considering drought is a complex trait, simultaneous selection using some methods would be more powerful.

IOP Conf. Series: Earth and Environmental Science **383** (2019) 012049 doi:10.1088/1755-1315/383/1/012049

5. Appendix

Tabel 5. Correlation among seedling characteristics, root patternd, and seedling stage response to drought of 60 rice genotypes simultaneously tested using mini pot and Seedling box methods into 60 rice genotypes.

	Shoot length	Tiller Number	Grean Leaf Number	Dry Leaf Number	Root Length	Root Number Zone 1	Root Number Zone 2	Root Number Zone 3	Total Root Number	Zone 3 Root Number Proportion	Leaf Rolling Score	Leaf Drying Score
Tiller Number	**0.524**											
	0											
Grean Leaf Number	**0.599**	**0.872**										
	0	0										
Dry Leaf Number	**-0.378**	-0.033	-0.072									
	0.003	0.804	0.585									
Root Length	**0.341**	**0.284**	**0.311**	**-0.344**								
	0.008	0.028	0.015	0.007								
Root Number Zone 1	**0.582**	**0.336**	**0.406**	0.033	0.133							
	0	0.009	0.001	0.802	0.311							
Root Number Zone 2	**0.531**	**0.489**	**0.625**	-0.126	**0.274**	**0.507**						
	0	0	0	0.337	0.034	0						
Root Number Zone 3	**0.393**	**0.458**	**0.415**	-0.193	0.106	0.177	0.198					
	0.002	0	0.001	0.14	0.422	0.175	0.129					
Total Root Number	**0.69**	**0.61**	**0.686**	-0.158	0.248	**0.701**	**0.813**	**0.649**				
	0	0	0	0.227	0.056	0	0	0				
Proportion of Zone 3	**-0.297**	-0.251	**-0.335**	-0.029	-0.133	**-0.518**	**-0.601**	**0.423**	**-0.304**			
	0.021	0.053	0.009	0.827	0.312	0	0	0.001	0.018			
Leaf Rolling Score	**0.237**	0.167	**0.299**	0.064	0.127	0.217	0.035	0.181	0.191	-0.006		
	0.068	0.202	0.02	0.628	0.333	0.097	0.788	0.165	0.144	0.966		
Leaf Drying Score	-0.068	0.005	-0.044	0.015	0.099	**-0.277**	-0.209	-0.029	-0.232	0.14	0.231	
	0.606	0.972	0.736	0.906	0.453	0.032	0.109	0.823	0.075	0.284	0.076	
Plant Recovery	0.232	0.184	**0.266**	-0.095	0.193	0.198	0.188	-0.094	0.11	**-0.276**	0.15	0.111
	0.075	0.159	0.04	0.471	0.14	0.129	0.15	0.476	0.403	0.033	0.253	0.397

6. References

[1] Serraj R, McNally K L, Slamet-Leodin I, Kohli A., Haefele S M, Atlin G, and Kumar A. 2011. Drought resistance improvement in rice: an integrated genetic and resources management strategy. Plant Prod. Sci. 14 (1): 1 - 14 .

[2] Wassmann, R., Jagadish, S.V.K., Sumfl eth, K., Pathak, H., Howell, G.,Ismail, A., Serraj, R., Redoña, E., Singh, R.K. and Heuer, S. 2009a. Regional vulnerability of climate change impacts on Asian rice production and scope for adaptation. Adv. Agron. 102: 91-133.

[3] Direktorat Perlindungan Tanaman Pangan. 2016. Laporan Kinerja Perlindungan Tanaman Pangan Tahun 2015. Direktorat Perlindungan Tanaman Pangan. Jakarta. Pp 54.

[4] Boer R. 2008. Pengembangan sistem prediksi perubahan iklim untuk ketahanan pangan. Laporan akhir konsorsium penelitian dan pengembangan perubahan iklim sector pertanian. Balai Besar Litbang Sumberdaya Pertanian. Bogor.

[5] Verulkar S B and Verma S K. 2014. Screening Protocols in Breeding for Drought Tolerance in Rice. Agricultural Research Volume 3, Issue 1, pp 32–40.

[6] IRRI. 2013. Standard Evaluation System 5th Edition, IRRI. Philippines. 55 pp

[7] Fukuai S and Cooper M. 1995. Development of drought-resistant cultivars using physiomorphological traits in rice. Field Crops Res, 40(2): 67–86.

[8] Efisue A A, Tonggoona P, Derera J, and Ubi B E. 2009. Screening Early-Generation Progenies of Interspecific Rice Genotypes for Drought-Stress Tolerance During Vegetative Phase. Journal of Crop Improvement, 23:174–193.

[9] Li Z and Ali J. 2017. Breeding green super rice (GSR varieties for sustainable rice cultivation. 10.19103/AS.2016.0003.05

[10] Zhang Q. 2007. Strategies for developing green super rice. Proceedings of the National Academy of Sciences of the United States of America 104: 16402–9

[11] Susanto U, Rohaeni W R, Prastika D, and Azkia F W. 2018. Observation of root architecture at vegetative stage of drought tolerant rice genotypes using mini pot method. IOP Conf. Series: Earth and Environmental Science 250 (2019) 012081. DOI. 10.1088/1755-1315/250/1/012081

[12] Zu X, Lub Y, Wang Q, Chu P, Miao W, Wanga H, La H. 2017. A new method for evaluating the drought tolerance of upland rice cultivars. The Crop Journal Vol 5: 4 8 8 – 4 9 8

[13] Kumar A, Verulkar S, Dixit S, Chauhan B, Bernier J, Venuprasad R, Zhao D, Shrivastava M N. 2009. Yield and yield-attributing traits of rice (Oryza sativa L.) under lowland drought and suitability of early vigor as a selection criterion. Field Crops Research. Vol. 114 (1): 99-107.

[14] Uga Y, Sugimoto K, Ogawa S, Rane J, Ishitani M, Hara N, Kitomi Y, Inukai Y, Ono K, Kanno N, Inoue H, Takehisa H, Motoyama R, Nagamura Y, Wu J, Matsumoto T, Takai T, Okuno K and Yano M. 2013. Control of root system architecture by DEEPER ROOTING 1 increases rice yield under drought conditions. Nature Genetics 45 (9): 1097 – 102

[15] Suwarto, Dinuriah I, Pramesthi R, and Soraya. 2019. Root growth dynamics and grain yield of ten new plant type of rice lines under aerobic and flooded condition. IOP Conf. Series: Earth and Environmental Science 250 (2019) 012082. DOI:10.1088/1755-1315/250/1/012082.

[16] Tomita A, Sato T, Uga Y, Obara M, and Fukuta Y. 2017. Genetic variation of root angle distribution in rice (Oryza sativa L.) seedlings. Breeding Science 67: 181–190. DOI:10.1270/jsbbs.16185

[17[Singh B, Reddy K R, Redona E D, Walker T. 2017. Screening of rice cultivars for morpho-physiolical responses to early –season soil moisture stress. Rice Science 24 (^): 322 – 335.

[18] Sourour A, Afef O, Mounir R, Mongi B Y. 2017. A review: Morphological, physiological, biochemical and molecular plant responses to water deficit stress. The International Journal Of Engineering And Science (IJES). Vol. 6 (1): 01-04

[19] Torres R O and Henry A. 2018. Yield stability of selected rice breeding lines and donors acrossconditions of mild to moderately severe drought stress. Field Crops Research 220: 37–45

[20] Kumar A, Shalabh D, Ram T, Yadaw R B, Mishra K K, Mandal N P. 2014. Breeding high-yielding drought-tolerant rice: genetic variations and conventional and molecular approaches. Journal of Experimental Botany. DOI:10.1093/jxb/eru363

[21] Sahebi M, Hanafi M M, Rafii M Y, Mahmud t M T, Azizi P, Osman M, Abiri R, Taheri S, kalhori N, Shabanimofrad M, Miah G, Atabaki N. 2018. Improvement of drought tolerant in rice (Oryza sativa L.): genetics, genomic tools, and the WRKY genes family. BioMed Research international Vol 2018. Article ID 3158474. Pp 20.

7. Acknowledgement

The research was funded by KKP3I research scheme of SMARTD Project at IAAR for the fiskal year of 2016. Special thanks is dedicated for Daud Heryanto and Desi Prastika who greatly assisting the experimen.

Characterization of chemical properties and color of starch from *Talas Beneng* (*Xanthosoma undipesh* K. Koch) extraction as a source of indigenous carbohydrate from Pandeglang regency, Banten province

W Nurtiana and V Y Pamela

Department of Food Technology, Faculty of Agriculture, University of Sultan Ageng Tirtayasa, Indonesia
E-mail winda@untirta.ac.id

Abstract. Indonesian are very dependent on wheat flour, even though wheat are subtropical plant. Therefore, the Indonesian government must import wheat continuously to supply the demand. While there are many potential local carbohydrates sources that have not been utilized such as tubers. One of local carbohydrates source is *Talas beneng* which only grows in Pandeglang Regency, Banten. This tuber grows wildly and is underutilized because it contains oxalate which causes itching when consumed. Oxalate reducing is done by soaking the taro in 1% NaCl for 60 minutes. To easier application on food products, this tuber can be made as starch. The aim of this study were to choose the best extraction method in producing starch and determine the color and chemical characteristics. The extraction method used was carried out with five attempts and two repetitions, with a ratio of taro and water 1:1, 1:2, 1:3.1:4, and 1:5. The results showed that the ratio that produces the biggest yield is 1:3 with the result 12.67%. Chemical characteristics of *talas beneng* starch are produce an L value 100, 00, a* 10.92, b* -15.54, and contains moisture 6.21%, ash 0.25%, fat 10.56%, protein 0.66%, starch 82.32%, amylose 28.91%, amylopectin 53.41%, and oxalate reduction 97.4%.

1. Introduction

Food is an important basic human needs besides clothing, education, and health. Without food, humans cannot live. Almost all snacks, traditional foods, and staple foods in Indonesia are made from wheat flour. Therefore, Indonesian people are very dependent on wheat flour. The Indonesian Flour Producers Association (Aptindo) revealed wheat imports in 2017 rose by around 9% to 11.48 million tons from the previous year. Dependence on certain foods can cause national food security to become fragile. Meanwhile there are still many potential food sources that have not been utilized optimally such as local food likes tubers. If Indonesian people can take advantage of local food this will undoubtedly be able to create resilient food security [1].

One method that can be used to reduce the use of wheat flour in food products is to substitute it with *talas beneng* flour or starch. *Talas beneng* (*Xanthosoma undipes* K. Koch.) is a local tuber that only grows in Pandeglang Regency, Banten Province. The size of this *talas beneng* is very large, grows wildly, and the color of tubers are yellowish [2]. *Talas beneng* has protein content of 6.29%, carbohydrates 84.88%, fat 1.12%, starch 75.62%, and calories 374.69 kcal [3]. With its large size and rich of nutrients, *talas beneng* is very potential to be applied in food products.

Talas beneng contains oxalate which can cause itching when consumed. Oxalate content in *talas beneng* are the highest compared to Bogor, Malang, and West Kalimantan taro [4]. The method used to reduce the oxalate content that is often used by the Pandeglang community is to soak it in a solution of

1% NaCl for 1 hour. For easier application of *talas beneng* in food products, it can be made as starch. Starch is homopolymer of glucose with α-glycosidic bonds. The characteristic of starch depends on the length of the carbon chain and the content of amylose and amylopectin. Each plant has a distinctive and different shape and size of the starch granules. Starch is composed of at least three main components, amylose, amylopectin and intermediate materials such as protein and fat. In general, starch contains 15-30% amylose, 70-85% amylopectin, and 5-10% intermediate material [5].

The aim of this research were to see the best extraction method to produce the largest yield of *talas beneng* starch and then determine of proximate, amylose, amylopectin, color, and oxalate contents in *talas beneng* starch.

2. Research methodology

2.1. Time and place of research

This research was conducted at the Laboratory of Food Technology and Agroecotechnology, Faculty of Agriculture, University of Sultan Ageng Tirtayasa in June-August 2019.

2.2. Tools and materials

The tools that used are set of kitchen tools (knives, cutting boards, basins, buckets), analytical balance, strainer, blender, baking pan, and oven for extracting and drying starch. Oven, furnace, cup, desiccator, soxhlet, kjeldahl apparatus, UV-Vis spectrophotometer (Perkin Elmer), HPLC UV-Vis (Agilent), chromameter (Minolta CR-300), and a set of glassware (Pyrex).

The materials that used used are *talas beneng* obtained from Juhut Village, Karang Tanjung District, Pandeglang Regency, Banten Province, NaCl salt, aquades, hexane, H_2SO_4 (merck), HgO (merck), NaOH (merck), H_3BO_3 (Sigma Aldrich), HCl (merck), aquabidest, phenolphtalein (Sigma Aldrich), ethanol 95%, amylose standard (Sigma Aldrich), acetic acid (merck), iodine (merck), and oxalate standard (sigma Aldrich).

2.3. Characterization of diameter, length, weight, color of talas beneng and extraction of starch

The process carried out is to observe the diameter, length, weight, skin color, and tuber color of *talas beneng*. The next step is extraction of *talas beneng* starch begins with washing, sorting, stripping, slicing, soaking in 1% NaCl salt solution, crushing with water for 5 minutes, filtering, settling for 24 hours, starch, and water separation, drying at 50°C for 24 hours, and size reduction. This extraction process was carried out in five attempts and two repetitions. The ratio of taro and water used for extraction is 1:1, 1:2, 1:3, 1:4, 1:5.

2.4. Determination of yield and proximate content

This extraction process will be chosen of the highest yield of starch. The yield is calculated using the formula:

$$\text{Yield} = \frac{\text{Weight of } Talas\ beneng}{\text{Weight of Dry Starch}} \text{ x } 100\% \quad (1)$$

Starch with the highest yield was analyse of proximate, amylose, amylopectin, degree of color, and oxalate contents. Proximate analysis was carried out of moisture content by gravimetric method (AOAC 2006), ash content by gravimetric method (AOAC 2006), protein content by Kjeldahl method (AOAC 2006), fat content by Soxhlet method (AOAC 2006) and carbohydrates (by difference).

2.5. Amylose and amylopectin content [7]

Analysis of amylose content using the spectrophotometric method with the principle of staining using iodine and calculated with blue value. 100 mg sample was placed in a test tube, then added with 1 mL of 95% ethanol and 9 mL of 1N NaOH. The mixture is heated in boiling water to form a gel and then the whole gel is transferred to a 100 mL measuring flask. The gel is added to distilled water and shaken, then adjusted to 100 mL with distilled water. A total of 5 mL of solution was put into a 100 mL measuring flask and added with 1 mL of 1 N acetic acid and 2 mL of iodine solution. The solution is fixed to 100 mL, then shaken and left for 20 minutes. The intensity of the blue color was measured by a spectrophotometer at a wavelength of 625 nm. Amylose content is calculated based on the standard amylose curve equation. The absorbance value obtained is entered into the standard amylose curve so the amylose % value will be obtained.

Carbohydrate content is considered as starch content, because the non-starch component of *talas beneng* has been removed during the extraction process. Starch is a combination of amylose and amylopectin. Amylopectin contents are calculated based on the starch content reduced by amylose content.

2.6. Degree of color

The degree of color is measured by chromameter. The color notation system used is a hunter system namely L* (brightness), a* (+ red, - green), b* (+ yellow, - blue). For calibration, chromameter is placed on white paper first. Chromameter is placed on the starch precisely without any gaps then the measurement button is pressed. The light source is on and the reflectance is measurable.

2.7 Oxalate content [8]

5 grams of *talas beneng* are crushed and placed in 100 mL Erlenmeyer flasks. To this, 50 Ml of HCl 2M (pH 0.08) was added for total oxalate analysis or 50 ml of distilled water (pH 7) for analysis of water-soluble oxalate. The mixture is homogenized and then covered with parafilm and placed in a waterbath (80 °C) for 20 minutes with periodic shaking. The sample is transferred to a 100 mL volumetric flask and water or acid is added to the pitch mark, then centrifuged at 1400 x g for 15 minutes. The supernatant was obtained and filtered with 0.45 μm cellulose acetate milipore and then analysed by HPLC. The column conditions that used were Aminex HPX-87H 300 mm x 7.8 mm, mobile phase 4 mM sulfuric acid, elution time 20 minutes, flow rate 0.6 ml/minute, column temperature 25°C, injection volume 20 μL, using multiwavelength detector and UV detection at 215 nm. The results obtained are compared to the oxalate standard.

3. Results and discussion

3.1. Physical appearance of talas beneng

Talas beneng is a unique taro because it has a large size compared to other taro. Table 1 shows that the weight of taro reached 832.23 grams, 5.64 grams in diameter, and 20.93 grams in length.

Table 1. Physical appearance of *Talas beneng*.

Physical Appearance	Characteristic
Length	20.93 cm
Diameter	5.64 cm
Weight	832.23 grams
Color of Skin	Reddish Brown
Color of Tuber	Yellowish white

The average weight of taro is 100 grams, so the table above shows that *talas beneng* has a very large size compared to other taro. The length and diameter of the large taro will affect the weight produced [9]. *Talas beneng* color is yellowish white, this is caused by the presence of β-carotene pigments. The level of β-carotene found in the yellowish taro tubers is around 0.0213 mg/100 g [10]. *Talas beneng* skin color is reddish brown caused by plant varieties, growing sites, nutrients obtained, and weather factors [3].

3.2. Talas beneng starch extraction
The extraction method used was wet extraction using water. The ratio of taro and water used was 1:1, 1:2, 1:3, 1:4, and 1:5.

Table 2. Comparison of taro and water ratio extraction and yield.

Taro: Water	Yield
1:1	3.51%
1:2	9.16%
1:3	12.67%
1:4	10.02%
1:5	12.10%

The extraction process is the most important thing in getting starch. On the table above, it can be seen that the ratio of 1:3 between taro and water produces the most yield. Yield starch resulting from the ratio of taro: 1:1 water extraction is the least, this is because the characteristics of *talas beneng* are hard and more difficult to crush, so the less water the more difficult to extract starch [2]. Ratio of 1:4 and 1:5 give smaller yield than 1:3 because some of the starch is dissolved and wasted in large amounts of water.

3.3. Color
Color is one of the important attributes that used to determine the quality of a food ingredient. Taro starch from 1:3 extracted was analysed for color attributes including L* value, indicating dark-light level with a range of 0-100, a value of 0 indicates black or very dark and value of 100 indicates a white or bright color tendency. a* value indicate the intensity of red (+) or green (-). b* value represents the intensity of the color yellow (+) or blue (-).

Talas beneng starch and tapioca starch are derived from the extraction of tubers. According to SNI 3451: 2011 about tapioca, the color quality requirements of starch are white with a minimum degree of white is 91. The color attributes of *talas beneng* starch are L* 100.00, a* 10.92, and b* -15.54. The L * value indicates that the white degree of *talas beneng* starch is perfect and exceeds the tapioca color standard. This perfect white color is obtained because there are no other components such as pigments that have been separated when the starch extraction process.

3.4. Chemical Characteristics
Chemical characteristics are a very important attribute in determining starch quality and its application in food products. *Talas beneng* starch with extraction ratio 1:3 contains 6.21% moisture content (Table 3), the moisture content requirements in tapioca according to SNI 3451: 2011 is maximum of 14%. This showed that the moisture content of *talas beneng* starch is better than tapioca standard. Therefore, the shelf life of *talas beneng* starch will be long because of moisture content less than 14%, the growth of microorganisms, especially mold will be inhibited [5].

The levels of ash in *talas beneng* starch reached 0.35% (Table 3), the quality requirements for tapioca ash according to SNI 3451: 2011 is maximum of 0.5%. This showed that the ash content of *talas beneng* starch is better than tapioca. The greater the ash content in flour, the worse the quality will be. In addition, if the flour or starch to be applied to food has high ash content, it can affect the fermentation activity, the strength of the dough, the color, and the final product, especially bread products [12].

Table 3. Chemical Characteristics of *Talas beneng* Starch

Characteristics	Content (%)
Moisture (%)	6.21
Ash (%)	0.25
Fat (%)	10.56
Protein (%)	0.66
Carbohydrate (%)	82.32
Amylose (%)	28.91
Amylopectin (%)	53.41

The fat content of *talas beneng* starch is 10.56% (Table 3). High fat content in food will accelerate rancidity so that it will reduce its quality [5]. Beside that, high fat in flour or starch will inhibit starch gelatinization due to the fat forming complex with amylose. The gelatinization process is inhibited because the fat is absorbed by the surface of the starch granules and forms a hydrophobic layer [2].

The protein content of *talas beneng* starch is 0.66% (Table 3). The quality standard of protein content in wheat flour according to SNI 3751: 2009 of wheat flour is a minimum of 7%. This low protein level results in *talas beneng* starch can be used for food ingredients in products that do not need development such as cookies, vermicelli, and glass noodles.

Carbohydrate content in *talas beneng* starch is considered as starch, which is 82.32%. These results are better than the standard tapioca according to SNI 3451: 2011 is 75%. The more starch content, the *talas beneng* starch can be used as a source of carbohydrates and substitute ingredients for wheat flour in various food products.

Amylose is part of the polymer with α- (1,4) bonds from 500-2000 D-glucose units and form a straight chain. Amylose content of *talas beneng* starch reached 28.91%, while amylose content in potatoes as much as 10%, wheat 25%, tapioca 15-18%, and rice 20%. Amylose content of 25-30% can provide character and texture strength of a compact gel [5].

Amylopectin is a branch chain polymer with α- (1,4)-glycosidic bonds and α- (1,6)-glycosidic bonds at its branching site. Each branch consists of 25-30 units of D-glucose. Amylopectin chain structure tends to form branches that make up 4-5% of all amylopectin structure [5]. Amylopectin content in *talas beneng* starch is 53.41%. In food products, amylopectin is stimulating the process of puffing, in which food products derived from starch with high amylopectin content will be mild, porous, and crisp [5].

Based on the composition of amylose and amylopectin, *talas beneng* starch contains higher amylose content than other ingredients. Because of its higher amylose content, *talas beneng* starch is suitable for use as an ingredient in food that does not need crisp, development, porous, and requires character and strength of gels such as vermicelli, glass noodles, noodles, and other traditional cakes.

3.5. Oxalate contents

Consumption of *talas beneng* can cause itching and irritation to the lips, mouth and throat. Itching that stimulates the oral cavity and skin is caused by the presence of small crystals in the form of fine needles composed of calcium oxalate called raphide. Oxalate ($C_2O_4^{2+}$) in taro is in the form of water-soluble (oxalate acid) and water-insoluble (calcium oxalate or oxalate salt) [4].

Talas beneng has the highest contents of oxalate compared to other taro. The oxalate content in Bogor, *talas beneng*, West Kalimantan, and Malang taro is 8.58 mg/g, 61.78 mg/g, 7.33 mg/g, and 10.887 mg/g [4]. The Pandeglang community used to soak the taro in 1% salt solution for 1 hour to reduce their oxalate content, so the itching will disappear when consuming *talas beneng* or its processed products.

In this study, *talas beneng* was soaked in a solution of 1% NaCl salt for 1 hour, the results showed that the oxalate content of *talas beneng* starch are 1.61 mg/g or oxalate reduction is 97.4%. This result is in accordance with Mayasari's study that soaked taro in a 10% NaCl solution for 1 hour showed an oxalate reduction of 96.83% [4]. Decreased contents of oxalate occur because of the reaction between sodium chloride (NaCl) and calcium oxalate (CaC_2O_4). Salt (NaCl) dissolved in water breaks down into Na^+ and Cl^- ions. The Na^+ ion attracts negatively charged ions and Cl^- ion attracts positively charged ions. In this reaction Na^+ ion bind $C_2O_4^{2-}$ ion to form sodium oxalate ($Na_2C_2O_4$) which can dissolve in water with a solubility of 3.7 g/100 ml at 20°C and 6.25 g/100 ml at 100°C, and Cl^- ion will bind ion Ca^{2+} forms a white precipitate of calcium dichloride ($CaCl_2$) which is easily soluble in water [4].

4. Conclusion

The best extraction method to get the highest yield of *talas beneng* starch is using a ratio of taro: water 1:3. Oxalate reduction on *talas beneng* by soaking taro in 1% NaCl salt solution for 1 hour and succeeded in reducing oxalate in taro starch by 97.4%. The water content, ash content, and starch content in *talas beneng* starch do not exceed the standards of SNI 3451: 2011 of tapioca. Amylose and amylopectin content of *talas beneng* starch are 28.91% and 53.41%, so that this starch is suitable for application in food products that require compact gel strength such as vermicelli, glass noodles, or noodles.

5. References

[1] Koswara S. 2014. *Processing Technology of Tubers, Chapter 1: Taro* (Bogor: USAID-Bogor Agricultural University)

[2] Rostianti T, Hakiki D N, Ariska A, Sumantri. 2018. Characterization of the physicochemical properties of beneng taro flour as local food biodiversity in Pandeglang regency. *Gorontalo Agriculture Technology Journal*. **1(2)** 1-7

[3] Apriani R R N, Setyadjit, Arpah M. 2011. Characterization of four varians of taro mentega, semir, hijau, beneng, and flour from the four varians of taro. *Jurnal Ilmiah dan Penelitian Ilmu Pangan*. **1(1)** 1-8

[4] Mayasari N. 2010. *The Effect of Acid And Salt Solutions Addition as Efforts to Reduce Oxalate Oo Taro Flour* (Bogor Agricultural University: Undergraduate Thesis)

[5] Kusnandar F. 2011. *Food Chemistry, Macro Components* (Jakarta: Dian Rakyat)

[6] Association of Official Analytical Chemist. 2006. *Official Methods of Analysis of The Association Analitycal Chemist* (Washington DC: AOAC)

[7] Association of Official Analytical Chemist. 1995. *Official Methods of Analysis of The Association Analitycal Chemist* (Washington DC: AOAC)

[8] Ross A B, Savage G P, Martin R J, Vanhanen L. 1999. Oxalates in oca (new zealand yam) (Oxalis tuberosa Mol.). *Journal of Agriculture and Food Chemistry*. **47** 5019–5022

[9] Setyowati M, Hanarida I, Sutoro. 2007. Characteristics of taro germplasm *(Colocasia esculenta)*. *Buletin Plasma Nutfah*. **13(2)** 49-55

[10] Tungriani D A, Karim A, Asmawati, Seniwati. 2012. Analysis of β-carotene content and vitamin c in various taro varieties (*Colocasia esculenta*). *Indonesia Chimica Acta*. **5** 1-10

[11] Indonesian National Standard. 2011. *SNI 3451:2011 Tapioca* (Jakarta: BSN)

[12] Rahmiati T M. 2015. *Characterization of Physicochemical Properties of Cassava (Manihot esculenta Crantz) Flour and Chips from Different Genotypes* (Bogor Agricultural University: Thesis)

[13] Indonesian National Standard. 2009. *SNI 3751:2009 Wheat Flour* (Jakarta: BSN)

5. Acknowledgement

This research was funded by Internal Research for Beginner Lecturer scheme, Faculty of Agriculture, University of Sultan Ageng Tirtayasa.

Technology and local knowledge of farmers in soybean seeds source management in Jambi province

Yardha, R Rubiana*, H Nugroho

Jambi Assessment Institute for Agricultural Technology, Indonesian Agency for Agricultural Research and Development, Jl. Samarinda Paal Lima, Kotabaru, Kota Jambi 36128.

*Corresponding author: ratna.rubiana@gmail.com

Abstract. The Ministry of Agriculture has optimism that soybean self-sufficiency can be realized in 2018. Consistency to stay on track that has been targeted has made Soybean Seed Management Unit (UPBS) activities accelerate with the technology of seed multiplication to supply quality seeds. This study examined the dissemination of technology and the existence of local knowledge, especially on consumption farmers to breeders as the output of UPBS activities. The method used is a descriptive qualitative method through literature studies and studies in the social field to explore the potential, roles, constraints and opportunities faced by farmers and soybean seed breeders related to local knowledge. The data collected consists of secondary and primary data. Benefit and Cost Ratio (B/C Ratio) analysis to determine the feasibility of changing consumption farmers to breeder farmers. The results show that farmers' local knowledge changes with technology dissemination. Changes occurred in the use of quality seeds of New Superior Varieties (VUB) with the application of soybean integrated crop management (PTT) models that were able to increase yields 16-27%. Local knowledge maintained in terms of attack control and infestation of plant pests (OPT) in the concept of Integrated Pest Management (IPM). In this case, local knowledge of breeder farmers can compensate for location-specific technology. At the field level, a combination of technology and local knowledge was followed by an increase in income from consumer farmers to B/C soybean growers. The ratio of consumption farmers to 1.45 increased to 1.69 for breeder farmers.

1. Introduction

Indonesian government must import around 1.9 million tons of soybeans each year to meet domestic needs [1]. This high import is a large enough foreign exchange expenditure, so efforts to increase domestic production need to be encouraged and one solution is to look for one with the availability of national seeds.

Seed is one of the production inputs that has a significant contribution to increasing productivity and quality of agricultural products. The availability of seeds with high yielding varieties and high quality, both physical, physiological, genetic and pathological quality is absolutely necessary in an agricultural production system. According to [2, 3] in modern agriculture, seed acts as a delivery mechanism that channels technological excellence to farmers and other consumers. Therefore, a good seed system is needed so that the technological superiority of a variety can be channeled.

Soybean seeds are seeds that quickly experience deterioration or decrease in viability and vigor, especially if stored in conditions that are less than optimal storage. According to [4] the use of low quality seeds with low viability and vigor will result in a low percentage of seedling emergence, seedlings that are less tolerant of abiotic stress and are more sensitive to disease attacks and will ultimately reduce yield.

Jambi province is one of the centers of soybeans on the island of Sumatra. However, the availability of soybean for public consumption in Jambi Province is still lacking. Until now, Jambi province has not been able to be independent in meeting the needs of seeds and soybean consumption. The cause of the low soybean production due to many farmers still use low quality saved seeds. Farmers continuously grow soybeans without selecting good seeds for cultivating in the next crop. In addition, the availability of seeds by farmers and private business entities and even relevant government agencies often misses when needed. Inadequate cultivation techniques, the attack of disturbing organisms and socio-economic factors are the causes of low soybean seed production.

Jambi Assessment Institute of Agricultural Technology (AIAT), which is the Technical Implementation Unit (TIU) of the Indonesian Agency for Agricultural Research and Development, has been given a mandate to resolve seedling problems in Jambi Province. To that end, Jambi AIAT through the Seed Source Management Unit (SSMU) is contributing in providing quality soybean seeds. SSMU soybean Jambi AIAT activities carried out through seed multiplication in the Integrated Plant Management Field School demonstration plot area and fostering farmer groups and establishing cooperation with Universities and Industrial Plantation Forest (IPF) to utilize the land between IPF plants. The activity aims to produce and distribute FS and SS class soybean seed sources and increase the ability and capacity of breeder farmers in producing soybean source seeds.

In this activity, SSMU also conducted a study on local knowledge of farmers in managing soybean fields and disseminated technology. Farmers 'responses to external knowledge and farmers' decision-making processes in adopting innovations are described in this paper. It was also stated that the factors which became an obstacle for farmers in implementing technological innovation in soybean seed multiplication.

2. Materials and Methods

Seed breeding activities were carried out in Dusun Baru Village, District VII Koto, Tebo Regency in January to December 2016, which had been prepared and determined by the local Food Crop Agriculture Office. Preparation for the location of activities included the selection of stretches as needed, carried out with soybean breeder farmer groups.

2.1. Technology and Local Knowledge

Farmer's local knowledge and technology dissemination was inventoryed through questions, then presented in the form of a comparison table. The data inventory process were carried out through in-depth interviews with farmers in each assistance that were conducted on their understanding of the cultivation process, IPM, ecological functions and interactions that occured in it until postharvest. Thus obtained a comprehensive description of the practice of soybean cultivation until postharvest.

There was two categories of farmers as sources of information, namely consumption farmers and soybean breeder farmers. Each of these groups of farmers had different experiences that had implications for land management patterns, so they could show relative differences in their knowledge and experience. Of the two categories above, approximately 30 farmers were interviewed and asked to articulate their knowledge and understanding of the process of managing soybean land.

2.2. Seed production of superior varieties of FS and SS class soybeans

Source seed production was carried out based on technical seed production, namely: 1) Determination of location, 2) determination of source seeds used, 3) land preparation, 4) planting, 5) fertilizing, 6) IPM and weed control, 7) rouging, 8) harvest, 9) drying, 10) seed processing, 11) and packaging. Certification and testing of seeds in the process of seed production is carried out in collaboration with BPSB Jambi Province.

Financial analysis and break-even point were carried out to determine the feasibility of seed multiplication efforts. Seed multiplication was considered feasible if the Gross B/C value was more than 1. The formulation of Gross B/C is [5]:

$$\text{Gross B/C} = \frac{P \times Q}{Bi}$$

where: P = Production price (Rp/Kg)
Q = Production result (kg/ha)
Bi = Production costs to i (Rp/ha)

2.3. Increasing the ability and capacity of breeder farmers in producing soybean seeds
One of the ways to meet the needs of quality soybean seeds was by increasing skills, knowledge, and the capacity of local seed breeders. Provision of seeds through local seed breeders was expected to be able to accelerate the dissemination and distribution of new superior varieties and quality seeds. Improvement of farmers' skills and knowledge was carried out through cooperation in seed production with cooperative farmers and through training by bringing competent sources in the field of seed.

3. Results and Discussion

3.1. Tillage
In order to obtain quality seeds from farmers' breeding results, breeder farmers must have knowledge about growing conditions, morphology, propagation technology and analysis of soybean seed farming. The Indonesian Agency for Agricultural Research and Development has issued recommendations for multiplication of soybean seeds. The soybean seed multiplication technology has been disseminated to soybean seed breeder groups.

Agroecology of Dusun Baru Village, District VII of Koto, Tebo Regency is non-acid dry land. Planting Season I is conducted in February - May (MH II) with a soybean cultivation technology package starting from preparing land by tillage till loose before the rainy season, by plowing 1 to 2 times then rake 1 time and leveled. The large seed variety, Anjasmoro, is the preference of farmers in Tebo, while the Wilis and Grobogan seed varieties are also planted. The total seed requirement is 40 kg / ha. Seed treatment with rhizobium source biological fertilizer at a dose of 20 g rhizobium / kg seed. Spacing of 40 x 15 cm with two seeds per planting hole. Fertilizer recommendations depend on conditions and soil fertility based on soil analysis, but in general in Dusun Baru Village the application of organic fertilizer or manure is recommended as much as 2 tons / hectare. Plant Pest Organisms (OPT) are controlled based on routine monitoring by farmers accompanied by researchers from BPTP Jambi. To overcome weeds as well as pests and plant diseases carried out based on the technical guidelines for Integrated Pest Management (IPM). Plants are ready to be harvested when the leaves have decayed and 95% of the pods are yellow-brown. Furthermore soybean seeding can be done manually.

Local knowledge refers to community-based agricultural technology and knowledge systems, in contrast to top-down knowledge and technology systems developed by modern research institutes. Because it is community based, in the process of disseminating technology, a suitable method is bottom-up. One way is to ask farmers to talk about traditional processes that are carried out. For example in fertilization, farmers are asked to explain the process of fertilization, the effects of fertilization. When the farmer explains the known stages, the researcher disseminates the technology. In this way farmers do not feel uprooted from the roots of their knowledge.

The farming system used by the community or traditional farmers is an agricultural system based on years of observation of the community's environment and the surrounding natural environment. Traditional societies have a variety of life-sustaining knowledge including farming knowledge of each type of plant. Based on observations, each type of plant has a variety of uses, so that in each variety there are various kinds of knowledge. The loss of one type of plant variety also means the loss of knowledge that follows it. In one variety contained knowledge about: cropping patterns, tillage, pest control, planting time, and others [6].

Farmers in Dusun Baru Village, District VII Koto, Tebo Regency have local knowledge that provides a specific method of hereditary cultivation. However, farmers who are members of the Harapan Mulya

IOP Conf. Series: Earth and Environmental Science **383** (2019) 012051 doi:10.1088/1755-1315/383/1/012051

Farmers Group are willing to accept outside influences selectively so that local knowledge can be maintained in accordance with the place. Their local knowledge is the result of thought based on good experience and has been passed on continuously, and now it is a local knowledge of the community that is sought again to be applied sustainably, because it has proven good results. Thus, in the process of dissemination of soybean seed culture cultivation technology packages there are parts that are changed and parts that are still maintained. The summary of the interview results illustrates the dissemination of technology and local knowledge of KT farmers. Mulya's expectations are as follows (Table 1).

Table 1. Technology Dissemination of Local Farmers KT. Harapan Mulya Dusun Baru Village, District VII Koto, Tebo Regency, Jambi Province, 2016.

Agriculture Sector	Technology[a]	Local Knowledge	Location Specific Technology
Tillage	The soil is processed until loose	The use of mulch straw and organic fertilizer	The soil is cultivated so that it reaches loose structures. After *tugal*, the seed is put into a hole, then the hole is sprinkled with organic fertilizer and covered with straw
Use of quality seeds	Anjasmoro, Baluran, Argopuro, Gumitir, Detam 1, Detam 2, Detam 3, Detam 4, Wilis, Kaba, Sinabung, Arjasari, Malika, Gema, Dering, Argomulyo, Burangrang, Malabar, Ijen, Gema, Dering	Preference for varieties with large and medium seeds, using seeds from the harvest	Anjasmoro, Grobogan Wilis
Seed Treatment	Rhizobium inoculation of seeds before planting, administration of growth regulators (PGR), and liquid supplementary fertilizers (PPC)	Using soil that has been planted with soybeans	Using artificial inoculation, for example: Legin, Rhizogin, and Rhizobin in the form of powder. , mixing is carried out evenly and in a shady place, and the seeds are planted immediately (no more than 6 hours).
Fertilization	Fertilizer recommendations include the main macro nutrients NPK, micro nutrients and manure as well as the use of straw according to the fertilizer reference	The main macro fertilizer of NPK, manure, use of straw	Recommendations for NPK main macro nutrients, micro nutrients, manure, straw according to fertilizer reference.

Agriculture Sector	Technology[a]	Local Knowledge	Location Specific Technology
Planting Distance Management	Adjusting plant population, 350,000 / hectare, spacing of 40 cm x 15 cm	Approximately when flowing, however, the spacing of related uniforms is more or less 40 cm x 15 cm	On land with enough water 40 cm x 15 cm On land with limited water 40 cm x 10 cm

[a]Balitkabi, 2015

In controlling pests in soybean cropping, farmers are still relying on pesticides. Nevertheless efforts to control pests by natural means are also carried out. For example, by setting traps in the form of cloth that has been soaked in an insecticide mixed with shrimp paste. The smell of shrimp paste attracts insect pests to perch on the fabric then the insects prick and suck the liquid that is in the fabric, and not how long the pest dies.

Important pests found at soybean cultivation sites are *Ophiomyia phaseoli*, known locally as bean flies, *Spodoptera litura* or grayak caterpillars, *Nezara viridula* or pale green ladybugs, *Riptortus linearis* or brown ladybugs and other pod borer pests. Currently the recommended technology for effective soybean pest control is the Sidamethrin insecticide for bean flies, Matador for leaf pests, and Deltamethrin for pod sucking pests, SlNPV-JTM 97C for controlling grayworm caterpillars and using entomopathogenic *Verticillum lecanii* for *R. linearis*. Alternative control using plant-based insecticide extracts of neem seed powder to control bean flies has not been done. Insects which are natural enemies of soybean pests are also commonly found such as *Harmonia conformis*, *Oxyopes macilentus*, *Chrysopogon sp.* and *Ischnura aurora*. So that this becomes an important note for the sustainability of soybean cultivation to be able to take pest control measures in a wise manner so that natural enemy populations are high and sufficient to play a role in suppressing pest populations.

3.2. Soybean Seed Production Production
The implementation of the activity involves two farmer groups, which are responsible for carrying out work starting on land preparation, planting, maintenance, harvesting and post-harvesting in accordance with the technology applied. Choosing an appropriate location is the first step that determines the success of soybean breeding activities. So that the risk of failure can be avoided or minimized, the location is directed at conditions suitable for soybean crop cultivation.

3.2.1. Seed Source Soy Classes FS.
Soybean SSMU used Anjasmoro varieties covering an area of 4 hectares, Wilis covering an area of 2 hectares and Grobogan covering an area of 1 hectare, carried out by 12 breeder farmers.

Soybean seed source used comes from Malang Balitkabi is a superior variety that has characteristics as: preferred by local farmers, can adapt to the environment, especially with acidity and aluminum levels that are not too high, its taste is liked and has a high price in the market local, high yield, tolerant of pests / diseases, and resistant to lodging. Variety choices are adjusted to the local location, and for the location of these activities the Anjasmoro, Wilis and Grobogan Superior Varieties are used.

The average range for growing soybeans is around 90%, the high percentage of growth is due to the high dormancy period and the level of nutrient availability in the soil. Growth of plants in the initial stage (age 14 days) until the time of growth, plants from the flowering period until the filling of the pods looks very good. The performance of soybean cultivation managed by recommended cultivation techniques is far better than the growth of soybean farmers, even though it has been fertilized with Urea, SP-36, and sometimes accompanied by KCl in the same dosage as recommended. The use of good quality seeds, as well as the addition of manure and dolomite are the main causes of differences in plant growth performance (Table 1).

Table 2. Appearance of FS class soybean breeding in Dusun Baru Village, District VII Koto, Tebo Regency, Jambi Province, 2016.

Blok	Variety	Land area (ha)	Date	Plant Height (cm)	Age of Flower (hr)	Number of Branches (piece)
I	Anjasmoro	1,0	10-4-2016	80,2	32	5
II	Grobogan	1,0	10-4-2016	80,4	33	4
III	Wilis	0,5	24-4-2016	86,1	34	4

Appearance of soybean plants in Anjsmoro, Wilis and Grobogan varieties when the plants are 30-50 HST shows good and good growth, with an attractive appearance, straight and clear lines and lines of plants. Appearance of plants against pest caterpillars is still below economic ambag/still relatively low (> 15%), so farmers are sufficiently guided by mechanical control, if the attack exceeds the threshold then control will be carried out using insecticides, so it does not affect the pod filling period.

Observation of the age of flowering Anjasmoro vareiats seen flowering plants with an age range of 32-35 HST, Wilis variety aged 35-36 HST and Grobogan Variety 28-29 HST. The occurrence of time that is too long for flowering out between Anjasmoro, Wilis and Grobogan varieties is caused by genetic traits of each variety.

The average height of Anjasmoro varieties ranged from 80.3 - 89.0 cm, Wilis varieties ranged from 79.3 - 81.4 cm, while Grobongan varieties ranged from 65.9 - 67.2 cm. It was seen that by fertilizing P and management of organic matter can increase the growth and productivity of soybean plants in a sustainable manner. Most of the soils in the tropics need a large amount of P needed by soybean plants is relatively small compared to the needs of the elements N, K and Ca, but some P research results show an increase in soybean yield. Besides lime will increase the need for plants for macro and micro elements. Therefore fertilizing N, P and K is needed. Application of technology in accordance with the recommendations will provide maximum results to the purity of soybean seeds.

Maintenance of the genetic quality of the varieties is carried out by means of rouging when the plants are aged 15-20 HST, the phases and cooking phases of physiology, namely removing plants whose morphological characteristics deviate from the characteristics of the varieties of plants whose seeds are produced. At the age of 15-20 HST the observed component is hypocotyl color, green hypokotyl will produce white flowers, while purple hypocotyl will produce purple flowers. In the flowering phase, flowers that are too deviant from the dominant plant then the plant is immediately discarded. Likewise in the physiological cooking phase, plants that deviate from the dominant plant are immediately uprooted.

All these activities are carried out by breeders with the guidance of Jambi AIAT researchers, field agricultural extension workers, and Agricultural Seed Supervision and Certification Center staff in the regions. Thus purity and later no longer have problems regarding labeling because from the beginning it was monitored by Agricultural Seed Supervision and Certification Center staff in the field.

The performance of the FS seed production shows that the Anjasmoro variety before the laboratory test was 6,300 kg, passed the test (certified seed) of 6.00 kg. Grobogan varieties as much as 800 production, the laboratory test did not pass, this is due to the quality of grobogan varieties are not suitable for planting in this location. The Wilis variety produces 500 kg of seeded seeds whereas the production of 1600 kg, the low percentage of passing the wilis variety certification is also influenced by the growing environment at the location of seed multiplication as presented in Table 3.

Table 3. Production of soybean seeds from BS to FS class, in Dusun Baru Village, District VII Koto, Tebo Regency, Jambi Province, 2016.

Variety	Blok	Land Area (ha)	Production (ton)		% Pass
			Test	Pass	

Anjasmoro	I	4,0	5.300	5.000	95,24
Grobogan	II	1,0	800	-	0
Wilis	III	2,0	1.600	500	31,25

The results of multiplication of BS-FS seeds with a target of 7 tons can only be reached as much as 5,500 kg, this is caused by constraints of environmental factors growing plants. The results of this FS propagation will be continued in the next season to get SS seeds.

3.2.2. Seed Source SS Soybean.

Propagation of FS-SS soybean seeds using Anjasmoro varieties covering an area of 15 hectares carried out by 22 breeder farmers. Activities are located at KT. Harapan Mulya, Dusun Baru Village District VII Koto, Tebo Regency. The implementation of the activity involved two farmer groups, who were responsible for carrying out work starting on land preparation, planting, maintenance, harvesting and post-harvesting in accordance with the technology applied.

Producing high-quality seeds, improving physical, physiological and genetic quality, is also carried out during post-harvest handling. Maintaining physical and genetic quality is mainly carried out during processing, while maintaining physiological quality of seeds is carried out from harvest to storage and even until the seeds are ready for planting by farmers.

The performance of soybean plants at the age of 30-45 days shows a fairly rapid growth and good, attractive appearance with a straight and clear plant flow. Appearance of plants also appears to have a caterpillar attack but still relatively low or below the economic threshold (> 15%), so farmers are sufficiently guided by mechanical control, if the attack exceeds the threshold then control is carried out using insecticides, so it does not affect the pod filling period (Table 4).

Table 4. Appearance of conditions for multiplication of FS to SS soybean seeds with Anjasmoro varieties in Dusun Baru Village, District VII Koto, Tebo Regency, Jambi Province, 2016.

Blok	Land Area (ha)	Date	Plant Height (cm)	Age of Flower (hr)	Number of Branches (piece)
IV	0,5	5-4-2016	95,2	31	5
V	1,0	6-4-2016	92,4	33	5
VI	0,5	7-4-2016	87,1	32	4
VII	0,5	8-4-2016	89,0	34	5

Furthermore, observations of flowering age showed that Anjasmoro varieties issued flowers at the age of 31 - 36 HST. The average plant height at 40 HST ranged from 87.3 to 99.9 cm (Table 2). It can be seen that the timely application of fertilizer and dosage is very influential on plant height. The purpose of fertilizing is to improve physical and chemical conditions in the soil. P fertilization and organic material management can increase productivity on an ongoing basis. Most of the soils in the tropics need a large amount of P needed by soybean plants is relatively small compared to the needs of the elements N, K and Ca, but some P research results show an increase in soybean yield [8].

Field performance of plants shows the growth and development of plants is very encouraging and is predicted to be able to produce seeds around 1.3 - 1.8 tons / ha of SS seeds. The multiplication of Anjasmoro variety seeds to get the SS seed class can be seen from the planting block which has quite encouraging production, with the percentage of certification passing ranging from 92 - 95% (Table 5).

Table 5. Production of soybean vareiats anjasmoro seeds from FS to SS seed classes, in Dusun Baru Village, District VII Koto, Tebo Regency, Jambi Province, 2016.

Blok	Land Area (ha)	Production (ton)		% Pass
		Test	Pass	
IV	5,0	6.500	6.000	92,30
V	5,0	7.400	7.000	94,60
VI	5,0	5.500	5.000	90,20
VII	5,0	5.300	5.000	94,34

3.2.3. Feasibility Analysis. The profit of soybean seed farming is more profitable than consumption of soybean farming. The high profitability of seed production farming compared to consumption production is caused by differences in selling prices. The selling price of soybean seeds is Rp 12,000 per kg, while the selling price of soybean consumption is only Rp 6000 to Rp 7,000 per kg or there is a difference in the selling price of Rp 5,000 to Rp 6,000 per kg.

Revenue from consumption farming is IDR 10,850,000 / ha / planting season or IDR 2,712,500 per month. Whereas, from the production of seeds, Rp 17,700,000 / ha / planting season or Rp 4,425,000 per month. The revenue from seed production farming is sourced from sales in the form of seeds amounting to Rp. 15,900,000 and in the form of consumption as much as Rp. 1,800,000. seed so it is sold in the form of consumption. Therefore, the profit gained from consumption farming is IDR 6,427,500 / ha / planting season, whereas in seed farming farming, it gives profit of IDR 11,113,500 / ha / planting season. B / C consumes 1.45 while B / C seeds 1.69.

In both of these farms, wages represent farm expenditures that are greater than those spent on material purchases. In farming, the consumption cost that must be spent for wages is 73.04%, while the costs for purchasing materials for production facilities are only 26.96%. Similarly, seed production farming, where the costs incurred for wages amounted to 63.87%, while the costs incurred for the purchase of production facilities amounted to 36.13%. The amount of wages incurred on existing farms compared to introductory technology farms is due to the low efficiency of farms on existing technology, especially on the expense of nursery costs that have moved twice.

3.3. Increasing the ability and capacity of breeder farmers in producing soybean seeds

With the implementation of site-specific technology that accommodates technology as a result of research combined with local knowledge, it is no wonder that at the farmer level there are still doubts about its success. So that the presence of the Jambi SSMU plays an important role so that farmers do not feel alone. SSMU becomes a supporting system if farmers experience failure.

The proper implementation of the activities so far with the pilot project is to make a pilot designed as a test or trial in order to demonstrate the effectiveness of a program implementation, find out the impact of program implementation. In other words, the area in the soybean center becomes a laboratory for researchers by sitting together with farmers. Researchers do not position themselves as superior to farmers but as colleagues or learners alike. Its application uses the participatory method, which is a method in which researchers go directly to the field following the activities under study. An example of the convention is by explaining the institutional strengthening of the corporation area through escorting, fostering and assisting in corn seed production techniques, then, provide assistance for production facilities, agricultural machinery, infrastructure, and market access [9].

Some requirements must be met when farmers become seed growers. In the context of strengthening the capacity of human resources (HR), farmers are equipped with technical capabilities of breeding soybean seeds. It is intended that farmers can make soybean seeds independently.

Some requirements, for example, have facilities for seed production, drying floors, processing facilities and warehouses. To facilitate the management of the seed breeder empowerment area, if possible it is placed on one stretch, but if it is not possible it can be implemented not on one stretch but still within 1 (one) subdistrict area. Seed breeders or breeder groups must carry out seed breeding in accordance with the target area of each commodity.

So that the implementation of seed breeder empowerment activities can run according to the provisions, it is necessary to provide guidance, monitoring and evaluation at the central level as well as provinces and districts / cities that are allocated the Seed Breeder Empowerment activities. Assistance is carried out from the beginning to the end of the process so that both farmers and researchers see and experience the implications of all stages of the process.

The sustainability of this program is also very much determined by the process of monitoring and evaluation of the relevant agencies collaborating with the relevant villages. It would be better if the seed multiplication program carried out in the related villages was included in the Village Development Planning. It is even possible to be grafted with a Financial Institution [7].

4. Conclusion

In the process of dissemination of soybean seed multiplication technology cultivation packages there are parts that are changed and parts that are still maintained. The production target of 7 tons of FS soybean seeds achieved was 5.5 tons, while the SS seed class with a target of 15 tons reached 23 tons or exceeded the target of 7 tons of SS seed. Through the guidance and assistance of breeder farmer groups regarding cultivation of seed, harvest, post-harvest technology and improvement of group administration, the supply of seeds from outside the region can be reduced or even eliminated. The capacity and capacity of breeder farmers has increased, this is seen from the success of producing seeds and getting an award as the best breeder group of Jambi Province. Soybean seeds from UPBS BPTP Jambi contributed to the supply of source seeds for the fulfillment of soybean seeds for breeder farmer groups and Provincial and District/City Government programs.

6. References

[1] Jambi Province Food Crop Agriculture Office. 2015. Food Crop Design Activities in 2015. Jambi Food Crop Agriculture Service.

[2] Nugraha US 2004. Legislation, policies and institutions for seedling development. *Technological development PRO*. 16 (1): 61-73.

[3] TeKrony DM 2006. Seeds: the delivery system for crop science. *Crop Sci*. 46: 2263-2269.

[4] Ilyas S 2006. Seed treatment using matriconditioning to improve vegetable seed quality. *Bul. Agron*. 34 (2): 124-132.

[5] Kadariah 2001. *Economic Analysis Project Evaluation.* Faculty of Economics Publisher Institute. University of Indonesia. Jakarta.

[6] Kastono D 2017. Application of Local Wisdom-Based Soy Cultivation Culture Culture in Supporting Food Security. *Development of a local soybean seed center.* https://pekakekal.org/

[7] Seed Supervision and Certification Center 2009. *Realization of Certification and Production of Rice, Maize, Soybean and Peanut Seeds*. Supervision and Certification of Seeds for Horticultural Crops in Jambi Province.

[8] Yardha, Adie M 2009. Performance of Soybean Hope Strain in Tidal Land. *Jurnal Agrista*. Faculty of Agriculture, Syiah Kuala University, Banda Aceh. Volume 13, Nomor 1. April 2009.

[9] Yardha. Nugroho H, Adri 2013. Acceleration of adoption of new high-yielding soybean varieties in tidal land. Proceedings of the 2013 National Seminar on the Result of Various Beans and Tuber Food Crops Research.

"Small farmer large field", a new model to developing and improving adoption of hybrid rice technology in Indonesia

Y Widyastuti[1]*, Satoto[1], I A Rumanti[1], B P Wibowo[1], A Ruskandar[1], Widyantoro[1], Rachmawati[1], and Z Zaini[2]

[1]Indonesian Center for Rice Research, Jalan Raya 9 Sukamandi, Ciasem Subang West Java 41256, Indonesia
[2]International Rice Research Institute Representative for Indonesia, Jalan Merdeka No 147, Central Bogor, Bogor City, 16111 West Java.

Corresponding author: yuniweicrr@gmail.com

Abstract. At present, the international market prefers to rice with some specific characters such as long slender shape, soft with 22-24% amylose content, a high percentage of head rice, and also aromatic. To fulfill rice with export standard, the farmers who involve have to apply good agricultural practices (GAP). Therefore need a high yielding variety, one of those varieties is hybrid rice variety possessing long slender shape, soft, a high percentage of head rice, and aromatic. Small Farmers, Large Field (SFLF) scheme was assumed précised and local specific for low land rice cultivation. This concept was integrated for farmers with small land holding to increase benefit collectively. The concept also integrated all cultivation system starting from planting up to marketing which was done together by some stakes holders. The objectives of this study are to (1) identify the hybrid rice productivity in large scale, (2) verify and validate the development of hybrid rice technology in small farmer large field's model, and (3) analyze the preference of farmers to hybrid rice varieties. The research activities consisted of SFLF approach by using hybrid rice varieties with Jarwo Super system in large scale. Adoption survey and preference test for the farmer in term of rice plant performance, quality of grain paddy and rice which was done simultaneously in subak Tajen, Penebel sub-district, Tabanan district, Bali province. Dem-farm was conducted at 50 ha with hybrid rice varieties from ICRR and private company. Average hybrid rice yield is about 7.5 ton/ha to 13.5 ton/ha. The results of participatory varietal selection showed that the yield potential, performance of the plant, tolerance to pest and disease, and quality of grain from hybrid rice varieties had a positive influence on farmers' preference.

1. Introduction

Rice has a special meaning in Asia, where about 90% of the rice is produced and consumed as a staple food. Indonesia, with the total population, predicted to reach 318.9 million in 2045, one of the urgent challenges is to fulfill the demand for rice as a major food crop. However, the ability to produce food is quite threatened due to limited natural and human resources, land conversion to modern industry, and global climate change because drought, flood, etc.

Considering the problems, then to increase rice production should be implemented by increasing rice productivity. Therefore, the development of high yielding varieties (HYVs) is a priority policy framework. Conventional technology to improve new HYVs using inbreed method becomes more difficult due to limited genetic differences. Responding to this challenge, hybrid rice technology should be developed as an alternative technology to increase national rice productivity. The term "hybrid" is used to refer to the first generation of a cross between two genetically diverse parents. The commercial hybrid refers to a superior F1, which not only outperforms the better parent but also shows significant (at least 1 ton ha^{-1}) yield superiority over the best high-yielding inbreed variety of similar duration and possesses acceptable grain quality.

Indonesian Center for Rice Research until 2019, has been released 21 new hybrid rice varieties. Hipa 18 and Hipa 19 are hybrid rice which has resistance to bacterial leaf blight (BLB) and brown planthopper (BPH) [1]. In 2019, Hipa 20 and Hipa 21 were released that having the potential for high yield, resistant to BLB and BPH, with a seed production potential of more than 2 t ha-1. Hybrids in Indonesia have the heterosis of 10-25% higher than the popular inbred varieties, such as Ciherang, Mekongga, IR64 [2]; [3].

The adoption of hybrid rice in Indonesia is still very low, around 0.34 – 3.39% during 2013 – 2017 (ICRR, 2017). Compared to other countries, this adoption rate is low, it is due to (a) gap yield between hybrid and non-hybrid is not significant, whereas hybrid rice seed price is much higher; (b) grain quality lower than non-hybrid variety; (c) seed production rate is still low which imply higher seed price; (d) limited information to the farmers about hybrid rice; (e) distribution channels that are not ready to absorb hybrid rice technology from producers to the farmers ([4]; [5]; [6]; [7]; dan [8].

The Ministry of Agriculture provides a grant of certified seed to smallholder farmers to improve adoption of hybrid rice varieties and accelerate the increase on rice production in major rice-producing provinces. In 2019 there was 100,000 hectare of paddy area which received the grant of hybrid rice seed with volume 1,500,000 Kg [9].

[10] reported that the Small Farmers, Large Field (SFLF) scheme have received attention in Vietnam. The SFLF is the solution to the problems of mechanization of smallholders and lack of bargaining power in the input and output markets. Through this scheme, farmers who organize themselves into groups and synchronize their operations by adopting a single rice variety to be planted, establishing group nurseries, and planting and harvesting in the same time, so that when changing. The "Small Farmers, Large Field" allows small farmers to benefit from economies of scale by organizing themselves and integrating their small rice areas into one large field. This way, they can increase their bargaining power input companies, such as fertilizer, pesticides, etc.

Therefore, the objectives of this article are: (1) identify the hybrid rice productivity in large scale, (2) verify and validate the development of hybrid rice technology in small farmer large field's model, and (3) analyze the preference of farmers to hybrid rice varieties.

2. Material and method

2.1. Demonstration farm

Demonstration farm of the Small Farmers, Large Field (SFLF) technology was conducted at Subak Tajen, Penebel, Tabanan Bali Province in Indonesia. The covered area of hybrid rice's planting field is about 50 ha consisted of 36 ha (dem-farm only for HIPA 18, Hipa 8, and Hipa 5 Ceva from ICRR), 2.5 ha (display of varieties), and 11.75 ha (dem-farm private company).

Seven hybrid rice varieties were tested in this dem-farm that presented in Table 1. The total area of dem-farm was about 12.6 hectares. Other inputs like fertilizer (NPK and organic fertilizer) were given based on crop need by soil tool kit for P and K, and leaf color chart for urea in integrated site-specific nutrient management (SSNM). While water supply to rice crop followed intermittent irrigation technique based on crop growth stages and synchronized with traditional water irrigation in Bali (Subak). Chemical to control pest and diseases was based on economic threshold unless there were endemic attacks.

Table 1. Type and source of rice varieties, and planted area in Tajen Penebel Tabanan Bali, 2017.

No	Variety	Area (acre)	Institution
1.	Hipa 18	40	ICRR licensed to Petrokimia Gresik
2.	Mapan P05	43	Primasid Andalan Utama
3.	SL 8 SHS	20	Syang Hyang Seri
4.	Sembada 188	37	Biogene Plantation
5.	Rejo	40	Agri Makmur Pertiwi
6.	Brang Biji	40	Padi Hibrida Nusantara
7.	Intani-602	3	BISI

2.2. Farmers preference of hybrid rice varieties

The preference test was conducted on the field, each panelist was positioned separately. A Hundred Balinese farmers from Subak Tajen were involved as a trained panelist. The panelists were allowed into the testing area in a smaller group (10-20 panelists) and had unlimited time to complete the test. They were asked to evaluate each hybrid rice variety based on its plant height, number of productive tillers, panicle performance, performance acceptability (PAcp), and overall performance of grain and plant.

2.3. Study of Small Farmers, Large Field scheme

The study of the was involved 100 farmers from irrigation society (Subak) Tajen in Penebel Sub-district, Tabanan District. They have planted 7 hybrid rice varieties in a total 50-hectare area. The technology applied was Jarwo Super which consists of several components including: cropping systems Jajar Legowo 2:1 and 4:1 ratio, hybrid rice varieties, decomposers, biological fertilizer as seed treatment and balanced fertilization based on Paddy Soil Test Kits (PUTS), monitoring of pest and disease using both biological and chemical application. There was used agricultural machinery and equipment, especially for planting (transplanters) and harvesting (combine harvester). Transfers information to the farmers were conducted by inviting 50 participants consisting of farmers, extension workers, and other stakeholders to field visit and got knowledge from the scientist and breeder from ICRR.

3. Result and Discussion

3.1. The location of demonstrasi area and culture of the farmers

Desa Tajen, which is the location of the Small Farmers, Large Field scheme, is located in Penebel sub-district, Tabanan district (picture 1). The population in 2016 was 2,921 and the population density was 744 persons / km^2. The area of the village is 300 ha used for SFLF is 50 ha. The majority of the people in Tajen Village work as farmers, both growers or smallholders. Subak Tajen (traditional agriculture association) is divided into 3 group farmers i.e. Dauh Yeh (1710 acre rice field), Lantang (2771 acre), and Ancut (1815 acre). The range ownership of rice fields is 6-90 acres from 340 farmers.

Balinese agriculture is a combination of science and art. The science, accumulated from hundreds of years of practical experience, the art from the attention to detail, and the value which the Balinese farmer places on the beauty of his fields. The organization community in Bali consisted of The subak (the irrigation society) and the banjar (the village). common features, e.g. leaders are elected from amongst the membership, the roles of these social institutions, vis-à-vis agriculture, are very different. The major role of the subak is to ensure an adequate and equitable supply of water for each farmer's rice crop; the role of the banjar is to provide labor for the cultivation of the rice crop. The rice farmers are members of both institutions, whilst hired labor and harvesting teams are members of the banjar only.

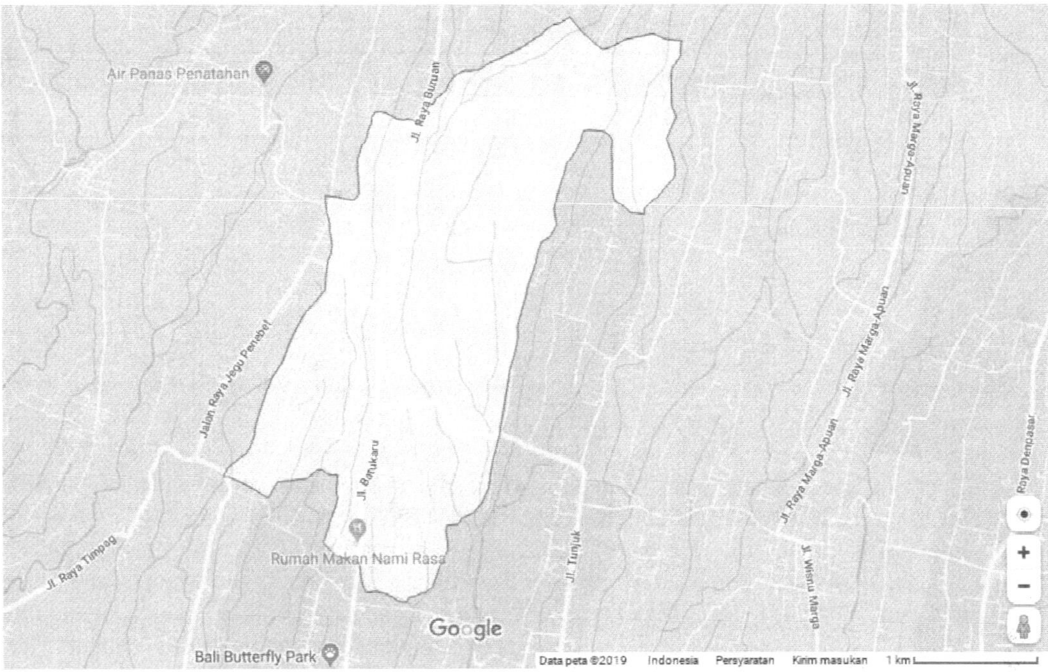

Picture 1. The location of Small Farmer Large Field scheme, Tajen Penebel Tabanan Bali Indonesia.

3.2. Farmers preference for performance of hybrid rice varieties

The farmers were asked to state the important traits of rice which have a bearing on the adoption of rice variety by the farmers. Several traits identified and were categorized in the nursery stage and plant growth stage, grain and eating quality. The traits having high significance for the adoption potential of rice variety were the crop duration, profitability, organoleptic traits. The traits having the least importance were disease vulnerability, threshing and germination vigor at the nursery stage.

The preference of the farmers to plant height and number of productive tillers (Table 2) showed that Rejo were most preferred by farmers. Other varieties that also got a higher percentage of liking were Mapan P05 and Sembada 188. Most farmers disliked of performance Brang Biji compared than hybrid rice varieties tested. The farmers like plants that have medium height. Plants that are too high are very vulnerable to lodging. The number of productive tillers is an important trait and related to grain yield. The farmers also like a hybrid rice varieties Rejo, Mapan P05, and Sembada 188 based on their quality of panicle and phenotypic acceptability. But most of the farmers dislike Hipa 18 and Brang Biji based on the above traits. It's the challenge for ICRR to improve their hybrid rice variety especially its performance in the field to compete with hybrids from the private sector.

Tabel 2. Preference of farmers to plant height and number of productive tillers of hybrid rice varieties.

No.	Hybrid rice varietis	Plant Height			Number of productive tillers		
		Like	Moderate	Dislike	Like	Moderate	Dislike
1	Mapan 05	52.5	47.5	0.0	45.8	54.2	0.0
2	Hipa 18	22.0	54.2	23.7	20.3	66.1	13.6
3	Rejo	72.9	27.1	0.0	62.7	37.3	0.0
4	SL-8	27.1	69.5	3.4	13.6	69.5	17.0
5	Intani-605	15.3	69.5	13.6	27.1	61.0	11.9
6	Brang Biji	11.9	55.9	32.2	11.9	59.3	28.8
7	Sembada 188	62.7	37.3	0.0	54.2	45.8	0.0

Tabel 3. Preference of farmers to panicle quality and PAcp of hybrid rice varieties.

No.	Hybrid rice varietis	Panicle quality (%)			Phenotypic Acceptability (PAcp) (%)		
		Like	Moderate	Dislike	Like	Moderate	Dislike
1	Mapan 05	49.2	50.9	0.0	35.6	64.4	0.0
2	Hipa 18	6.8	52.5	40.7	8.5	59.3	32.2
3	Rejo	61.0	39.0	0.0	54.2	45.8	0.0
4	SL-8	18.6	62.7	18.6	15.3	74.6	10.2
5	Intani-605	10.2	55.9	33.9	11.9	66.1	22.0
6	Brang Biji	3.4	25.4	71.2	5.1	59.3	35.6
7	Sembada 188	57.6	40.7	1.7	52.5	47.5	0.0

The relative yield performances of the hybrid rice varieties are presented in Table 4. Overall, rice hybrid performed excellently with an average yield of 10.50 ton ha^{-1}. However, Sembada 188 a hybrid rice variety from Biogene Plantation showed highest grain yield than others (13.5 ton ha-1), whereas Hipa 18 from ICRR showed lowest grain yield (7.5 ton ha-1). The result of grain yield the same as the level preference farmers for the overall performance of the plant. Most of the farmers (50.9%) were prefer Sembada 188 as their favorite varieties (Table 4).

Tabel 4. Preference of farmers to overall hybrid rice varieties and its grain yield.

No.	Hybrid rice varieties	Overall plant (%)			Grain yield (t ha-1)
		Like	Moderate	Dislike	
1	HIPA 18	5.08	49.15	45.76	7.5
2	SL 8 SHS	15.3	67.8	16.95	8.7
3	MAPAN P.05	35.6	62.71	1.69	12.4
4	Sembada 188	50.9	47.46	1.69	13.5
5	Rejo	49.2	49.15	1.69	12.1
6	BRANG BIJI	3.39	54.24	42.37	8.2
7	Intani-602	6.78	72.88	20.34	11.0

[11] reported that most farmers prefer Sembada, Hipa 18, Mapan P 05, Intani 602 based on its cooked rice had moist and tender texture while Hipa 19 cooked rice had a dry and hard texture. Grain physical and milling quality determined farmers' preference. [12] also had shown in their study that technology-specific variables (e.g. yield potential and acceptability) are significant for explaining adoption behavior, implying that it is important to take farmers' preferences to varietal characteristics into consideration in the design of a research and development program.

3.3. The studi of Small Farmers, Large Field scheme

The development of hybrid rice, public-private partnership is very important to improve the adoption level in the farmers. The government should be facilitating public-private partnership for sustainable hybrid rice growth and development. The activites i.e encouraging farmers to join the field production experiment, organize farmers participation in the hybrid rice production model, establish high-yield cultivation model, organize a field visit and product evaluation for the farmers, develop training on seed production technique, cultivation techniques, and new varieties selection transfer for farmers.

The Small Farmers, Large Field scheme in Bali was a program that supported by various institution i.e. government (IAARD-ICFORD, ICRR, Bali's AIAT, local government), a private company, and international organization (IRRI). ICRR as coordinator this program was sharing the qualified seeds, technology for agriculture, and high-tech harvesting tools. That variety has been planted in Subak Tajen. Tajen headman, Gusti Putu Sumerta Yasa reported rice harvest previously only reaches to the

6-7 ton, but after planting a hybrid seed it increases to 9-10 ton of rice with planting seven seeds of hybrid as a model. It is certainly supporting because the income of the farmer is increasing. Besides that, farmer no longer burning straw, as usual, they make fertilizer with that straw and it is certainly useful for rice field. The comparison of income between using non-hybrid seed and hybrid seed is about Rp. 10 million per hectare, and of course it is a good opportunity for attracting society back to being farmer.

The SFLF schema on 50 ha of farmers' land has shown that Balinese farmers who have a characteristic of small land, when joined by farmer groups through joint management with one type of variety, the same cultivation system, integrated pest control in groups is very influential besides increasing togetherness and cooperation between farmers also increases their income. This approach is expected to be carried out in other regions, especially for the development of hybrid rice which has been considered slow in terms of development.

4. Conclusion

The SFLF schema outputs are generally comply most of the technology components and standard operational procedure (SOP) in the implementation of Best Management Practices (BMP) in large scale mechanized rice farming. Early warning system must be continuously exercised to avoid pest and diseases outbreak. Brown plant hopper and rat are two main pests that must be closely monitor and immediately controlled. In average, potential yield of Sembada 188 hybrid rice variety is highest compared with other hybrid rice varieties or about 13.5 ton/ha average.

The SFLF schema on 50 ha of farmers' land has shown that Balinese farmers who have a characteristic of small land, when joined by farmer groups through joint management with one type of variety, the same cultivation system, integrated pest control in groups is very influential besides increasing togetherness and cooperation between farmers also increases their income. This approach is expected to be carried out in other regions, especially for the development of hybrid rice which has been considered slow in terms of development.

5. References

[1] Satoto, Rumanti I A, and Widyastuti Y 2016a Yield stability of new hybrid rice across locations Agrivita Journal of Agricultural Science 38(1):33-39.

[2] Widyastuti Y, Satoto, and Rumanti I A 2015 Performance of promising hybrid rice in two different elevations of irrigated lowland in Indonesia *Agrivita Journal of Agricultural Science* **37**(2):169-177.

[3] Kartina N, Wibowo B P, Widyastuti Y, Rumanti I A, and Satoto 2016 Korelasi dan sidik lintas karakter agronomi padi hibrida *Jurnal Ilmu Pertanian Indonesia* (Bahasa) **21**(2):76-83.

[4] Ruskandar A 2010 Persepsi petani dan identifikasi faktor penentu pengembangan dan adopsi varietas padi hibrida *Iptek Tanaman Pangan* (Bahasas) **5** (2):113-125.

[5] Sumarno, Wargiono J, Kartasasmita U G, Ismail I G, and Soejitno J 2008. Pemahaman dan kesiapan petani mengadopsi padi hibrida *Iptek Tanaman Pangan* (Bahasa) **3**(2):167-183.

[6] Firohmatillah A R, Nurmalina R 2012 Pengembangan padi varietas unggul hibrida: pendekatan metode *quality function development* dan *sensitivity price analysis. Jurnal Ekonomi Pembangunan* (Bahasa) **13**(1):29-45.

[7] Satoto, Widyastuti Y, Kartina N, and Wibowo B P 2016b Analisis adopsi pengembangan padi hibrida di Indonesia *Iptek Tanaman Pangan* (Bahasa) **12**(1):1-8.

[8] Mottaleb K A, Mohanti S, and Nelson A 2014 Factors influencing hybrid rice adoption: a Bangladesh case *Australian Journal of Agricultural and Resource Economics* **59**:258-274.

[9] Directorate General of Crops 2019 *Petunjuk Pelaksanaan Kegiatan Perbenihan Tanaman Pangan* Ministry of Agriculture.

[10] Rosellon E 2015 "Small farmers, large field" scheme gaining success in Vietnam http://news.irri.org/2015/05/small-farmers-large-field-scheme.html. Friday, May 15, 2015.

[11] Ardhiyanti S D, Aurum F S, Handoko D, Liyanan, Widyastuti Y, and Satoto 2018 Farmers' preferences for grain quality of hybrid rice: A study case in Bali *Preprint.*

[12] Ghimire R, Wen-Chi Huang, and Shrestha R B 2015 Factors Affecting Adoption of Improved Rice Varieties among Rural Farm Households in Central Nepal *Rice Science* **22**(1):35-43.

AUTHOR INDEX

AUTHOR INDEX

Institute of Physics
Dirac House, Temple Back
Bristol BS1 6BE UK

ISSN: 1755-1307
ISBN 978-1-7138-0251-8

CURRAN ASSOCIATES INC.
proceedings.com